BIG IDEAS MATH®

Advanced 1

A Bridge to Success

TEACHING EDITION

Ron Larson
Laurie Boswell

BIG IDEAS LEARNING®
Erie, Pennsylvania
BigIdeasLearning.com

Big Ideas Learning, LLC
1762 Norcross Road
Erie, PA 16510-3838
USA

For product information and customer support, contact Big Ideas Learning at **1-877-552-7766** or visit us at ***BigIdeasLearning.com***.

Cover Image
Steve Horsley/Shutterstock.com

Printed in the U.S.A.

ISBN 13: 978-1-68033-129-5
ISBN 10: 1-68033-129-9

2 3 4 5 6 7 8 9 10 WEB 17 16 15

AUTHORS

Ron Larson is a professor of mathematics at Penn State Erie, The Behrend College, where he has taught since receiving his Ph.D. in mathematics from the University of Colorado. Dr. Larson is well known as the lead author of a comprehensive program for mathematics that spans middle school, high school, and college courses. His high school and Advanced Placement books are published by Holt McDougal. Ron's numerous professional activities keep him in constant touch with the needs of students, teachers, and supervisors. Ron and Laurie Boswell began writing together in 1992. Since that time, they have authored over two dozen textbooks. In their collaboration, Ron is primarily responsible for the pupil edition and Laurie is primarily responsible for the teaching edition of the text.

Laurie Boswell is the Head of School and a mathematics teacher at the Riverside School in Lyndonville, Vermont. Dr. Boswell received her Ed.D. from the University of Vermont in 2010. She is a recipient of the Presidential Award for Excellence in Mathematics Teaching. Laurie has taught math to students at all levels, elementary through college. In addition, Laurie was a Tandy Technology Scholar, and served on the NCTM Board of Directors from 2002 to 2005. She currently serves on the board of NCSM, and is a popular national speaker. Along with Ron, Laurie has co-authored numerous math programs.

ABOUT THE BOOK

Big Ideas Math: A Bridge to Success was developed using the same research-based strategy of a balanced approach that has become synonymous with the *Big Ideas Math* series. This approach opens doors to abstract thought, reasoning, and inquiry as students persevere to answer the Essential Questions that introduce each section.

From start to finish, this program was designed with the student in mind. Students are subtly introduced to "Habits of Mind" that help them internalize concepts for a greater depth of understanding. These habits serve students well not only in mathematics, but across all curricula throughout their academic careers. *Big Ideas Math: A Bridge to Success* exposes students to highly motivating and relevant problems. Woven throughout the program are the depth and rigor students need to prepare for career-readiness and other college-level courses.

We consider the Big Ideas Math series to be the crowning jewel of 30 years of achievement in writing educational materials.

Ron Larson   Laurie Boswell

TEACHER REVIEWERS

- Lisa Amspacher
  Milton Hershey School
  Hershey, PA
- Mary Ballerina
  Orange County Public Schools
  Orlando, FL
- Lisa Bubello
  School District of Palm Beach County
  Lake Worth, FL
- Sam Coffman
  North East School District
  North East, PA
- Kristen Karbon
  Troy School District
  Rochester Hills, MI
- Laurie Mallis
  Westglades Middle School
  Coral Springs, FL
- Dave Morris
  Union City Area School District
  Union City, PA
- Bonnie Pendergast
  Tolleson Union High School District
  Tolleson, AZ
- Valerie Sullivan
  Lamoille South Supervisory Union
  Morrisville, VT
- Becky Walker
  Appleton Area School District
  Appleton, WI
- Zena Wiltshire
  Dade County Public Schools
  Miami, FL

STUDENT REVIEWERS

- Mike Carter
- Matthew Cauley
- Amelia Davis
- Wisdom Dowds
- John Flatley
- Nick Ganger
- Hannah Iadeluca
- Paige Lavine
- Emma Louie
- David Nichols
- Mikala Parnell
- Jordan Pashupathi
- Stephen Piglowski
- Robby Quinn
- Michael Rawlings
- Garrett Sample
- Andrew Samuels
- Addie Sedelmyer
- Tyler Steffy
- Erin Taylor
- Reid Wilson

CONSULTANTS

- Patsy Davis
  Educational Consultant
  Knoxville, Tennessee
- Bob Fulenwider
  Mathematics Consultant
  Bakersfield, California
- Linda Hall
  Mathematics Assessment Consultant
  Norman, Oklahoma
- Ryan Keating
  Special Education Advisor
  Gilbert, Arizona
- Michael McDowell
  Project-Based Instruction Specialist
  Fairfax, California
- Sean McKeighan
  Interdisciplinary Advisor
  Norman, Oklahoma
- Bonnie Spence
  Differentiated Instruction Consultant
  Missoula, Montana

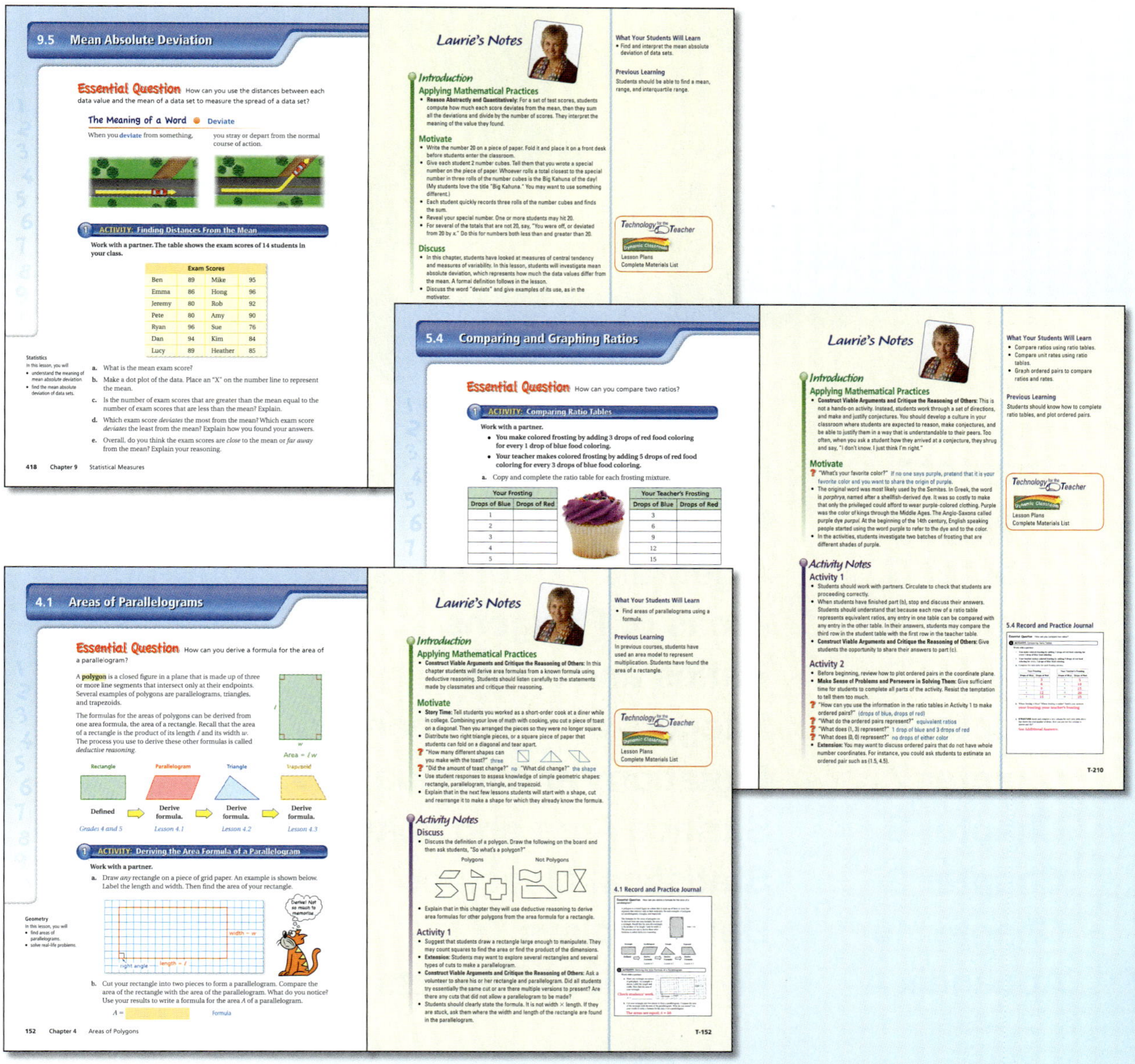

BIG IDEAS MATH

The *Big Ideas Math Advanced* series allows students to complete the material for grades 6, 7, and 8 in two years without skipping any learning objectives.

ADVANCED 1

Using the *Big Ideas Math Advanced* series, students can complete the Advanced Pathway and have the opportunity for conceptual understanding, procedural fluency, and application through the use of focus, coherence, and rigor.

Chapter 11	Integers
Chapter 12	Rational Numbers
Chapter 13	Expressions and Equations
Chapter 14	Ratios and Proportions
Chapter 15	Percents

11 Integers

"Before my school had Big Ideas Math I would always lose test points because I left units off my answers. Now I see why they are so important."

Rational Numbers

12

"Word problems used to confuse me. Now I understand how to look for patterns and what the question is asking!"

13

Expressions and Equations

“I like the Big Ideas Math Tutorials because they help explain the math when I am at home.”

14 Ratios and Proportions

“I really like the Graphic Organizers because they show me another way to take notes.”

15 Percents

"Using the Interactive Manipulatives from the Dynamic Student Edition helps me to see the mathematics that I am learning."

LOOKING FORWARD TO ADVANCED 2

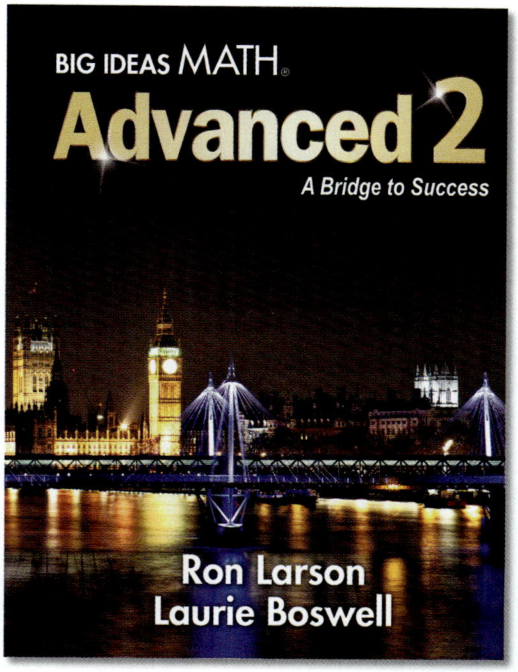

Chapter 1	Equations
Chapter 2	Transformations
Chapter 3	Angles and Triangles
Chapter 4	Graphing and Writing Linear Equations
Chapter 5	Systems of Linear Equations
Chapter 6	Functions
Chapter 7	Real Numbers and the Pythagorean Theorem
Chapter 8	Volume and Similar Solids
Chapter 9	Data Analysis and Displays
Chapter 10	Exponents and Scientific Notation
Chapter 11	Inequalities
Chapter 12	Constructions and Scale Drawings
Chapter 13	Circles and Area
Chapter 14	Surface Area and Volume
Chapter 15	Probability and Statistics
Appendix A	My Big Ideas Projects

PROGRAM OVERVIEW

Print *Also available online*

Pupil Edition

Teaching Edition

- Laurie's Notes

Record and Practice Journal

Available in English and Spanish

Resources by Chapter and Assessment Book

- Resources by Chapter
  - Family and Community Involvement: English and Spanish
  - Start Thinking! and Warm Up
  - Extra Practice
  - Enrichment and Extension
  - Puzzle Time
  - Technology Connection
- Assessment Book
  - Quizzes
  - Chapter Tests
  - Cumulative Assessment
  - Alternative Assessment
  - End-of-Course Tests

Technology

Student Edition

With complete English and Spanish Audio

- Dynamic eBook App
  - Lesson Tutorial Videos
  - Dynamic Investigations
- Online Home Edition
  - Multi-Language Glossary
  - Basic Skills Handbook
  - Skills Review Handbook
  - Game Closet

Dynamic Classroom

- Interactive Manipulatives
- Answer Presentation Tool
- Mini-Assessment
- Closure Activity

Dynamic Assessment and Progress Monitoring Tool

- Homework and Assessment Creation
- Progress Monitoring
- Direct Ties to Remediation
- All-In-One Reporting
- Online Chat Tutor

Dynamic Teaching Tools

- Interactive Whiteboard Lesson Library
  - Includes standard and customizable lessons
  - Compatible with SMART®, Promethean®, and Mimeo® technology
- **Exam**View® Assessment Suite
- Real-Life STEM Videos
- Editable Online Resources
  - Lesson Plans
  - Pacing Guides
  - Assessment Book
  - Resources by Chapter

SCOPE AND

Regular Pathway

Grade 6

Ratios and Proportional Relationships	– Understand Ratio Concepts; Use Ratio Reasoning
The Number System	– Perform Fraction and Decimal Operations; Understand Rational Numbers
Expressions and Equations	– Write, Interpret, and Use Expressions, Equations, and Inequalities
Geometry	– Solve Problems Involving Area, Surface Area, and Volume
Statistics and Probability	– Summarize and Describe Distributions; Understand Variability

Grade 7

Ratios and Proportional Relationships	– Analyze Proportional Relationships
The Number System	– Perform Rational Number Operations
Expressions and Equations	– Generate Equivalent Expressions; Solve Problems Using Linear Equations and Inequalities
Geometry	– Understand Geometric Relationships; Solve Problems Involving Angles, Surface Area, and Volume
Statistics and Probability	– Analyze and Compare Populations; Find Probabilities of Events

Grade 8

The Number System	– Approximate Real Numbers; Perform Real Number Operations
Expressions and Equations	– Use Radicals and Integer Exponents; Connect Proportional Relationships and Lines; Solve Systems of Linear Equations
Functions	– Define, Evaluate, and Compare Functions; Model Relationships
Geometry	– Understand Congruence and Similarity; Apply the Pythagorean Theorem; Apply Volume Formulas
Statistics and Probability	– Analyze Bivariate Data

SEQUENCE

Advanced Pathway

Grade 6 Advanced

Domain	Topics
Ratios and Proportional Relationships	– Understand Ratio Concepts; Use Ratio Reasoning; Analyze Proportional Relationships
The Number System	– Perform Fraction and Decimal Operations; Understand Rational Numbers; Perform Rational Number Operations
Expressions and Equations	– Write, Interpret, and Use Expressions, Equations, and Inequalities; Generate Equivalent Expressions; Solve Problems Using Linear Equations
Geometry	– Solve Problems Involving Area, Surface Area, and Volume
Statistics and Probability	– Summarize and Describe Distributions; Understand Variability

Grade 7 Advanced

Domain	Topics
The Number System	– Approximate Real Numbers; Perform Real Number Operations
Expressions and Equations	– Solve Problems Using Linear Inequalities; Use Radicals and Integer Exponents; Connect Proportional Relationships and Lines; Solve Systems of Linear Equations
Functions	– Define, Evaluate, and Compare Functions; Model Relationships
Geometry	– Understand Geometric Relationships; Solve Problems Involving Angles, Surface Area, and Volume; Understand Congruence and Similarity; Apply the Pythagorean Theorem
Statistics and Probability	– Analyze and Compare Populations; Find Probabilities of Events; Analyze Bivariate Data

LEARNING PROGRESSION

Advanced Pathway

Domain	Advanced 1	Advanced 2	Algebra 1
Number and Quantity			
Ratios and Proportional Relationships	Use ratios to solve problems. *Chapter 5* Use proportional relationships to solve problems. *Chapters 14, 15*		
The Number System	Perform operations with multi-digit numbers and find common factors and multiples. *Chapter 1* Divide fractions by fractions. *Chapter 2* Extend understanding of numbers to the rational number system. *Chapter 6* Perform operations with rational numbers. *Chapter 12*	Extend understanding of numbers to the real number system. *Chapter 7*	Use properties of rational exponents, rational numbers, and irrational numbers. *Chapters 6, 9, 10*
Algebra			
Expressions and Equations	Perform arithmetic with algebraic expressions. *Chapter 3* Solve one-variable equations and inequalities. *Chapter 7* Analyze relationships between dependent and independent variables. *Chapter 7* Write equivalent expressions. *Chapter 13* Use numerical and algebraic expressions and equations to solve problems. *Chapter 13*	Understand the connections between proportional relationships, lines, and linear equations. *Chapter 4* Solve linear equations and systems of linear equations. *Chapter 5* Work with radicals and integer exponents. *Chapters 7, 10* Use numerical and algebraic inequalities to solve problems. *Chapter 11*	Perform arithmetic on linear and quadratic polynomials. *Chapters 7–9* Solve linear and absolute value equations and inequalities, and quadratic and exponential equations. *Chapters 1–4, 6–10* Solve linear systems of equations and inequalities, and nonlinear systems of equations. *Chapters 5, 9*

MIDDLE SCHOOL MATH

Domain	Advanced 1	Advanced 2	Algebra 1
Functions			
Functions		Define, evaluate, and compare functions, and use functions to model relationships between quantities. *Chapter 6*	Interpret and analyze linear, quadratic, exponential, absolute value, step, and piecewise functions. *Chapters 3, 4, 6, 8–10* Build linear, quadratic, exponential, and absolute value functions from existing functions. *Chapters 3, 6, 8, 10*
Geometry			
Geometry	Solve real-world and mathematical problems involving area, surface area, and volume. *Chapters 4, 8*	Understand congruence and similarity. *Chapters 2, 3* Use the Pythagorean Theorem. *Chapter 7* Solve problems involving volume of cylinders, cones, and spheres. *Chapter 8* Draw, construct, and describe geometrical figures and describe the relationships between them. *Chapters 12–14* Solve problems involving angle measure, area, surface area, and volume. *Chapters 12–14*	
Statistics and Probability			
Statistics and Probability	Develop understanding of statistical variability and summarize and describe distributions. *Chapters 9, 10*	Investigate patterns of association in bivariate data. *Chapter 9* Make inferences about a population, compare two populations, and use probability models. *Chapter 15*	Summarize, represent, and interpret data using measures of center and spread. *Chapter 11*

PACING GUIDE FOR ADVANCED 1

Chapters 1–15 **145 Days**

Scavenger Hunt (1 Day)

Chapter 1 (11 Days)

Chapter Opener	1 Day
Section 1.1	1 Day
Section 1.2	1 Day
Section 1.3	1 Day
Study Help/Quiz	1 Day
Section 1.4	1 Day
Section 1.5	1 Day
Section 1.6	1 Day
Extension 1.6	1 Day
Chapter Review/Chapter Tests	2 Days

Chapter 2 (14 Days)

Chapter Opener	1 Day
Section 2.1	2 Days
Section 2.2	2 Days
Section 2.3	1 Day
Study Help/Quiz	1 Day
Section 2.4	1 Day
Section 2.5	2 Days
Section 2.6	2 Days
Chapter Review/Chapter Tests	2 Days

Chapter 3 (8 Days)

Chapter Opener	1 Day
Section 3.1	1 Day
Section 3.2	1 Day
Section 3.3	1 Day
Section 3.4	1 Day
Extension 3.4	1 Day
Chapter Review/Chapter Tests	2 Days

Chapter 4 (8 Days)

Chapter Opener	1 Day
Section 4.1	1 Day
Section 4.2	1 Day
Section 4.3	1 Day
Extension 4.3	1 Day
Section 4.4	1 Day
Chapter Review/Chapter Tests	2 Days

Chapter 5 (11 Days)

Chapter Opener	1 Day
Section 5.1	1 Day
Section 5.2	1 Day
Section 5.3	1 Day
Section 5.4	1 Day
Study Help/Quiz	1 Day
Section 5.5	1 Day
Section 5.6	1 Day
Section 5.7	1 Day
Chapter Review/Chapter Tests	2 Days

Chapter 6 (10 Days)

Chapter Opener	1 Day
Section 6.1	1 Day
Section 6.2	1 Day
Section 6.3	1 Day
Study Help/Quiz	1 Day
Section 6.4	1 Day
Section 6.5	1 Day
Extension 6.5	1 Day
Chapter Review/Chapter Tests	2 Days

Chapter 7 (11 Days)

Chapter Opener	1 Day
Section 7.1	1 Day
Section 7.2	1 Day
Section 7.3	1 Day
Section 7.4	1 Day
Study Help/Quiz	1 Day
Section 7.5	1 Day
Section 7.6	1 Day
Section 7.7	1 Day
Chapter Review/Chapter Tests	2 Days

Chapter 8 (7 Days)

Chapter Opener	1 Day
Section 8.1	1 Day
Section 8.2	1 Day
Section 8.3	1 Day
Section 8.4	1 Day
Chapter Review/Chapter Tests	2 Days

Chapter 9 (9 Days)

Chapter Opener	1 Day
Section 9.1	1 Day
Section 9.2	1 Day
Section 9.3	1 Day
Study Help/Quiz	1 Day
Section 9.4	1 Day
Section 9.5	1 Day
Chapter Review/Chapter Tests	2 Days

Chapter 10 (8 Days)

Chapter Opener	1 Day
Section 10.1	1 Day
Section 10.2	1 Day
Section 10.3	1 Day
Extension 10.3	1 Day
Section 10.4	1 Day
Chapter Review/Chapter Tests	2 Days

Chapter 11 (8 Days)

Chapter Opener	1 Day
Section 11.1	1 Day
Section 11.2	1 Day
Section 11.3	1 Day
Section 11.4	1 Day
Section 11.5	1 Day
Chapter Review/Chapter Tests	2 Days

Chapter 12 (7 Days)

Chapter Opener	1 Day
Section 12.1	1 Day
Section 12.2	1 Day
Section 12.3	1 Day
Section 12.4	1 Day
Chapter Review/Chapter Tests	2 Days

Chapter 13 (10 Days)

Chapter Opener	1 Day
Section 13.1	1 Day
Section 13.2	1 Day
Extension 13.2	1 Day
Study Help/Quiz	1 Day
Section 13.3	1 Day
Section 13.4	1 Day
Section 13.5	1 Day
Chapter Review/Chapter Tests	2 Days

Chapter 14 (11 Days)

Chapter Opener	1 Day
Section 14.1	1 Day
Section 14.2	1 Day
Extension 14.2	1 Day
Section 14.3	1 Day
Study Help/Quiz	1 Day
Section 14.4	1 Day
Section 14.5	1 Day
Section 14.6	1 Day
Chapter Review/Chapter Tests	2 Days

Chapter 15 (11 Days)

Chapter Opener	1 Day
Section 15.1	1 Day
Section 15.2	1 Day
Section 15.3	1 Day
Section 15.4	1 Day
Study Help/Quiz	1 Day
Section 15.5	1 Day
Section 15.6	1 Day
Section 15.7	1 Day
Chapter Review/Chapter Tests	2 Days

Mathematical Practices

Make sense of problems and persevere in solving them.

- Multiple representations are presented to help students move from concrete to representative and into abstract thinking
- *Essential Questions* help students focus and analyze
- *In Your Own Words* provide opportunities for students to look for meaning and entry points to a problem

Reason abstractly and quantitatively.

- Visual problem solving models help students create a coherent representation of the problem
- Opportunities for students to decontextualize and contextualize problems are presented in every lesson

Construct viable arguments and critique the reasoning of others.

- *Error Analysis*; *Different Words, Same Question*; and *Which One Doesn't Belong* features provide students the opportunity to construct arguments and critique the reasoning of others
- *Inductive Reasoning* activities help students make conjectures and build a logical progression of statements to explore their conjecture

Model with mathematics.

- Real-life situations are translated into diagrams, tables, equations, and graphs to help students analyze relations and to draw conclusions
- Real-life problems are provided to help students learn to apply the mathematics that they are learning to everyday life

Use appropriate tools strategically.

- *Graphic Organizers* support the thought process of what, when, and how to solve problems
- A variety of tool papers, such as graph paper, number lines, and manipulatives, are available as students consider how to approach a problem
- Opportunities to use the web, graphing calculators, and spreadsheets support student learning

Attend to precision.

- *On Your Own* questions encourage students to formulate consistent and appropriate reasoning
- Cooperative learning opportunities support precise communication

Look for and make use of structure.

- *Inductive Reasoning* activities provide students the opportunity to see patterns and structure in mathematics
- Real-world problems help students use the structure of mathematics to break down and solve more difficult problems

Look for and express regularity in repeated reasoning.

- Opportunities are provided to help students make generalizations
- Students are continually encouraged to check for reasonableness in their solutions

Go to *BigIdeasMath.com* for more information on the Mathematical Practices.

Mathematical Content

Chapter Coverage

Chapters: 1 2 3 4 **5** 6 7 8 9 10 11 12 13 **14** **15**

Ratios and Proportional Relationships

- Understand ratio concepts and use ratio reasoning to solve problems.
- Analyze proportional relationships and use them to solve real-world and mathematical problems.

Chapters: **1** **2** 3 4 5 **6** 7 8 9 10 **11** **12** 13 14 15

The Number System

- Apply and extend previous understandings of multiplication and division to divide fractions by fractions.
- Compute fluently with multi-digit numbers and find common factors and multiples.
- Apply and extend previous understandings of numbers to the system of rational numbers.
- Apply and extend previous understandings of operations with fractions to add, subtract, multiply, and divide rational numbers.

Chapters: **1** 2 **3** 4 5 6 **7** 8 9 10 11 12 **13** 14 15

Expressions and Equations

- Apply and extend previous understandings of arithmetic to algebraic expressions.
- Reason about and solve one-variable equations and inequalities.
- Represent and analyze quantitative relationships between dependent and independent variables.
- Use properties of operations to generate equivalent expressions.
- Solve real-life and mathematical problems using numerical and algebraic expressions and equations.

Chapters: 1 2 3 **4** 5 6 7 **8** 9 10 11 12 13 14 15

Geometry

- Solve real-world and mathematical problems involving area, surface area, and volume.

Chapters: 1 2 3 4 5 6 7 8 **9** **10** 11 12 13 14 15

Statistics and Probability

- Develop understanding of statistical variability.
- Summarize and describe distributions.

11 Integers

"Look, subtraction is not that difficult. Imagine that you have five squeaky mouse toys."
5 + (-1) = 4
"After your friend Fluffy comes over for a visit, you notice that one of the squeaky toys is missing."
4 - (-1) = 5
I'm going to count my toys
"Now, you go over to Fluffy's and retrieve the missing squeaky mouse toy. It's easy."

"Dear Sir: You asked me to 'find' the opposite of −1."
That's not the only thing that is missing.
"I didn't know it was missing."

What Your Students Have Learned

- Fluently multiply.
- Divide whole numbers, find quotients with and without remainders.
- Create and solve problems with whole number operations.
- Fluently divide.
- Identify and represent integers.
- Order and compare integers.
- Identify and describe absolute values of integers.

What Your Students Will Learn

- Use and justify rules for addition, subtraction, multiplication, and division of integers.
- Find the absolute value of integers.
- Add, subtract, multiply, and divide integers.

Pacing Guide for Chapter 11

Chapter Opener Advanced	1 Day
Section 1 Advanced	1 Day
Section 2 Advanced	1 Day
Section 3 Advanced	1 Day
Section 4 Advanced	1 Day
Section 5 Advanced	1 Day
Chapter Review/ Chapter Tests Advanced	2 Days
Total Chapter 11 Advanced	8 Days
Year-to-Date Advanced	106 Days

Technology for the Teacher

BigIdeasMath.com
Chapter at a Glance
Complete Materials List
Parent Letters: English and Spanish

What Your Students Have Learned

- Use the Commutative and Associative properties to simplify expressions.
- Use the properties of zero and one to simplify expressions.

Additional Topics for Review

- Compare and Order Integers
- Operations with Whole Numbers
- Distributive Property
- Mental Math Strategies

Try It Yourself

1–6. See Additional Answers.

Record and Practice Journal Fair Game Review

1. $7 + y$;
$2 + (5 + y) = (2 + 5) + y$ Assoc. Prop. of Add.
$= 7 + y$ Add 2 and 5.

2. $c + 10$;
$(c + 1) + 9 = c + (1 + 9)$ Assoc. Prop. of Add.
$= c + 10$ Add 1 and 9.

3. $n + 3.7$;
$(2.3 + n) + 1.4$
$= (n + 2.3) + 1.4$ Comm. Prop. of Add.
$= n + (2.3 + 1.4)$ Assoc. Prop. of Add.
$= n + 3.7$ Add 2.3 and 1.4.

4. $12 + d$;
$7 + (d + 5) = 7 + (5 + d)$ Comm. Prop. of Add.
$= (7 + 5) + d$ Assoc. Prop. of Add.
$= 12 + d$ Add 7 and 5.

5. $70t$;
$10(7t) = (10 \cdot 7)t$ Assoc. Prop. of Mult.
$= 70t$ Multiply 10 and 7.

6–12. See Additional Answers.

Math Background Notes

Vocabulary Review

- Commutative Property of Addition
- Commutative Property of Multiplication
- Associative Property of Addition
- Associative Property of Multiplication
- Addition Property of Zero
- Multiplication Property of Zero
- Multiplication Property of One

Commutative and Associative Properties

- Discuss the meaning of the word *commute*. Talk about *commuters* who go to work and back home, switching the direction.
- Discuss the meaning of the word *associate*, putting two things together. You *associate* dogs with barking, Arizona with the Grand Canyon, and sports cars with speed.
- Remind students that it is okay to *just combine numbers* for only the operations of addition and multiplication. If the first problem was $6 - (14 - x)$, combining the numbers would result in a wrong answer.
- Remind students that grouping symbols, such as parentheses, tell you to perform the operation inside them first.
- Ask students to write the Commutative and Associative Properties using variables.

Properties of Zero and One

- Multiplying by zero, in any form, produces a product of 0.
- Multiplying by one, in any form, does not change the value of the quantity. Remind students that 1 can be represented in different ways: 1, $\left(\frac{1}{2} + \frac{1}{2}\right)$, or $\frac{3}{3}$. This is also called the Multiplicative Identity.
- Ask students to write the Properties of Zero and One using variables.

Reteaching and Enrichment Strategies

If students need help. . .	If students got it. . .
Record and Practice Journal • Fair Game Review Skills Review Handbook Lesson Tutorials	Game Closet at *BigIdeasMath.com* Start the next section

What You Learned Before

"I liked it because it is the opposite of the freezing point on the Fahrenheit temperature scale."

Commutative and Associative Properties

Example 1 **a. Simplify the expression $6 + (14 + x)$.**

$6 + (14 + x) = (6 + 14) + x$	Associative Property of Addition
$= 20 + x$	Add 6 and 14.

b. Simplify the expression $(3.1 + x) + 7.4$.

$(3.1 + x) + 7.4 = (x + 3.1) + 7.4$	Commutative Property of Addition
$= x + (3.1 + 7.4)$	Associative Property of Addition
$= x + 10.5$	Add 3.1 and 7.4.

c. Simplify the expression $5(12y)$.

$5(12y) = (5 \cdot 12)y$	Associative Property of Multiplication
$= 60y$	Multiply 5 and 12.

Try It Yourself

Simplify the expression. Explain each step.

1. $3 + (b + 8)$ **2.** $(d + 4) + 6$ **3.** $6(5p)$

Properties of Zero and One

Example 2 **a. Simplify the expression $6 \cdot 0 \cdot q$.**

$6 \cdot 0 \cdot q = (6 \cdot 0) \cdot q$	Associative Property of Multiplication
$= 0 \cdot q = 0$	Multiplication Property of Zero

b. Simplify the expression $3.6 \cdot s \cdot 1$.

$3.6 \cdot s \cdot 1 = 3.6 \cdot (s \cdot 1)$	Associative Property of Multiplication
$= 3.6 \cdot s$	Multiplication Property of One
$= 3.6s$	

Try It Yourself

Simplify the expression. Explain each step.

4. $13 \cdot m \cdot 0$ **5.** $1 \cdot x \cdot 29$ **6.** $(n + 14) + 0$

11.1 Integers and Absolute Value

Essential Question How can you use integers to represent the velocity and the speed of an object?

On these two pages, you will investigate vertical motion (up or down).

- Speed tells how fast an object is moving, but it does not tell the direction.
- Velocity tells how fast an object is moving, and it also tells the direction.

  When velocity is positive, the object is moving up.

  When velocity is negative, the object is moving down.

1 ACTIVITY: Falling Parachute

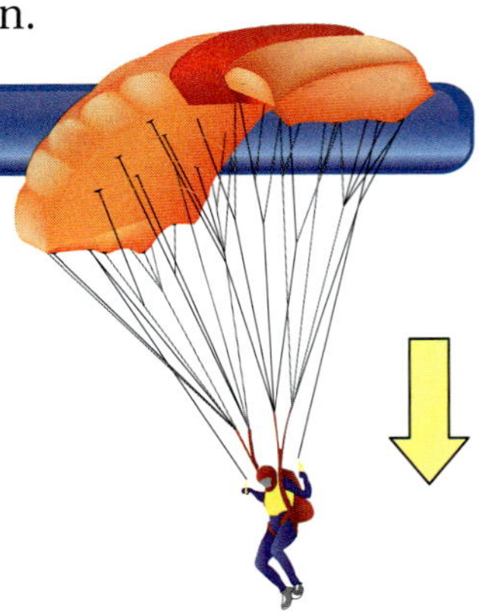

Work with a partner. You are gliding to the ground wearing a parachute. The table shows your height above the ground at different times.

Time (seconds)	0	1	2	3
Height (feet)	90	75	60	45

a. Describe the pattern in the table. How many feet do you move each second? After how many seconds will you land on the ground?

b. What integer represents your speed? Give the units.

c. Do you think your velocity should be represented by a positive or negative integer? Explain your reasoning.

d. What integer represents your velocity? Give the units.

2 ACTIVITY: Rising Balloons

Work with a partner. You release a group of balloons. The table shows the height of the balloons above the ground at different times.

Time (seconds)	0	1	2	3
Height (feet)	8	12	16	20

a. Describe the pattern in the table. How many feet do the balloons move each second? After how many seconds will the balloons be at a height of 40 feet?

b. What integer represents the speed of the balloons? Give the units.

c. Do you think the velocity of the balloons should be represented by a positive or negative integer? Explain your reasoning.

d. What integer represents the velocity of the balloons? Give the units.

Integers

In this lesson, you will
- define the absolute value of a number.
- find absolute values of numbers.
- solve real-life problems.

Laurie's Notes

Introduction

Applying Mathematical Practices

- **Attend to Precision.** In this lesson, students will use integers to describe the velocity of an object. Although integers have not been defined formally, students have heard of negative numbers. In discussing speed and velocity, ask students for examples, being sure that appropriate units are stated.

Motivate

- Ask students if they have ever watched the launch of a NASA shuttle. The velocity of the shuttle describes both its speed and its direction. Today's activity looks at the velocities of various objects.
- **Model:** Use a handkerchief to make a simple parachute by stapling or taping a piece of string or yarn to each corner of the handkerchief. Tie a paper clip to the loose ends of the string or yarn.
- Have two students stand on chairs. One student drops a paper clip without the parachute and the other drops the paper clip with the opened parachute.
- Both paper clips have speed, which can be measured in ft/sec. They also have velocity, moving down, so the velocity is negative and the units would still be ft/sec.
- Write the definitions of speed and velocity on the board and ask for (or share) examples of each.
  - **Examples of speed:** a pitcher throws a 94 mi/h fastball; a manatee swims 4 mi/h; a football travels 15 ft/sec
  - **Examples of velocity:** a NASA shuttle has an orbital velocity of 17,500 mi/h; you walk to school at 8 ft/sec; a feather falls at -2 ft/sec

Activity Notes

Activity 1

? "Do you see any pattern(s) in the table? Describe the pattern(s)." The heights are going down by 15. "Is there a name for this collection of numbers: 90, 75, 60, 45, ...?" multiples of 15

? "When will you land on the ground? Explain how you know." You land after 6 seconds. Students may say that $6 \times 15 = 90$ so after 6 seconds the person is on the ground.

- Discuss the difference between the speed of 15 feet per second and the velocity of -15 feet per second.
- **Attend to Precision:** Be sure to point out that in the table, each of the quantities have units associated with them. The labeling of units is a good habit to develop.

Activity 2

- This activity is similar to Activity 1 except the balloons are rising versus a parachute falling. The speed and velocity will be positive.

? "What was the height of the balloons when they were released? Explain your reasoning." 8 feet; the height when time equals 0 seconds

What Your Students Will Learn

- Define the absolute value of a number.
- Find and compare absolute values of numbers.

Previous Learning

Students should know how to compare, order, and graph integers.

Technology for the Teacher

Dynamic Classroom

Lesson Plans
Complete Materials List

11.1 Record and Practice Journal

Essential Question How can you use integers to represent the velocity and the speed of an object?

On these three pages, you will investigate vertical motion (up or down).

- Speed tells how fast an object is moving, but it does not tell the direction.
- Velocity tells how fast an object is moving, and it also tells the direction.

When velocity is positive, the object is moving up.

When velocity if negative, the object is moving down.

1 ACTIVITY: Falling Parachute

Work with a partner. You are gliding to the ground wearing a parachute. The table shows your height above the ground at different times.

Time (seconds)	0	1	2	3
Height (feet)	90	75	60	45

a. Describe the pattern in the table. How many feet do you move each second? After how many seconds will you land on the ground?
After each second, your height decreases by 15 feet; 15 ft; 6 sec

b. What integer represents your speed? Give the units.
15 ft/sec

c. Do you think your velocity should be represented by a positive or negative integer? Explain your reasoning.
negative

d. What integer represents your velocity? Give the units.
-15 ft/sec

English Language Learners

Vocabulary

On a long strip of paper, mark integers from −10 through 10, with zero in the middle. Hold the strip vertically so that −10 touches the ground. Discuss that when an object is moving from 0 to 10, the *velocity* is positive. Point out the numbers 1 through 10 on the strip are positive integers. Then discuss that an object moving from zero down to the ground has a negative velocity. Again refer to the number strip, pointing out that −1 through −10 are negative integers.

11.1 Record and Practice Journal

2 ACTIVITY: Rising Balloons

Work with a partner. You release a group of balloons. The table shows the height of the balloons above the ground at different times.

Time (seconds)	0	1	2	3
Height (feet)	8	12	16	20

a. Describe the pattern in the table. How many feet do the balloons move each second? After how many seconds will the balloons be at a height of 40 feet?

After each second, the height increases by 4 feet; 4 ft; 8 sec

b. What integer represents the speed of the balloons? Give the units.

4 ft/sec

c. Do you think the velocity of the balloons should be represented by a positive or negative integer? Explain your reasoning.

positive

d. What integer represents the velocity of the balloons? Give the units.

4 ft/sec

3 ACTIVITY: Firework Parachute

Work with a partner. The table shows the height of a firework's parachute above the ground at different times.

Time (seconds)	Height (feet)
0	480
1	360
2	240
3	120
4	0

a. Describe the pattern in the table. How many feet does the parachute move each second? After each second, the height decreases by 120 feet; 120 ft

b. What integer represents the speed of the parachute? What integer represents the velocity? How are these integers similar in their relation to 0 on a number line?

120 ft/sec; −120 ft/sec; same distance from zero

Inductive Reasoning

4. Complete the table.

Velocity (feet per second)	−14	20	−2	0	25	−15
Speed (feet per second)	14	20	2	0	25	15

5. Find two different velocities for which the speed is 16 feet per second.

−16 ft/sec; 16 ft/sec

6. Which number is greater: −4 or 3? Use a number line to explain your reasoning. 3

7. One object has a velocity of −4 feet per second. Another object has a velocity of 3 feet per second. Which object has the greater speed? Explain your answer.

The object with the velocity of −4 ft/sec

8. **IN YOUR OWN WORDS** How can you use integers to represent the velocity and the speed of an object?

speed: positive integers
velocity: positive and negative integers

What Is Your Answer?

9. **LOGIC** In this lesson, you will study **absolute value**. Here are some examples:

Absolute value of $|-16| = 16$ Absolute value of $|16| = 16$

Absolute value of $|0| = 0$ Absolute value of $|-2| = 2$

Which of the following is a true statement? Explain your reasoning.

$|\text{velocity}| = \text{speed}$ $|\text{speed}| = \text{velocity}$

$|\text{velocity}| = \text{speed}$

Laurie's Notes

Activity 3

- **Think-Pair-Share:** Students should read each question independently and then work in pairs to answer the questions. When they have answered the questions, the pair should compare their answers with another group and discuss any discrepancies.
- Use curved arrows to show the change in time and the change in height.

Change	Time (in seconds)	Height (in feet)	Change
	0	480	
+1	1	360	−120
+1	2	240	−120
+1	3	120	−120
+1	4	0	−120

Inductive Reasoning

- In Question 4, if students have connected the relationship between speed and velocity, they will be able to complete the table. To reinforce labeling of answers, have students write *ft/sec* with each entry in the table.
- Questions 5–7 are developing the notion of absolute value—there are two velocities that have a speed of 16 ft/sec, namely 16 ft/sec and −16 ft/sec.
- In Question 6, it is common for students to say −4 > 3. Remind students that a number farther to the right on a number line is greater, so 3 > −4.
- In Question 7, the sign (negative or positive) is not considered because the question concerns speed and the direction does not matter. An object moving 4 ft/sec has a greater speed than an object moving 3 ft/sec.

What Is Your Answer?

- In Question 9, speed is the **absolute value** of velocity.

Words of Wisdom

- A formal definition of absolute value will be presented in the lesson.
- Do not let students suggest that absolute value simply means to take away the negative sign. This could cause problems in the future when students work with variables or variable expressions within the absolute value symbols. Direct discussions toward the idea that you want to know how far (distance) a number is from zero.

Closure

- **Communication:** Give examples of velocities of two objects (A and B), where Velocity A > Velocity B but Speed A < Speed B.
- Examples are shown below.

Velocity Object A	Velocity Object B	Compare Velocities	Compare Speeds
4 ft/sec	−5 ft/sec	Vel (A) > Vel (B)	Sp (A) < Sp (B)
10 ft/sec	−15 ft/sec	Vel (A) > Vel (B)	Sp (A) < Sp (B)
−5 ft/sec	−15 ft/sec	Vel (A) > Vel (B)	Sp (A) < Sp (B)

3 ACTIVITY: Firework Parachute

Work with a partner. The table shows the height of a firework's parachute above the ground at different times.

Time (seconds)	Height (feet)
0	480
1	360
2	240
3	120
4	0

Math Practice

Use Clear Definitions

What information can you use to support your answer?

a. Describe the pattern in the table. How many feet does the parachute move each second?

b. What integer represents the speed of the parachute? What integer represents the velocity? How are these integers similar in their relation to 0 on a number line?

Inductive Reasoning

4. Copy and complete the table.

Velocity (feet per second)	-14	20	-2	0	25	-15
Speed (feet per second)						

5. Find two different velocities for which the speed is 16 feet per second.

6. Which number is greater: -4 or 3? Use a number line to explain your reasoning.

7. One object has a velocity of -4 feet per second. Another object has a velocity of 3 feet per second. Which object has the greater speed? Explain your answer.

What Is Your Answer?

8. **IN YOUR OWN WORDS** How can you use integers to represent the velocity and the speed of an object?

9. **LOGIC** In this lesson, you will study *absolute value*. Here are some examples:

$|-16| = 16$ $\quad$ $|16| = 16$ $\quad$ $|0| = 0$ $\quad$ $|-2| = 2$

Which of the following is a true statement? Explain your reasoning.

$|\text{velocity}| = \text{speed}$ $\qquad$ $|\text{speed}| = \text{velocity}$

Practice

Use what you learned about absolute value to complete Exercises 4–11 on page 480.

11.1 Lesson

Check It Out
Lesson Tutorials
BigIdeasMath.com

Key Vocabulary
integer, *p. 478*
absolute value, *p. 478*

The following numbers are **integers:**

$\ldots, -3, -2, -1, 0, 1, 2, 3, \ldots$

Key Idea

Absolute Value

Words The **absolute value** of an integer is the distance between the number and 0 on a number line. The absolute value of a number a is written as $|a|$.

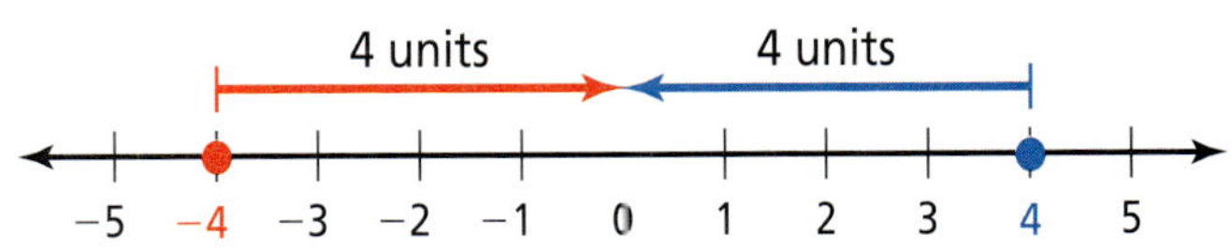

Numbers $|-4| = 4$ $|4| = 4$

EXAMPLE 1 Finding Absolute Value

Find the absolute value of 2.

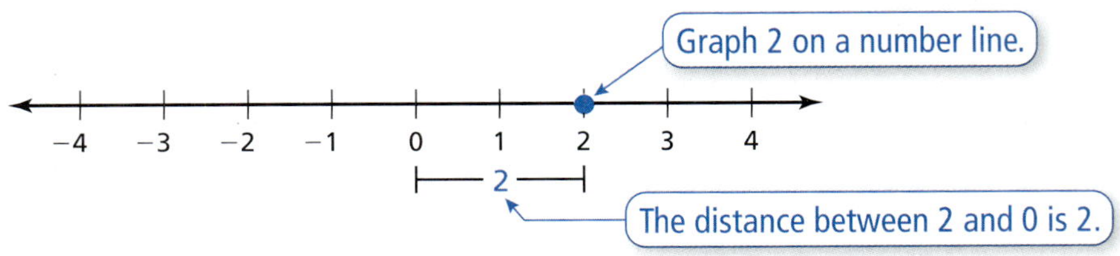

So, $|2| = 2$.

EXAMPLE 2 Finding Absolute Value

Find the absolute value of −3.

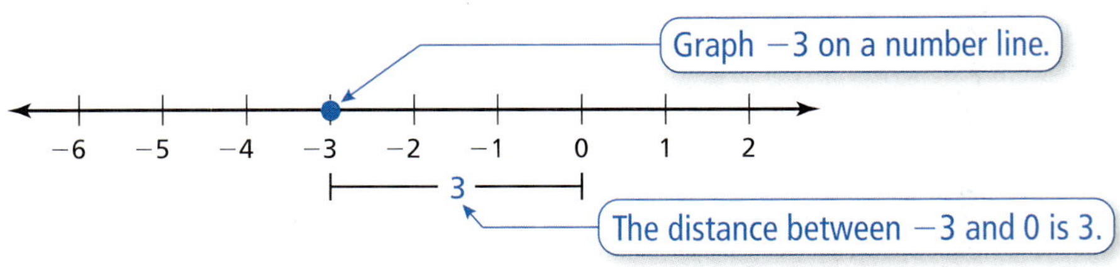

So, $|-3| = 3$.

On Your Own

Now You're Ready
Exercises 4–19

Find the absolute value.

1. $|7|$ **2.** $|-1|$ **3.** $|-5|$ **4.** $|14|$

Laurie's Notes

Introduction

Connect

- **Yesterday:** Students explored speed and velocity in the activity. They used the relationship between speed and velocity to find speed (the absolute value of velocity) when the velocity was known.
- **Today:** Have students think about the distance a number is from 0. This distance is always positive, just like the speed of an object.

Motivate

- Have two students (A and B) stand at the front of the room with a piece of string between them. You hold a piece of paper with the number 0 written on it. State that the distance between A and B is 10 units.

? Position yourself at the midpoint of A and B.
  - "If A is 5, what number does B represent?" −5
  - "Who is closer to me (meaning 0)?" Neither; both are the same distance.
  - "How far away from me is each person?" 5 units

? Move closer to A so that if A is 3, B would be approximately −7.
  - "If A is 3, what number does B represent?" −7
  - "Who is closer to me (meaning 0)?" A
  - "How far away from me is each person?" A is 3 units. B is 7 units.

? Move closer to B so that if B is −2, A would be approximately 8.
  - "If B is −2, what number does A represent?" 8
  - "Who is closer to me (meaning 0)?" B
  - "How far away from me is each person?" A is 8 units. B is 2 units.

? Without the string, ask students, "What number or numbers are 6 units from 0?" 6 and −6

Lesson Notes

Key Idea

- When students state that "absolute values are always positive," try to clarify their statement. Students may make incorrect assumptions when there are numeric or variable expressions within absolute value symbols.
- At this stage of development, stress the geometric definition of absolute value. **Absolute value** is the distance a number is from zero.

Example 1 and Example 2

- Work through Examples 1 and 2 and then ask the following:
  - ? "What is the absolute value of 12?" 12
  - ? "What is the absolute value of −8?" 8
- **Common Error:** When students plot 2 on the number line, they may make a scale on the number line and fail to put a dot on the number 2. Remind students that plotting a point involves actually putting a dot on the number line at the point they are plotting. Remind them as well that the numbers go below the number line, not above it.

Goal Today's lesson is finding the **absolute value** of an **integer**.

Technology for the Teacher

Dynamic Classroom

Lesson Tutorials
Lesson Plans
Answer Presentation Tool

Extra Example 1

Find the absolute value of 6. 6

Extra Example 2

Find the absolute value of −11. 11

On Your Own

1. 7 **2.** 1

3. 5 **4.** 14

Extra Example 3

Compare $|-9|$ and 7. $|-9| > 7$

On Your Own

5. $|-2| > -1$
6. $-7 < |6|$
7. $|10| < 11$
8. $9 = |-9|$

Extra Example 4

Seawater freezes at −2°C. Is the freezing point of honey (from Example 4) or seawater closer to the freezing point of water, 0°C? seawater

On Your Own

9. airplane fuel; Because $|-53| < |55|$, the freezing point of airplane fuel is closer to 0°C, the freezing point of water.

Differentiated Instruction

Kinesthetic

Stand in front of the classroom with two students, one on each side. Tell the students that you represent zero, your left (the students' right) is the positive direction, and your right (the students' left) is the negative direction. Have the student on your left walk three paces away from you. Say, "This student represents +3." Have the student on the right walk three paces from you. Say, "This student represents −3." Ask, "How far is each student from me?" (3 paces away). Say, "Positive 3 and negative 3 are the same distance from zero, so they have the same absolute value, which is 3."

Laurie's Notes

Example 3

- Students may mix up or forget the inequality symbols.
- **Common Error:** Students may label the number line left-to-right for both positive and negative integers as shown below. Explain that negative integers are labeled from 0, such that 1 and −1 are both one unit from 0.

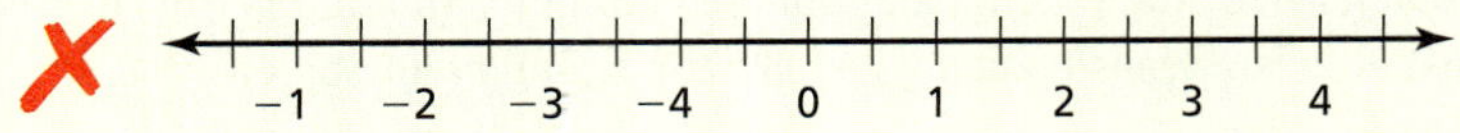

- **Attend to Precision:** When students write their answers, check to see that the absolute value symbols are included. It would be wrong to write $1 < -4$. The correct answer is $1 < |-4|$.

On Your Own

- Make sure that students understand that when you write the notation for the absolute value, it means *take the absolute value of the number inside the symbols.*

Example 4

- Have students discuss different liquids that they know freeze, such as water, ice cream, and chocolate. Probe to see if students know that water freezes at 0°C and 32°F.
- **FYI:** Citrus fruit trees can sustain damage from low temperatures. Depending upon the temperature, the leaves, wood, or fruit can be damaged, causing economic problems for both the growers and the consumers.
- In part (a), discuss the substances listed in the table.
  - ? "Why is the point representing −3 closer to 0 than −10?" −3 is 3 units to the left of 0, and −10 is 10 units to the left of 0. Because 3 is less than 10, −3 is closer to 0 than −10.
  - ? "Why is it important for airplane fuel to have a very low freezing point?" Planes flying in very cold temperatures need fuel to remain in liquid form.
- In part (b), to answer the question "Which substance has a freezing point closer to the freezing point of water?" be sure students understand that the absolute value of the freezing point is how you measure the distance to 0.

Words of Wisdom

- **Construct Viable Arguments and Critique the Reasoning of Others:** Too often students will say, "Because it is," or "It's obvious." Look for a reference to absolute value when students explain their reasoning.

Closure

- **Exit Ticket:** The freezing point of vinegar is −2°C. Is the freezing point of vinegar or honey closer to the freezing point of water? Explain your reasoning. vinegar; Because $|-2| < |-3|$, the freezing point of vinegar is closer to 0°C, the freezing point of water.

EXAMPLE 3 Comparing Values

Remember

A number line can be used to compare and order integers. Numbers to the left are less than numbers to the right. Numbers to the right are greater than numbers to the left.

Compare 1 and $|-4|$.

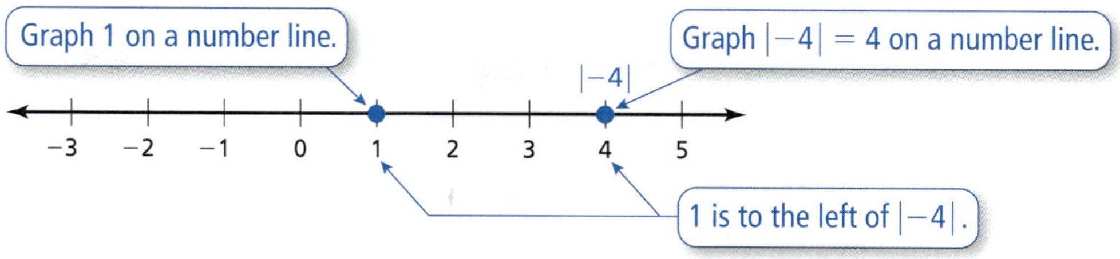

So, $1 < |-4|$.

On Your Own

Now You're Ready Exercises 20–25

Copy and complete the statement using <, >, or =.

5. $|-2|$ ▢ -1

6. -7 ▢ $|6|$

7. $|10|$ ▢ 11

8. 9 ▢ $|-9|$

EXAMPLE 4 Real-Life Application

Substance	Freezing Point (°C)
Butter	35
Airplane fuel	−53
Honey	−3
Mercury	−39
Candle wax	55

The *freezing point* is the temperature at which a liquid becomes a solid.

a. Which substance in the table has the lowest freezing point?

b. Is the freezing point of mercury or butter closer to the freezing point of water, 0°C?

a. Graph each freezing point.

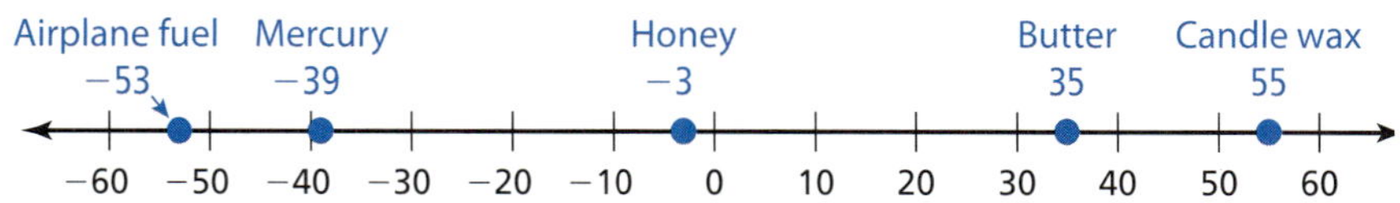

Airplane fuel has the lowest freezing point, −53°C.

b. The freezing point of water is 0°C, so you can use absolute values.

Mercury: $|-39| = 39$ **Butter:** $|35| = 35$

Because 35 is less than 39, the freezing point of butter is closer to the freezing point of water.

On Your Own

9. Is the freezing point of airplane fuel or candle wax closer to the freezing point of water? Explain your reasoning.

11.1 Exercises

Check It Out
Help with Homework
BigIdeasMath.com

Vocabulary and Concept Check

1. **VOCABULARY** Which of the following numbers are integers?

$$9, 3.2, -1, \frac{1}{2}, -0.25, 15$$

2. **VOCABULARY** What is the absolute value of an integer?

3. **WHICH ONE DOESN'T BELONG?** Which expression does *not* belong with the other three? Explain your reasoning.

$\lvert 6 \rvert$	6	-6	$\lvert -6 \rvert$

Practice and Problem Solving

Find the absolute value.

1 2 4. $\lvert 9 \rvert$ 5. $\lvert -6 \rvert$ 6. $\lvert -10 \rvert$ 7. $\lvert 10 \rvert$

8. $\lvert -15 \rvert$ 9. $\lvert 13 \rvert$ 10. $\lvert -7 \rvert$ 11. $\lvert -12 \rvert$

12. $\lvert 5 \rvert$ 13. $\lvert -8 \rvert$ 14. $\lvert 0 \rvert$ 15. $\lvert 18 \rvert$

16. $\lvert -24 \rvert$ 17. $\lvert -45 \rvert$ 18. $\lvert 60 \rvert$ 19. $\lvert -125 \rvert$

Copy and complete the statement using <, >, or =.

3 20. $2 \;\square\; \lvert -5 \rvert$ 21. $\lvert -4 \rvert \;\square\; 7$ 22. $-5 \;\square\; \lvert -9 \rvert$

23. $\lvert -4 \rvert \;\square\; -6$ 24. $\lvert -1 \rvert \;\square\; \lvert -8 \rvert$ 25. $\lvert 5 \rvert \;\square\; \lvert -5 \rvert$

ERROR ANALYSIS Describe and correct the error.

26. ✗ $\lvert 10 \rvert = -10$

27. ✗ $\lvert -5 \rvert < 4$

28. **SAVINGS** You deposit \$50 in your savings account. One week later, you withdraw \$20. Write each amount as an integer.

29. **ELEVATOR** You go down 8 floors in an elevator. Your friend goes up 5 floors in an elevator. Write each amount as an integer.

Order the values from least to greatest.

30. $8, \lvert 3 \rvert, -5, \lvert -2 \rvert, -2$ 31. $\lvert -6 \rvert, -7, 8, \lvert 5 \rvert, -6$

32. $-12, \lvert -26 \rvert, -15, \lvert -12 \rvert, \lvert 10 \rvert$ 33. $\lvert -34 \rvert, 21, -17, \lvert 20 \rvert, \lvert -11 \rvert$

Simplify the expression.

34. $\lvert -30 \rvert$ 35. $-\lvert 4 \rvert$ 36. $-\lvert -15 \rvert$

Assignment Guide and Homework Check

Level	Assignment	Homework Check
Advanced	1–11, 20–44 even 46–50	24, 30, 38, 40

Common Errors

- **Exercises 4–19** Students may think that the absolute value of a number is its opposite and say $|6| = -6$. Use a number line to show them that the absolute value is a number's distance from 0, so it is always a positive number or zero.
- **Exercises 20–27** When comparing absolute values of negative integers, students may not find the absolute values and instead compare the integers themselves.
- **Exercise 22** A student may find $-5 > |-9|$. The student likely did not find the absolute value of -9 to be 9 and instead compared -5 to -9.
- **Exercise 41** Students may write 14 and 18, rather than -14 and -18, for the diver's positions. These students did not account for the fact that the diver is *below* sea level. Use a vertical scale by turning a number line so that it runs vertically. Point out to students that the exercise defines sea level as 0 on the number line, so the diver's positions are negative.

11.1 Record and Practice Journal

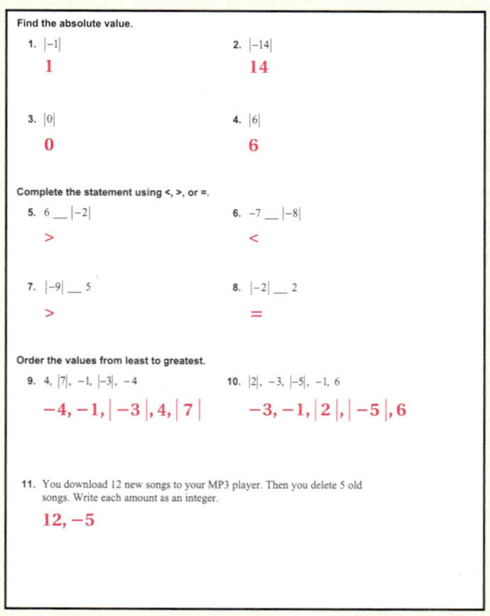

Find the absolute value.

1. $|-1|$ **1**
2. $|-14|$ **14**
3. $|0|$ **0**
4. $|6|$ **6**

Complete the statement using <, >, or =.

5. $6 \_\_ |-2|$ **>**
6. $-7 \_\_ |-8|$ **<**
7. $|-9| \_\_ 5$ **>**
8. $|-2| \_\_ 2$ **=**

Order the values from least to greatest.

9. $4, |7|, -1, |-3|, -4$ **$-4, -1, |-3|, 4, |7|$**
10. $|2|, -3, |-5|, -1, 6$ **$-3, -1, |2|, |-5|, 6$**

11. You download 12 new songs to your MP3 player. Then you delete 5 old songs. Write each amount as an integer.
12, -5

Vocabulary and Concept Check

1. $9, -1, 15$
2. the distance between the integer and zero on a number line
3. -6; All of the other expressions are equal to 6.

Practice and Problem Solving

4. 9
5. 6
6. 10
7. 10
8. 15
9. 13
10. 7
11. 12
12. 5
13. 8
14. 0
15. 18
16. 24
17. 45
18. 60
19. 125
20. $2 < |-5|$
21. $|-4| < 7$
22. $-5 < |-9|$
23. $|-4| > -6$
24. $|-1| < |-8|$
25. $|5| = |-5|$
26. The absolute value of a number cannot be negative. $|10| = 10$
27. Because $|-5| = 5$, the statement is incorrect. $|-5| > 4$
28. $50, -20$
29. $-8, 5$
30. $-5, -2, |-2|, |3|, 8$
31. $-7, -6, |5|, |-6|, 8$
32. $-15, -12, |10|, |-12|, |-26|$
33. $-17, |-11|, |20|, 21, |-34|$
34. 30
35. -4
36. -15

Practice and Problem Solving

37. a. MATE

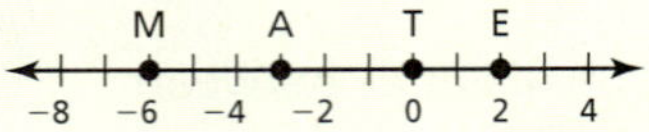

b. TEAM

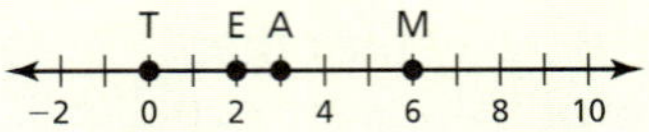

38. *Sample answer:* -4

39. $n \geq 0$

40. $n \leq 0$

41. See *Taking Math Deeper.*

42. Loihi

43. a. Player 3

b. Player 2

c. Player 1

44. true; If a number x is negative, then its absolute value is its opposite, $-x$.

45. false; The absolute value of zero is zero, which is neither positive nor negative.

Fair Game Review

46. 51 **47.** 144

48. 398 **49.** 3170

50. A

Mini-Assessment

Find the absolute value of the integer.

1. 6 6 **2.** −13 13

3. −17 17 **4.** 0 0

5. You deposit $125 in your checking account. One month later, you withdraw $65. Write each amount as an integer. 125; −65

Taking Math Deeper

Exercise 41

In this problem, negative numbers are used on a vertical scale to indicate *positions* that are below sea level. The numbers on the number line describe *position* (a negative number), not *depth* below sea level (a positive number).

1 **a.** Draw and label a vertical number line.

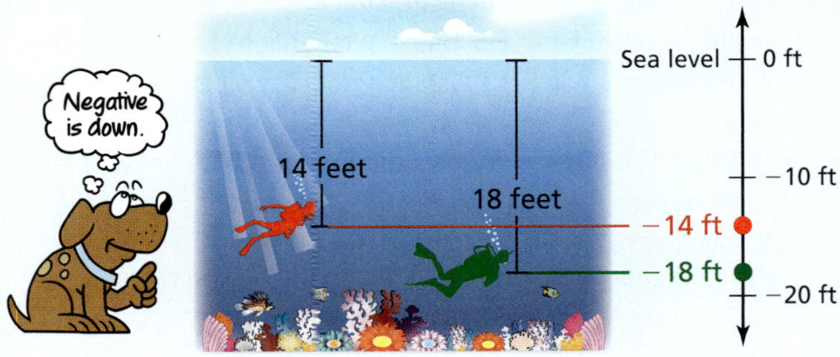

2 **b.** Which integer is greater?

Of the two integers −14 and −18, −14 is greater because it is *higher* on a vertical number line.

c. Which integer has the greater absolute value?

$|-18| = 18$ is greater than $|-14| = 14$.

Compare this absolute value with the depth of that diver. The numbers are the same. The diver is 18 feet below sea level.

3 In mathematics, many concepts require understanding. Some, however, simply require acceptance. No one really knows why negative means "left" or "down" on a number line. We are not sure of the reason for choosing "right" to be positive, but it could have been something as simple as the fact that René Descartes was right-handed.

Project

Draw a picture that illustrates a real-life use of negative integers. Write a paragraph that explains how negative numbers are used in your picture.

Reteaching and Enrichment Strategies

If students need help...	If students got it...
Resources by Chapter • Practice A and Practice B • Puzzle Time Record and Practice Journal Practice Differentiating the Lesson Lesson Tutorials Skills Review Handbook	Resources by Chapter • Enrichment and Extension • Technology Connection Start the next section

37. **PUZZLE** Use a number line.

a. Graph and label the following points on a number line: $A = -3$, $E = 2$, $M = -6$, $T = 0$. What word do the letters spell?

b. Graph and label the absolute value of each point in part (a). What word do the letters spell now?

38. **OPEN-ENDED** Write a negative integer whose absolute value is greater than 3.

REASONING **Determine whether $n \geq 0$ or $n \leq 0$.**

39. $n + |-n| = 2n$

40. $n + |-n| = 0$

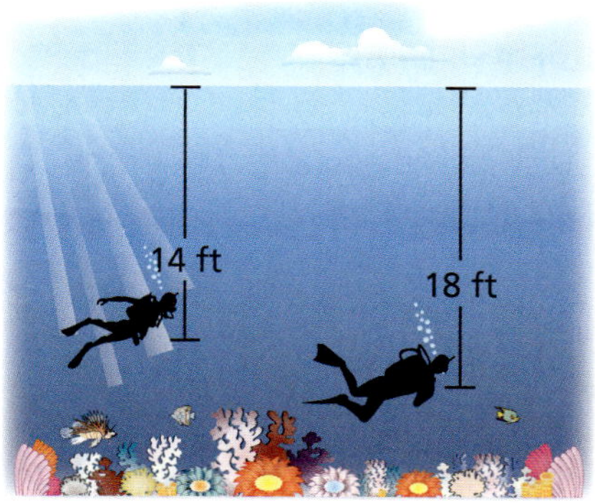

41. **CORAL REEF** The depths of two scuba divers exploring a living coral reef are shown.

a. Write an integer for the position of each diver relative to sea level.

b. Which integer in part (a) is greater?

c. Which integer in part (a) has the greater absolute value? Compare this absolute value with the depth of that diver.

42. **VOLCANOES** The *summit elevation* of a volcano is the elevation of the top of the volcano relative to sea level. The summit elevation of the volcano Kilauea in Hawaii is 1277 meters. The summit elevation of the underwater volcano Loihi in the Pacific Ocean is −969 meters. Which summit is closer to sea level?

43. **MINIATURE GOLF** The table shows golf scores, relative to *par*.

a. The player with the lowest score wins. Which player wins?

b. Which player is at par?

c. Which player is farthest from par?

Player	Score
1	+5
2	0
3	−4
4	−1
5	+2

True or False? **Determine whether the statement is *true* or *false*. Explain your reasoning.**

44. If $x < 0$, then $|x| = -x$.

45. The absolute value of every integer is positive.

Fair Game Review What you learned in previous grades & lessons

Add. *(Section 1.1)*

46. $19 + 32$

47. $50 + 94$

48. $181 + 217$

49. $1149 + 2021$

50. **MULTIPLE CHOICE** Which value is *not* a whole number? *(Skills Review Handbook)*

Ⓐ −5 Ⓑ 0 Ⓒ 4 Ⓓ 113

11.2 Adding Integers

Essential Question Is the sum of two integers *positive*, *negative*, or *zero*? How can you tell?

1 ACTIVITY: Adding Integers with the Same Sign

Work with a partner. Use integer counters to find $-4 + (-3)$.

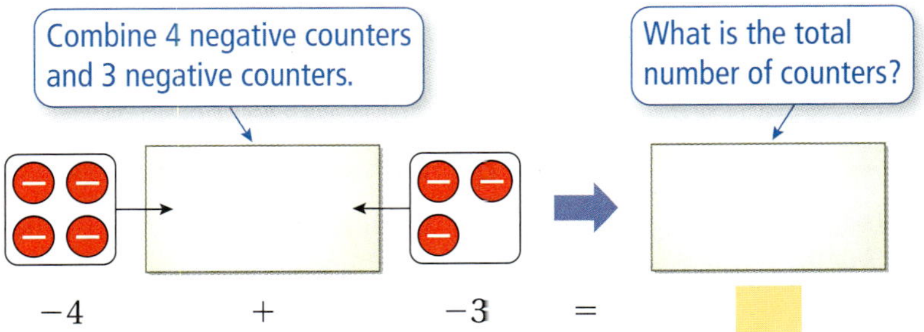

So, $-4 + (-3) =$ ______.

2 ACTIVITY: Adding Integers with Different Signs

Work with a partner. Use integer counters to find $-3 + 2$.

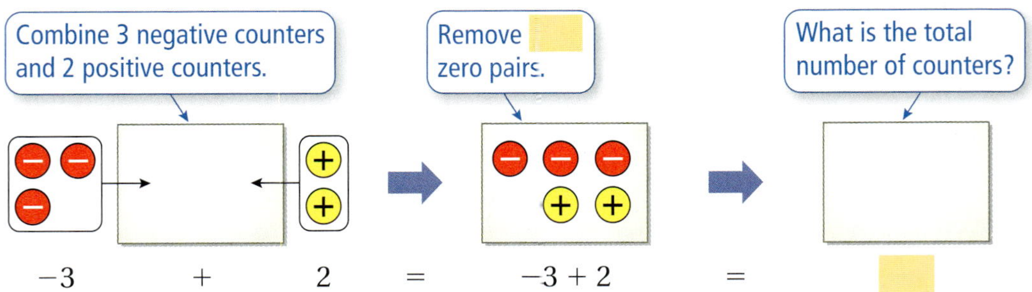

So, $-3 + 2 =$ ______.

Integers

In this lesson, you will

- add integers.
- show that the sum of a number and its opposite is 0.
- solve real-life problems.

3 ACTIVITY: Adding Integers with Different Signs

Work with a partner. Use a number line to find $5 + (-3)$.

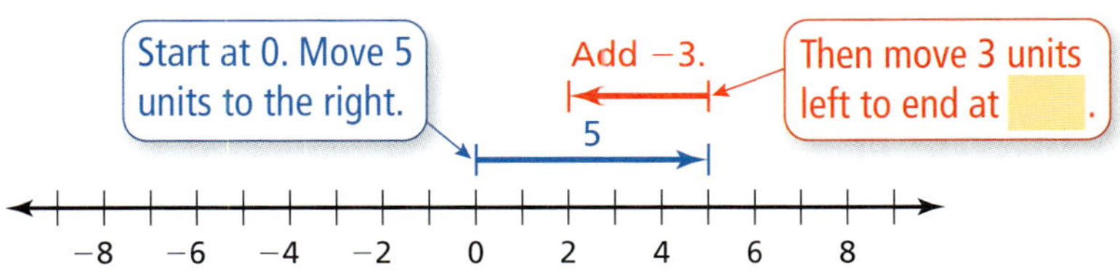

So, $5 + (-3) =$ ______.

Laurie's Notes

Introduction

Applying Mathematical Practices

- **Look For and Express Regularity in Repeated Reasoning.** In this lesson, students will use color integer counters to develop a conceptual understanding of integer addition. When the signs are the same, it is a combining process and the sum has the same sign. When the signs are different, it is a combining process *and* you remove some zero pairs.

Motivate

? "What is the net result of an 8-yard loss in football followed by a 10-yard gain?" 2-yard gain

? "What is the net result of scoring 25 points in a video game then losing 40 points?" a loss of 15 points

- Today's activity is about how integers are added.

Demonstrate

- If this is the student's first experience with integer counters, define a *yellow* counter as positive 1 $(+1)$ and a *red* counter as negative 1 (-1).
- Counters of opposite color "neutralize" each other, so the net result of such a pair is zero. This is called a *zero pair.*
- **Model:** Show students that there are many ways to model a single integer. For example, the number 2 can be represented by two yellow counters or by three yellow counters with one red counter (2 plus 1 zero pair).

Activity Notes

Activity 1 and Activity 2

- **Management Tip:** Store integer counters in self-locking bags. Put 15–20 counters in each bag.
- Students should use counters even if they say they know the answer.
- **Model:** A student volunteer could model Activity 1 and Activity 2 at the overhead projector saying aloud what he or she is doing with the counters.
- **Common Error:** You may hear students say that -3 is greater than 2. You should respond "Gee, 2 is farther to the right on the number line than -3. Are you sure -3 is greater?" Remind students that the number farther to the right on the number line is greater.
- The use of parentheses around the integer -3 is for clarity. Sometimes people write $-4 + {}^{-}3$, with the raised negative sign.

Activity 3

- Numbers are being represented by *directed line segments.* Positive numbers point to the right and negative numbers point to the left.
- **Connection:** The amount that the two directed line segments overlap is the same as the number of zero pairs that would result if the same problem were modeled using integer counters.

What Your Students Will Learn

- Add integers with the same sign or with different signs.
- Show that the sum of a number and its opposite is 0 using the Additive Inverse Property.

Previous Learning

Students need to know how to add and subtract whole numbers.

Technology for the Teacher

Dynamic Classroom

Lesson Plans
Complete Materials List

11.2 Record and Practice Journal

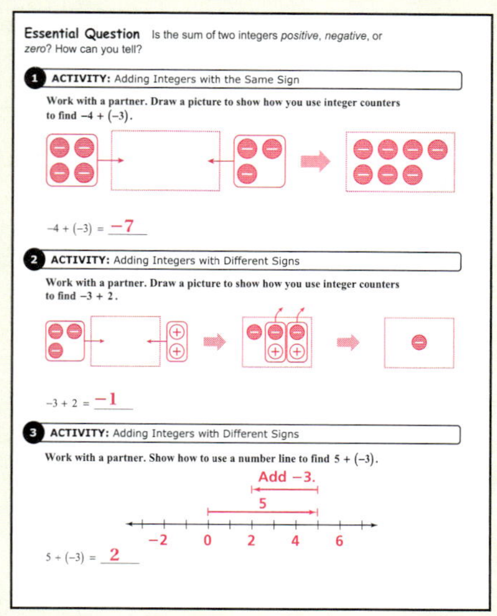

Essential Question Is the sum of two integers *positive, negative,* or *zero*? How can you tell?

1 ACTIVITY: Adding Integers with the Same Sign

Work with a partner. Draw a picture to show how you use integer counters to find $-4 + (-3)$.

$-4 + (-3) = -7$

2 ACTIVITY: Adding Integers with Different Signs

Work with a partner. Draw a picture to show how you use integer counters to find $-3 + 2$.

$-3 + 2 = -1$

3 ACTIVITY: Adding Integers with Different Signs

Work with a partner. Show how to use a number line to find $5 + (-3)$.

$5 + (-3) = 2$

Differentiated Instruction

Visual

Use the number line to review and reinforce the Commutative Property of Addition. Draw a number line on the board. Model the expressions $-5 + 2$ and $2 + (-5)$. Students should see from the movement on the number line that the sum (result) is the same. The initial direction of movement does not matter.

11.2 Record and Practice Journal

4 ACTIVITY: Adding Integers with Different Signs

Work with a partner. Write the addition expression shown. Then find the sum. How are the integers in the expression related to 0 on a number line?

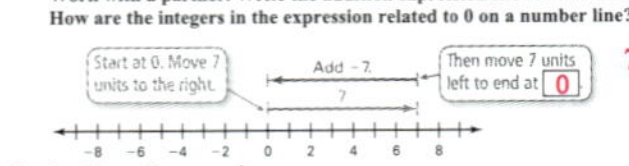

$7 + (-7) = 0$

Inductive Reasoning

Work with a partner. Use integer counters or a number line to complete the table.

	Exercise	Type of Sum	Sum	Sum: Positive, Negative, or Zero
1	5. $-4 + (-3)$	same sign	-7	Negative
2	6. $-3 + 2$	different signs	-1	Negative
3	7. $5 + (-3)$	different signs	2	Positive
4	8. $7 + (-7)$	different signs	0	Zero
	9. $2 + 4$	same sign	6	Positive
	10. $-6 + (-2)$	same sign	-8	Negative
	11. $-5 + 9$	different signs	4	Positive
	12. $15 + (-9)$	different signs	6	Positive
	13. $-10 + 10$	different signs	0	Zero
	14. $-6 + (-6)$	same sign	-12	Negative
	15. $13 + (-13)$	different signs	0	Zero

What Is Your Answer?

16. **IN YOUR OWN WORDS** Is the sum of two integers *positive*, *negative*, or *zero*? How can you tell?
Sample answer: The sum will have the same sign as the integer with the greater absolute value.

17. **STRUCTURE** Write a general rule for adding Sample answers given.
a. two integers with the same sign.
To add two positive integers, add normally. To add two negative integers, ignore the signs and add. Then make the answer negative.
b. two integers with different signs.
Subtract the lesser absolute value from the greater absolute value. Use the sign of the integer with the greater absolute value.
c. two integers that vary in sign.
The sum is zero.

Laurie's Notes

Activity 4

? "How would you use the number line to show $-7 + 7$?" Ask a student volunteer to show the solution. Start at zero. Move 7 units to the left to get to -7. Then move 7 units to the right. End at 0.

- Students should recognize that the order in which you add two numbers does not matter (Commutative Property of Addition). You can start at 7 and then move 7 units to the left.

Words of Wisdom

- You can use a standard deck of playing cards to generate random addition problems. Define red = negative, black = positive, Jacks = 11, Queens = 12, Kings = 13, Aces = 1, Jokers = 0.

Inductive Reasoning

- Students should work with partners to complete the table. Note that Questions 5–8 are the problems completed in Activities 1–4.
- The goal of Questions 5–15 is to develop some understanding about the two types of addition problems: integers with the same sign and integers with different signs.
- **Look for and Express Regularity in Repeated Reasoning:** It is important for students to record the *Type of Sum*. Mathematically proficient students will observe that when the signs are the same, you "just add." When the signs are different, the strategy is different, and students may not be articulate in describing the pattern.
- If time permits, you might give a problem such as $47 + (-58)$. Clearly, you do not want to model this with color integer counters. Students should be able to describe the strategy of putting 47 yellow and 58 red together. There would be 47 zero pairs with 11 red remaining representing a sum of -11.

What Is Your Answer?

- In Questions 16 and 17, the sum of two integers can be positive, negative, or zero. If the two integers have the same sign, the sign of the sum is the same as the integers. If the integers have different signs, then the sum is the sign of the integer with the greater absolute value. (Formal rules will be presented in the lesson.)

? **Extension:** Use the integer counters or a number line to model the sum of three integers.

- "What is the sum of $3 + (-2) + 5$?" 6
- "What is the sum of $(-4) + 2 + (-5)$?" -7

Closure

- "If the sum of two integers is negative, are both integers negative? How do you know?" Both integers could be negative, but they may not be. The sum of 4 and -5 is negative, but both integers are not negative.
- "If the sum of two integers is positive, are both integers positive? How do you know?" Both integers could be positive, but they may not be. The sum of -4 and 5 is positive, but both integers are not positive.

4 ACTIVITY: Adding Integers with Different Signs

Math Practice

Make Conjectures

How can the relationship between the integers help you write a rule?

Work with a partner. Write the addition expression shown. Then find the sum. How are the integers in the expression related to 0 on a number line?

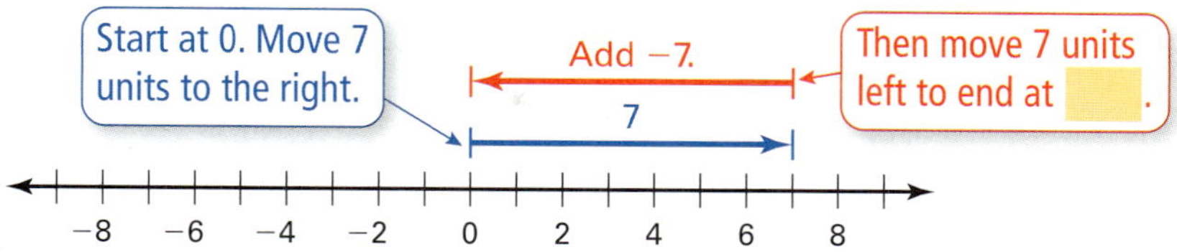

Inductive Reasoning

Work with a partner. Use integer counters or a number line to complete the table.

	Exercise	Type of Sum	Sum	Sum: Positive, Negative, or Zero
1	**5.** $-4 + (-3)$	Integers with the same sign		
2	**6.** $-3 + 2$			
3	**7.** $5 + (-3)$			
4	**8.** $7 + (-7)$			
	9. $2 + 4$			
	10. $-6 + (-2)$			
	11. $-5 + 9$			
	12. $15 + (-9)$			
	13. $-10 + 10$			
	14. $-6 + (-6)$			
	15. $13 + (-13)$			

What Is Your Answer?

16. **IN YOUR OWN WORDS** Is the sum of two integers *positive*, *negative*, or *zero*? How can you tell?

17. **STRUCTURE** Write general rules for adding (a) two integers with the same sign, (b) two integers with different signs, and (c) two integers that vary only in sign.

Practice Use what you learned about adding integers to complete Exercises 8–15 on page 486.

11.2 Lesson

Check It Out
Lesson Tutorials
BigIdeasMath.com

Key Vocabulary

opposites, *p. 484*

additive inverse, *p. 484*

Key Idea

Adding Integers with the Same Sign

Words Add the absolute values of the integers. Then use the common sign.

Numbers $2 + 5 = 7$ $\quad -2 + (-5) = -7$

EXAMPLE 1 Adding Integers with the Same Sign

Find $-2 + (-4)$. Use a number line to check your answer.

$-2 + (-4) = -6$ Add $|-2|$ and $|-4|$.

Use the common sign.

The sum is -6.

Check

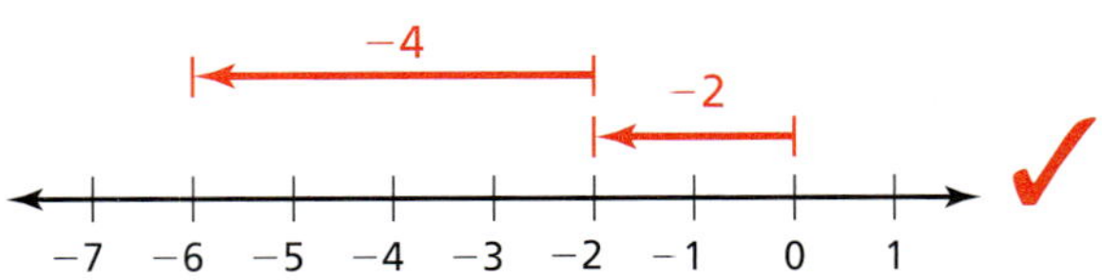

On Your Own

Add.

1. $7 + 13$
2. $-8 + (-5)$
3. $-20 + (-15)$

The Meaning of a Word

Opposite

When you walk across a street, you are moving to the **opposite** side of the street.

Two numbers that are the same distance from 0, but on opposite sides of 0, are called **opposites.** For example, -3 and 3 are opposites.

Key Ideas

Adding Integers with Different Signs

Words Subtract the lesser absolute value from the greater absolute value. Then use the sign of the integer with the greater absolute value.

Numbers $8 + (-10) = -2$ $\quad -13 + 17 = 4$

Additive Inverse Property

Words The sum of an integer and its **additive inverse,** or opposite, is 0.

Numbers $6 + (-6) = 0$ $\quad -25 + 25 = 0$ **Algebra** $a + (-a) = 0$

Multi-Language Glossary at BigIdeasMath.com

Laurie's Notes

Introduction

Connect

- **Yesterday:** Students used integer counters and a number line to add integers of the same sign and of different signs.
- **Today:** Students will add integers without the use of a visual or concrete model.

Motivate

? "Is the sum of 58 and −72 positive or negative? How do you know?"
Negative. *Sample answers:*
Using Counters: Some students may say that there would be more red counters (72) than yellow counters (58), so the sum is negative.
Using a Number Line: Some students may describe a number line: if you go back (left) 72 units and then forward (right) 58 units, you won't get back to 0, so the sum is negative.
Using Definitions: Some students may remember that the sign of the integer with the greater absolute value (72) is negative, so the sum is negative.

Lesson Notes

Example 1

- **Model with Mathematics:** As you discuss the example, refer to the models from the Activity.
  - When the signs are the *same,* the counters will be the *same color.*
  - When the signs are the *same,* both directed line segments will be going in the *same direction.*

Key Ideas

- Discuss the definition of opposites.

? **Construct Viable Arguments and Critique the Reasoning of Others:** "When you add two integers with different signs, how do you know if the sum is positive or negative?" Students should be using the concept of absolute value even if they don't use the precise language. You want to hear something about the size of the number, meaning its absolute value.

? **Look for and Express Regularity in Repeated Reasoning:** Write these problems on the board: $14 + (-8) = ?$ and $(-14) + 8 = ?$. Ask, "How are the problems alike? How are they different?" *Sample answer:*
Alike: They each use the numbers 14 and 8. They both consist of two different signs, which are being added together.
Different: In the first problem, 14 is positive and 8 is negative. In the second problem, 14 is negative and 8 is positive.

- Define how to add integers with different signs.
- Define the Additive Inverse Property. This is a special case of adding integers with different signs.

? "How many zero pairs are there when you add $(-5) + 5$?" 5

Goal Today's lesson is finding the sum of two or more integers.

Technology for the Teacher

Dynamic Classroom

Lesson Tutorials
Lesson Plans
Answer Presentation Tool

Extra Example 1

Find $-3 + (-12)$. -15

On Your Own

1. 20
2. −13
3. −35

Extra Example 2

Add.

a. $-11 + 6$ -5

b. $12 + (-5)$ 7

c. $4 + (-4)$ 0

Extra Example 3

Find the change in the account balance for August.

August Transactions	
Deposit	\$35
Deposit	\$40
Withdrawal	$-$\$25

increased \$50

On Your Own

4. 9 **5.** -1

6. 0 **7.** $-\$20$

Differentiated Instruction

Vocabulary

Write a table of opposites on the board. Encourage students to add to the list.

Word	Opposite
little	big
forward	backward

Ask the English learners to write the words in their native languages in another column and to share them with the class. Explain to the class that in mathematics, every nonzero number has an opposite. Every pair of opposites consists of a positive number and a negative number. Ask students to name some pairs of opposite numbers. Write opposite numbers and the words that represent them in the table.

Laurie's Notes

Example 2

- For each part of this example, have student volunteers explain how it was computed.
- **Part (a):** Because 5 has the lesser absolute value, you subtract it from the absolute value of -10. In general, use the sign of the number with the greater absolute value. In this case, the answer will be negative.

Words of Wisdom

- **Model with Mathematics:** Be sure students understand that the subtraction of the two absolute values is connected to the zero pairs that get removed when using integer counters, or it is the overlapping distance when the number line is used.
- If students are not getting correct answers, use integer counters or the number line to model several additional examples.

Example 3

- Review the properties of addition (associative, commutative, zero). Explain to students that these rules also apply to integers.
- **Financial Literacy:** Provide a brief description of what *deposit* and *withdrawal* mean in banking. A *deposit* is when you add money to an account, and a *withdrawal* is when you take money out of an account. Checkbooks are one context where addition of opposites occurs. Depositing \$100 and writing a check for \$100 results in a zero change in the balance.
- **Model:** Using *play* money and a student volunteer, act out the following.
  Hand \$50 to a banker (*deposit*).
  Ask for \$40 back (*withdrawal*).
  Hand the banker \$75 (*deposit*).
  Ask for \$50 back (*withdrawal*).

? "What is the change in the balance in your account?" \$35

? "Would the change in the balance be the same if you added the two deposits first, added the two withdrawals next, and then found the sum of the two answers?" yes; $(50 + 75) + (-40 + (-50)) = 125 + (-90) = 35$

- **FYI:** The Addition Property of Zero states that the sum of any number and zero is that number.

Closure

- What do you know about the sum $A + B$? Explain your reasoning.

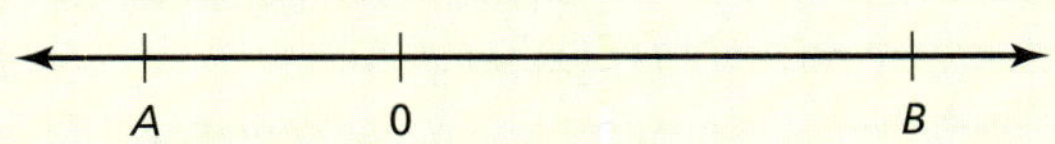

- Two integers have different signs. Their sum is -8. What are possible values for the two integers? *Sample answers:* -9 and 1, -10 and 2, 3 and -11, 4 and -12

EXAMPLE 2 Adding Integers with Different Signs

a. Find $5 + (-10)$.

$5 + (-10) = -5$ — $|-10| > |5|$. So, subtract $|5|$ from $|-10|$.

Use the sign of -10.

The sum is -5.

b. Find $-3 + 7$.

$-3 + 7 = 4$ — $|7| > |-3|$. So, subtract $|-3|$ from $|7|$.

Use the sign of 7.

The sum is 4.

c. Find $-12 + 12$.

$-12 + 12 = 0$ — The sum is 0 by the Additive Inverse Property.

-12 and 12 are opposites.

The sum is 0.

EXAMPLE 3 Adding More Than Two Integers

The list shows four bank account transactions in July. Find the change C in the account balance.

JULY TRANSACTIONS	
Withdrawal	-$40
Deposit	$50
Deposit	$75
Withdrawal	-$50

Study Tip

A deposit of $50 and a withdrawal of $50 represent opposite quantities, $+50$ and -50, which have a sum of 0.

Find the sum of the four transactions.

$C = -40 + 50 + 75 + (-50)$ — Write the sum.

$= -40 + 75 + 50 + (-50)$ — Commutative Property of Addition

$= -40 + 75 + [50 + (-50)]$ — Associative Property of Addition

$= -40 + 75 + 0$ — Additive Inverse Property

$= 35 + 0$ — Add -40 and 75.

$= 35$ — Addition Property of Zero

Because $C = 35$, the account balance increased $35 in July.

On Your Own

Now You're Ready
Exercises 8–23 and 28–39

Add.

4. $-2 + 11$

5. $9 + (-10)$

6. $-31 + 31$

7. WHAT IF? In Example 3, the deposit amounts are $30 and $40. Find the change C in the account balance.

11.2 Exercises

Check It Out
Help with Homework
BigIdeasMath.com

Vocabulary and Concept Check

1. **WRITING** How do you find the additive inverse of an integer?
2. **NUMBER SENSE** Is $3 + (-4)$ the same as $-4 + 3$? Explain.

Tell whether the sum is *positive*, *negative*, or *zero* without adding. Explain your reasoning.

3. $-8 + 20$
4. $30 + (-30)$
5. $-10 + (-18)$

Tell whether the statement is *true* or *false*. Explain your reasoning.

6. The sum of two negative integers is always negative.
7. An integer and its absolute value are always opposites.

Practice and Problem Solving

Add.

1 2
8. $6 + 4$
9. $-4 + (-6)$
10. $-2 - (-3)$
11. $-5 + 12$
12. $5 + (-7)$
13. $8 + (-8)$
14. $9 + (-11)$
15. $-3 + 13$
16. $-4 + (-16)$
17. $-3 + (-1)$
18. $14 + (-5)$
19. $0 + (-11)$
20. $-10 + (-15)$
21. $-13 + 9$
22. $18 + (-18)$
23. $-25 + (-9)$

ERROR ANALYSIS **Describe and correct the error in finding the sum.**

24. ✗ $9 + (-6) = -3$
25. ✗ $-10 + (-10) = 0$

26. **TEMPERATURE** The temperature is −3°F at 7:00 A.M. During the next 4 hours, the temperature increases 21°F. What is the temperature at 11:00 A.M.?
27. **BANKING** Your bank account has a balance of −\$12. You deposit \$60. What is your new balance?

Tell how the Commutative and Associative Properties of Addition can help you find the sum mentally. Then find the sum.

3
28. $9 + 6 + (-6)$
29. $-8 + 13 + (-13)$
30. $9 + (-17) + (-9)$
31. $7 + (-12) + (-7)$
32. $-12 + 25 + (-15)$
33. $6 + (-9) + 14$

Add.

34. $13 + (-21) + 16$
35. $22 + (-14) + (-35)$
36. $-13 + 27 + (-18)$
37. $-19 + 26 + 14$
38. $-32 + (-17) + 42$
39. $-41 + (-15) + (-29)$

Assignment Guide and Homework Check

Level	Assignment	Homework Check
Advanced	1–15, 24–48 even, 49–54	30, 36, 40, 46, 48

Common Errors

- **Exercises 8–23, 28–39** Students may try to ignore the signs and just add the integers. Remind them of the meaning of absolute value. Make sure they understand that they should use the sign of the number that is farther from zero. Also remind them of the Key Ideas, and how the signs of the integers determine if they need to add or subtract the integers.
- **Exercise 48** Students may not realize that each height measurement is given in reference to the previous point. Tell them to determine the measurement in relation to point *A*, which would be zero on a number line.

11.2 Record and Practice Journal

Add.

1. $-9 + 2$ — -7
2. $5 + (-5)$ — 0
3. $-12 + (-6)$ — -18
4. $-10 + 19 + 5$ — 14
5. $-11 + (-20) + 9$ — -22
6. $-7 + 7 + (-8)$ — -8

Use mental math to solve the equation.

7. $x + (-5) = 4$ — $x = 9$
8. $y + 6 = -2$ — $y = -8$
9. $-10 = -7 + z$ — $z = -3$

10. The table shows the change in your hair length over a year.

Month	January	February	August	September	December
Change in hair length (inches)	2	−1	3	−4	3

a. What is the total change in your hair length at the end of the year?
3 inches

b. Is your hair longer in January or December? Explain your reasoning.
December

c. When is your hair the longest? Explain your reasoning.
August

Vocabulary and Concept Check

1. Change the sign of the integer.
2. yes; The sums are the same by the Commutative Property of Addition.
3. positive; 20 has the greater absolute value and is positive.
4. zero; 30 and -30 are additive inverses.
5. negative; The common sign is a negative sign.
6. true; To add integers with the same sign, add the absolute values and use the common sign.
7. false; A positive integer and its absolute value are equal, not opposites.

Practice and Problem Solving

8. 10
9. -10
10. -5
11. 7
12. -2
13. 0
14. -2
15. 10
16. -20
17. -4
18. 9
19. -11
20. -25
21. -4
22. 0
23. -34
24. The wrong sign is used.
$9 + (-6) = 3$
25. -10 and -10 are not opposites.
$-10 + (-10) = -20$
26. 18°F
27. $48
28. Use the Associative Property to add 6 and -6 first. 9
29. Use the Associative Property to add 13 and -13 first. -8

Practice and Problem Solving

30. *Sample answer:* Use the Commutative Property to switch the last two terms. -17

31. *Sample answer:* Use the Commutative Property to switch the last two terms. -12

32. *Sample answer:* Use the Commutative Property to switch the last two terms. -2

33. *Sample answer:* Use the Commutative Property to switch the last two terms. 11

34. 8 **35.** -27

36. -4 **37.** 21

38. -7 **39.** -85

40. 0

41. *Sample answer:* $-26 + 1$; $-12 + (-13)$

42. -1 **43.** -3

44. 9 **45.** $d = -10$

46. $b = 2$ **47.** $m = -7$

48. See Additional Answers.

49. See *Taking Math Deeper*.

Fair Game Review

50. 31 **51.** 8

52. 114 **53.** 183

54. D

Mini-Assessment

Add.

1. $10 + (-12)$ -2
2. $-7 + (-5)$ -12
3. $-17 + 25$ 8
4. $65 + (-99)$ -34
5. The temperature is -2°F at 6 A.M. During the next three hours, the temperature increases 15°F. What is the temperature at 9 A.M.? 13°F

Taking Math Deeper

Exercise 49

In this puzzle, students get a chance to apply integer addition with *Guess, Check, and Revise.*

1. Solve the straightforward part of the puzzle.

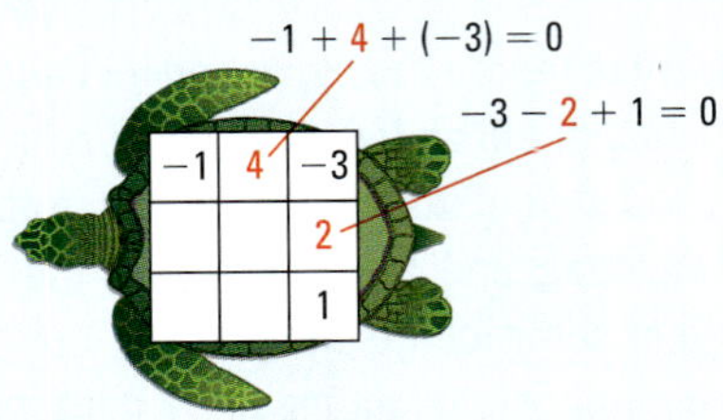

2. Make a list of the numbers from -4 to 4. Cross off the numbers you have used.

-4 ~~-3~~ -2 ~~-1~~ 0 ~~1~~ ~~2~~ 3 ~~4~~

3. Use the strategy *Guess, Check, and Revise* to complete the square.

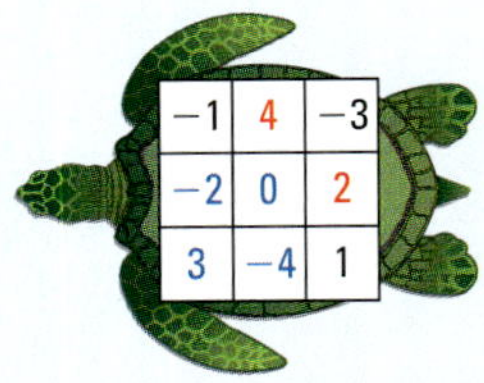

You can create other magic squares by repeating the same digit in each number in the magic square (-11, 44, -33, etc.).

The emperor Yu-Huang was a legendary figure in China, in the same sense that King Arthur was a legendary figure in Europe. He was called the Jade Emperor, and there are many stories about him. In addition to the story of the magic square and the turtle, the Jade Emperor is credited with creating the Chinese Zodiac in which each sequence of 12 years is given the name of an animal, such as the "Year of the Snake" or the "Year of the Rat."

Project

Create a new magic square using integers. Decide which squares will contain numbers and which squares will be blank. Complete your magic square. Then switch puzzles with a classmate and complete the puzzle you receive. Check your answers.

Reteaching and Enrichment Strategies

If students need help. . .	If students got it. . .
Resources by Chapter • Practice A and Practice B • Puzzle Time Record and Practice Journal Practice Differentiating the Lesson Lesson Tutorials Skills Review Handbook	Resources by Chapter • Enrichment and Extension • Technology Connection Start the next section

40. **SCIENCE** A lithium atom has positively charged protons and negatively charged electrons. The sum of the charges represents the charge of the lithium atom. Find the charge of the atom.

Lithium Atom

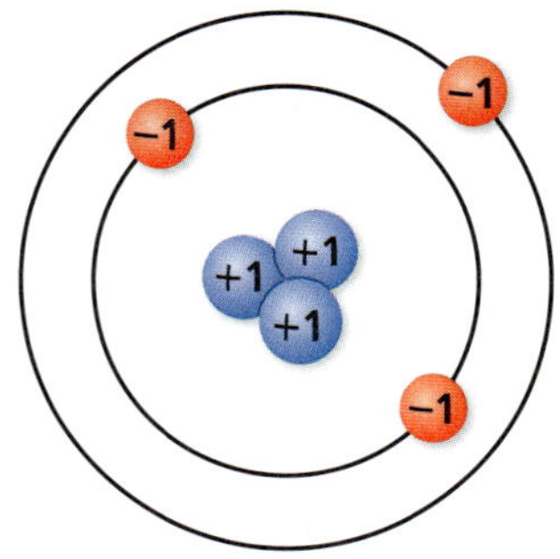

41. **OPEN-ENDED** Write two integers with different signs that have a sum of -25. Write two integers with the same sign that have a sum of -25.

ALGEBRA **Evaluate the expression when $a = 4$, $b = -5$, and $c = -8$.**

42. $a + b$ **43.** $-b + c$ **44.** $|a + b + c|$

MENTAL MATH **Use mental math to solve the equation.**

45. $d + 12 = 2$ **46.** $b + (-2) = 0$ **47.** $-8 + m = -15$

48. **PROBLEM SOLVING** Starting at point A, the path of a dolphin jumping out of the water is shown.

a. Is the dolphin deeper at point C or point E? Explain your reasoning.

b. Is the dolphin higher at point B or point D? Explain your reasoning.

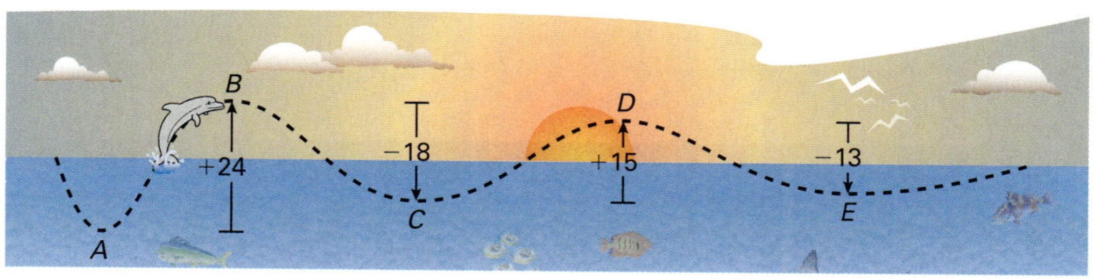

49. **Puzzle** According to a legend, the Chinese Emperor Yu-Huang saw a magic square on the back of a turtle. In a *magic square*, the numbers in each row and in each column have the same sum. This sum is called the *magic sum*.

Copy and complete the magic square so that each row and each column has a magic sum of 0. Use each integer from -4 to 4 exactly once.

Fair Game Review What you learned in previous grades & lessons

Subtract. *(Section 1.1)*

50. $69 - 38$ **51.** $82 - 74$ **52.** $177 - 63$ **53.** $451 - 268$

54. **MULTIPLE CHOICE** What is the range of the numbers below? *(Section 9.4)*

12, 8, 17, 12, 15, 18, 30

Ⓐ 12 Ⓑ 15 Ⓒ 18 Ⓓ 22

11.3 Subtracting Integers

Essential Question How are adding integers and subtracting integers related?

1 ACTIVITY: Subtracting Integers

Work with a partner. Use integer counters to find $4 - 2$.

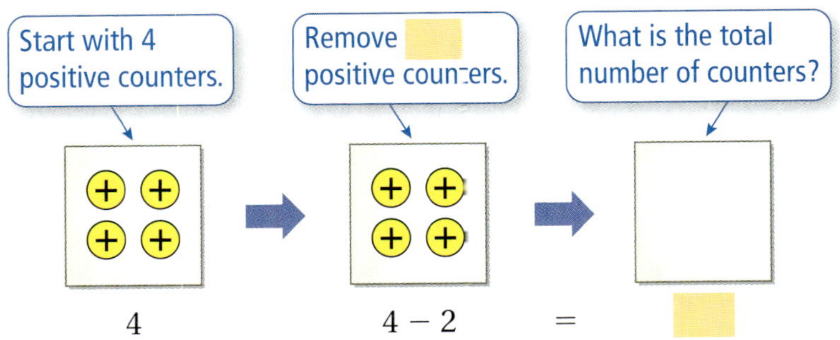

So, $4 - 2 =$ ____.

2 ACTIVITY: Adding Integers

Work with a partner. Use integer counters to find $4 + (-2)$.

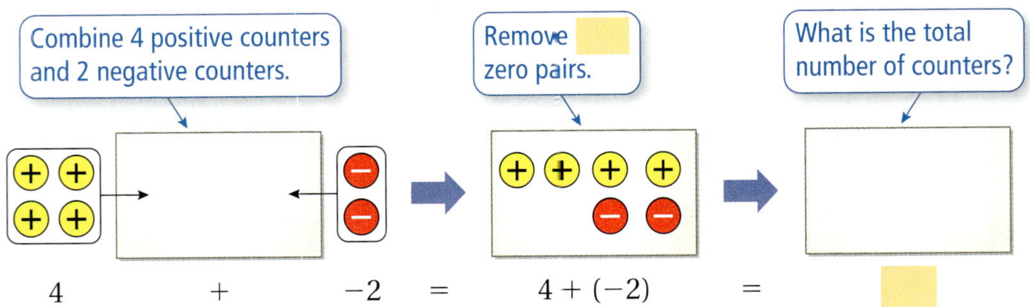

So, $4 + (-2) =$ ____.

Integers

In this lesson, you will

- subtract integers.
- solve real-life problems.

3 ACTIVITY: Subtracting Integers

Work with a partner. Use a number line to find $-3 - 1$.

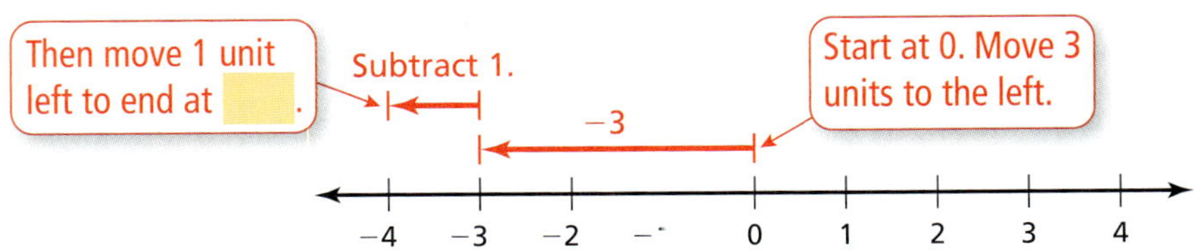

So, $-3 - 1 =$ ____.

Laurie's Notes

Introduction

Applying Mathematical Practices

- **Look For and Express Regularity in Repeated Reasoning:** Students will develop a conceptual understanding of integer subtraction. Each subtraction problem is followed by a related addition problem. Modeling each problem helps students make sense of subtraction.

Motivate

- ? Hand a student a collection of objects (8 pencils, 12 index cards, 9 paper clips) and ask another student to take some of the objects (5 pencils, 7 index cards, 3 paper clips). "What expressions represent this situation?" $8 - 5, 12 - 7, 9 - 3$
- ? "One way to think about subtraction: you have some amount and you take away another amount. Does this still work when you begin with negative amounts like -3 (owe a friend \$3)?"
- Today's activity investigates subtraction of integers.

Activity Notes

Activity 1 and Activity 2

- ? "How would you model $4 - 2$ using integer counters?" Subtraction means that you model the first number, and then take the second number from that original collection. Students should model the problem using their counters even if they say they know the answer.
- "Now let's see how $4 - 2$ is like the addition problem you did a few days ago." Have students work through Activity 2 with partners.

Words of Wisdom

- Before trying the number line examples, try $2 - 4$ with integer counters. It will remind students that subtraction is *not* commutative.
- ? "How would you model $2 - 4$?" Some students may say that this is not possible, because you should subtract the lesser number from the greater number. Show the model below.
- **Model:** Show the class two yellow counters.
- ? "How can you take 4 yellow counters away?" Add two zero pairs and then take away 4 yellow counters. Two red counters are left. $2 - 4 = -2$
- ? "$4 - 2$ had the related problem $4 + (-2)$. What do you think the related addition problem would be for $2 - 4$?" $2 + (-4)$

Activity 3

- **Reason Abstractly and Quantitatively:** To subtract a positive number, move to the left. To subtract a negative number, move to the right. To help students make sense of subtracting a negative quantity, use the context of money and owing someone \$2. If you subtract away a debt of \$2, you are moving 2 units in the positive direction, to the right.
- **Model:** To model $-3 - 1$, draw an arrow pointing to the left from zero to -3. Then move *left one* because you are *subtracting positive 1*.

What Your Students Will Learn

- Subtract integers by adding their opposties.

Previous Learning

Students need to know how to add and subtract whole numbers.

Technology for the Teacher

Dynamic Classroom

Lesson Plans
Complete Materials List

11.3 Record and Practice Journal

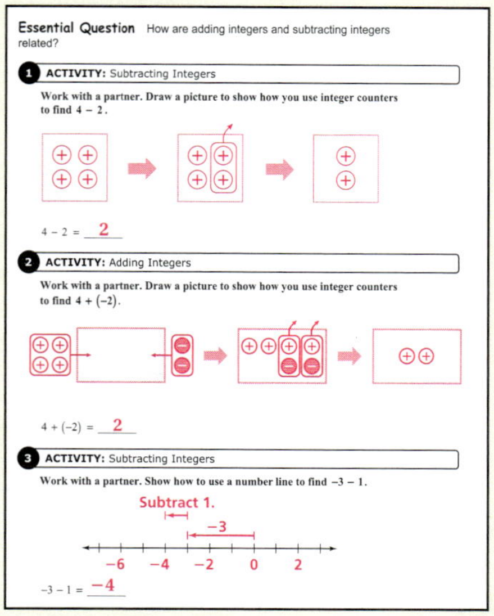

Essential Question How are adding integers and subtracting integers related?

1 ACTIVITY: Subtracting Integers

Work with a partner. Draw a picture to show how you use integer counters to find $4 - 2$.

$4 - 2 = 2$

2 ACTIVITY: Adding Integers

Work with a partner. Draw a picture to show how you use integer counters to find $4 + (-2)$.

$4 + (-2) = 2$

3 ACTIVITY: Subtracting Integers

Work with a partner. Show how to use a number line to find $-3 - 1$.

$-3 - 1 = -4$

Differentiated Instruction

Auditory

When you subtract a number, you add its opposite. So when you subtract a number using a number line, you move in the opposite direction you would move if adding the number. This is why you move left when subtracting a positive number and move right when subtracting a negative number.

11.3 Record and Practice Journal

4 ACTIVITY: Adding Integers

Work with a partner. Write the addition expression shown. Then find the sum.

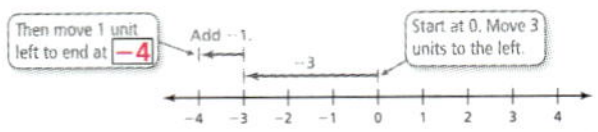

$-3 + (-1) = -4$

Inductive Reasoning

Work with a partner. Use integer counters or a number line to complete the table.

	Exercise	Operation: Add or Subtract	Answer
1	5. $4 - 2$	Subtract 2	2
2	6. $4 + (-2)$	Add −2	2
3	7. $-3 - 1$	Subtract 1	−4
4	8. $-3 + (-1)$	Add −1	−4
	9. $3 - 8$	Subtract 8	−5
	10. $3 + (-8)$	Add −8	−5
	11. $9 - 13$	Subtract 13	−4
	12. $9 + (-13)$	Add −13	−4
	13. $-6 - (-3)$	Subtract −3	−3
	14. $-6 + 3$	Add 3	−3
	15. $-5 - (-12)$	Subtract −12	7
	16. $-5 + 12$	Add 12	7

What Is Your Answer?

17. **IN YOUR OWN WORDS** How are adding integers and subtracting integers related?
Subtracting an integer is the same as adding its opposite.

18. **STRUCTURE** Write a general rule for subtracting integers.
To subtract an integer, add its opposite.

19. Use a number line to find the value of the expression $-4 + 4 - 9$. What property can you use to make your calculation easier? Explain.

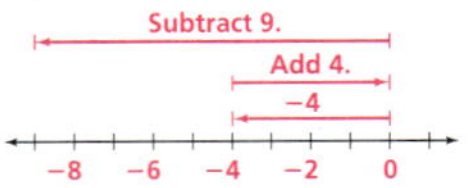

Additive Inverse Property; $-4 + 4 = 0$, so the answer is the remaining number −9.

Laurie's Notes

Activity 4

- Look at the related addition problem $-3 + (-1)$. Draw an arrow from 0 to −3 to represent −3. Now move to the left one because you are adding 1 in the negative direction (−1). Draw the arrow and write, "Add −1."
- Have students work with partners to write an addition expression.
- You just wrote sum and difference expressions that meant the same thing.

Inductive Reasoning

- Students should work with partners to find the sums. Note that Questions 5–8 are the problems completed in Activities 1–4.
- The goal is to develop some understanding about subtraction and the related addition problem.
- **Look for and Express Regularity in Repeated Reasoning:** It is important for students to record the *Operation: Add or Subtract*. Mathematically proficient students will observe that there is a pattern in the table. There are pairs of problems, one subtraction and one addition.

What Is Your Answer?

- In Questions 17 and 18, subtraction is the same as adding the opposite.
- ? **Extension:** "Use the integer counters or a number line to model $(8 - 4) - 2$ and $8 - (4 - 2)$. Are the results the same?" No, subtraction is not associative.
- **FYI:** The Associative Property of Addition states that the value of a sum does not depend on how the numbers are grouped. This does not apply for subtraction, as illustrated by the example above.

Closure

- Explain how you would use integer counters to model $4 - 6$.
Add 6 red counters to the 4 yellow counters and remove the 4 zero pairs. The result is 2 red counters.

4 ACTIVITY: Adding Integers

Math Practice

Make Sense of Quantities

What integers will you use in your addition expression?

Work with a partner. Write the addition expression shown. Then find the sum.

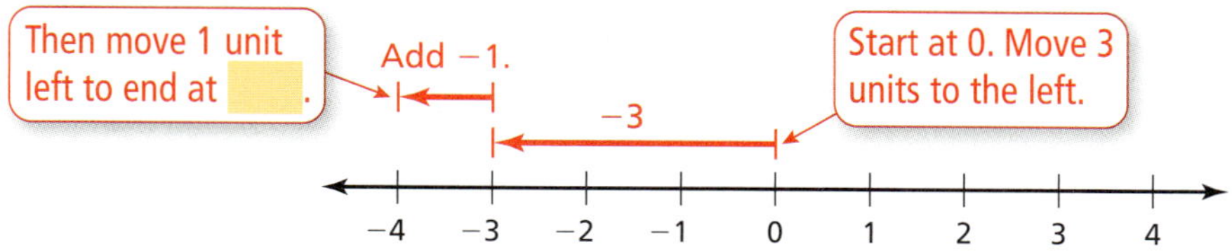

Inductive Reasoning

Work with a partner. Use integer counters or a number line to complete the table.

	Exercise	Operation: Add or Subtract	Answer
1	**5.** $4 - 2$	Subtract 2	
2	**6.** $4 + (-2)$		
3	**7.** $-3 - 1$		
4	**8.** $-3 + (-1)$		
	9. $3 - 8$		
	10. $3 + (-8)$		
	11. $9 - 13$		
	12. $9 + (-13)$		
	13. $-6 - (-3)$		
	14. $-6 + 3$		
	15. $-5 - (-12)$		
	16. $-5 + 12$		

What Is Your Answer?

17. **IN YOUR OWN WORDS** How are adding integers and subtracting integers related?

18. **STRUCTURE** Write a general rule for subtracting integers.

19. Use a number line to find the value of the expression $-4 + 4 - 9$. What property can you use to make your calculation easier? Explain.

Practice Use what you learned about subtracting integers to complete Exercises 8–15 on page 492.

11.3 Lesson

Check It Out
Lesson Tutorials
BigIdeasMath.com

Key Idea

Subtracting Integers

Words To subtract an integer, add its opposite.

Numbers $3 - 4 = 3 + (-4) = -1$

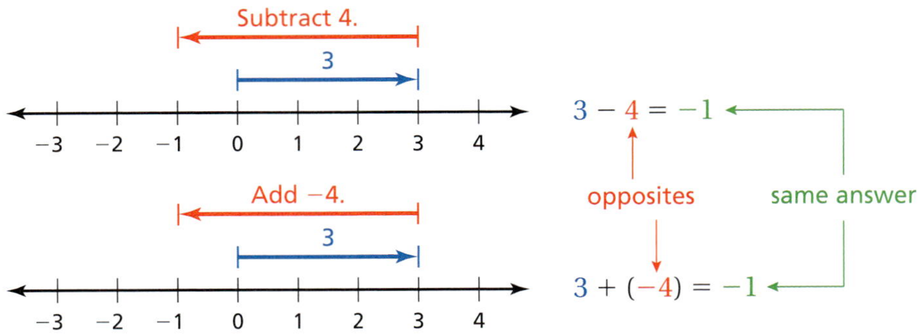

EXAMPLE 1 Subtracting Integers

a. Find 3 − 12.

$3 - 12 = 3 + (-12)$ Add the opposite of 12.

$= -9$ Add.

The difference is −9.

b. Find −8 − (−13).

$-8 - (-13) = -8 + 13$ Add the opposite of −13.

$= 5$ Add.

The difference is 5.

c. Find 5 − (−4).

$5 - (-4) = 5 + 4$ Add the opposite of −4.

$= 9$ Add.

The difference is 9.

On Your Own

Now You're Ready
Exercises 8–23

Subtract.

1. $8 - 3$	**2.** $9 - 17$	**3.** $-3 - 3$
4. $-14 - 9$	**5.** $9 - (-8)$	**6.** $-12 - (-12)$

Laurie's Notes

Introduction

Connect

- **Yesterday:** Students used integer counters and a number line to subtract integers.
- **Today:** Students will use the idea that a subtraction problem can be rewritten as an addition problem.

Motivate

? "Is my age (teacher) minus your age (point to a student) the same as your age minus my age?" Most students will understand that an older person's age minus a younger person's age is a positive number and subtracting in the other order is a negative number.

- Draw the two number lines shown.

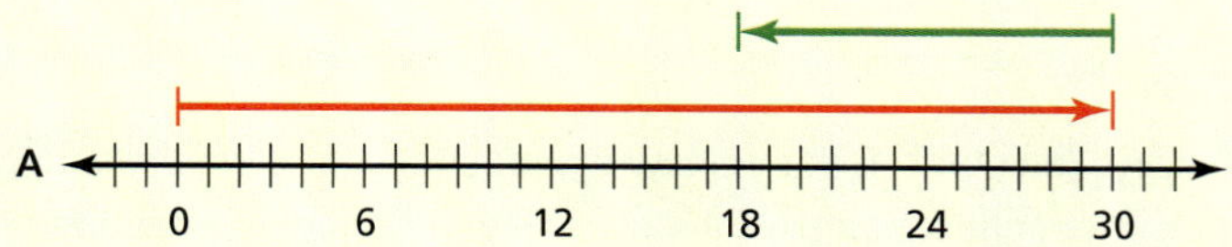

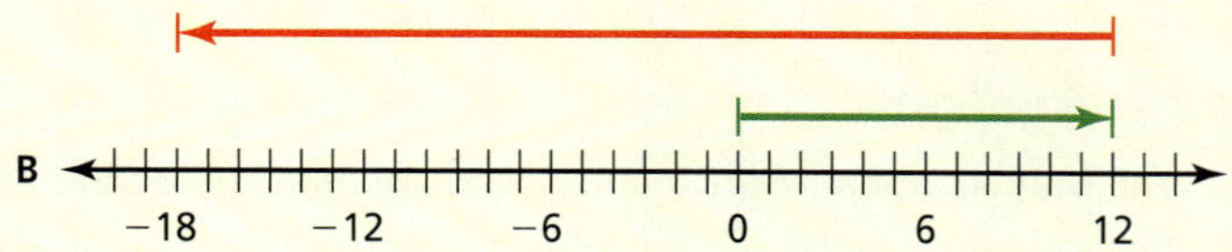

- Assume the teacher is 30 and the student is 12. Take away the context (age of two people) and simply look at the two number lines.

? "What two addition problems are modeled in each diagram?" A: $30 + (-12)$; B: $12 + (-30)$

- Now add the context of the problem and write the following on the board.

$$30 - 12 = 30 + (-12) = 18$$
$$12 - 30 = 12 + (-30) = -18$$

- This introduction states the relationship between a subtraction problem and its related addition problem. It reminds students that subtraction is not commutative and that they need to be careful of order when subtracting.
- **Vocabulary Review:** Ask students what the word *opposite* means.

Lesson Notes

Key Idea

- Each subtraction problem can be rewritten as an addition problem. Remind students that they already know how to add integers.

Example 1

- Work through each part of the example. Pointing to a classroom number line may be helpful.
- Use a different color when rewriting the problem as "add the opposite."

$$3 - 12 = 3 + (-12)$$

Goal Today's lesson is finding the difference of integers.

Technology for the Teacher

Dynamic Classroom

Lesson Tutorials
Lesson Plans
Answer Presentation Tool

Extra Example 1

Subtract.

a. $8 - 10$ -2
b. $-3 - (-6)$ 3
c. $2 - (-4)$ 6

On Your Own

1. 5 **2.** −8
3. −6 **4.** −23
5. 17 **6.** 0

Extra Example 2

Evaluate $-11 - 5 - (-8)$. -8

On Your Own

7. -33 8. -33
9. -4 10. -2
11. 15 12. -59

Extra Example 3

Which continent has the greater range of elevations? South America

	South America	Europe
Highest Elevation	6960 m	5642 m
Lowest Elevation	−40 m	−92 m

On Your Own

13. 5710 meters

English Language Learners

Class Activity

Reinforce the meaning of the words *difference*, *subtract*, *positive*, and *negative*.

1. Ask students to name two numbers whose difference is 0. Students should realize that the only possibility is a number subtracted from itself.
2. Ask students to name two numbers whose difference is 3. Students should realize that if they start with a number greater than 3, they need to subtract a positive number. If they begin with a number less than 3, they need to subtract a negative number.
3. Ask students to name two numbers whose difference is −3. Starting with a number greater than −3 means you need to subtract a positive number. Starting with a number less than −3 means you need to subtract a negative number.

Laurie's Notes

On Your Own

- **Questions 1–6:**
  - For students who are having difficulty, have them record the problem on the board. They should say aloud, "Add the opposite," and state what that means for the particular problem.
  - **Construct Viable Arguments and Critique the Reasoning of Others:** Ask students if it s possible to determine when the difference of two negative numbers will be positive and when the difference of two negative numbers will be negative.

Example 2

- Caution students to work slowly.
- Subtraction must be performed in order from left to right.

Example 3

- **Vocabulary:** You may need to review the meanings of *elevation* and *range*.
- **Fun Fact:** The highest point in Hawaii is Mauna Kea at 4208 meters above sea level. The lowest points in Hawaii are at sea level, where the coast of Hawaii meets the Pacific Ocean.

Words of Wisdom

- **Model with Mathematics:** Make a colorful number line that stretches the length of your board. Use two 3-inch wide strips of different colored paper. Positive integers are on one color and negative integers are on the other. Label only the integers, but make a hash mark at $\frac{1}{2}$ between each of the integers. This will be useful later in the year.

Closure

- **Writing:** Your friend is home sick today. Imagine you are on the telephone with him or her. How would you explain how to subtract integers? Be sure to use an example.

EXAMPLE 2 Subtracting Integers

Evaluate $-7 - (-12) - 14$.

$-7 - (-12) - 14 = -7 + 12 - 14$	Add the opposite of –12.
$= 5 - 14$	Add –7 and 12.
$= 5 + (-14)$	Add the opposite of 14.
$= -9$	Add.

So, $-7 - (-12) - 14 = -9$.

On Your Own

Now You're Ready
Exercises 27–32

Evaluate the expression.

7. $-9 - 16 - 8$

8. $-4 - 20 - 9$

9. $0 - 9 - (-5)$

10. $-8 - (-6) - 0$

11. $15 - (-20) - 20$

12. $-14 - 9 - 36$

EXAMPLE 3 Real-Life Application

Which continent has the greater range of elevations?

	North America	Africa
Highest Elevation	6198 m	5895 m
Lowest Elevation	–86 m	–155 m

To find the range of elevations for each continent, subtract the lowest elevation from the highest elevation.

North America

$$\begin{aligned} \text{range} &= 6198 - (-86) \\ &= 6198 + 86 \\ &= 6284 \text{ m} \end{aligned}$$

Africa

$$\begin{aligned} \text{range} &= 5895 - (-155) \\ &= 5895 + 155 \\ &= 6050 \text{ m} \end{aligned}$$

Because 6284 is greater than 6050, North America has the greater range of elevations.

On Your Own

13. The highest elevation in Mexico is 5700 meters, on Pico de Orizaba. The lowest elevation in Mexico is –10 meters, in Laguna Salada. Find the range of elevations in Mexico.

11.3 Exercises

Check It Out
Help with Homework
BigIdeasMath.com

Vocabulary and Concept Check

1. **WRITING** How do you subtract one integer from another?
2. **OPEN-ENDED** Write two integers that are opposites.
3. **DIFFERENT WORDS, SAME QUESTION** Which is different? Find "both" answers.

Find the difference of 3 and -2.	What is 3 less than -2?
How much less is -2 than 3?	Subtract -2 from 3.

MATCHING Match the subtraction expression with the corresponding addition expression.

4. $9 - (-5)$
5. $-9 - 5$
6. $-9 - (-5)$
7. $9 - 5$

A. $-9 + 5$ B. $9 + (-5)$ C. $-9 + (-5)$ D. $9 + 5$

Practice and Problem Solving

Subtract.

1
8. $4 - 7$
9. $8 - (-5)$
10. $-6 - (-7)$
11. $-2 - 3$
12. $5 - 8$
13. $-4 - 6$
14. $-8 - (-3)$
15. $10 - 7$
16. $-8 - 13$
17. $15 - (-2)$
18. $-9 - (-13)$
19. $-7 - (-8)$
20. $-6 - (-6)$
21. $-10 - 12$
22. $32 - (-6)$
23. $0 - 20$

24. **ERROR ANALYSIS** Describe and correct the error in finding the difference $7 - (-12)$.

✗ $7 - (-12) = 7 + (-12) = -5$

25. **SWIMMING POOL** The floor of the shallow end of a swimming pool is at -3 feet. The floor of the deep end is 9 feet deeper. Which expression can be used to find the depth of the deep end?

$-3 + 9$ $-3 - 9$ $9 - 3$

26. **SHARKS** A shark is at -80 feet. It swims up and jumps out of the water to a height of 15 feet. Write a subtraction expression for the vertical distance the shark travels.

Evaluate the expression.

2
27. $-2 - 7 + 15$
28. $-9 + 6 - (-2)$
29. $12 - (-5) - 8$
30. $-87 - 5 - 13$
31. $-6 - (-8) - 6$
32. $-15 - 7 - (-11)$

Assignment Guide and Homework Check

Level	Assignment	Homework Check
Advanced	1–15, 24–48 even, 50–56	30, 32, 38, 42, 44

For Your Information

- **Exercise 3** In the *Different Words, Same Question* exercise, three of the four choices pose the same question using different words. The remaining choice poses a different question. So there are two answers.

Common Errors

- **Exercises 8–23** Students may change the sign of the first number or forget to change the problem from subtraction to addition when changing the sign of the second number. Remind them that the first number is a starting point and will never change. Also remind students that the sign of the second number and the operation change.
- **Exercises 27–32** Students may try to do the addition first instead of working left to right. Remind them that the order of operations does not put addition before subtraction, but that addition *and* subtraction are performed from left to right.
- **Exercise 40** Students may try to add $-4 + 11$ instead of subtract $-4 - 11$ because they do not recognize that *change in elevation* means a range (subtraction). Use a number line rotated vertically to help students see the meaning of change in elevation.

11.3 Record and Practice Journal

Subtract.

1. $3 - 8$ **−5**
2. $6 - (-7)$ **13**
3. $-10 - 9$ **−19**
4. $-5 - (-4)$ **−1**

Evaluate the expression.

5. $11 - (-2) + 14$ **27**
6. $-16 - (-12) + (-8)$ **−12**
7. $6 - 17 - 4$ **−15**

Use mental math to solve the equation.

8. $6 - x = 10$ **$x = -4$**
9. $y - (-10) = 2$ **$y = -8$**
10. $z - 17 = -14$ **$z = 3$**

11. You begin a hike in Death Valley, California, at an elevation of −86 meters. You hike to a point of elevation at 45 meters. What is your change in elevation?
131 meters

12. You sell T-shirts for a fundraiser. It costs $112 to have the T-shirts made. You make $98 in sales. What is your profit?
−$14

Vocabulary and Concept Check

1. You add the integer's opposite.
2. *Sample answer:* 3, −3
3. What is 3 less than −2?; −5; 5
4. D
5. C
6. A
7. B

Practice and Problem Solving

8. −3
9. 13
10. 1
11. −5
12. −3
13. −10
14. −5
15. 3
16. −21
17. 17
18. 4
19. 1
20. 0
21. −22
22. 38
23. −20
24. The *opposite* of −12 should be added.
$7 - (-12) = 7 + 12 = 19$
25. $-3 - 9$
26. $15 - (-80)$
27. 6
28. −1
29. 9
30. −105
31. 8
32. −11

Practice and Problem Solving

33. $m = 14$ **34.** $w = 4$

35. $c = 15$ **36.** -5

37. 2 **38.** -17

39. 3 **40.** -15 m

41. *Sample answer:* $x = -2$, $y = -1$; $x = -3$, $y = -2$

42. See *Taking Math Deeper.*

43. sometimes; It's positive only if the first integer is greater.

44. sometimes; It's positive only if the first integer is greater.

45. always; It's always positive because the first integer is always greater.

46. never; It's never positive because the first integer is never greater.

47. all values of a and b

48. when a and b both have the same sign, or $a = 0$, or $b = 0$

49. when a and b have the same sign and $|a| \geq |b|$ or $b = 0$

Fair Game Review

50. -20 **51.** -45

52. 40 **53.** 468

54. 1476 **55.** 2378

56. C

Mini-Assessment

Subtract.

1. $6 - 10$ -4
2. $-14 - 16$ -30
3. $-9 - (-4)$ -5
4. $-26 - (-35)$ 9
5. The top of a flag pole is 15 feet high. The base is at -3 feet. Find the length of the flag pole. 18 feet

Taking Math Deeper

Exercise 42

The exercise reviews the concept of the range of a data set. Students should know that the range of a set is the difference between the greatest number and the least number in the set.

1 **a.** Find the range of temperatures for each month.

	Jan	Feb	Mar	Apr	May	Jun	Jul	Aug	Sep	Oct	Nov	Dec
High (°F)	56	57	56	72	82	92	84	85	73	64	62	53
Low (°F)	−35	−38	−24	−15	1	29	34	31	19	−6	−21	−36
Range (°F)	91	95	80	87	81	63	50	54	54	70	83	89

$56 - (-35) = 91$

2 Help me see it.

$$56 - (-35) = 56 + 35 = 91$$

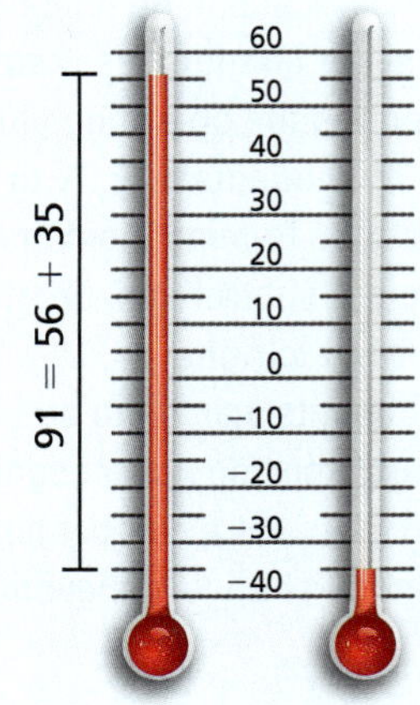

3 **b.** Find the all-time high and the all-time low temperatures.

All-time high: 92°F

All-time low: −38°F

c. Find the range of the all-time high and the all-time low.

$$92 - (-38) = 92 + 38 = 130°F$$

Project

Create a chart showing the high and low temperatures in your town for each month of the year. Give the range of temperatures for each month.

Reteaching and Enrichment Strategies

If students need help...	If students got it...
Resources by Chapter • Practice A and Practice B • Puzzle Time Record and Practice Journal Practice Differentiating the Lesson Lesson Tutorials Skills Review Handbook	Resources by Chapter • Enrichment and Extension • Technology Connection Start the next section

MENTAL MATH **Use mental math to solve the equation.**

33. $m - 5 = 9$ **34.** $w - (-3) = 7$ **35.** $6 - c = -9$

ALGEBRA **Evaluate the expression when $k = -3$, $m = -6$, and $n = 9$.**

36. $4 - n$ **37.** $m - (-8)$

38. $-5 + k - n$ **39.** $|m - k|$

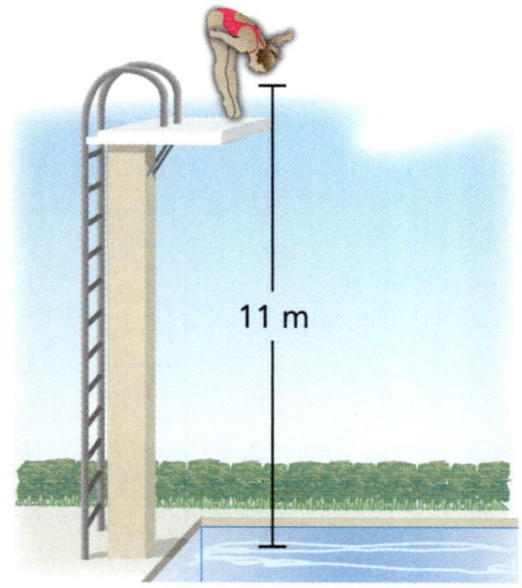

40. **PLATFORM DIVING** The figure shows a diver diving from a platform. The diver reaches a depth of 4 meters. What is the change in elevation of the diver?

41. **OPEN-ENDED** Write two different pairs of negative integers, x and y, that make the statement $x - y = -1$ true.

42. **TEMPERATURE** The table shows the record monthly high and low temperatures for a city in Alaska.

	Jan	Feb	Mar	Apr	May	Jun	Jul	Aug	Sep	Oct	Nov	Dec
High (°F)	56	57	56	72	82	92	84	85	73	64	62	53
Low (°F)	−35	−38	−24	−15	1	29	34	31	19	−6	−21	−36

a. Find the range of temperatures for each month.

b. What are the all-time high and all-time low temperatures?

c. What is the range of the temperatures in part (b)?

REASONING **Tell whether the difference between the two integers is *always*, *sometimes*, or *never* positive. Explain your reasoning.**

43. two positive integers **44.** two negative integers

45. a positive integer and a negative integer **46.** a negative integer and a positive integer

Number Sense **For what values of a and b is the statement true?**

47. $|a - b| = |b - a|$ **48.** $|a + b| = |a| + |b|$ **49.** $|a - b| = |a| - |b|$

Fair Game Review What you learned in previous grades & lessons

Add. *(Section 11.2)*

50. $-5 + (-5) + (-5) + (-5)$ **51.** $-9 + (-9) + (-9) + (-9) + (-9)$

Multiply. *(Section 1.1)*

52. 8×5 **53.** 6×78 **54.** 36×41 **55.** 82×29

56. **MULTIPLE CHOICE** Which value of n makes the value of the expression $4n + 3$ a composite number? *(Skills Review Handbook)*

Ⓐ 1 Ⓑ 2 Ⓒ 3 Ⓓ 4

11 Study Help

Check It Out
Graphic Organizer
BigIdeasMath.com

You can use an **idea and examples chart** to organize information about a concept. Here is an example of an idea and examples chart for absolute value.

Absolute Value: the distance between a number and 0 on the number line

- Example: $|3| = 3$
- Example: $|-5| = 5$
- Example: $|0| = 0$

On Your Own

Make idea and examples charts to help you study these topics.

1. integers
2. adding integers
   a. with the same sign
   b. with different signs
3. Additive Inverse Property
4. subtracting integers

After you complete this chapter, make idea and examples charts for the following topics.

5. multiplying integers
   a. with the same sign
   b. with different signs
6. dividing integers
   a. with the same sign
   b. with different signs

"I made an idea and examples chart to give my owner ideas for my birthday next week."

Sample Answers

1\.

Integers: ..., −3, −2, −1, 0, 1, 2, 3, ...

- Example: −586
- Example: 0
- Example: 16

2a\.

Adding integers with the same sign: Add the absolute values of the integers. Then use the common sign.

- Example: 16 + 17 = 33
- Example: −5 + (−4) = −9
- Example: −55 + (−45) = −100

2b\.

Adding integers with different signs: Subtract the lesser absolute value from the greater absolute value. Then use the sign of the integer with the greater absolute value.

- Example: 8 + (−2) = 6
- Example: −8 + 2 = −6
- Example: −97 + 19 = −78

3\.

Additive Inverse Property: The sum of an integer and its *additive inverse*, or opposite, is 0.

- Example: 5 + (−5) = 0
- Example: −100 + 100 = 0
- Example: 16 + (−16) = 0

4\. Available at *BigIdeasMath.com.*

List of Organizers

Available at *BigIdeasMath.com*

Comparison Chart
Concept Circle
Example and Non-Example Chart
Formula Triangle
Four Square
Idea (Definition) and Examples Chart
Information Frame
Information Wheel
Notetaking Organizer
Process Diagram
Summary Triangle
Word Magnet
Y Chart

About this Organizer

An **Idea and Examples Chart** can be used to organize information about a concept. Students fill in the top rectangle with a term and its definition or description. Students fill in the rectangles that follow with examples to illustrate the term. Each sample answer shows 3 examples, but students can show more or fewer examples. Idea and examples charts are useful for concepts that can be illustrated with more than one type of example.

Technology for the Teacher

Editable Graphic Organizer

Answers

1. $|-8| > 3$
2. $7 = |-7|$
3. $-6, -4, 3, |-4|, |-5|$
4. $-10, -8, |-9|, 12, |-15|$
5. -11
6. 12
7. -6
8. 0
9. 1
10. 13
11. **a.** $-10, -7$

    b. -7

    c. -10
12. yes; They raised $1129.
13. 130°F

Technology for the Teacher

Online Assessment
Assessment Book
ExamView® Assessment Suite

Alternative Quiz Ideas

100% Quiz
Error Notebook
Group Quiz
Homework Quiz
Math Log
Notebook Quiz
Partner Quiz
Pass the Paper

Partner Quiz

- Partner quizzes are to be completed by students working in pairs. Student pairs can be selected by the teacher, by students, through a random process, or any way that works for your class.
- Students are permitted to use their notebooks and other appropriate materials.
- Each pair submits a draft of the quiz for teacher feedback. Then they revise their work and turn it in for a grade.
- When the pair is finished they can submit one paper, or each can submit their own.
- Teachers can give feedback in a variety of ways. It is important that the teacher does not reteach or provide the solution. The teacher can tell students which questions they have answered correctly, if they are on the right track, or if they need to rethink a problem.

Reteaching and Enrichment Strategies

If students need help. . .	If students got it. . .
Resources by Chapter • Practice A and Practice B • Puzzle Time Lesson Tutorials *BigIdeasMath.com*	Resources by Chapter • Enrichment and Extension • Technology Connection Game Closet at *BigIdeasMath.com* Start the next section

11.1–11.3 Quiz

Check It Out
Progress Check
BigIdeasMath.com

Copy and complete the statement using <, >, or =. *(Section 11.1)*

1. $\lvert -8 \rvert \ \square\ 3$

2. $7\ \square\ \lvert -7 \rvert$

Order the values from least to greatest. *(Section 11.1)*

3. $-4, \lvert -5 \rvert, \lvert -4 \rvert, 3, -6$

4. $12, -8, \lvert -15 \rvert, -10, \lvert -9 \rvert$

Evaluate the expression. *(Section 11.2 and Section 11.3)*

5. $-3 + (-8)$

6. $-4 + 16$

7. $3 - 9$

8. $-5 - (-5)$

Evaluate the expression when $a = -2$, $b = -8$, and $c = 5$. *(Section 11.2 and Section 11.3)*

9. $4 - a - c$

10. $\lvert b - c \rvert$

11. EXPLORING Two climbers explore a cave. *(Section 11.1)*

- **a.** Write an integer for the position of each climber relative to the surface.
- **b.** Which integer in part (a) is greater?
- **c.** Which integer in part (a) has the greater absolute value?

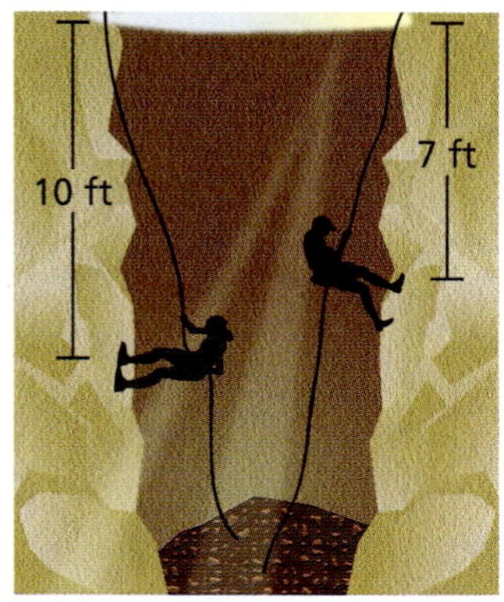

12. SCHOOL CARNIVAL The table shows the income and expenses for a school carnival. The school's goal was to raise \$1100. Did the school reach its goal? Explain. *(Section 11.2)*

Games	Concessions	Donations	Flyers	Decorations
\$650	\$530	\$52	−\$28	−\$75

13. TEMPERATURE Temperatures in the Gobi Desert reach −40°F in the winter and 90°F in the summer. Find the range of the temperatures. *(Section 11.3)*

11.4 Multiplying Integers

Essential Question Is the product of two integers *positive*, *negative*, or *zero*? How can you tell?

1 ACTIVITY: Multiplying Integers with the Same Sign

Work with a partner. Use repeated addition to find 3 • 2.

Recall that multiplication is repeated addition. 3 • 2 means to add 3 groups of 2.

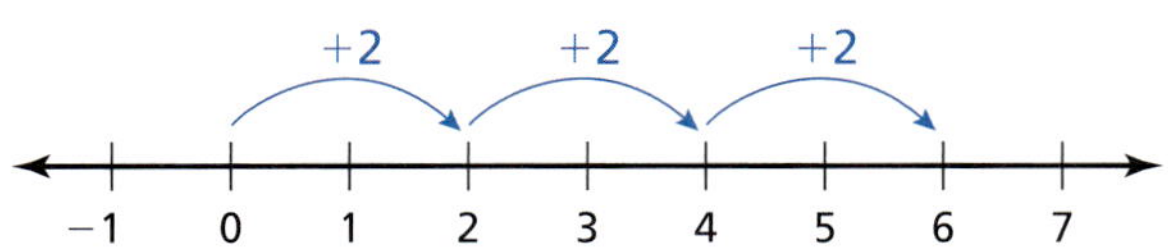

Now you can write

$3 \cdot 2 =$ ____ + ____ + ____

= ____.

So, $3 \cdot 2 =$ ____.

2 ACTIVITY: Multiplying Integers with Different Signs

Work with a partner. Use repeated addition to find 3 • (−2).

$3 \cdot (-2)$ means to add 3 groups of −2.

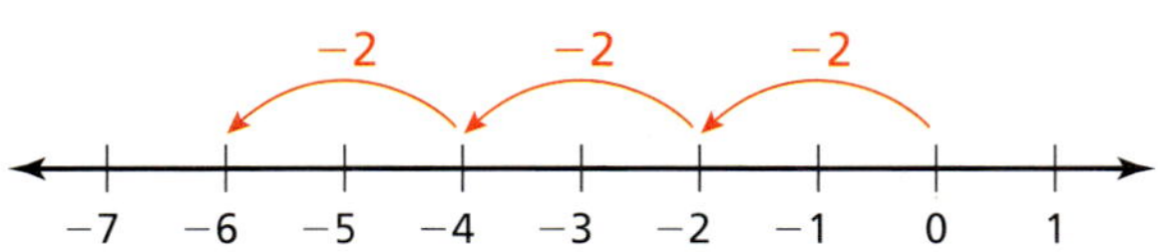

Now you can write

$3 \cdot (-2) =$ ____ + ____ + ____

= ____.

So, $3 \cdot (-2) =$ ____.

3 ACTIVITY: Multiplying Integers with Different Signs

Work with a partner. Use a table to find −3 • 2.

Describe the pattern of the products in the table. Then complete the table.

2	•	2	=	4
1	•	2	=	2
0	•	2	=	0
−1	•	2	=	____
−2	•	2	=	____
−3	•	2	=	____

So, $-3 \cdot 2 =$ ____.

Integers

In this lesson, you will

- multiply integers.
- solve real-life problems.

Laurie's Notes

Introduction

Applying Mathematical Practices

- **Look for and Express Regularity in Repeated Reasoning:** In this lesson, students will use the repeated addition model of multiplication and inductive reasoning to develop a conceptual understanding of integer multiplication.

Motivate

- Play *Guess My Rule.* Write the first 4 terms of a sequence on the board. Ask students to give the next few terms and guess the rule. Here are some possibilities.
  - 2, 4, 8, 16, . . . 32, 64, 128; The rule is to multiply by 2, powers of 2, or doubling (any of those 3 answers is acceptable).
  - 0, 4, 8, 12, . . . 16, 20, 24; The rule is adding 4, multiples of 4, or counting by 4s (any of those 3 answers is acceptable).
  - $-6, -3, 0, 3, \ldots$ 6, 9, 12; The rule is adding 3.
- ? Ask students if any of the patterns seem different than the others. In the sequences above, the third one involves negative numbers.

Discuss

- ? "Do you remember *skip counting* in elementary school?" If no one remembers, explain to students that skip counting is a fast way to count by a number other than 1.
- Skip counting is one way to show multiplication. For example, skip counting by 5s yields 5, 10, 15, 20, 25, ... You can think of the terms of the sequence formed by skip counting by 5s as $5 \times 1, 5 \times 2, 5 \times 3$, etc.

Activity Notes

Activity 1 and Activity 2

- **Model with Mathematics:** This is the *repeated addition model* of multiplication.
- Have students draw a number line to represent 3 groups of 2.
- ? **Connection:** "I noticed that $3 \times 2 = 2 \times 3$. What property is this?" The Commutative Property of Multiplication
- In Activity 2, make sure students understand why the arrows are moving to the left, instead of moving to the right.
- Ask a student to read the last result in Activity 2, namely that $3 \cdot (-2) = -6$. So, a positive number times a negative number is a negative product.

Activity 3

- Make sure that students recognize the pattern—the first factor is decreasing by 1, the second factor is constant, and the product is decreasing by 2.
- Ask a student to read the last result, namely that $-3 \cdot 2 = -6$. So, a negative number times a positive number is a negative product.

What Your Students Will Learn

- Multiply integers with the same sign and with different signs.
- Evaluate expressions with whole number exponents by using repeated multiplication.

Previous Learning

Students need to know how to multiply whole numbers.

Technology for the Teacher

Dynamic Classroom

Lesson Plans
Complete Materials List

11.4 Record and Practice Journal

Essential Question Is the product of two integers *positive, negative,* or *zero*? How can you tell?

1 ACTIVITY: Multiplying Integers with the Same Sign

Work with a partner. Use repeated addition to find $3 \cdot 2$.

Recall that multiplication is repeated addition. $3 \cdot 2$ means to add 3 groups of 2.

$3 \cdot 2 = 2 + 2 + 2 = 6$

$3 \cdot 2 =$ 6

2 ACTIVITY: Multiplying Integers with Different Signs

Work with a partner. Use repeated addition to find $3 \cdot (-2)$.

$3 \cdot (-2) = (-2) + (-2) + (-2) = -6$

$3 \cdot (-2) =$ −6

3 ACTIVITY: Multiplying Integers with Different Signs

Work with a partner. Use a table to find $-3 \cdot 2$.

Describe the pattern of the products in the table. Then complete the table.

$2 \cdot 2 = 4$
$1 \cdot 2 = 2$
$0 \cdot 2 = 0$
$-1 \cdot 2 = -2$
$-2 \cdot 2 = -4$
$-3 \cdot 2 = -6$

The products decrease by 2 in each row.

$-3 \cdot 2 =$ −6

Differentiated Instruction

Visual

Use integer counters to demonstrate that the product of two integers with different signs is negative.

$$3(-4) = (-4) + (-4) + (-4) = -12$$

11.4 Record and Practice Journal

4 ACTIVITY: Multiplying Integers with the Same Sign

Work with a partner. Use a table to find $-3 \cdot (-2)$.

Describe the pattern of the products in the table. Then complete the table.

−3	•	3	=	−9
−3	•	2	=	−6
−3	•	1	=	−3
−3	•	0	=	0
−3	•	−1	=	3
−3	•	−2	=	6

The products increase by 3 in each row.

$-3 \cdot (-2) =$ 6

Inductive Reasoning

Work with a partner. Complete the table.

	Exercise	Type of Product	Product	Product: Positive or Negative
1	5. 3 • 2	same sign	6	Positive
2	6. 3 • (−2)	different signs	−6	Negative
3	7. −3 • 2	different signs	−6	Negative
4	8. −3 • (−2)	same sign	6	Positive
	9. 6 • 3	same sign	18	Positive
	10. 2 • (−5)	different signs	−10	Negative
	11. −6 • 5	different signs	−30	Negative
	12. −5 • (−3)	same sign	15	Positive

What Is Your Answer?

13. Write two integers whose product is 0.

Sample answer: 3 and 0

14. **IN YOUR OWN WORDS** Is the product of two integers *positive*, *negative*, or *zero*? How can you tell?

It can be positive, negative, or zero.

15. **STRUCTURE** Write a general rule for multiplying

a. two integers with the same sign.

Multiply the absolute values and make the product positive.

b. two integers with different signs.

Multiply the absolute values and make the product negative.

Laurie's Notes

Activity 4

- **Connection:** Activity 1 showed that the product of two positive integers is positive. Activity 2 and Activity 3 showed that the product of a positive and a negative (or a negative and a positive) is negative.
- ? "Are there any other combinations to consider?" the product of two negatives
- Tell students: "Let's look at the product of two negatives." Students should recognize the patterns: the first factor is constant, the second factor is decreasing by 1, and the product is increasing by 3.
- **Extension:** Use the patterns developed to find the product of three numbers, such as $3(-2)(-4)$. 24

Inductive Reasoning

- Students should work with partners to find the products. The goal is for the students to recognize the bigger pattern. When the factors have the same signs, the product is positive. When the factors have different signs, the product is negative.
- Note that Questions 5–8 are the problems completed in Activities 1–4.
- **Look for and Express Regularity in Repeated Reasoning:** It is important for students to record the *Type of Product*. Mathematically proficient students will observe that there is a pattern in the table. When both factors have the same sign, the product is positive. When the factors have different signs, the product is negative.

Words of Wisdom

- **Common Error:** Students may make mistakes with addition. Review with them that a negative integer added to a negative integer has a negative sum. (Remind students that red counters added to red counters equal red counters.)

What Is Your Answer?

- Students may have a good sense of how to predict the sign of the product; however, they often use language such as: "two positives make a positive and two negatives make a positive." This language should be avoided.

Closure

- "Today we learned that a negative integer multiplied by a negative integer is positive. Be sure to mentally check all of your steps so that you are not confusing anything."

4 ACTIVITY: Multiplying Integers with the Same Sign

Work with a partner. Use a table to find $-3 \cdot (-2)$.

Describe the pattern of the products in the table. Then complete the table.

-3	$\cdot$	3	$=$	-9
-3	$\cdot$	2	$=$	-6
-3	$\cdot$	1	$=$	-3
-3	$\cdot$	0	$=$	
-3	$\cdot$	-1	$=$	
-3	$\cdot$	-2	$=$	

So, $-3 \cdot (-2) =$ ______.

Math Practice

Look for Patterns

How can you use the pattern to complete the table?

Inductive Reasoning

Work with a partner. Complete the table.

	Exercise	Type of Product	Product	Product: Positive or Negative
1	**5.** $3 \cdot 2$	Integers with the same sign		
2	**6.** $3 \cdot (-2)$			
3	**7.** $-3 \cdot 2$			
4	**8.** $-3 \cdot (-2)$			
	9. $6 \cdot 3$			
	10. $2 \cdot (-5)$			
	11. $-6 \cdot 5$			
	12. $-5 \cdot (-3)$			

What Is Your Answer?

13. Write two integers whose product is 0.

14. **IN YOUR OWN WORDS** Is the product of two integers *positive*, *negative*, or *zero*? How can you tell?

15. **STRUCTURE** Write general rules for multiplying (a) two integers with the same sign and (b) two integers with different signs.

Practice Use what you learned about multiplying integers to complete Exercises 8–15 on page 500.

11.4 Lesson

Check It Out
Lesson Tutorials
BigIdeasMath.com

Key Ideas

Multiplying Integers with the Same Sign

Words The product of two integers with the same sign is positive.

Numbers $2 \cdot 3 = 6$ $\qquad -2 \cdot (-3) = 6$

Multiplying Integers with Different Signs

Words The product of two integers with different signs is negative.

Numbers $2 \cdot (-3) = -6$ $\qquad -2 \cdot 3 = -6$

EXAMPLE 1 Multiplying Integers with the Same Sign

Find $-5 \cdot (-6)$.

The integers have the same sign.

$$-5 \cdot (-6) = 30$$

The product is positive.

The product is 30.

EXAMPLE 2 Multiplying Integers with Different Signs

Multiply.

a. $3(-4)$ **b.** $-7 \cdot 4$

The integers have different signs.

$$3(-4) = -12 \qquad -7 \cdot 4 = -28$$

The product is negative.

The product is -12. The product is -28.

On Your Own

Now You're Ready
Exercises 8–23

Multiply.

1. $5 \cdot 5$

2. $4(11)$

3. $-1(-9)$

4. $-7 \cdot (-8)$

5. $12 \cdot (-2)$

6. $4(-6)$

7. $-10(-6)(0)$

8. $-7 \cdot (-5) \cdot (-4)$

Laurie's Notes

Introduction

Connect

- **Yesterday:** Students used repeated addition on a number line to develop a sense of integer multiplication.
- **Today:** Students will find products of integers with the same sign and products of integers with different signs.

Motivate

? "Will someone summarize what we learned yesterday?"

- Listen for informal language such as "two negatives make a positive." While it is understood that the remark was made about multiplying two negative numbers, some students may incorrectly remember the comment when they are adding two negatives later on.
- **Vocabulary Review:** Ask students to define *factor, product,* and *Commutative Property of Multiplication.*

Lesson Notes

Key Ideas

- **Attend to Precision:** Write the rules for the two cases of multiplying integers. Discuss how multiplication can be represented. The multiplication dot is shown in the book, and parentheses are used to surround a negative integer. The parentheses are used for clarity so that the negative sign is not confused with the operation of subtraction.

Example 1

- Work through each part of the example.
- Say, "You know that 5 times 6 is 30, and because both integers in this example are negative (-5 and -6), the product is 30."
- **Attend to Precision:** Students should use correct language in reading the problems. They should say, "Negative 5 times negative 6 equals 30." If students say "minus 5," remind students that minus is an operation.

Example 2

- **Look for and Make Use of Structure:** Point out to students how multiplication is represented differently in the two problems. Before doing part (a), ask if there is another way the problem could be written.
- The goal is for students to be comfortable with all of the ways in which multiplication is represented.

On Your Own

- Students should work independently and check their work.
- Alternately, you could write the problems on index cards. Ask two students to sort the cards into two piles: integers with the same sign, and integers with different signs. Ask, "What is true about all of the products (or answers) in this pile?" Point to one of the piles and then repeat the same question for the other pile. Ask for volunteers to do the problems aloud.
- When multiplying more than two numbers, remind students that they can rearrange factors using properties of multiplication.

Goal Today's lesson is finding the product of integers.

Technology for the Teacher

Dynamic Classroom

Lesson Tutorials
Lesson Plans
Answer Presentation Tool

Extra Example 1

Find $-8 \cdot (-12)$. 96

Extra Example 2

Multiply.

a. $9(-7)$ -63

b. $-6 \cdot 6$ -36

On Your Own

1. 25	**2.** 44
3. 9	**4.** 56
5. -24	**6.** -24
7. 0	**8.** -140

Laurie's Notes

Extra Example 3

Evaluate the expression.

a. $(-8)^2$ 64

b. -9^2 -81

c. -5^3 -125

On Your Own

9. 9 **10.** -8

11. -49 **12.** -216

Extra Example 4

A football jersey is marked down \$10 each week for 3 weeks. Find the total change in the price of the football jersey. $-\$30$

On Your Own

13. -45 manatees

English Language Learners

Vocabulary

Make sure that the students understand the mathematical meanings of the words *positive* and *negative*. In math, *positive* means a number greater than zero and *negative* means a number less than zero. Explain that positive and negative do not mean good and bad.

Example 3

- Students should know the meaning of exponents. Write the expression 5^2 on the board and ask students to tell you what it means.
- **Vocabulary Review:** 5 is the *base* and 2 is the *exponent*. The exponent tells you how many factors of the base (how many times you will see the base number) will be multiplied.
- So, $5 \times 5 = 25$. It is read "5 raised to the second power" or "5 squared."
- **Common Error:** When a negative number is raised to a power, the number must be written within parentheses. In part (b), the example is read "the opposite of 5 squared." If you wanted to raise -5 to the second power, it would be written $(-5)^2$. For the given problem, the order of operations says to square the number and then take its opposite. Part (c) shows how to raise a negative integer to a power.

? **Extension:** "When you raise a negative number to a power, is the answer always positive?" No. If the exponent is odd, the answer is negative.

On Your Own

- Students should work with partners.
- **Construct Viable Arguments and Critique the Reasoning of Others:** Caution students about Questions 11 and 12.
- **Common Error:** Students sometimes multiply the exponent by the base, particularly if the exponent is greater than 2.

Example 4

- There is no scale written on the vertical axis.

? "Is it possible to determine the number of taxis the company began with?" From the graph, you could estimate 300 taxis to start: each horizontal line is 50 taxis, and the first bar is 6 increments tall.

- Read the verbal model:
  total change = change per year $\times$ number of years.

? **Extension:** "At the same rate, how many years before there are no taxis?" 6

On Your Own

- Note that you do *not* need to know the initial population of manatees in order to answer the question.

Closure

- Write the number -4 on the board. Ask students to write each of the following and then share their responses.
  - "Write a multiplication problem that has -4 as one of the factors, and has a negative product." *Sample answer:* $(-4)(-1)(-1) = -4$
  - "Write a second multiplication problem that has -4 as one of the factors, and has a positive product." *Sample answer:* $(-4)(-1)(1) = 4$

EXAMPLE 3 Using Exponents

Study Tip

Place parentheses around a negative number to raise it to a power.

a. Evaluate $(-2)^2$.

$(-2)^2 = (-2) \cdot (-2)$ Write $(-2)^2$ as repeated multiplication.

$= 4$ Multiply.

b. Evaluate -5^2.

$-5^2 = -(5 \cdot 5)$ Write 5^2 as repeated multiplication.

$= -25$ Multiply.

c. Evaluate $(-4)^3$.

$(-4)^3 = (-4) \cdot (-4) \cdot (-4)$ Write $(-4)^3$ as repeated multiplication.

$= 16 \cdot (-4)$ Multiply.

$= -64$ Multiply.

On Your Own

Now You're Ready
Exercises 32–37

Evaluate the expression.

9. $(-3)^2$ **10.** $(-2)^3$ **11.** -7^2 **12.** -6^3

EXAMPLE 4 Real-Life Application

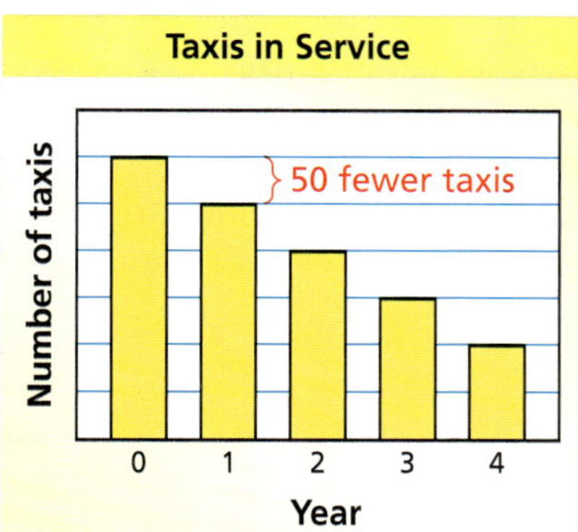

The bar graph shows the number of taxis a company has in service. The number of taxis decreases by the same amount each year for 4 years. Find the total change in the number of taxis.

The bar graph shows that the number of taxis in service decreases by 50 each year. Use a model to solve the problem.

total change = change per year • number of years

$= -50 \cdot 4$

$= -200$

Use -50 for the change per year because the number *decreases* each year.

The total change in the number of taxis is -200.

On Your Own

13. A manatee population decreases by 15 manatees each year for 3 years. Find the total change in the manatee population.

11.4 Exercises

Check It Out
Help with Homework
BigIdeasMath.com

Vocabulary and Concept Check

1. **WRITING** What can you conclude about the signs of two integers whose product is (a) positive and (b) negative?
2. **OPEN-ENDED** Write two integers whose product is negative.

Tell whether the product is *positive* or *negative* without multiplying. Explain your reasoning.

3. $4(-8)$
4. $-5(-7)$
5. $-3 \cdot 12$

Tell whether the statement is *true* or *false*. Explain your reasoning.

6. The product of three positive integers is positive.
7. The product of three negative integers is positive.

Practice and Problem Solving

Multiply.

1 2
8. $6 \cdot 4$
9. $7(-3)$
10. $-2(8)$
11. $-3(-4)$
12. $-6 \cdot 7$
13. $3 \cdot 9$
14. $8 \cdot (-5)$
15. $-1 \cdot (-12)$
16. $-5(10)$
17. $-13(0)$
18. $-9 \cdot 9$
19. $15(-2)$
20. $-10 \cdot 11$
21. $-6 \cdot (-13)$
22. $7(-14)$
23. $-11 \cdot (-11)$

24. **JOGGING** You burn 10 calories each minute you jog. What integer represents the change in your calories after you jog for 20 minutes?
25. **WETLANDS** About 60,000 acres of wetlands are lost each year in the United States. What integer represents the change in wetlands after 4 years?

Multiply.

26. $3 \cdot (-8) \cdot (-2)$
27. $6(-9)(-1)$
28. $-3(-5)(-4)$
29. $(-5)(-7)(-20)$
30. $-6 \cdot 3 \cdot (-2)$
31. $3 \cdot (-12) \cdot 0$

Evaluate the expression.

3
32. $(-4)^2$
33. $(-1)^3$
34. -8^2
35. -6^2
36. $-5^2 \cdot 4$
37. $-2 \cdot (-3)^3$

ERROR ANALYSIS **Describe and correct the error in evaluating the expression.**

38. ✗ $-2(-7) = -14$
39. ✗ $-10^2 = 100$

Assignment Guide and Homework Check

Level	Assignment	Homework Check
Advanced	1–15, 30–46 even, 47–53	7, 34, 40, 47

Common Errors

- **Exercises 8–23** Students may not remember that a negative number multiplied by a negative number is positive. Tell them that it is similar to multiplying by -1, which means to take the opposite. For example, $-6(-13) = (-1 \cdot 6)(-13) = -1[6 \cdot (-13)] = -1(-78) = 78$.
- **Exercises 26–31** Students may multiply all the numbers together ignoring the signs and then place the incorrect sign in front. For example, a student might say $-7(-3)(-5) = 105$. Tell them to multiply only two integers at a time, determine the sign, and then multiply by the last number.
- **Exercises 32–37** Students may erroneously interpret -8^2 as $(-8)(-8)$ instead of $-1(8^2)$. Remind them that the negative sign means multiplication by -1 and that exponents are evaluated before multiplication.

11.4 Record and Practice Journal

Multiply.

1. $8 \cdot 9$ **72**
2. $7(-7)$ **−49**
3. $-10 \cdot 4$ **−40**
4. $-5(-6)$ **30**
5. $12 \cdot (-1) \cdot (-2)$ **24**
6. $-10(-3)(-7)$ **−210**
7. $-20 \cdot 0 \cdot (-4)$ **0**
8. $-4 \cdot 8 \cdot 3$ **−96**

Evaluate the expression.

9. $(-8)^2$ **64**
10. -11^2 **−121**
11. $9 \cdot (-5)^2$ **225**
12. $(-2)^3 \cdot (-6)$ **48**
13. You lose 5 points for every wrong answer in a trivia game. What integer represents the change in your points after answering 8 questions wrong? **−40**

Vocabulary and Concept Check

1. **a.** They are the same.
   b. They are different.
2. *Sample answer:* 2, −3
3. negative; different signs
4. positive; same signs
5. negative; different signs
6. true; The product of the first two positive integers is positive. The product of the result and the third positive integer is positive.
7. false; The product of the first two negative integers is positive. The product of the positive result and the third negative integer is negative.

Practice and Problem Solving

8. 24	**9.** −21
10. −16	**11.** 12
12. −42	**13.** 27
14. −40	**15.** 12
16. −50	**17.** 0
18. −81	**19.** −30
20. −110	**21.** 78
22. −98	**23.** 121
24. −200	**25.** −240,000
26. 48	**27.** 54
28. −60	**29.** −700
30. 36	**31.** 0
32. 16	**33.** −1
34. −64	**35.** −36
36. −100	**37.** 54

Practice and Problem Solving

38. The product should be positive. $-2(-7) = 14$

39. The answer should be negative. $-10^2 = -(10 \cdot 10) = -100$

40. -6 **41.** 32

42. 38

43. -7500, 37,500

44. 1792, -7168

45. -12

46. See *Taking Math Deeper*.

47. **a.** 153; 141; 129

b. The price drops \$12 every month.

c. no; yes; In August, you have \$135 but the cost is \$141. In September, you have \$153 and the cost is only \$129.

48. -25

Fair Game Review

49. 3 **50.** 8

51. 14 **52.** 17

53. D

Mini-Assessment

Multiply.

1. $-4(-5)$ 20
2. $3(-3)$ -9
3. $-1(-12)$ 12
4. $-2(15)$ -30
5. You have \$900 in a checking account. You pay a \$60 cell phone bill each month using this account. The account balance is given by $900 + (-60t)$, where t is the time in months. What is the balance of the account after 4 months? \$660

Taking Math Deeper

Exercise 46

This is a classic type of problem in mathematics. A real-life measurement (such as height) is modeled by an expression. The height h (in feet) depends on the time t (in minutes).

$$h = 22{,}000 + (-480t) = 22{,}000 - 480t$$

1 First, help students understand the model.

$h = 22{,}000 - 480t$

Starting height is 22,000 feet.

Plane descends 480 feet each minute.

2 **a.** Copy and complete the table.

Time (minutes)	5	10	15	20
Height (feet)	19,600	17,200	14,800	12,400

3 **b.** When does the plane land?

$$\frac{22{,}000}{480} \approx 45.83 \text{ minutes}$$

The **height** of a plane is called its **altitude**.

Descent rates vary greatly. The rate of 480 feet per minute in this problem is low. Descent rates between 500 and 1500 feet per minute are more common. After take-off, an ascent rate of 1000 to 2000 feet per minute is common.

Project

Draw a graph showing the height of the plane in 5-minute intervals from the time it begins the descent at 22,000 feet until it lands.

Reteaching and Enrichment Strategies

If students need help...	If students got it...
Resources by Chapter • Practice A and Practice B • Puzzle Time Record and Practice Journal Practice Differentiating the Lesson Lesson Tutorials Skills Review Handbook	Resources by Chapter • Enrichment and Extension • Technology Connection Start the next section

ALGEBRA Evaluate the expression when $a = -2$, $b = 3$, and $c = -8$.

40. ab

41. $|a^2c|$

42. $-ab^3 - ac$

NUMBER SENSE Find the next two numbers in the pattern.

43. $-12, 60, -300, 1500, \ldots$

44. $7, -28, 112, -448, \ldots$

45. **GYM CLASS** You lose four points each time you attend gym class without sneakers. You forget your sneakers three times. What integer represents the change in your points?

46. **MODELING** The height of an airplane during a landing is given by $22{,}000 + (-480t)$, where t is the time in minutes.

a. Copy and complete the table.

b. Estimate how many minutes it takes the plane to land. Explain your reasoning.

Time (minutes)	5	10	15	20
Height (feet)				

47. **INLINE SKATES** In June, the price of a pair of inline skates is \$165. The price changes each of the next 3 months.

a. Copy and complete the table.

Month	Price of Skates
June	$165 = \$165$
July	$165 + (-12) = \$\_\_\_\_$
August	$165 + 2(-12) = \$\_\_\_\_$
September	$165 + 3(-12) = \$\_\_\_\_$

b. Describe the change in the price of the inline skates for each month.

c. The table at the right shows the amount of money you save each month to buy the inline skates. Do you have enough money saved to buy the inline skates in August? September? Explain your reasoning.

Amount Saved	
June	\$35
July	\$55
August	\$45
September	\$18

48. **Reasoning** Two integers, a and b, have a product of 24. What is the least possible sum of a and b?

Fair Game Review What you learned in previous grades & lessons

Divide. *(Section 1.1)*

49. $27 \div 9$

50. $48 \div 6$

51. $56 \div 4$

52. $153 \div 9$

53. **MULTIPLE CHOICE** What is the prime factorization of 84? *(Section 1.4)*

Ⓐ $2^2 \times 3^2$ Ⓑ $2^3 \times 7$ Ⓒ $3^3 \times 7$ Ⓓ $2^2 \times 3 \times 7$

11.5 Dividing Integers

Essential Question Is the quotient of two integers *positive*, *negative*, or *zero*? How can you tell?

1 ACTIVITY: Dividing Integers with Different Signs

Work with a partner. Use integer counters to find $-15 \div 3$.

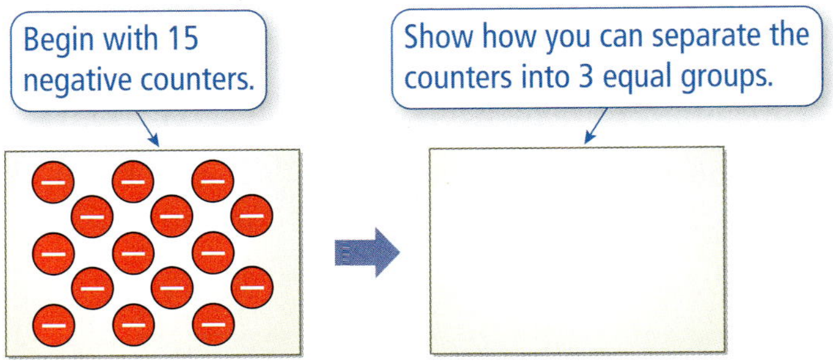

Because there are ____ negative counters in each group, $-15 \div 3 =$ ____.

2 ACTIVITY: Rewriting a Product as a Quotient

Work with a partner. Rewrite the product $3 \cdot 4 = 12$ as a quotient in two different ways.

First Way

12 is equal to 3 groups of ____.

So, $12 \div 3 =$ ____.

Second Way

12 is equal to 4 groups of ____.

So, $12 \div 4 =$ ____.

3 ACTIVITY: Dividing Integers with Different Signs

Work with a partner. Rewrite the product $-3 \cdot (-4) = 12$ as a quotient in two different ways. What can you conclude?

First Way

$12 \div ($ ____ $) =$ ____

Second Way

$12 \div ($ ____ $) =$ ____

In each case, when you divide a ____ integer by a ____ integer, you get a ____ integer.

Integers

In this lesson, you will

- divide integers.
- solve real-life problems.

Laurie's Notes

Introduction

Applying Mathematical Practices

- **Look for and Express Regularity in Repeated Reasoning:** In this lesson, students will use the relationship between multiplication and division to develop a conceptual understanding of integer division.

Motivate

- ? "What do you know about football?" Guide students to discuss the length of the field. A football field is 100 yards long, plus two 10-yard end zones, for a total length of 120 yards.
- ? "If I told you the area of the football field, could you tell me the width of the football field?" The goal is to have students think about the area formula ($A = \ell w$) and realize that if they know the area and one dimension, they can divide to find the other dimension.
- ? "The area is 6400 yd^2 and the length is 120 yd. What is the width?" $53\frac{1}{3}$ yd

Discuss

- ? "What are fact families? Give some examples for multiplication and division." Fact families show the inverse relationship between multiplication and division.

  Sample answers: $2 \times 3 = 6, 3 \times 2 = 6, 6 \div 2 = 3$, and $6 \div 3 = 2$
  $6 \times 8 = 48, 8 \times 6 = 48, 48 \div 6 = 8$, and $48 \div 8 = 6$

Activity Notes

Activity 1

- ? Place 15 red integer counters on the overhead, arranged in a 3×5 array. "What integer is being modeled?" -15
- ? Use a ruler to separate the counters into 3 groups of 5. "What division problem does this suggest?" $-15 \div 3$
- There are 5 red counters in every group. The quotient is -5. This is the *grouping model* of division.

Activity 2

- **Look for and Make Use of Structure:** Relate this example back to the football problem and fact families, because length $\times$ width = area, area $\div$ length = width, and area $\div$ width = length.

Activity 3

- ? "What is the product of two negative integers?" a positive integer
- **Look for and Make Use of Structure:** You can rewrite a multiplication problem as a division problem, but be sure to pay attention to the signs.

What Your Students Will Learn

- Divide integers with the same sign and with different signs.
- Evaluate expressions involving division by using substitution and the order of operations.

Previous Learning

Students need to know how to divide whole numbers.

Technology for the Teacher

Dynamic Classroom

Lesson Plans
Complete Materials List

11.5 Record and Practice Journal

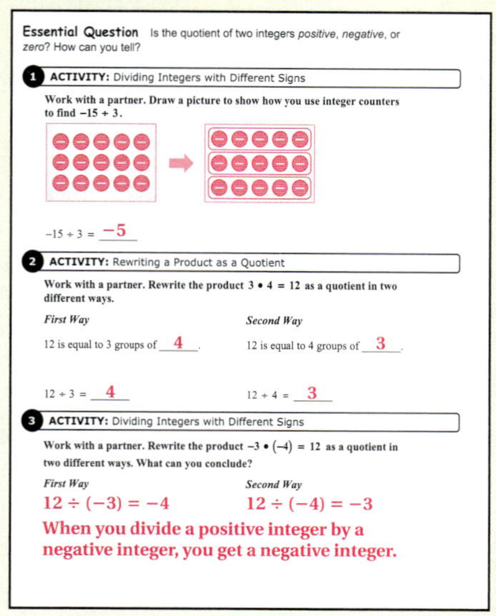

Essential Question Is the quotient of two integers *positive*, *negative*, or *zero*? How can you tell?

1 ACTIVITY: Dividing Integers with Different Signs

Work with a partner. Draw a picture to show how you use integer counters to find $-15 \div 3$.

$-15 \div 3 =$ -5

2 ACTIVITY: Rewriting a Product as a Quotient

Work with a partner. Rewrite the product $3 \bullet 4 = 12$ as a quotient in two different ways.

First Way 12 is equal to 3 groups of 4.
Second Way 12 is equal to 4 groups of 3.

$12 \div 3 =$ 4 $\quad$ $12 \div 4 =$ 3

3 ACTIVITY: Dividing Integers with Different Signs

Work with a partner. Rewrite the product $-3 \bullet (-4) = 12$ as a quotient in two different ways. What can you conclude?

First Way $12 \div (-3) = -4$
Second Way $12 \div (-4) = -3$

When you divide a positive integer by a negative integer, you get a negative integer.

Differentiated Instruction

Visual

Have students find the mean of two negative integers. Then tell the students to graph the two numbers and the mean on a number line. Have students share their results with the class. If any student has a mean that is zero or positive, identify the error.

11.5 Record and Practice Journal

4 ACTIVITY: Dividing Negative Integers

Work with a partner. Rewrite the product $3 \bullet (-4) = -12$ as a quotient in two different ways. What can you conclude?

First Way

$-12 \div (-4) = 3$

Second Way

$-12 \div 3 = -4$

The quotient of integers with the same sign is positive. The quotient of integers with different signs is negative.

Inductive Reasoning

Work with a partner. Complete the table.

	Exercise	Type of Quotient	Quotient	Quotient: Positive, Negative, or Zero
1	**5.** $-15 \div 3$	different signs	-5	Negative
2	**6.** $12 \div 4$	same sign	3	Positive
3	**7.** $12 \div (-3)$	different signs	-4	Negative
4	**8.** $-12 \div (-4)$	same sign	3	Positive
	9. $-6 \div 2$	different signs	-3	Negative
	10. $-21 \div (-7)$	same sign	3	Positive
	11. $10 \div (-2)$	different signs	-5	Negative
	12. $12 \div (-6)$	different signs	-2	Negative
	13. $0 \div (-15)$	first integer zero	0	Zero
	14. $0 \div 4$	first integer zero	0	Zero

What Is Your Answer?

15. IN YOUR OWN WORDS Is the quotient of two integers *positive*, *negative*, or *zero*? How can you tell?

It could be positive, negative, or zero.

16. STRUCTURE Write a general rule for dividing

a. two integers with the same sign.

Divide the absolute values and make the quotient positive.

b. two integers with different signs.

Divide the absolute values and make the quotient negative.

Laurie's Notes

Activity 4

- **Look for and Make Use of Structure:** Let's look at one last related problem, $3 \bullet (-4) = -12$. In this example, it is the first way (negative ÷ negative) that is new. The second way (dividing integers with different signs) is similar to Activity 3.

Words of Wisdom

? "How are Activity 2 and Activity 3 alike? How are Activity 2 and Activity 4 alike?" In each problem, you rewrote the integer multiplication problem as an integer division problem.

Inductive Reasoning

- Students should work with partners to find the quotients. The goal is for the students to recognize the bigger pattern. When the dividend and divisor have the same signs, the quotient is positive. When the dividend and divisor have different signs, the quotient is negative.
- Note that Questions 5–8 are the problems completed in Activities 1–4.
- **Look for and Express Regularity in Repeated Reasoning:** It is important for students to record the *Type of Quotient*. Mathematically proficient students will observe that there is a pattern in the table. When the dividend and divisor have the same sign, the quotient is positive. When the dividend and divisor have different signs, the quotient is negative.
- **Extension:**
  - ? "Is division commutative, meaning do 18 ÷ 9 and 9 ÷ 18 have the same quotient?" no
  - ? "What is the relationship between the two solutions?" They are reciprocals.

What Is Your Answer?

- Students may have a good sense of how to predict the sign of the quotient; however, they often use language (as they do with multiplication) such as: two positives make a positive and two negatives make a positive. This language should be avoided.

Closure

- "Today you learned that a negative integer divided by a negative integer is positive. Be sure to mentally check all of your steps so that you are not confusing anything."

4 ACTIVITY: Dividing Negative Integers

Math Practice

Maintain Oversight

How do you know what the sign will be when you divide two integers?

Work with a partner. Rewrite the product $3 \cdot (-4) = -12$ as a quotient in two different ways. What can you conclude?

First Way

$-12 \div (\ \ ) = \ \ $

Second Way

$-12 \div (\ \ ) = \ \ $

When you divide a ______ integer by a ______ integer, you get a ______ integer. When you divide a ______ integer by a ______ integer, you get a ______ integer.

Inductive Reasoning

Work with a partner. Complete the table.

	Exercise	Type of Quotient	Quotient	Quotient: Positive, Negative, or Zero
1	**5.** $-15 \div 3$	Integers with different signs		
2	**6.** $12 \div 4$			
3	**7.** $12 \div (-3)$			
4	**8.** $-12 \div (-4)$			
	9. $-6 \div 2$			
	10. $-21 \div (-7)$			
	11. $10 \div (-2)$			
	12. $12 \div (-6)$			
	13. $0 \div (-15)$			
	14. $0 \div 4$			

What Is Your Answer?

15. IN YOUR OWN WORDS Is the quotient of two integers *positive*, *negative*, or *zero*? How can you tell?

16. STRUCTURE Write general rules for dividing (a) two integers with the same sign and (b) two integers with different signs.

Practice Use what you learned about dividing integers to complete Exercises 8–15 on page 506.

11.5 Lesson

Check It Out
Lesson Tutorials
BigIdeasMath.com

Key Ideas

Remember

Division by 0 is undefined.

Dividing Integers with the Same Sign

Words The quotient of two integers with the same sign is positive.

Numbers $8 \div 2 = 4$ $\quad -8 \div (-2) = 4$

Dividing Integers with Different Signs

Words The quotient of two integers with different signs is negative.

Numbers $8 \div (-2) = -4$ $\quad -8 \div 2 = -4$

EXAMPLE 1 Dividing Integers with the Same Sign

Find $-18 \div (-6)$.

The integers have the same sign.

$$-18 \div (-6) = 3$$

The quotient is positive.

The quotient is 3.

EXAMPLE 2 Dividing Integers with Different Signs

Divide.

a. $75 \div (-25)$ **b.** $\frac{-54}{6}$

The integers have different signs.

$$75 \div (-25) = -3 \qquad \frac{-54}{6} = -9$$

The quotient is negative.

The quotient is -3. The quotient is -9.

On Your Own

Now You're Ready
Exercises 8–23

Divide.

1. $14 \div 2$ **2.** $-32 \div (-4)$ **3.** $-40 \div (-8)$

4. $0 \div (-6)$ **5.** $\frac{-49}{7}$ **6.** $\frac{21}{-3}$

Laurie's Notes

Introduction

Connect

- **Yesterday:** Students used fact families and what they knew about multiplication of integers to develop a sense of integer division.
- **Today:** Students will use the idea that when the signs of the dividend and divisor are the same, the quotient is positive; when the signs of the dividend and divisor are different, the quotient is negative.

Motivate

? "Will someone summarize what we learned yesterday?"

- **Listen:** Again, watch for informal language such as "two negatives make a positive." While it is understood that the remark was made about dividing two negative numbers, some students may incorrectly remember the comment when they are adding or subtracting two negatives later on.
- **Vocabulary Review:** Ask students to define *dividend, divisor, quotient, Commutative Property,* and *division involving zero.*

Lesson Notes

Key Ideas

- **Attend to Precision:** Students should know that to check a division problem, you multiply the quotient by the divisor, and the answer is the dividend.
- Summary of division involving zero: You can divide 0 by a nonzero number and the answer is 0. You cannot divide a number by 0. Later in this lesson, connect this concept to "0 cannot be in the denominator when division is represented in fraction form."

Example 1

- Work through the example.
- Say, "We know that 18 divided by 6 is 3, and because both integers are negative for this example (-18 and -6), the quotient is positive 3."
- **Attend to Precision:** Be sure that students use correct language in reading the problems. When they read the problem they should say, "Negative 18 divided by negative 6 equals 3." If students say "minus 18," remind them that minus is an operation.

Example 2

- **Look for and Make Use of Structure:** Point out to students how division is represented differently in the two problems. Before doing part (a), you may want to ask if the problem could be written another way.
- The goal is for students to be comfortable with all of the ways in which division is represented.

On Your Own

- Students should work independently and check their work.

Goal Today's lesson is dividing integers.

Technology for the Teacher

Dynamic Classroom

Lesson Tutorials
Lesson Plans
Answer Presentation Tool

Extra Example 1

Find $-48 \div (-6)$. 8

Extra Example 2

Divide.

a. $\dfrac{84}{-4}$ -21

b. $-39 \div 3$ -13

On Your Own

1. 7 **2.** 8

3. 5 **4.** 0

5. -7 **6.** -7

Extra Example 3

Evaluate $\frac{x-8}{y^2}$ when $x = 4$ and $y = 2$.
−1

On Your Own

7. 3
8. −4
9. 2

Extra Example 4

The morning high tide at a beach is 57 inches. Six hours later, the afternoon low tide is 12 inches. What is the mean hourly change in the height? −7.5 in.

On Your Own

10. −6 ft/h

English Language Learners

Visual
Show students that the rules for multiplication can be used to understand the rules for division. For example, to evaluate $16 \div (-8) = ?$, rewrite it using multiplication, $-8 \times ? = 16$. Students should be able to determine that the answer to the multiplication problem is −2, and the answer to the division problem is −2. So, the quotient of two integers with different signs is negative. Use the same approach to demonstrate the other three cases: $-16 \div 8 = ?$, $16 \div 8 = ?$, and $-16 \div (-8) = ?$.

Laurie's Notes

Example 3

- Students should know the order of operations and the meaning of exponents. Students will need to use order of operations. Instead of telling students "remember the order of operations," you want to see what the students remember without prompting.
- ? Write the problem and ask what it means to "evaluate." To evaluate a numerical expression means to perform the operations to find the value of the expression.
- **Common Error:** If students forget the order of operations, they will perform the operations left to right. Solicit responses as to what operations should be done, in the correct order, and why.

On Your Own

- Students should work with partners.
- ? **Extension:** "Can Question 8 be rewritten as $a + 6/2$?" No. There is an implied order of operations by the division bar and, therefore, it would need to be written as $(a + 6)/2$.

Example 4

- Discuss how the word *mean* is used in this context. Students often only think of computing a mean by adding values and then dividing by the number of values. So, if students think of *mean* as adding values, it could lead to a problem.
- In this problem, the total change in height is found by finding the difference of the final height and the initial height.

On Your Own

- There is a decrease in the water level, so the hourly change in height is negative. If the tide is coming in, the height would increase, and so the hourly change would be positive.

Closure

- How are the rules for multiplication and division of integers related? Why? The rules are the same because the operations are inverses. You can use fact families to rewrite a division problem as a multiplication problem, and vice versa.

EXAMPLE 3 Evaluating an Expression

Evaluate $10 - x^2 \div y$ when $x = 8$ and $y = -4$.

> **Remember**
> Use order of operations when evaluating an expression.

$10 - x^2 \div y = 10 - 8^2 \div (-4)$	Substitute 8 for x and -4 for y.
$= 10 - 8 \cdot 8 \div (-4)$	Write 8^2 as repeated multiplication.
$= 10 - 64 \div (-4)$	Multiply 8 and 8.
$= 10 - (-16)$	Divide 64 by -4.
$= 26$	Subtract.

On Your Own

Now You're Ready
Exercises 28–31

Evaluate the expression when $a = -18$ and $b = -6$.

7. $a \div b$
8. $\dfrac{a + 6}{3}$
9. $\dfrac{b^2}{a} + 4$

EXAMPLE 4 Real-Life Application

You measure the height of the tide using the support beams of a pier. Your measurements are shown in the picture. What is the mean hourly change in the height?

Use a model to solve the problem.

$$\text{mean hourly change} = \frac{\text{final height} - \text{initial height}}{\text{elapsed time}}$$

$= \dfrac{8 - 59}{6}$	Substitute. The elapsed time from 2 P.M. to 8 P.M. is 6 hours.
$= \dfrac{-51}{6}$	Subtract.
$= -8.5$	Divide.

∴ The mean change in the height of the tide is -8.5 inches per hour.

On Your Own

10. The height of the tide at the Bay of Fundy in New Brunswick decreases 36 feet in 6 hours. What is the mean hourly change in the height?

11.5 Exercises

Check It Out
Help with Homework
BigIdeasMath.com

Vocabulary and Concept Check

1. **WRITING** What can you tell about two integers when their quotient is positive? negative? zero?
2. **VOCABULARY** A quotient is undefined. What does this mean?
3. **OPEN-ENDED** Write two integers whose quotient is negative.
4. **WHICH ONE DOESN'T BELONG?** Which expression does *not* belong with the other three? Explain your reasoning.

$\frac{10}{-5}$ $\frac{-10}{5}$ $\frac{-10}{-5}$ $-\left(\frac{10}{5}\right)$

Tell whether the quotient is *positive* or *negative* without dividing.

5. $-12 \div 4$
6. $\frac{-6}{-2}$
7. $15 \div (-3)$

Practice and Problem Solving

Divide, if possible.

① ② 8. $4 \div (-2)$
9. $21 \div (-7)$
10. $-20 \div 4$
11. $-18 \div (-3)$
12. $\frac{-14}{7}$
13. $\frac{0}{6}$
14. $\frac{-15}{-5}$
15. $\frac{54}{-9}$
16. $-33 \div 11$
17. $-49 \div (-7)$
18. $0 \div (-2)$
19. $60 \div (-6)$
20. $\frac{-56}{14}$
21. $\frac{18}{0}$
22. $\frac{65}{-5}$
23. $\frac{-84}{-7}$

ERROR ANALYSIS **Describe and correct the error in finding the quotient.**

24. ✗ $\frac{-63}{-9} = -7$

25. ✗ $0 \div (-5) = -5$

26. **ALLIGATORS** An alligator population in a nature preserve in the Everglades decreases by 60 alligators over 5 years. What is the mean yearly change in the alligator population?

27. **READING** You read 105 pages of a novel over 7 days. What is the mean number of pages you read each day?

ALGEBRA **Evaluate the expression when $x = 10$, $y = -2$, and $z = -5$.**

③ 28. $x \div y$
29. $\frac{10y^2}{z}$
30. $\left|\frac{xz}{-y}\right|$
31. $\frac{-x^2 + 6z}{y}$

Assignment Guide and Homework Check

Level	Assignment	Homework Check
Advanced	1–15, 24–40 even, 41–45	28, 30, 36, 38

Common Errors

- **Exercises 8–23** In problems involving zero, students may just say that the quotient is undefined. Remind students that when 0 is the dividend, it means $\square \cdot -2 = 0$, where $\square = 0$. Also, when 0 is the divisor, it means $\square \cdot 0 = 18$, where the answer is undefined.
- **Exercises 28–31 and 34–35** Students may forget to follow the order of operations. Review the order of operations, especially the left-to-right rule in evaluating multiplication/division and addition/subtraction.
- **Exercises 32 and 33** Students may not remember how to find the mean of several numbers. They may get confused by the negative numbers and subtract instead of add. Remind students of the definition of mean.

11.5 Record and Practice Journal

Divide, if possible.

1. $3 \div (-1)$ **−3**
2. $8 \div 2$ **4**
3. $-10 \div 5$ **−2**
4. $-21 \div (-7)$ **3**
5. $\frac{48}{-6}$ **−8**
6. $\frac{-13}{-13}$ **1**
7. $\frac{0}{3}$ **0**
8. $\frac{-55}{11}$ **−5**

Evaluate the expression.

9. $-63 \div (-7) + 6$ **15**
10. $-5 - 12 \div 3$ **−9**
11. $-8 \cdot 7 + 33 \div (-11)$ **−59**

12. The table shows the number of yards a football player runs in each quarter of a game. Find the mean number of yards the player runs per quarter.

Quarter	1	2	3	4
Yards	−2	14	−18	−6

−3 yards

Vocabulary and Concept Check

1. They have the same sign. They have different signs. The dividend is zero.
2. The divisor is zero.
3. *Sample answer:* −4, 2
4. $\frac{-10}{-5}$, which equals 2.

   All the others equal −2.
5. negative
6. positive
7. negative

Practice and Problem Solving

8. −2
9. −3
10. −5
11. 6
12. −2
13. 0
14. 3
15. −6
16. −3
17. 7
18. 0
19. −10
20. −4
21. undefined
22. −13
23. 12
24. The quotient should be positive. $\frac{-63}{-9} = 7$
25. The quotient should be 0. $0 \div (-5) = 0$
26. −12 alligators
27. 15 pages
28. −5
29. −8
30. 25
31. 65

Practice and Problem Solving

32. 3 **33.** 5

34. -10 **35.** 4

36. $-8, 4$; Divide the previous number by -2 to obtain the next number.

37. -400 ft/min

38. See *Taking Math Deeper*.

39. 5

40. 20 people

41. *Sample answer:* $-20, -15, -10, -5, 0$; Start with -10, then pair -15 with -5 and -20 with 0. The sum of the integers must be $5(-10) = -50$.

Fair Game Review

42–44. See Additional Answers for number lines.

42. $-6, -1, |2|, 4, |-10|$

43. $-8, -3, |0|, 3, |-4|$

44. $-7, -5, -2, |-2|, |5|$

45. B

Mini-Assessment

Divide.

1. $-16 \div (-4)$ 4
2. $-22 \div 11$ -2
3. $35 \div (-5)$ -7
4. $-36 \div (-6)$ 6
5. You play a video game for 15 minutes. You lose 75 points. What integer represents the mean change in points per minute? -5

Taking Math Deeper

Exercise 38

This problem reviews the concept of mean (average) that students learned last year. The difference here is that the data set contains negative numbers. When students studied mean previously, all of the numbers in the data set were positive.

Scorecard	
Round 1	-2
Round 2	-6
Round 3	-7
Round 4	-3

1. **a.** Find the total score.

$$-2 + (-6) + (-7) + (-3) = -18$$

2. **b.** Find the mean score per round.

Divide the total score by the number of rounds.

$$\frac{-18}{4} = -4.5$$

3. In golf, par is the number of strokes it should take to complete a hole. An 18-hole course may contain four par 3s, ten par 4s, and four par 5s with a total par score of 72.

If a course has a par of 72 and a golfer takes 75 strokes to complete the course, the golfer's score is $+3$, or "three over par." If a golfer takes 70 strokes, the score is -2, or "two under par."

Golf tournaments usually have 4 rounds, for a total par of $4(72) = 288$. As an extension, ask students to determine the golfer's total score.
$288 + (-18) = 270$

Project

Make a table showing the winning scores for each Masters Tournament since 2000. Graph the winning scores for each year. Describe any patterns you notice.

Reteaching and Enrichment Strategies

If students need help. . .	If students got it. . .
Resources by Chapter • Practice A and Practice B • Puzzle Time Record and Practice Journal Practice Differentiating the Lesson Lesson Tutorials Skills Review Handbook	Resources by Chapter • Enrichment and Extension • Technology Connection Start the next section

Find the mean of the integers.

32. 3, −10, −2, 13, 11

33. −26, 39, −10, −16, 12, 31

Evaluate the expression.

34. $-8 - 14 \div 2 + 5$

35. $24 \div (-4) + (-2) \cdot (-5)$

36. PATTERN Find the next two numbers in the pattern −128, 64, −32, 16, ... Explain your reasoning.

37. SNOWBOARDING A snowboarder descends a 1200-foot hill in 3 minutes. What is the mean change in elevation per minute?

38. GOLF The table shows a golfer's score for each round of a tournament.

a. What was the golfer's total score?

b. What was the golfer's mean score per round?

Scorecard	
Round 1	−2
Round 2	−6
Round 3	−7
Round 4	−3

39. TUNNEL The Detroit-Windsor Tunnel is an underwater highway that connects the cities of Detroit, Michigan, and Windsor, Ontario. How many times deeper is the roadway than the bottom of the ship?

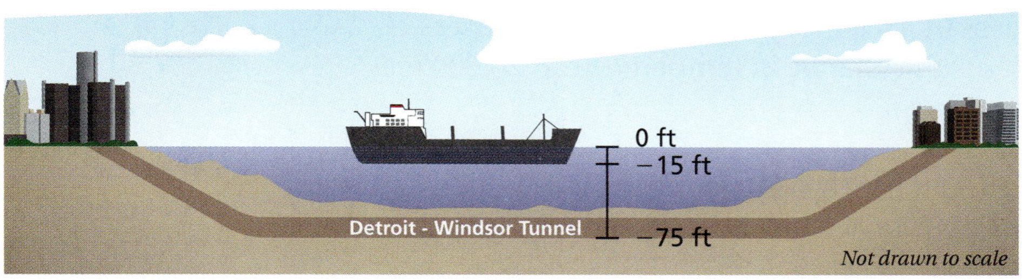

40. AMUSEMENT PARK The regular admission price for an amusement park is \$72. For a group of 15 or more, the admission price is reduced by \$25. How many people need to be in a group to save \$500?

41. Number Sense Write five different integers that have a mean of −10. Explain how you found your answer.

Fair Game Review What you learned in previous grades & lessons

Graph the values on a number line. Then order the values from least to greatest. *(Section 11.1)*

42. $-6, 4, |2|, -1, |-10|$

43. $3, |0|, |-4|, -3, -8$

44. $|5|, -2, -5, |-2|, -7$

45. MULTIPLE CHOICE What is the value of $4 \cdot 3 + (12 \div 2)^2$? *(Section 1.3)*

Ⓐ 15 Ⓑ 48 Ⓒ 156 Ⓓ 324

11.4–11.5 Quiz

Check It Out
Progress Check
BigIdeasMath.com

Evaluate the expression. *(Section 11.4 and Section 11.5)*

1. $-7(6)$

2. $-1(-10)$

3. $\dfrac{-72}{-9}$

4. $-24 \div 3$

5. $-3 \cdot 4 \cdot (-6)$

6. $(-3)^3$

Evaluate the expression when $a = 4$, $b = -6$, and $c = -12$. *(Section 11.4 and Section 11.5)*

7. c^2

8. bc

9. $\dfrac{ab}{c}$

10. $\dfrac{|c - b|}{a}$

11. SPEECH In speech class, you lose 3 points for every 30 seconds you go over the time limit. Your speech is 90 seconds over the time limit. What integer represents the change in your points? *(Section 11.4)*

12. MOUNTAIN CLIMBING On a mountain, the temperature decreases by 18°F every 5000 feet. What integer represents the change in temperature at 20,000 feet? *(Section 11.4)*

13. GAMING You play a video game for 15 minutes. You lose 165 points. What is the mean change in points per minute? *(Section 11.5)*

14. DIVING You dive 21 feet from the surface of a lake in 7 seconds. *(Section 11.4 and Section 11.5)*

a. What is the mean change in your position in feet per second?

b. You continue diving. What is your position relative to the surface after 5 more seconds?

15. HIBERNATION A female grizzly bear weighs 500 pounds. After hibernating for 6 months, she weighs only 200 pounds. What is the mean change in weight per month? *(Section 11.5)*

Alternative Assessment Options

Math Chat	Student Reflective Focus Question
Structured Interview	Writing Prompt

Math Chat

- Have students work in pairs. One student describes the rule for multiplying two integers with the same sign and the rule for multiplying two integers with different signs. The student should include examples. The other student should probe for more information. Students then switch roles and repeat the process for dividing two integers with the same sign and dividing two integers with different signs.
- The teacher should walk around the classroom listening to the pairs and asking questions to ensure understanding.

Study Help Sample Answers

Remind students to complete Graphic Organizers for the rest of the chapter.

5a. Multiplying integers with the same sign: The product of two integers with the same sign is positive.

- Example: $6 \cdot 3 = 18$
- Example: $-6 \cdot (-3) = 18$
- Example: $-18(-16) = 288$

5b. Multiplying integers with different signs: The product of two integers with different signs is negative.

- Example: $6 \cdot (-3) = -18$
- Example: $-6 \cdot 3 = -18$
- Example: $18(-16) = -288$

6. Available at *BigIdeasMath.com.*

Reteaching and Enrichment Strategies

If students need help. . .	If students got it. . .
Resources by Chapter • Practice A and Practice B • Puzzle Time Lesson Tutorials *BigIdeasMath.com*	Resources by Chapter • Enrichment and Extension • Technology Connection Game Closet at *BigIdeasMath.com* Start the Chapter Review

Answers

1. -42

2. 10

3. 8

4. -8

5. 72

6. -27

7. 144

8. 72

9. 2

10. $\frac{3}{2}$

11. -9

12. -72

13. -11

14. **a.** -3

b. -36 ft

15. -50 lb/mo

Technology for the Teacher

Online Assessment
Assessment Book
ExamView® Assessment Suite

For the Teacher
Additional Review Options

- *BigIdeasMath.com*
- Online Assessment
- Game Closet at *BigIdeasMath.com*
- Vocabulary Help
- Resources by Chapter

Answers

1. 3
2. 9
3. 17
4. 8
5. Mississippi River in Illinois
6. −27
7. −10
8. 25
9. −34

Review of Common Errors

Exercises 1–5

- Students may think they can find the absolute value of a number by changing its sign and incorrectly find $|6| = -6$.

Exercises 6–9

- Students may ignore the signs and just add the integers.
- Remind students of the Key Ideas and how the signs of the integers determine if the student needs to add or subtract the integers.

Exercises 10–14

- Students may change the sign of the first number or forget to change the problem from subtraction to addition when changing the sign of the second number.

Exercises 15–22

- Students may not remember that a negative number multiplied or divided by a negative number is positive.

Exercises 23–25

- Remind students of the order of operations.

Exercises 26–28

- Remind students of the definition of mean.

Exercise 29

- Students may not recognize that division should be used to find the answer.
- Point out to students that all of the shirts are priced the same and that the total price of the shirts is $30.60.

11 Chapter Review

Check It Out
Vocabulary Help
BigIdeasMath .com

Review Key Vocabulary

integer, *p. 478*
absolute value, *p. 478*
opposites, *p. 484*
additive inverse, *p. 484*

Review Examples and Exercises

11.1 Integers and Absolute Value *(pp. 476–481)*

Find the absolute value of −2.

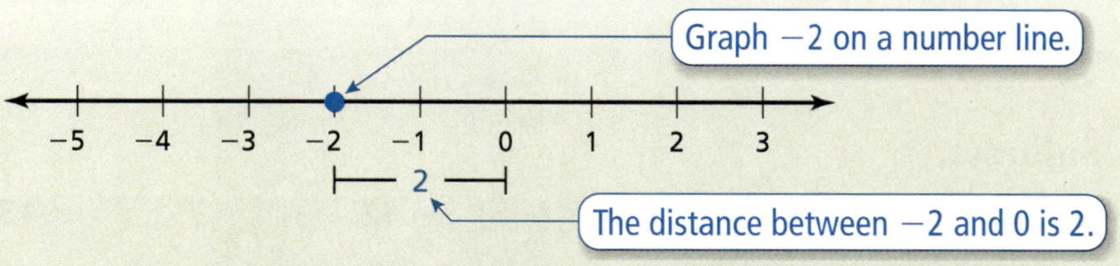

So, $|-2| = 2$.

Exercises

Find the absolute value.

1. $|3|$
2. $|-9|$
3. $|-17|$
4. $|8|$
5. **ELEVATION** The elevation of Death Valley, California, is −282 feet. The Mississippi River in Illinois has an elevation of 279 feet. Which is closer to sea level?

11.2 Adding Integers *(pp. 482–487)*

Find 6 + (−14).

$6 + (-14) = -8$ — $|-14| > |6|$. So, subtract $|6|$ from $|-14|$.

Use the sign of −14.

The sum is −8.

Exercises

Add.

6. $-16 + (-11)$
7. $-15 + 5$
8. $100 + (-75)$
9. $-32 + (-2)$

11.3 Subtracting Integers *(pp. 488–493)*

Subtract.

a. $7 - 19 = 7 + (-19)$ — Add the opposite of 19.

$= -12$ — Add.

∴ The difference is -12.

b. $-6 - (-10) = -6 + 10$ — Add the opposite of -10.

$= 4$ — Add.

∴ The difference is 4.

Exercises

Subtract.

10. $8 - 18$ **11.** $-16 - (-5)$ **12.** $-18 - 7$ **13.** $-12 - (-27)$

14. **GAME SHOW** Your score on a game show is -300. You answer the final question incorrectly, so you lose 400 points. What is your final score?

11.4 Multiplying Integers *(pp. 496–501)*

a. Find $-7 \cdot (-9)$.

The integers have the same sign.

$$-7 \cdot (-9) = 63$$

The product is positive.

∴ The product is 63.

b. Find $-6(14)$.

The integers have different signs.

$$-6(14) = -84$$

The product is negative.

∴ The product is -84.

Exercises

Multiply.

15. $-8 \cdot 6$ **16.** $10(-7)$ **17.** $-3 \cdot (-6)$ **18.** $-12(5)$

Review Game

Integer Operations

Materials per Group

- 52 index cards, numbered 1 through 52
- paper for each group member
- pencil for each group member

Directions

This game is like the card game *War*. Divide the class into equally sized groups. One person in each group shuffles and deals the index cards for that group. Eight rounds are played. For each round, each person is dealt two cards. A different operation is used in each round, as follows.

Round 1—addition
Round 2—subtraction
Round 3—multiplication
Round 4—division
Round 5—addition
Round 6—subtraction
Round 7—multiplication
Round 8—division

Using the two cards dealt for the round, in Rounds 1 through 4, players evaluate an expression using the number on the first card, the operation for the round, and the number on the second card. In Rounds 5 through 8, players evaluate an expression using the number on the first card, the operation for the round, and *the negative of* the number on the second card.

The person with the greatest result in their group wins the round and takes all the other players' cards. In the event of a tie, the players involved receive one additional card, and the operation for the round is performed using *all three* cards. Players continue in this manner until there is one winner.

If a group runs out of cards before the end of Round 8, each player records how many cards he or she has collected, and the cards are collected from the players. The cards are then reshuffled and reused.

Who Wins?

The player with the most cards after Round 8 wins. In the event of a tie, the rounds are repeated, starting with Round 1, until there is a winner.

For the Student Additional Practice

- Lesson Tutorials
- Multi-Language Glossary
- Self-Grading Progress Check
- *BigIdeasMath.com*
  Dynamic Student Edition
  Student Resources

Answers

10. -10 **11.** -11

12. -25 **13.** 15

14. -700 points

15. -48 **16.** -70

17. 18 **18.** -60

19. -2 **20.** 7

21. -5 **22.** -12

23. -2 **24.** 2

25. 3 **26.** -1

27. -48 **28.** $-\$48$

29. 5 shirts

My Thoughts on the Chapter

Teacher Tip
Not allowed to write in your teaching edition? Use sticky notes to record your thoughts.

What worked. . .

What did not work. . .

What I would do differently. . .

11.5 Dividing Integers *(pp. 502–507)*

a. Find 30 ÷ (−10).

The integers have different signs.

$$30 \div (-10) = -3$$

The quotient is negative.

The quotient is −3.

b. Find $\frac{-72}{-9}$.

The integers have the same sign.

$$\frac{-72}{-9} = 8$$

The quotient is positive.

The quotient is 8.

Exercises

Divide.

19. $-18 \div 9$ **20.** $\frac{-42}{-6}$ **21.** $\frac{-30}{6}$ **22.** $84 \div (-7)$

Evaluate the expression when $x = 3$, $y = -4$, and $z = -6$.

23. $z \div x$ **24.** $\frac{xy}{z}$ **25.** $\frac{z - 2x}{y}$

Find the mean of the integers.

26. −3, −8, 12, −15, 9 **27.** −54, −32, −70, −25, −65, −42

28. PROFITS The table shows the weekly profits of a fruit vendor. What is the mean profit for these weeks?

Week	1	2	3	4
Profit	−$125	−$86	$54	−$35

29. RETURNS You return several shirts to a store. The receipt shows that the amount placed back on your credit card is −$30.60. Each shirt is −$6.12. How many shirts did you return?

11 Chapter Test

Check It Out
Test Practice
BigIdeasMath.com

Find the absolute value.

1. $|-9|$
2. $|64|$
3. $|-22|$

Copy and complete the statement using <, >, or =.

4. $4 \;\square\; |-8|$
5. $|-7| \;\square\; -12$
6. $-7 \;\square\; |3|$

Evaluate the expression.

7. $-6 + (-11)$
8. $2 - (-9)$
9. $-9 \cdot 2$
10. $-72 \div (-3)$

Evaluate the expression when $x = 5$, $y = -3$, and $z = -2$.

11. $\dfrac{y + z}{x}$
12. $\dfrac{x - 5z}{y}$

Find the mean of the integers.

13. 11, −7, −14, 10, −5
14. −32, −41, −39, −27, −33, −44

15. **NASCAR** A driver receives −25 points for each rule violation. What integer represents the change in points after 4 rule violations?

16. **GOLF** The table shows your scores, relative to *par*, for nine holes of golf. What is your total score for the nine holes?

Hole	1	2	3	4	5	6	7	8	9	Total
Score	+1	−2	−1	0	−1	+3	−1	−3	+1	?

17. **VISITORS** In a recent 10-year period, the change in the number of visitors to U.S. national parks was about −11,150,000 visitors.
   a. What was the mean yearly change in the number of visitors?
   b. During the seventh year, the change in the number of visitors was about 10,800,000. Explain how the change for the 10-year period can be negative.

Test Item References

Chapter Test Questions	Section to Review
1–6	11.1
7, 16	11.2
8	11.3
9, 15	11.4
10–14, 17	11.5

Test-Taking Strategies

Remind students to quickly look over the entire test before they start so that they can budget their time. They should not spend too much time on any single problem. Urge students to try to work on a part of each problem because partial credit is better than none. Teach students to use the Stop and Think strategy before answering. **Stop** and carefully read the question, and **Think** about what the answer should look like.

Common Errors

- **Exercise 2** Students may think that the absolute value of a number is always its opposite and write $|64| = -64$. Use a number line to show students that absolute value is a number's distance from 0, so it is always positive or zero.
- **Exercises 4 and 5** When comparing absolute values of negative integers, students may not find absolute values first and instead just compare the integers. Remind students to find the absolute values first.
- **Exercises 7–12** Students may ignore the signs of the integers when simplifying. Remind students of the Key Ideas for addition, subtraction, multiplication, and division of integers and that the signs of the integers will affect their answers.
- **Exercises 13 and 14** Students may not remember how to find the mean of several numbers. They may get confused by the negative numbers and subtract instead of add. Remind students of the definition of mean.
- **Exercise 16** Students may not know the meaning of the word *par*, so explain it to them.

Reteaching and Enrichment Strategies

If students need help. . .	If students got it. . .
Resources by Chapter • Practice A and Practice B • Puzzle Time Record and Practice Journal Practice Differentiating the Lesson Lesson Tutorials *BigIdeasMath.com* Skills Review Handbook	Resources by Chapter • Enrichment and Extension • Technology Connection Game Closet at *BigIdeasMath.com* Start Cumulative Assessment

Answers

1. 9
2. 64
3. 22
4. $4 < |-8|$
5. $|-7| > -12$
6. $-7 < |3|$
7. -17
8. 11
9. -18
10. 24
11. -1
12. -5
13. -1
14. -36
15. -100
16. -3
17. a. $-1{,}115{,}000$ visitors

    b. During other years, there were more significant changes in visitors in the negative direction.

Technology for the Teacher

Online Assessment
Assessment Book
ExamView® Assessment Suite

Test-Taking Strategies

Available at *BigIdeasMath.com*

After Answering Easy Questions, Relax
Answer Easy Questions First
Estimate the Answer
Read All Choices before Answering
Read Question before Answering
Solve Directly or Eliminate Choices
Solve Problem before Looking at Choices
Use Intelligent Guessing
Work Backwards

About this Strategy

When taking a multiple choice test, be sure to read each question carefully and thoroughly. Before answering a question, determine exactly what is being asked, then eliminate the wrong answers and select the best choice.

Answers

1. C
2. H
3. C
4. 25
5. G

Technology for the Teacher

Performance Tasks
Online Assessment
Assessment Book
ExamView® Assessment Suite

Item Analysis

1. **A.** The student treats all numbers as gains and finds their sum.

   B. The student finds the correct difference but thinks it is a gain instead of a loss.

   C. Correct answer

   D. The student treats all numbers as losses and finds their sum.

2. **F.** The student does not perform the operation correctly.

   G. The student does not perform the operation correctly.

   H. Correct answer

   I. The student does not perform the operation correctly.

3. **A.** The student does not evaluate a^2 correctly and does not find the absolute value.

   B. The student does not find the absolute value.

   C. Correct answer

   D. The student does not evaluate a^2 correctly.

4. **Gridded Response:** Correct answer: 25

   Common Error: The student thinks that $17 - (-8)$ is equivalent to $17 - 8$, getting an answer of 9.

5. **F.** The student thinks that $-(-5) = -5$.

   G. Correct answer

   H. The student does not follow the order of operations.

   I. The student does not follow the order of operations.

Cumulative Assessment Icons

Gridded Response

Short Response (2-point rubric)

Extended Response (4-point rubric)

11 Cumulative Assessment

Test-Taking Strategy
Solve Directly or Eliminate Choices

"You can eliminate A and C. Then, solve directly to determine that the correct answer is D."

1. A football team gains 2 yards on the first play, loses 5 yards on the second play, loses 3 yards on the third play, and gains 4 yards on the fourth play. What is the team's overall gain or loss for all four plays?

A. a gain of 14 yards

B. a gain of 2 yards

C. a loss of 2 yards

D. a loss of 14 yards

2. Which expression is *not* equal to the number 0?

F. $5 - 5$

G. $-7 + 7$

H. $6 - (-6)$

I. $-8 - (-8)$

3. What is the value of the expression below when $a = -2$, $b = 3$, and $c = -5$?

$$|a^2 - 2ac + 5b|$$

A. -9

B. -1

C. 1

D. 9

4. What is the value of the expression below?

$$17 - (-8)$$

5. Sam was evaluating an expression in the box below.

$$\begin{aligned}(-2)^3 \cdot 3 - (-5) &= 8 \cdot 3 - (-5)\\ &= 24 + 5\\ &= 29\end{aligned}$$

What should Sam do to correct the error that he made?

F. Subtract 5 from 24 instead of adding.

G. Rewrite $(-2)^3$ as -8.

H. Subtract -5 from 3 before multiplying by $(-2)^3$.

I. Multiply -2 by 3 before raising the quantity to the third power.

6. What is the value of the expression below when $x = 6$, $y = -4$, and $z = -2$?

$$\frac{x - 2y}{-z}$$

A. -7

B. -1

C. 1

D. 7

7. What is the missing number in the sequence below?

$$39, 24, 9, \_\_\_, -21$$

8. You are playing a game using the spinner shown. You start with a score of 0 and spin the spinner four times. When you spin blue or green, you add the number to your score. When you spin red or orange, you subtract the number from your score. Which sequence of colors represents the greatest score?

F. red, green, green, red

G. orange, orange, green, blue

H. red, blue, orange, green

I. blue, red, blue, red

9. Which expression represents a negative integer?

A. $5 - (-6)$

B. $(-3)^3$

C. $-12 \div (-6)$

D. $(-2)(-4)$

10. Which expression has the greatest value when $x = -2$ and $y = -3$?

F. $-xy$

G. xy

H. $x - y$

I. $-x - y$

Item Analysis (continued)

6. **A.** The student thinks that $-(-2) = -2$.

   B. The student thinks that $(-2)(-4) = -8$.

   C. The student thinks that $(-2)(-4) = -8$ and $-(-2) = -2$.

   D. Correct answer

7. **Gridded Response:** Correct answer: -6

   Common Error: The student incorrectly finds a number halfway between 9 and -21 by subtracting 9 from -21 to get -30, half of which is -15.

8. **F.** The student adds the red value instead of subtracting.

   G. Correct answer

   H. The student adds the absolute value of each number.

   I. The student chooses the value with the greatest absolute value, not the greatest value.

9. **A.** The student does not perform the operation correctly.

   B. Correct answer

   C. The student does not perform the operation correctly.

   D. The student does not perform the operation correctly.

10. **F.** The student does not perform the operation correctly.

    G. Correct answer

    H. The student does not perform the operation correctly.

    I. The student does not perform the operation correctly.

Answers

6. D
7. −6
8. G
9. B
10. G

Answers

11. B

12. G

13. D

14. *Part A* Start at 0. Then move 2 to the left and then 3 more to the left, which results in a position of −5.

Part B Start at 0. Then move 2 to the right and then 5 to the left, which results in a position of −3.

15. H

Item Analysis (continued)

11. **A.** The student thinks that $-(-3) = -3$.

B. Correct answer

C. The student thinks that $-5 \cdot (-4)^2 = 80$ and $-(-3) = -3$.

D. The student thinks that $-5 \cdot (-4)^2 = 80$.

12. **F.** The student does not have a clear understanding of the properties.

G. Correct answer

H. The student does not have a clear understanding of the properties.

I. The student does not have a clear understanding of the properties.

13. **A.** The student finds the mode instead of the mean.

B. The student finds the range instead of the mean. The student also makes multiple errors in finding the range, using −8 and 1 instead of −8 and 4, thinking that the range between −8 and 1 is 7, and thinking that the range is negative.

C. The student finds the median instead of the mean.

D. Correct answer

14. **2 points** The student demonstrates a thorough understanding of adding and subtracting integers using a number line. The student demonstrates how to add using a number line, writes the expression, and gets the correct answer $-2 + (-3) = -5$. The student demonstrates how to subtract using a number line, writes the expression, and gets the correct answer $2 - 5 = -3$.

1 point The student demonstrates a partial understanding of adding and subtracting integers using a number line. The student shows some knowledge of adding and subtracting integers, but is not successful in determining the correct answers.

0 points The student demonstrates insufficient understanding of adding and subtracting integers using a number line.

15. **F.** The student incorrectly evaluates the numerator as $(-3 - 2)^2$ instead of $-3 - (2)^2$.

G. The student incorrectly evaluates the power as $(-2)^2$ instead of 2^2.

H. Correct answer

I. The student incorrectly evaluates the numerator as $(-3 - 2)^2$ instead of $-3 - (2)^2$ and uses the wrong sign in the answer.

11. What is the value of the expression below?

$$-5 \cdot (-4)^2 - (-3)$$

A. -83

B. -77

C. 77

D. 83

12. Which property does the equation below represent?

$$-80 + 30 + (-30) = -80 + [30 + (-30)]$$

F. Commutative Property of Addition

G. Associative Property of Addition

H. Additive Inverse Property

I. Addition Property of Zero

13. What is the mean of the data set in the box below?

$$-8, -6, -2, 0, -6, -8, 4, -7, -8, 1$$

A. -8

B. -7

C. -6

D. -4

14. Consider the number line shown below.

Think Solve Explain

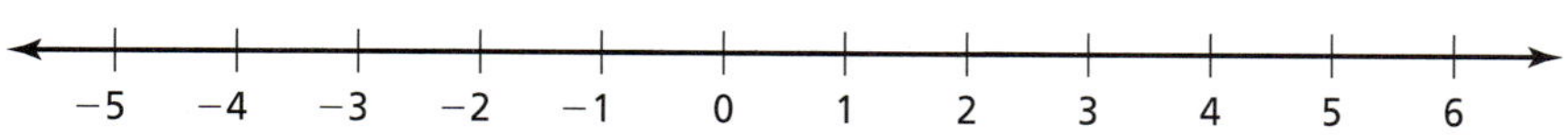

Part A Use the number line to explain how to add -2 and -3.

Part B Use the number line to explain how to subtract 5 from 2.

15. What is the value of the expression below?

$$\frac{-3 - 2^2}{-1}$$

F. -25

G. -1

H. 7

I. 25

12 Rational Numbers

What Your Students Have Learned

- Fluently multiply and divide whole numbers.
- Add and subtract fractions with unlike denominators.
- Multiply fractions or whole numbers by fractions.
- Divide unit fractions by whole numbers and whole numbers by unit fractions.
- Add, subtract, multiply, and divide decimals up to hundredths.
- Divide fractions.
- Fluently add, subtract, multiply, and divide decimals.
- Describe quantities with positive and negative numbers.

What Your Students Will Learn

- Add, subtract, multiply, and divide rational numbers.
- Apply properties of operations as strategies to perform operations with rational numbers.
- Convert rational numbers to decimals using long division.

Pacing Guide for Chapter 12

Chapter Opener Advanced	1 Day
Section 1 Advanced	1 Day
Section 2 Advanced	1 Day
Section 3 Advanced	1 Day
Section 4 Advanced	1 Day
Chapter Review/ Chapter Tests Advanced	2 Days
Total Chapter 12 Advanced	7 Days
Year-to-Date Advanced	113 Days

Technology for the Teacher

BigIdeasMath.com
Chapter at a Glance
Complete Materials List
Parent Letters: English and Spanish

What Your Students Have Learned

- Write decimals as fractions and fractions as decimals.
- Add and subtract fractions with unlike denominators.
- Multiply and divide fractions with unlike denominators.

Additional Topics for Review

- Place Value
- Writing Mixed Numbers as Improper Fractions
- Dividing Numbers Using the Standard Algorithm
- Simplifying Fractions

Try It Yourself

1. $\frac{51}{100}$ **2.** $\frac{731}{1000}$

3. 0.6 **4.** 0.875

5. $\frac{9}{10}$ **6.** $\frac{3}{5}$

7. $\frac{27}{70}$ **8.** $\frac{17}{20}$

Record and Practice Journal Fair Game Review

1. $\frac{13}{50}$ **2.** $\frac{79}{100}$

3. $\frac{571}{1000}$ **4.** $\frac{423}{500}$

5. 0.375 **6.** 0.4

7. 0.6875 **8.** 0.85

9. $\frac{3}{5}$ **10.** $\frac{17}{72}$

11–18. See Additional Answers.

Math Background Notes

Vocabulary Review

- Denominator
- Least Common Multiple
- Common Denominator
- Least Common Denominator (LCD)
- Reciprocal
- Divisor

Writing Decimals and Fractions

- Students should know how to convert between decimals and fractions.
- You may need to review place values to the right of the decimal place with students prior to completing Example 1.

Adding and Subtracting Fractions

- Students should know how to add and subtract fractions.
- Remind students that adding and subtracting fractions requires a common denominator.
- You should review the least common multiple with students. This concept will help some students to find a common denominator.
- Using the least common multiple of the denominators will produce the least common denominator. Remind students that there are many common denominators to choose from. Some choices will require students to simplify the fraction at the end.

Multiplying and Dividing Fractions

- Students should know how to multiply and divide fractions.
- Remind students that the rules for multiplying and dividing fractions are different from the rules for adding and subtracting fractions. Multiplying and dividing fractions does not require a common denominator.
- **Teaching Tip:** Most students will remember the process to divide fractions. If your students are comfortable with the process, encourage them to describe it using math vocabulary Instead of "change the sign and flip the second fraction," encourage "multiply by the reciprocal of the divisor."

Reteaching and Enrichment Strategies

If students need help...	If students got it...
Record and Practice Journal • Fair Game Review Skills Review Handbook Lesson Tutorials	Game Closet at *BigIdeasMath.com* Start the next section

What You Learned Before

"Let's play a game. The goal is to say a positive rational number that is less than the other pet's number... You go first."

Writing Decimals and Fractions

Example 1 Write 0.37 as a fraction.

$$0.37 = \frac{37}{100}$$

Example 2 Write $\frac{2}{5}$ as a decimal.

$$\frac{2}{5} = \frac{2 \cdot 2}{5 \cdot 2} = \frac{4}{10} = 0.4$$

Try It Yourself

Write the decimal as a fraction or the fraction as a decimal.

1. 0.51 **2.** 0.731 **3.** $\frac{3}{5}$ **4.** $\frac{7}{8}$

Adding and Subtracting Fractions

Example 3 Find $\frac{1}{3} + \frac{1}{5}$.

$$\frac{1}{3} + \frac{1}{5} = \frac{1 \cdot 5}{3 \cdot 5} + \frac{1 \cdot 3}{5 \cdot 3}$$
$$= \frac{5}{15} + \frac{3}{15}$$
$$= \frac{8}{15}$$

Example 4 Find $\frac{1}{4} - \frac{2}{9}$.

$$\frac{1}{4} - \frac{2}{9} = \frac{1 \cdot 9}{4 \cdot 9} - \frac{2 \cdot 4}{9 \cdot 4}$$
$$= \frac{9}{36} - \frac{8}{36}$$
$$= \frac{1}{36}$$

Multiplying and Dividing Fractions

Example 5 Find $\frac{5}{6} \cdot \frac{3}{4}$.

$$\frac{5}{6} \cdot \frac{3}{4} = \frac{5 \cdot \cancel{3}^{1}}{_{2}\cancel{6} \cdot 4}$$
$$= \frac{5}{8}$$

Example 6 Find $\frac{2}{3} \div \frac{9}{10}$.

$$\frac{2}{3} \div \frac{9}{10} = \frac{2}{3} \cdot \frac{10}{9}$$

Multiply by the reciprocal of the divisor.

$$= \frac{2 \cdot 10}{3 \cdot 9}$$
$$= \frac{20}{27}$$

Try It Yourself

Evaluate the expression.

5. $\frac{1}{4} + \frac{13}{20}$ **6.** $\frac{14}{15} - \frac{1}{3}$ **7.** $\frac{3}{7} \cdot \frac{9}{10}$ **8.** $\frac{4}{5} \div \frac{16}{17}$

12.1 Rational Numbers

Essential Question How can you use a number line to order rational numbers?

The Meaning of a Word ● Rational

The word **rational** comes from the word *ratio*. Recall that you can write a ratio using fraction notation.

If you sleep for 8 hours in a day, then the ratio of your sleeping time to the total hours in a day can be written as $\dfrac{8\text{ h}}{24\text{ h}}$.

A **rational number** is a number that can be written as the ratio of two integers.

$$2 = \frac{2}{1} \qquad -3 = \frac{-3}{1} \qquad -\frac{1}{2} = \frac{-1}{2} \qquad 0.25 = \frac{1}{4}$$

1 ACTIVITY: Ordering Rational Numbers

Work in groups of five. Order the numbers from least to greatest.

- Use masking tape and a marker to make a number line on the floor similar to the one shown.

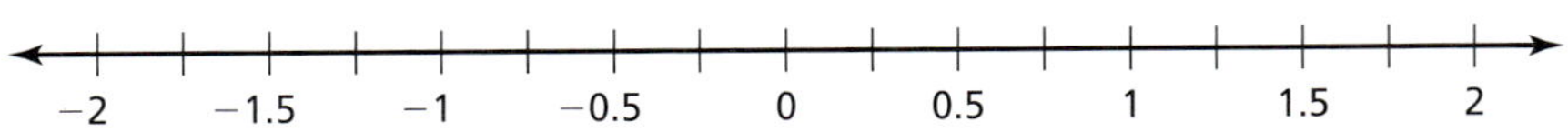

- Write the numbers on pieces of paper. Then each person should choose one.
- Stand on the location of your number on the number line.
- Use your positions to order the numbers from least to greatest.

a. $-0.5,\ 1.25,\ -\frac{1}{3},\ 0.5,\ -\frac{5}{3}$

b. $-\frac{7}{4},\ 1.1,\ \frac{1}{2},\ -\frac{1}{10},\ -1.3$

c. $-1.4,\ -\frac{3}{5},\ \frac{9}{2},\ \frac{1}{4},\ 0.9$

d. $\frac{5}{4},\ 0.75,\ -\frac{5}{4},\ -0.8,\ -1.1$

Rational Numbers

In this lesson, you will
- understand that a rational number is an integer divided by an integer.
- convert rational numbers to decimals.

Laurie's Notes

Introduction

Applying Mathematical Practices

- **Make Sense of Problems and Persevere in Solving Them:** In this lesson, students will write fractions as decimals and vice versa. Students should always check the reasonableness of their answers. For instance, $\frac{7}{11}$ is greater than $\frac{1}{2}$. So, when you write $\frac{7}{11}$ as a decimal, the result should be greater than 0.5.

Motivate

- A key skill for both activities today will be the ability to compare fractions and decimals. Try a warm up where students need to fill in the following table.

Fraction	$\frac{1}{2}$		$\frac{3}{5}$		$\frac{3}{4}$	
Decimal		0.1		0.8		1.4

- Check for understanding of the process of converting between these two forms of numbers.
- Students have studied operations with whole numbers, fractions, decimals, and integers.
- ? "Do you think there is a number halfway between −3 and −4? What is that number?"
- Explain that in this chapter, they will perform operations on numbers such as −3.5. Define rational numbers.

Activity Notes

Activity 1

- **Model with Mathematics:** In preparing for this activity, be sure to leave sufficient space between the number line marks so that students are able to stand at their locations comfortably. If there is enough space in the classroom, make multiple number lines on the floor. Consider different orientations so that students may see some as vertical number lines and some as horizontal number lines. If space is limited, pairs of students could do the same problem on the board or on a piece of paper.
- **Construct Viable Arguments:** When the first set of numbers has been located, spend time having students give their reasoning as to why they located the numbers as they did. For instance, how did they know that $-\frac{5}{3}$ was to the left of −0.5 versus to the right of it?
- So that all students have an opportunity to use the number line on the floor, rotate groups at the end of each set of numbers.
- **Extension:** If time permits, ask students to name a decimal between two fractions $\left(-\frac{1}{2} \text{ and } -\frac{3}{4}\right)$ and to name a fraction between two decimals (−0.6 and −0.7).

What Your Students Will Learn

- Understand that a rational number is an integer divided by an integer.
- Convert rational numbers to decimals using long division.
- Write decimals as fractions in simplest form using equivalent fractions.
- Order rational numbers by converting them to decimals and then graphing them on a number line.

Previous Learning

Students should know how to convert between common fractions (halves, fourths, fifths, and tenths) and decimals. They should also be able to graph common fractions and decimals on a number line.

Technology for the Teacher

Dynamic Classroom

Lesson Plans
Complete Materials List

12.1 Record and Practice Journal

Essential Question How can you use a number line to order rational numbers?

A **rational number** is a number that can be written as a ratio of two integers.

$2 = \frac{2}{1}$ $\quad -3 = \frac{-3}{1}$ $\quad -\frac{1}{2} = \frac{-1}{2}$ $\quad 0.25 = \frac{1}{4}$

1 ACTIVITY: Ordering Rational Numbers

Work in groups of five. Order the numbers from least to greatest.

- Use masking tape and a marker to make a number line on the floor similar to the one shown.

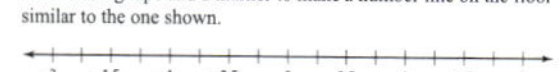

- Write the numbers on pieces of paper. Then each person should choose one piece of paper.
- Stand on the location of your number on the number line.
- Use your positions to order the numbers from least to greatest.

The numbers from least to greatest are

_____, _____, _____, _____, and _____.

a. $-0.5, 1.25, -\frac{1}{3}, 0.5, -\frac{5}{3}$

$-\frac{5}{3}, -0.5, -\frac{1}{3}, 0.5, 1.25$

b. $-\frac{7}{4}, 1.1, \frac{1}{2}, -\frac{1}{10}, -1.3$

$-\frac{7}{4}, -1.3, -\frac{1}{10}, \frac{1}{2}, 1.1$

c. $-1.4, -\frac{3}{5}, \frac{9}{2}, \frac{1}{4}, 0.9$

$-1.4, -\frac{3}{5}, \frac{1}{4}, 0.9, \frac{9}{2}$

d. $\frac{5}{4}, 0.75, -\frac{5}{4}, -0.8, -1.1$

$-\frac{5}{4}, -1.1, -0.8, 0.75, \frac{5}{4}$

English Language Learners

Vocabulary

Let English language learners know that the pronunciation of rational numbers that end in –th such as fourths and fifths is sometimes difficult for native English speakers.

12.1 Record and Practice Journal

2 ACTIVITY: The Game of Math Card War

Preparation:

- Cut index cards to make 40 playing cards.*
- Write each number in the table on a card.

$-\frac{3}{2}$	$\frac{3}{10}$	$-\frac{3}{4}$	−0.6	1.25	−0.15	$\frac{5}{4}$	$\frac{3}{5}$	−1.6	−0.3
$\frac{3}{20}$	$\frac{8}{5}$	−1.2	$\frac{19}{10}$	0.75	−1.5	$-\frac{6}{5}$	$-\frac{3}{5}$	1.2	0.3
1.5	1.9	−0.75	−0.4	$\frac{3}{4}$	$-\frac{5}{4}$	−1.9	$\frac{2}{5}$	$-\frac{3}{20}$	$\frac{19}{10}$
$\frac{6}{5}$	$-\frac{3}{10}$	1.6	$-\frac{2}{5}$	0.6	0.15	$\frac{3}{2}$	−1.25	0.4	$-\frac{8}{5}$

To Play:

- Play with a partner.
- Deal 20 cards to each player facedown.
- Each player turns one card faceup. The player with the greater number wins. The winner collects both cards and places them at the bottom of his or her cards.
- Suppose there is a tie. Each player lays three cards facedown, then a new card faceup. The player with the greater of these new cards wins. The winner collects all ten cards and places them at the bottom of his or her cards.
- Continue playing until one player has all the cards. This player wins the game.

*Cut-outs are available in the back of the Record and Practice Journal.

What Is Your Answer?

3. **IN YOUR OWN WORDS** How can you use a number line to order rational numbers? Give an example.

Sample answer: A number line can be used to organize rational numbers from least to greatest based on their order from left to right on the line.

The numbers are in order from least to greatest. Fill in the blank spaces with rational numbers. Sample answers are given.

4. $-\frac{1}{2}$, $\boxed{-\frac{1}{4}}$, $\frac{1}{3}$, $\boxed{\frac{3}{4}}$, $\frac{7}{5}$, $\boxed{2}$

5. $-\frac{5}{2}$, $\boxed{-2}$, −1.9, $\boxed{-\frac{3}{2}}$, $-\frac{2}{3}$, $\boxed{\frac{1}{2}}$

6. $-\frac{1}{3}$, $\boxed{-\frac{1}{5}}$, −0.1, $\boxed{0.1}$, $\frac{4}{5}$, $\boxed{\frac{5}{4}}$

7. −3.4, $\boxed{}$, −1.5, $\boxed{1.1}$, 2.2, $\boxed{2.5}$

−2.1

Laurie's Notes

Activity 2

- You may want to make the game cards ahead of time or have students create the game cards.
- **Management Tip:** To preserve cards for multiple uses, make cards on colored cardstock and store individual sets in sealable plastic bags.
- The card game *War* is familiar to many students. The question asked each play is, "Which number is greater?" The player with the greater value collects both cards. If the cards have an equivalent value, there is a tie. As stated in the text, each player lays 3 cards face down and then 1 card face up. The player with the card of greater value collects all of the cards.
- **Comparing Cards:** The key component of this activity is when students actually compare the two rational numbers. Discuss with students how they will compare the numbers. When both numbers are positive or the signs are different, students will have less difficulty. If both numbers are negative, students need to remember that the farther the number is to the right on the number line, the greater its value.

  For example, $-\frac{3}{5} > -0.75$ because $-\frac{3}{5}$ is to the right of -0.75 on a number line.
- To start play, give students the opportunity to preview the cards. Explain the rules and let students begin. If one group finishes early, have them shuffle the cards and play again.
- **Extension:** The cards can also be used to play the game *Memory*. Put the fraction cards in one group and the decimal cards in another group. Place all cards face down in two grids. Students select one card from each group. If the cards match, (meaning they are equivalent), then the student keeps the cards. If they do not match, the cards are put back face down. A deck of 40 cards is too many! Reduce the deck to 24 (12 in each group). Make sure the equivalent decimals and fractions are in each deck.

What Is Your Answer?

- Listen for the big idea, namely that the farther to the right the number is on the number line, the greater the value of that number.
- **Construct Viable Arguments and Critique the Reasoning of Others:** For Questions 4–7, students should work with partners. Have students share their results and their reasoning. Answers will vary, so the explanation is important to hear.

Closure

- Which is greater: *A* or *B*? All have *B* as the greater number.

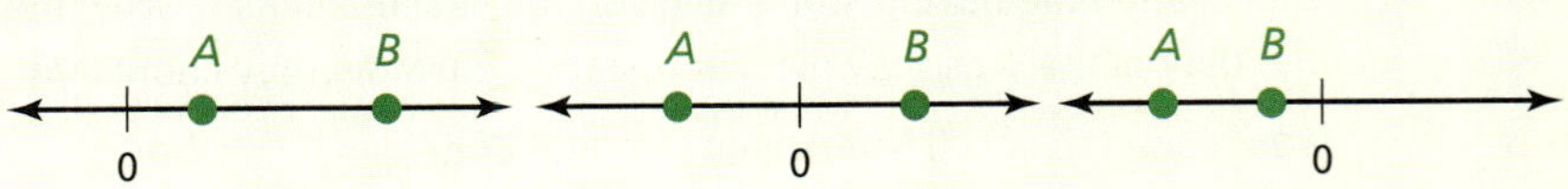

2 ACTIVITY: The Game of Math Card War

Math Practice

Consider Similar Problems

What are some ways to determine which number is greater?

Preparation:

- Cut index cards to make 40 playing cards.
- Write each number in the table on a card.

To Play:

- Play with a partner.
- Deal 20 cards to each player facedown.
- Each player turns one card faceup. The player with the greater number wins. The winner collects both cards and places them at the bottom of his or her cards.
- Suppose there is a tie. Each player lays three cards facedown, then a new card faceup. The player with the greater of these new cards wins. The winner collects all ten cards and places them at the bottom of his or her cards.
- Continue playing until one player has all the cards. This player wins the game.

$-\frac{3}{2}$	$\frac{3}{10}$	$-\frac{3}{4}$	-0.6	1.25	-0.15	$\frac{5}{4}$	$\frac{3}{5}$	-1.6	-0.3
$\frac{3}{20}$	$\frac{8}{5}$	-1.2	$\frac{19}{10}$	0.75	-1.5	$-\frac{6}{5}$	$-\frac{3}{5}$	1.2	0.3
1.5	1.9	-0.75	-0.4	$\frac{3}{4}$	$-\frac{5}{4}$	-1.9	$\frac{2}{5}$	$-\frac{3}{20}$	$-\frac{19}{10}$
$\frac{6}{5}$	$-\frac{3}{10}$	1.6	$-\frac{2}{5}$	0.6	0.15	$\frac{3}{2}$	-1.25	0.4	$-\frac{8}{5}$

What Is Your Answer?

3. **IN YOUR OWN WORDS** How can you use a number line to order rational numbers? Give an example.

The numbers are in order from least to greatest. Fill in the blank spaces with rational numbers.

4. $-\frac{1}{2}$, ____, $\frac{1}{3}$, ____, $\frac{7}{5}$, ____

5. $-\frac{5}{2}$, ____, -1.9, ____, $-\frac{2}{3}$, ____

6. $-\frac{1}{3}$, ____, -0.1, ____, $\frac{4}{5}$, ____

7. -3.4, ____, -1.5, ____, 2.2, ____

Practice Use what you learned about ordering rational numbers to complete Exercises 28–30 on page 522.

12.1 Lesson

Check It Out
Lesson Tutorials
BigIdeasMath.com

Key Vocabulary
rational number, *p. 520*
terminating decimal, *p. 520*
repeating decimal, *p. 520*

Key Idea

Rational Numbers

A **rational number** is a number that can be written as $\frac{a}{b}$ where a and b are integers and $b \neq 0$.

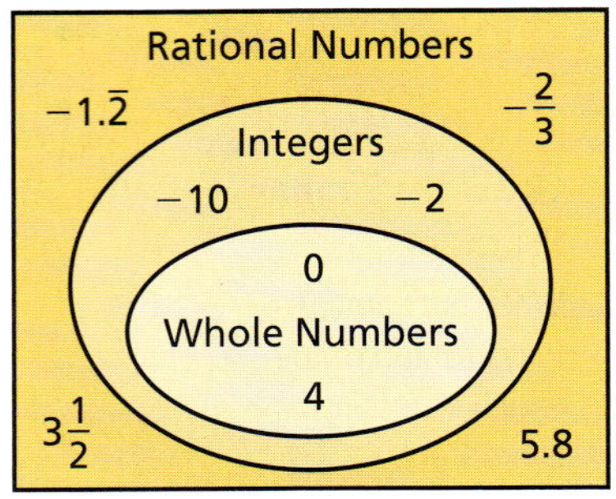

Because you can divide any integer by any nonzero integer, you can use long division to write fractions and mixed numbers as decimals. These decimals are also rational numbers and will either *terminate* or *repeat.*

A **terminating decimal** is a decimal that ends.

$1.5, -0.25, 10.625$

A **repeating decimal** is a decimal that has a pattern that repeats.

$-1.333\ldots = -1.\overline{3}$

$0.151515\ldots = 0.\overline{15}$

Use *bar notation* to show which of the digits repeat.

EXAMPLE 1 Writing Rational Numbers as Decimals

a. Write $-2\frac{1}{4}$ as a decimal.

Notice that $-2\frac{1}{4} = -\frac{9}{4}$.

Divide 9 by 4.

$$4\overline{)9.00} = 2.25$$

```
    2.25
 4)9.00
  -8
   1 0
  -  8
     20
   - 20
      0
```

The remainder is 0. So, it is a terminating decimal.

So, $-2\frac{1}{4} = -2.25$.

b. Write $\frac{5}{11}$ as a decimal.

Divide 5 by 11.

```
    0.4545
11)5.0000
  - 4 4
      60
    - 55
      50
    - 44
      60
    - 55
       5
```

The remainder repeats. So, it is a repeating decimal.

So, $\frac{5}{11} = 0.\overline{45}$.

On Your Own

Now You're Ready
Exercises 11–18

Write the rational number as a decimal.

1. $-\frac{6}{5}$ **2.** $-7\frac{3}{8}$ **3.** $-\frac{3}{11}$ **4.** $1\frac{5}{27}$

Multi-Language Glossary at BigIdeasMath.com

Laurie's Notes

Introduction

Connect

- **Yesterday:** Students ordered rational numbers.
- **Today:** Students will extend this knowledge to include repeating decimals.

Motivate

- Ask students to form a "name fraction," where the numerator is the number of letters in their first name and the denominator is the number of letters in their last name.
- Before class, go through your class roster and select two students whose name fractions are nearly equivalent, but one is a terminating decimal and the other is a repeating decimal. Discuss writing the fractions as decimals.
- When you look at the repeating decimal, share that today's lesson is about writing rational numbers, which may be repeating decimals.

Lesson Notes

Key Idea

- **Discuss:** Students have worked with fractions and decimals before. Explain that when negative fractions and decimals are included, we refer to these numbers as rational numbers. Also point out that the definition includes the word *can*, meaning that the rational numbers do not have to be written in the form $\frac{a}{b}$, but they *can* be.
- Define terminating and repeating decimals. Give examples of each.
- **Common Error:** Some students will write $\frac{1}{3}$ as 0.333 and think that is sufficient. They do not realize what the repeat bar represents.
- **Big Idea:** In this lesson students should gain an understanding that every quotient of integers (with a non-zero divisor) is a rational number.

Example 1

? "How do you write a fraction as a decimal?" Listen for 3 methods: 1) benchmark fractions you know, 2) write the fraction as an equivalent fraction with a denominator as a power of 10 and use the place value, or 3) divide the numerator by the denominator.

- **Make Sense of Problems:** Mathematically proficient students are able to plan a solution. Choosing between methods may help students be more efficient and accurate when writing fractions as decimals.
- Complete part (a) as a class. The first step is to write the mixed number as the equivalent improper fraction. Then divide the numerator by the denominator. Point out that the negative sign is simply placed in the answer after the calculations are complete.
- Complete part (b) as a class. Remind students that you always divide the numerator by the denominator, regardless of the size of the numbers!

On Your Own

- **Neighbor Check:** Have students work independently and then have their neighbors check their work. Have students discuss any discrepancies.

Goal Today's lesson is writing fractions as decimals, including **repeating decimals**, and writing decimals as fractions in simplest form.

Technology for the Teacher

Dynamic Classroom

Lesson Tutorials
Lesson Plans
Answer Presentation Tool

Extra Example 1

Write the rational number as a decimal.

a. $4\frac{3}{16}$ 4.1875

b. $-3\frac{4}{9}$ $-3.\overline{4}$

On Your Own

1. -1.2 **2.** -7.375

3. $-0.\overline{27}$ **4.** $1.\overline{185}$

Extra Example 2

Write -2.625 as a mixed number in simplest form. $-2\frac{5}{8}$

On Your Own

5. $-\frac{7}{10}$
6. $\frac{1}{8}$
7. $-3\frac{1}{10}$
8. $-10\frac{1}{4}$

Extra Example 3

Order the rational numbers $-\frac{5}{9}$, $-1\frac{3}{4}$, $-\frac{13}{8}$, and -0.6 from least to greatest. $-1\frac{3}{4}$, $-\frac{13}{8}$, -0.6, $-\frac{5}{9}$

On Your Own

9. All of the sea creatures (anglerfish, squid, shark, and whale) are deeper than the dolphin.

Differentiated Instruction

Auditory

Writing terminating decimals as rational numbers is easier if the students read the decimal using place value as opposed to reading the digits.

Terminating decimal:
-0.26

Read as place value:
negative twenty-six hundredths

Read as digits:
negative zero point two six

Laurie's Notes

Example 2

- ? "How do you write a decimal as a fraction?" Look at the place value of the last digit in the decimal and that will be the denominator.
- Work through Example 2.
- ? "How was the fraction simplified?" Both the numerator and the denominator were divided by a common factor of 2.
- Be sure to discuss the Study Tip.
- **Extension:** Write -0.026 and -2.6 as fractions. This helps students focus on the importance of place value and where the last digit is located.

On Your Own

- **Neighbor Check:** Have students work independently and then have their neighbors check their work. Have students discuss any discrepancies.
- In Questions 7 and 8, the whole number portion of the decimal can be a problem.

Example 3

- Discuss the unit of measure, kilometers.
- Work through the problem. When doing this problem in class, draw the number line vertically and identify sea level.

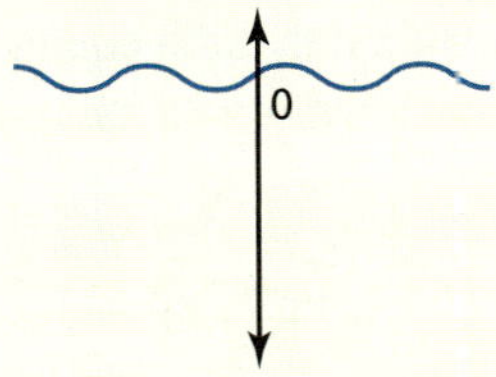

On Your Own

- **Neighbor Check:** Have students work independently and then have their neighbors check their work. Have students discuss any discrepancies.
- **Extension:** If calculators are available to students, explore the repeating patterns for certain sets of fractions (thirds, ninths, elevenths, etc.).

Closure

- **Exit ticket:**
  - Write $-\frac{5}{6}$ as a decimal. $-0.8\overline{3}$
  - Write -0.56 as a fraction. $-\frac{56}{100} = -\frac{14}{25}$

EXAMPLE 2 Writing a Decimal as a Fraction

Write -0.26 as a fraction in simplest form.

> **Study Tip**
>
> If p and q are integers, then $-\frac{p}{q} = \frac{-p}{q} = \frac{p}{-q}$.

$$-0.26 = -\frac{26}{100}$$

Write the digits after the decimal point in the numerator.

The last digit is in the hundredths place. So, use 100 in the denominator.

$$= -\frac{13}{50}$$

Simplify.

On Your Own

Now You're Ready Exercises 20–27

Write the decimal as a fraction or a mixed number in simplest form.

5. -0.7 **6.** 0.125 **7.** -3.1 **8.** -10.25

EXAMPLE 3 Ordering Rational Numbers

Creature	Elevation (kilometers)
Anglerfish	$-\frac{13}{10}$
Squid	$-2\frac{1}{5}$
Shark	$-\frac{2}{11}$
Whale	-0.8

The table shows the elevations of four sea creatures relative to sea level. Which of the sea creatures are deeper than the whale? Explain.

Write each rational number as a decimal.

$$-\frac{13}{10} = -1.3$$

$$-2\frac{1}{5} = -2.2$$

$$-\frac{2}{11} = -0.\overline{18}$$

Then graph each decimal on a number line.

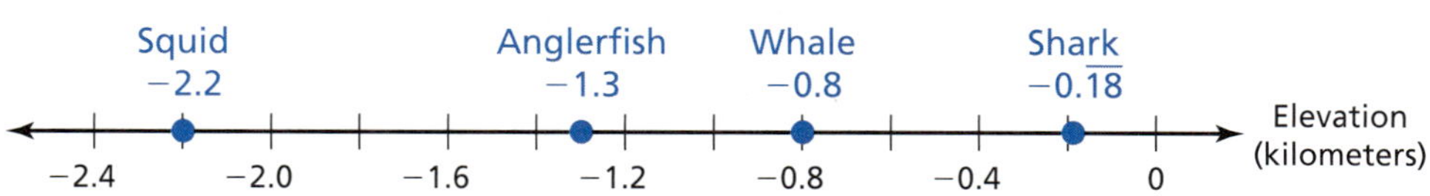

∴ Both -2.2 and -1.3 are less than -0.8. So, the squid and the anglerfish are deeper than the whale.

On Your Own

Now You're Ready Exercises 28–33

9. WHAT IF? The elevation of a dolphin is $-\frac{1}{10}$ kilometer. Which of the sea creatures in Example 3 are deeper than the dolphin? Explain.

12.1 Exercises

Check It Out
Help with Homework
BigIdeasMath.com

Vocabulary and Concept Check

1. **VOCABULARY** Is the quotient of two integers always a rational number? Explain.

2. **WRITING** Are all terminating and repeating decimals rational numbers? Explain.

Tell whether the number belongs to each of the following number sets: *rational numbers, integers, whole numbers.*

3. -5 4. $-2.1\overline{6}$ 5. 12 6. 0

Tell whether the decimal is *terminating* or *repeating*.

7. $-0.4848\ldots$ 8. -0.151 9. 72.72 10. $-5.2\overline{36}$

Practice and Problem Solving

Write the rational number as a decimal.

(1) 11. $\frac{7}{8}$ 12. $\frac{1}{11}$ 13. $-\frac{7}{9}$ 14. $-\frac{17}{40}$

15. $1\frac{5}{6}$ 16. $-2\frac{17}{18}$ 17. $-5\frac{7}{12}$ 18. $8\frac{15}{22}$

19. **ERROR ANALYSIS** Describe and correct the error in writing the rational number as a decimal.

$$-\frac{7}{11} = -0.6\overline{3}$$

Write the decimal as a fraction or a mixed number in simplest form.

(2) 20. -0.9 21. 0.45 22. -0.258 23. -0.312

24. -2.32 25. -1.64 26. 6.012 27. -12.405

Order the numbers from least to greatest.

(3) 28. $-\frac{3}{4}, 0.5, \frac{2}{3}, -\frac{7}{3}, 1.2$ 29. $\frac{9}{5}, -2.5, -1.1, -\frac{4}{5}, 0.8$ 30. $-1.4, -\frac{8}{5}, 0.6, -0.9, \frac{1}{4}$

31. $2.1, -\frac{6}{10}, -\frac{9}{4}, -0.75, \frac{5}{3}$ 32. $-\frac{7}{2}, -2.8, -\frac{5}{4}, \frac{4}{3}, 1.3$ 33. $-\frac{11}{5}, -2.4, 1.6, \frac{15}{10}, -2.25$

34. **COINS** You lose one quarter, two dimes, and two nickels.
   a. Write the amount as a decimal.
   b. Write the amount as a fraction in simplest form.

35. **HIBERNATION** A box turtle hibernates in sand at $-1\frac{5}{8}$ feet. A spotted turtle hibernates at $-1\frac{16}{25}$ feet. Which turtle is deeper?

Assignment Guide and Homework Check

Level	Assignment	Homework Check
Advanced	1–10, 19, 28–42 even, 45–52	28, 36, 42, 46

Common Errors

- **Exercises 11–18** Students may forget to carry the negative sign through the division operation. Tell them to create a space for the final answer and to write the sign of the number in the space at the beginning.
- **Exercises 20–27** Students may try to put the decimal number over the denominator. Remind them to remove the decimal point before they write it as a fraction. They can also write the whole number in front of the fraction while they are reducing it.
- **Exercises 28–33** Students may just order the fractions or decimals without the negative signs. Remind them that some numbers are negative and will be less than the positive numbers.

12.1 Record and Practice Journal

Write the rational number as a decimal.

1. $-\frac{9}{10}$ **-0.9**
2. $-4\frac{2}{3}$ **$-4.\overline{6}$**
3. $1\frac{7}{16}$ **1.4375**

Write the decimal as a fraction or mixed number in simplest form.

4. -0.84 **$-\frac{21}{25}$**
5. 5.22 **$5\frac{11}{50}$**
6. -1.716 **$-1\frac{179}{250}$**

Order the numbers from least to greatest.

7. $\frac{1}{5}, 0.1, -\frac{1}{2}, -0.25, 0.3$ **$-\frac{1}{2}, -0.25, 0.1, \frac{1}{5}, 0.3$**
8. $-1.6, \frac{5}{2}, -\frac{7}{8}, 0.9, -\frac{6}{5}$ **$-1.6, -\frac{6}{5}, -\frac{7}{8}, 0.9, \frac{5}{2}$**
9. $-\frac{2}{3}, \frac{5}{9}, 0.5, -1.3, -\frac{10}{3}$ **$-\frac{10}{3}, -1.3, -\frac{2}{3}, 0.5, \frac{5}{9}$**

10. The table shows the position of each runner relative to when the first place finisher crossed the finish line. Who finished in second place? Who finished in fifth place?

Runner	A	B	C	D	E	F
Meters	−1.264	$-\frac{5}{4}$	−1.015	−0.480	$-\frac{14}{25}$	$-\frac{13}{8}$

Runner D; Runner B

Vocabulary and Concept Check

1. no; The denominator cannot be 0.
2. yes; These decimals can be written as $\frac{a}{b}$ where a and b are integers and $b \neq 0$.
3. rational numbers, integers
4. rational numbers
5. rational numbers, integers, whole numbers
6. rational numbers, integers, whole numbers
7. repeating
8. terminating
9. terminating
10. repeating

Practice and Problem Solving

11. 0.875
12. $0.\overline{09}$
13. $-0.\overline{7}$
14. -0.425
15. $1.8\overline{3}$
16. $-2.9\overline{4}$
17. $-5.58\overline{3}$
18. $8.\overline{681}$
19. The bar should be over both digits to the right of the decimal point.

    $-\frac{7}{11} = -0.\overline{63}$
20. $-\frac{9}{10}$
21. $\frac{9}{20}$
22. $-\frac{129}{500}$
23. $-\frac{39}{125}$
24. $-2\frac{8}{25}$
25. $-1\frac{16}{25}$
26. $6\frac{3}{250}$
27. $-12\frac{81}{200}$
28. $-\frac{7}{3}, -\frac{3}{4}, 0.5, \frac{2}{3}, 1.2$

Practice and Problem Solving

29. $-2.5, -1.1, -\frac{4}{5}, 0.8, \frac{9}{5}$

30. $-\frac{8}{5}, -1.4, -0.9, \frac{1}{4}, 0.6$

31. $-\frac{9}{4}, -0.75, -\frac{6}{10}, \frac{5}{3}, 2.1$

32. $-\frac{7}{2}, -2.8, -\frac{5}{4}, 1.3, \frac{4}{3}$

33. $-2.4, -2.25, -\frac{11}{5}, \frac{15}{10}, 1.6$

34. **a.** -0.55

b. $-\frac{11}{20}$

35. spotted turtle

36. $-2.2 > -2.42$

37. $-1.82 < -1.81$

38. $\frac{15}{8} = 1\frac{7}{8}$ **39.** $-4\frac{6}{10} > -4.65$

40–45. See Additional Answers.

46. See *Taking Math Deeper.*

47. See Additional Answers.

Fair Game Review

48. $\frac{31}{35}$ **49.** $\frac{7}{30}$

50. 4.72 **51.** 21.15

52. D

Mini-Assessment

Write the rational number as a decimal.

1. $\frac{8}{9}$ $0.\overline{8}$ **2.** $-\frac{11}{10}$ -1.1

3. $\frac{4}{125}$ 0.032 **4.** $-\frac{13}{15}$ $-0.8\overline{6}$

5. When your cousin was born, she was $21\frac{4}{5}$ inches long. When your friend was born, he was $21\frac{5}{6}$ inches long. Who was longer at birth? your friend

Taking Math Deeper

Exercise 46

Students have already learned that it is easier to order numbers in decimal form than in fraction form. This problem gives students practice with this skill using negative numbers. The challenge in this problem is that students need to decide what place values to use for all four decimals.

1 Write each number as a decimal.

Week	1	2	3	4
Change (inches)	$-\frac{7}{5}$	$-1\frac{5}{11}$	-1.45	$-1\frac{91}{200}$
Decimal	-1.4000	$-1.45\overline{45}$	-1.4500	-1.4550

2 Graph the numbers on a number line.

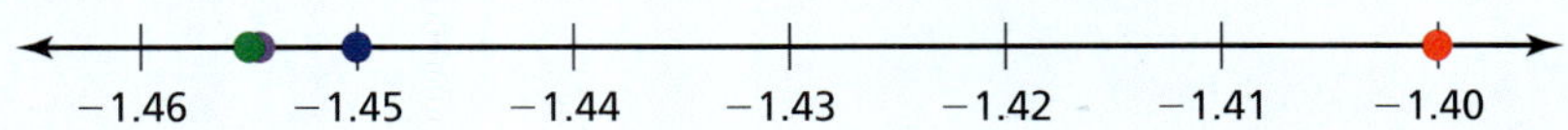

This problem is more difficult than it appears.

3 Write the numbers in order from least to greatest.

$-1\frac{91}{200}$ $-1\frac{5}{11}$ -1.45 $-\frac{7}{5}$

The U.S. Geological Survey (USGS) records the water levels at various locations in the United States. You can track these measurements by going to www.usgs.org.

Project

Create a chart showing the water levels at various locations in the Great Lakes during the same week. What is the range in water levels? Why do you think the levels vary?

Reteaching and Enrichment Strategies

If students need help. . .	If students got it. . .
Resources by Chapter • Practice A and Practice B • Puzzle Time Record and Practice Journal Practice Differentiating the Lesson Lesson Tutorials Skills Review Handbook	Resources by Chapter • Enrichment and Extension • Technology Connection Start the next section

Copy and complete the statement using <, >, or =.

36. $-2.2 \ \square \ -2.42$ **37.** $-1.82 \ \square \ -1.81$ **38.** $\frac{15}{8} \ \square \ 1\frac{7}{8}$

39. $-4\frac{6}{10} \ \square \ -4.65$ **40.** $-5\frac{3}{11} \ \square \ -5.\overline{2}$ **41.** $-2\frac{13}{16} \ \square \ -2\frac{11}{14}$

42. **OPEN-ENDED** Find one terminating decimal and one repeating decimal between $-\frac{1}{2}$ and $-\frac{1}{3}$.

Player	Hits	At Bats
Eva	42	90
Michelle	38	80

43. **SOFTBALL** In softball, a batting average is the number of hits divided by the number of times at bat. Does Eva or Michelle have the higher batting average?

44. **PROBLEM SOLVING** You miss 3 out of 10 questions on a science quiz and 4 out of 15 questions on a math quiz. Which quiz has a higher percent of correct answers?

45. **SKATING** Is the half pipe deeper than the skating pool? Explain.

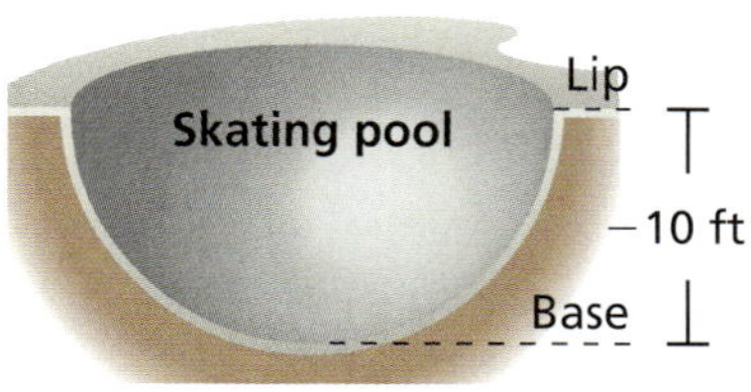

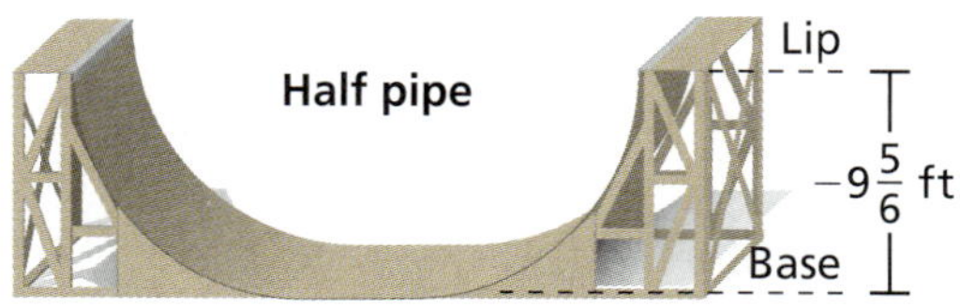

46. **ENVIRONMENT** The table shows the changes from the average water level of a pond over several weeks. Order the numbers from least to greatest.

Week	1	2	3	4
Change (inches)	$-\frac{7}{5}$	$-1\frac{5}{11}$	-1.45	$-1\frac{91}{200}$

47. Critical Thinking Given: a and b are integers.

a. When is $-\frac{1}{a}$ positive? **b.** When is $\frac{1}{ab}$ positive?

Fair Game Review What you learned in previous grades & lessons

Add or subtract. *(Section 1.6 and Section 2.4)*

48. $\frac{3}{5} + \frac{2}{7}$ **49.** $\frac{9}{10} - \frac{2}{3}$ **50.** $8.79 - 4.07$ **51.** $11.81 + 9.34$

52. **MULTIPLE CHOICE** In one year, a company has a profit of −$2 million. In the next year, the company has a profit of $7 million. How much more profit did the company make the second year? *(Section 11.3)*

Ⓐ $2 million Ⓑ $5 million Ⓒ $7 million Ⓓ $9 million

12.2 Adding Rational Numbers

Essential Question How can you use what you know about adding integers to add rational numbers?

1 ACTIVITY: Adding Rational Numbers

Work with a partner. Use a number line to find the sum.

a. $2.7 + (-3.4)$

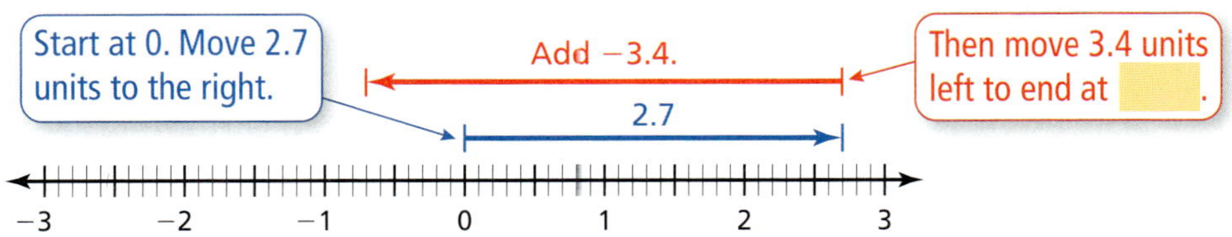

So, $2.7 + (-3.4) = $ ▭.

b. $1.3 + (-1.5)$

c. $-2.1 + 0.8$

d. $-1\frac{1}{4} + \frac{3}{4}$

e. $\frac{3}{10} + \left(-\frac{3}{10}\right)$

2 ACTIVITY: Adding Rational Numbers

Work with a partner. Use a number line to find the sum.

a. $-1\frac{2}{5} + \left(-\frac{4}{5}\right)$

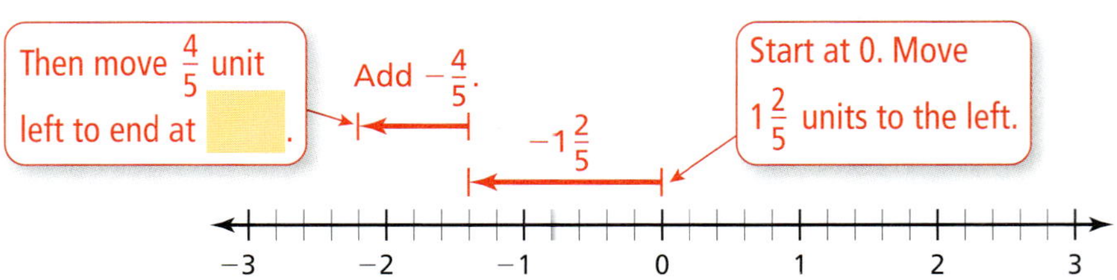

So, $-1\frac{2}{5} + \left(-\frac{4}{5}\right) = $ ▭.

b. $-\frac{7}{10} + \left(-1\frac{7}{10}\right)$

c. $-1\frac{2}{3} + \left(-1\frac{1}{3}\right)$

d. $-0.4 + (-1.9)$

e. $-2.3 + (-0.6)$

Rational Numbers

In this lesson, you will

- add rational numbers.
- solve real-life problems.

Laurie's Notes

Introduction

Applying Mathematical Practices

- **Reason Abstractly and Quantitatively:** Mathematically proficient students are able to use a number line to represent the sum of two rational numbers. Representing a sum on a number line means students must attend to the meaning of addition and to the meaning of a rational number.

Motivate

? Pose a series of contextual questions that will help students think about negative rational numbers. These questions should suggest why you need to be able to add rational numbers. Examples:
  - "If finding a dollar and a quarter represented \$1.25, how would losing a dollar and a quarter be represented?" $-\$1.25$
  - "If a half mile above sea level is represented as $\frac{1}{2}$, how would a half mile below sea level be represented?" $-\frac{1}{2}$
  - "Represent a loss of $5\frac{1}{2}$ yards on a play in football." $-5\frac{1}{2}$
  - "Represent a drop in temperature of 4.2°." -4.2
- Discuss the use of number lines as a model for addition. Ask students to describe how addition is modeled on a number line.

Activity Notes

Activity 1

- **Reason Abstractly and Quantitatively:** The number line helps students see that the rules for adding rational numbers shouldn't be different from the rules for adding integers.
- Remind students that the first number is represented on the number line by a ray starting at 0. The second number starts at the end of that ray.
- In part (a), you can move 3.4 units to the left in stages. Moving 2 units to the left puts you at 0.7. Moving 1 more unit to the left puts you at -0.3 (not -0.7). Finally, another 0.4 unit left puts you at -0.7.
- **Teaching Tip:** Suggest to students that they use the tick marks on the number line to help them perform the moves in stages.

? "When adding rational numbers with different signs, can you predict the sign of the sum? Explain." Yes; the sum has the same sign as the number with the greater absolute value.

Activity 2

- This activity involves adding rational numbers with the same sign. Students should recall that when adding two negative numbers using the number line model, both rays will be drawn in the same direction.

? "What is $-1\frac{2}{5}$ as a decimal?" -1.4 "What is $-\frac{4}{5}$ as a decimal?" -0.8

? "When adding rational numbers with the same sign, can you predict the sign of the sum? Explain." Yes; the sign is the same.

What Your Students Will Learn

- Add rational numbers with the same sign or with different signs.
- Evaluate expressions involving rational numbers by using substitution.

Previous Learning

Students should be comfortable adding positive fractions, positive decimals, and integers.

Technology for the Teacher

Dynamic Classroom

Lesson Plans
Complete Materials List

12.2 Record and Practice Journal

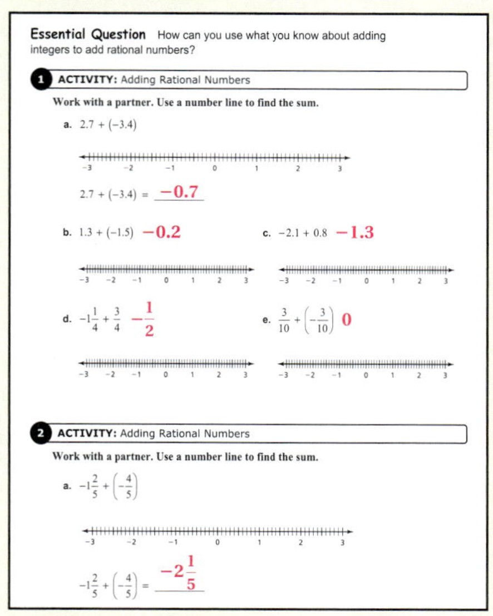

Essential Question How can you use what you know about adding integers to add rational numbers?

1 ACTIVITY: Adding Rational Numbers

Work with a partner. Use a number line to find the sum.

a. $2.7 + (-3.4)$

$2.7 + (-3.4) = $ **-0.7**

b. $1.3 + (-1.5)$ **-0.2**

c. $-2.1 + 0.8$ **-1.3**

d. $-1\frac{1}{4} + \frac{3}{4}$ **$-\frac{1}{2}$**

e. $\frac{3}{10} + \left(-\frac{3}{10}\right)$ **0**

2 ACTIVITY: Adding Rational Numbers

Work with a partner. Use a number line to find the sum.

a. $-1\frac{2}{5} + \left(-\frac{4}{5}\right)$

$-1\frac{2}{5} + \left(-\frac{4}{5}\right) = $ **$-2\frac{1}{5}$**

English Language Learners

Vocabulary

Use color-coded examples to help students understand the vocabulary used in this section: denominator, common denominator, least common denominator, improper fraction, and mixed number.

12.2 Record and Practice Journal

b. $-\frac{7}{10} + \left(-1\frac{7}{10}\right)$ $-2\frac{2}{5}$

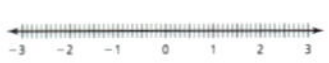

c. $-1\frac{2}{3} + \left(-1\frac{1}{3}\right)$ -3

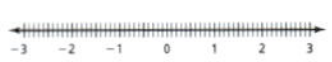

d. $-0.4 + (-1.9)$ -2.3

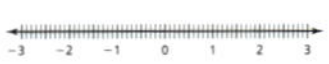

e. $-2.3 + (-0.6)$ -2.9

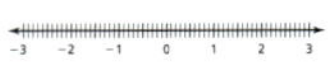

3 ACTIVITY: Writing Expressions

Work with a partner. Write the addition expression shown. Then find the sum.

a.

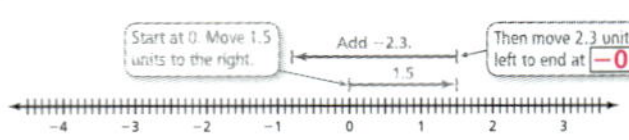

$1.5 + (-2.3) = -0.8$

b.

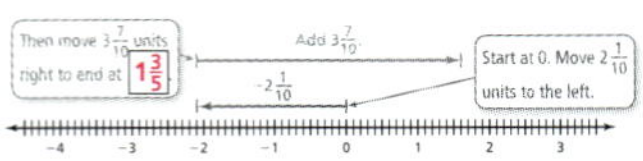

$-2\frac{1}{10} + 3\frac{7}{10} = 1\frac{3}{5}$

c.

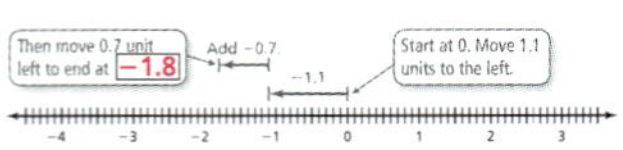

$-1.1 + (-0.7) = -1.8$

What Is Your Answer?

4. **IN YOUR OWN WORDS** How can you use what you know about adding integers to add rational numbers?

To add rational numbers, use the same rules for signs used for integers.

PUZZLE **Find a path through the table so that the numbers add up to the sum. You can move horizontally or vertically.**

5. Sum: $\frac{3}{4}$

Start → ... ← End

$\frac{1}{2}$	$\frac{2}{3}$	$-\frac{5}{7}$
$-\frac{1}{8}$	$-\frac{3}{4}$	$\frac{1}{3}$

$\frac{1}{2} + \frac{2}{3} + \left(-\frac{3}{4}\right) + \frac{1}{3} = \frac{3}{4}$

6. Sum: -0.07

Start → ... ← End

2.43	1.75	−0.98
−1.09	3.47	−4.88

$2.43 + (-1.09) + 3.47 + (-4.88) = -0.07$

Laurie's Notes

Activity 3

- Have students write the problem modeled on each number line. Ask what clues helped them figure out the problem. State the solution.
- **Common Error:** Students say that $1.5 + (-2.3) = -1.2$, meaning that students will subtract the lesser digit from the greater digit regardless of how the problem is written. (They subtract 1 from 2 and 0.3 from 0.5.) Take time to look at the number line model.

What Is Your Answer?

- For Question 5, students should work with partners. Let students wrestle with the question first, then offer a hint if needed. Five of the six fractions have a common denominator of 24, as does the desired sum of $\frac{3}{4}$. Write each of the fractions as an equivalent fraction with a denominator of 24. The fraction $\left(-\frac{5}{7}\right)$ is not needed to solve the puzzle, so it is not rewritten.

Closure

- Write an addition problem using two rational numbers with different signs whose sum is positive. Find the sum. *Sample answer:* $-3.6 + 7.5 = 3.9$
- Write an addition problem using two rational numbers with different signs whose sum is negative. Find the sum. *Sample answer:* $\frac{1}{4} + \left(-\frac{3}{8}\right) = -\frac{1}{8}$

3 ACTIVITY: Writing Expressions

Work with a partner. Write the addition expression shown. Then find the sum.

Math Practice

Use Operations

What operation is represented in each number line? How does this help you write an expression?

a.

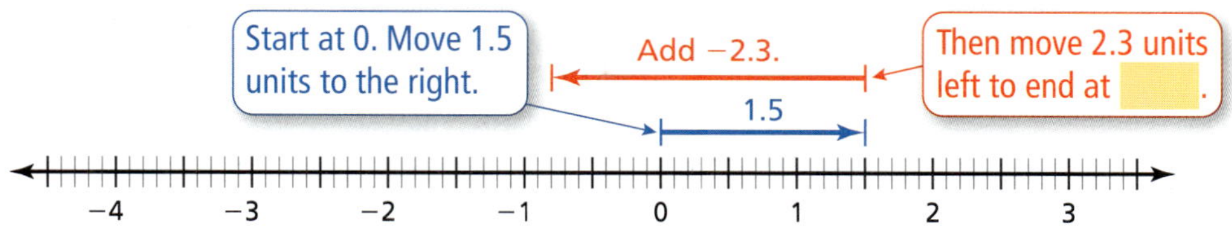

b.

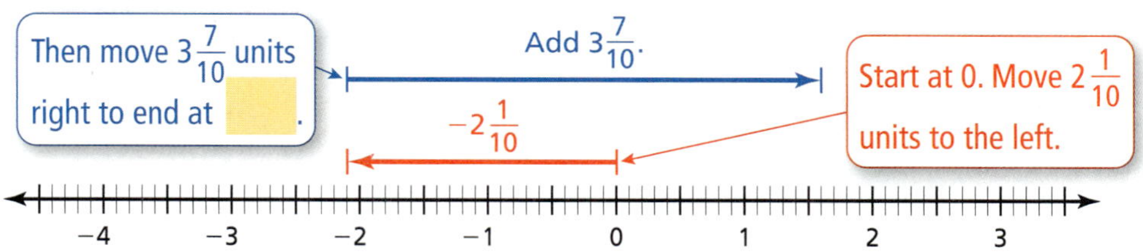

c.

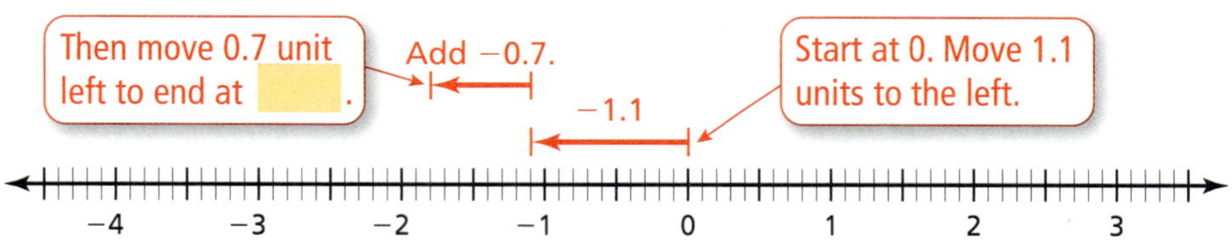

What Is Your Answer?

4. **IN YOUR OWN WORDS** How can you use what you know about adding integers to add rational numbers?

PUZZLE Find a path through the table so that the numbers add up to the sum. You can move horizontally or vertically.

5. Sum: $\frac{3}{4}$

Start →

$\frac{1}{2}$	$\frac{2}{3}$	$-\frac{5}{7}$
$-\frac{1}{8}$	$-\frac{3}{4}$	$\frac{1}{3}$

← End

6. Sum: −0.07

Start →

2.43	1.75	−0.98
−1.09	3.47	−4.88

← End

Practice Use what you learned about adding rational numbers to complete Exercises 4–6 on page 528.

12.2 Lesson

Check It Out
Lesson Tutorials
BigIdeasMath.com

Key Idea

Adding Rational Numbers

Words To add rational numbers, use the same rules for signs as you used for integers.

Numbers $-\frac{1}{3}+\frac{1}{6}=\frac{-2}{6}+\frac{1}{6}=\frac{-2+1}{6}=\frac{-1}{6}=-\frac{1}{6}$

EXAMPLE 1 Adding Rational Numbers

Study Tip

In Example 1, notice how $-\frac{8}{3}$ is written as $-\frac{8}{3}=\frac{-8}{3}=\frac{-16}{6}$.

Find $-\frac{8}{3}+\frac{5}{6}$.

Estimate $-3+1=-2$

$-\frac{8}{3}+\frac{5}{6}=\frac{-16}{6}+\frac{5}{6}$ Rewrite using the LCD (least common denominator).

$=\frac{-16+5}{6}$ Write the sum of the numerators over the common denominator.

$=\frac{-11}{6}$ Add.

$=-1\frac{5}{6}$ Write the improper fraction as a mixed number.

The sum is $-1\frac{5}{6}$.

Reasonable? $-1\frac{5}{6}\approx-2$ ✓

EXAMPLE 2 Adding Rational Numbers

Find $-4.05+7.62$.

$-4.05+7.62=3.57$ $|7.62|>|-4.05|$. So, subtract $|-4.05|$ from $|7.62|$.

Use the sign of 7.62.

The sum is 3.57.

On Your Own

Now You're Ready
Exercises 4–12

Add.

1. $-\frac{7}{8}+\frac{1}{4}$ **2.** $-6\frac{1}{3}+\frac{20}{3}$ **3.** $2+\left(-\frac{7}{2}\right)$

4. $-12.5+15.3$ **5.** $-8.15+(-4.3)$ **6.** $0.65+(-2.75)$

Laurie's Notes

Introduction

Connect

- **Yesterday:** Students explored how to add rational numbers.
- **Today:** Students will formalize the process completed yesterday, and they will add rational numbers.

Motivate

? Ask students whether the following questions are *true* or *false*.

$\frac{\Diamond}{8} + \frac{\triangle}{8} = \frac{\Diamond + \triangle}{16}$ false, unless the symbols are opposites

$\frac{\Diamond}{8} + \frac{\triangle}{8} = \frac{\Diamond + \triangle}{8}$ true

$0.4 + 0.34 = 0.38$ false

Lesson Notes

Key Idea

- **Representation:** Take time to talk about how negative fractions are represented, meaning where the negative sign is written. All of the following are equivalent: $-\frac{2}{3} = \frac{-2}{3} = \frac{2}{-3}$.
- **Discuss:** Emphasize the intermediate step: $\frac{-2}{6} + \frac{1}{6} = \frac{-2+1}{6}$. This will help the students a great deal.

Example 1

? "What type of fraction is $-\frac{8}{3}$?" improper

? "What would $-\frac{8}{3}$ be as a mixed number?" $-2\frac{2}{3}$

- Note the *Study Tip* when working through the example.
- Be sure to tell students to check for reasonableness of their answers.

Example 2

? Write the problem and ask, "Should the final answer be *positive* or *negative*? Why?"

On Your Own

- Have three pairs of students complete one of the first three fraction problems at the board, while the other students try the problems at their desks. Have students explain their work at the board.

? Ask questions such as, "How do you know your answer is reasonable?"

- Have three different pairs of students complete one of the last three problems at the boards, while the other students try the problems at their desks.

Goal Today's lesson is adding rational numbers.

Technology for the Teacher

Dynamic Classroom

Lesson Tutorials
Lesson Plans
Answer Presentation Tool

Extra Example 1

Find $\frac{4}{5} + \left(-\frac{3}{10}\right)$. $\frac{1}{2}$

Extra Example 2

Find $-3.92 + (-6.89)$. -10.81

On Your Own

1. $-\frac{5}{8}$
2. $\frac{1}{3}$
3. $-1\frac{1}{2}$
4. 2.8
5. -12.45
6. -2.1

Extra Example 3

Evaluate $x + 2y$ when $x = -3.5$ and $y = 1.7$. -0.1

Extra Example 4

The table shows the changes in the value of a stock during one week. Find the stock's gain or loss for the week.

Day	Change in value (dollars)
Monday	3.45
Tuesday	−12.90
Wednesday	−5.02
Thursday	29.31
Friday	−9.44

gain of $5.40

On Your Own

7. $-\frac{1}{2}$ **8.** 2

9. gain of $770 million

Differentiated Instruction

Auditory

Have students verbally describe the process for adding rational numbers, as well as rewriting and simplifying fractions and mixed numbers. Students should use the word *negative* when referring to the opposite of a positive number. The words *subtract* and *minus* refer to the arithmetic operation. Encourage students to use the words *numerator, denominator, least common denominator,* and *improper fraction.*

Laurie's Notes

Example 3

? "What does it mean to *evaluate* an expression?" Substitute the given value for the variable(s) and do the arithmetic.

- **Reason Abstractly and Quantitatively:** When substituting $\frac{1}{4}$ for *x*, make sure students understand that $2\left(\frac{1}{4}\right)$ is a multiplication problem. It is not the mixed number $2\frac{1}{4}$.

Example 4

- Ask a volunteer to read the problem. Check to see that students are comfortable with the vocabulary.
- Before beginning, tell students that the properties of addition also apply to rational numbers.

? "What is the relationship between 1.7 and −1.7?" They are opposites.

- Explain how 1.7 and −1.7 represent a gain and a loss of $1.7 billion, which combine to make $0. Discuss other quantities that combine to make 0 for standard 7.NS.1a.
- **Attend to Precision:** When students finish the computation they will have −0.3 billion. Students need to convert −0.3 billion to millions.

On Your Own

- **Neighbor Check:** Have students work independently and then have their neighbors check their work. Have students discuss any discrepancies.

Closure

- Ask students to explain how addition of rational numbers is similar to addition of integers. *Sample answer:* The sign of the sum is the sign of the number with the greater absolute value.

EXAMPLE 3 Evaluating Expressions

Evaluate $2x + y$ when $x = \frac{1}{4}$ and $y = -\frac{3}{2}$.

$$2x + y = 2\left(\frac{1}{4}\right) + \left(-\frac{3}{2}\right)$$ Substitute $\frac{1}{4}$ for x and $-\frac{3}{2}$ for y.

$$= \frac{1}{2} + \left(\frac{-3}{2}\right)$$ Multiply.

$$= \frac{1 + (-3)}{2}$$ Write the sum of the numerators over the common denominator.

$$= -1$$ Simplify.

EXAMPLE 4 Real-Life Application

Year	Profit (billions of dollars)
2008	−1.7
2009	−4.75
2010	1.7
2011	0.85
2012	3.6

The table shows the annual profits (in billions of dollars) of a financial company from 2008 to 2012. Positive numbers represent *gains*, and negative numbers represent *losses*. Which statement describes the profit over the five-year period?

Ⓐ gain of \$0.3 billion Ⓑ gain of \$30 million

Ⓒ loss of \$3 million Ⓓ loss of \$300 million

To determine whether there was a gain or a loss, find the sum of the profits.

five-year profit $= -1.7 + (-4.75) + 1.7 + 0.85 + 3.6$ Write the sum.

$= -1.7 + 1.7 + (-4.75) + 0.85 + 3.6$ Comm. Prop. of Add.

$= 0 + (-4.75) + 0.85 + 3.6$ Additive Inv. Prop.

$= -4.75 + 0.85 + 3.6$ Add. Prop. of Zero

$= -3.9 + 3.6$ Add -4.75 and 0.85.

$= -0.3$ Add -3.9 and 3.6.

The five-year profit is −\$0.3 billion. So, the company has a five-year loss of \$0.3 billion, or \$300 million.

∴ The correct answer is Ⓓ.

On Your Own

Now You're Ready
Exercises 15–17

Evaluate the expression when $a = \frac{1}{2}$ and $b = -\frac{5}{2}$.

7. $b + 4a$

8. $|a + b|$

9. **WHAT IF?** In Example 4, the 2013 profit is \$1.07 billion. State the company's gain or loss over the six-year period in millions of dollars.

12.2 Exercises

Check It Out
Help with Homework
BigIdeasMath.com

Vocabulary and Concept Check

1. **WRITING** Explain how to find the sum $-8.46 + 5.31$.

2. **OPEN-ENDED** Write an addition expression using fractions that equals $-\frac{1}{2}$.

3. **DIFFERENT WORDS, SAME QUESTION** Which is different? Find "both" answers.

Add -4.5 and 3.5.	What is the distance between -4.5 and 3.5?
What is -4.5 increased by 3.5?	Find the sum of -4.5 and 3.5.

Practice and Problem Solving

Add. Write fractions in simplest form.

1 2 4. $\frac{11}{12} + \left(-\frac{7}{12}\right)$

5. $-1\frac{1}{5} + \left(-\frac{3}{5}\right)$

6. $-4.2 + 3.3$

7. $-\frac{9}{14} + \frac{2}{7}$

8. $4 + \left(-1\frac{2}{3}\right)$

9. $\frac{15}{4} + \left(-4\frac{1}{3}\right)$

10. $-3.1 + (-0.35)$

11. $12.48 + (-10.636)$

12. $20.25 + (-15.711)$

ERROR ANALYSIS **Describe and correct the error in finding the sum.**

13. ✗

$$\begin{array}{r} -3.7 \\ +\ (-0.25) \\ \hline -0.62 \end{array}$$

14. ✗ $-\frac{5}{8} + \frac{1}{8} = \frac{-5+1}{8} = \frac{-6}{8} = -\frac{3}{4}$

Evaluate the expression when $x = \frac{1}{3}$ and $y = -\frac{7}{4}$.

3 15. $x + y$

16. $3x + y$

17. $-x + |y|$

18. **BANKING** Your bank account balance is –\$20.85. You deposit \$15.50. What is your new balance?

19. **HOT DOGS** You eat $\frac{3}{10}$ of a pack of hot dogs. Your friend eats $\frac{1}{5}$ of the pack of hot dogs. What fraction of the pack of hot dogs do you and your friend eat?

Assignment Guide and Homework Check

Level	Assignment	Homework Check
Advanced	1–6, 8–22 even, 23–33	8, 12, 16, 20, 24

Common Errors

- **Exercises 7–9** Students may try to identify the sign of the answer before finding a common denominator. Remind them that they need to find the common denominator first.
- **Exercises 10–12** Students may forget to line up the decimal points when they add decimals. Remind them that the decimal points must be lined up before adding. Students may want to use half-inch graph paper to help keep the numbers and decimal points aligned.

12.2 Record and Practice Journal

Add. Write fractions in simplest form.

1. $-\frac{4}{5} + \frac{3}{20}$ — $-\frac{13}{20}$
2. $-8 + \left(-\frac{6}{7}\right)$ — $-8\frac{6}{7}$
3. $1\frac{2}{15} + \left(-3\frac{1}{2}\right)$ — $-2\frac{11}{30}$
4. $-\frac{1}{6} + \left(-\frac{5}{12}\right)$ — $-\frac{7}{12}$
5. $\frac{9}{10} + (-3)$ — $-2\frac{1}{10}$
6. $-5\frac{3}{4} + \left(-4\frac{5}{6}\right)$ — $-10\frac{7}{12}$
7. $0.46 + (-0.642)$ — -0.182
8. $0.13 + (-5.7)$ — -5.57
9. $-2.57 + (-3.48)$ — -6.05
10. Before a race, you start $4\frac{5}{8}$ feet behind your friend. At the halfway point, you are $3\frac{2}{3}$ feet ahead of your friend. What is the change in distance between you and your friend from the beginning of the race?

    $8\frac{7}{24}$ feet

Vocabulary and Concept Check

1. Because $|-8.46| > |5.31|$, subtract $|5.31|$ from $|-8.46|$ and the sign is negative.
2. *Sample answer:* $-\frac{1}{4} + \left(-\frac{1}{4}\right)$
3. What is the distance between -4.5 and 3.5?; 8; -1

Practice and Problem Solving

4. $\frac{1}{3}$
5. $-1\frac{4}{5}$
6. -0.9
7. $-\frac{5}{14}$
8. $2\frac{1}{3}$
9. $-\frac{7}{12}$
10. -3.45
11. 1.844
12. 4.539
13. The decimals are not lined up correctly; Line up the decimals; -3.95
14. The sum of the numerators is incorrect.
    $$-\frac{5}{8} + \frac{1}{8} = \frac{-5+1}{8} = \frac{-4}{8} = -\frac{1}{2}$$
15. $-1\frac{5}{12}$
16. $-\frac{3}{4}$
17. $1\frac{5}{12}$
18. $-\$5.35$
19. $\frac{1}{2}$

Practice and Problem Solving

20. $-\frac{7}{8}$

21. $-9\frac{3}{4}$

22. -2.6

23. The sum is an integer when the sum of the fractional parts of the numbers adds up to an integer.

24. See Additional Answers.

25. less than; The water level for the three-month period compared to the normal level is $-1\frac{7}{16}$.

26. \$450

27. no; This is only true when a and b have the same sign.

28. See *Taking Math Deeper.*

Fair Game Review

29. Commutative Property of Addition; 7

30. Associative Property of Multiplication; 81

31. Associative Property of Addition; $1\frac{1}{8}$

32. Commutative Property of Multiplication; $\frac{8}{45}$

33. A

Mini-Assessment

Add. Write fractions in simplest form.

1. $2\frac{4}{5}+\left(-\frac{12}{15}\right)$ 2
2. $-\frac{3}{4}+\left(-\frac{8}{9}\right)$ $-1\frac{23}{36}$
3. $15.48+(-17.23)$ -1.75
4. $-3.89+(-5.34)$ -9.23
5. Your bank account balance is $-\$15.50$. You deposit \$75. What is your new balance? \$59.50

Taking Math Deeper

Exercise 28

You can evaluate this expression by adding each of the 19 fractions as written. However, this would be tedious. To save time, notice that all the denominators are the same. So, you could find the sum of the numerators and write it over the common denominator, 20.

1 One way to quickly find the sum of the numerators is by finding the sums of consecutive terms (1st and 2nd terms, 3rd and 4th terms, and so on.)

$$\overbrace{19+(-18)}^{1}+\overbrace{17+(-16)}^{1}+\overbrace{15+(-14)}^{1}+\overbrace{13+(-12)}^{1}+$$
$$\underbrace{11+(10)}_{1}+\underbrace{9+(-8)}_{1}+\underbrace{7+(-6)}_{1}+\underbrace{5+(-4)}_{1}+\underbrace{3+(-2)}_{1}+\underbrace{1}_{1}$$

There are nine pairs whose sum is 1, and the term 1 is remaining. So, the sum of the numerators is 10.

2 Another way to quickly find the sum of the numerators is by pairing terms at the beginning of the expression with terms at the end (1st term and 19th term, 2nd term and 18th term, and so on).

$$\overbrace{19+1+(-18)+(-2)}^{0}+\overbrace{17+3+(-16)+(-4)}^{0}+$$
$$\underbrace{15+5+(-14)+(-6)}_{0}+\underbrace{13+7+(-12)+(-8)}_{0}+\underbrace{11+9+(-10)}_{0}$$

This also shows that the sum of the numerators is 10.

3 Answer the question.

So, the value of the expression is $\frac{10}{20}$, or $\frac{1}{2}$.

Project

Write an addition problem with more than 10 terms that can be more easily solved by rearranging terms. Trade problems with a friend and solve.

Reteaching and Enrichment Strategies

If students need help. . .	If students got it. . .
Resources by Chapter • Practice A and Practice B • Puzzle Time Record and Practice Journal Practice Differentiating the Lesson Lesson Tutorials Skills Review Handbook	Resources by Chapter • Enrichment and Extension • Technology Connection Start the next section

Add. Write fractions in simplest form.

20. $6 + \left(-4\frac{3}{4}\right) + \left(-2\frac{1}{8}\right)$

21. $-5\frac{2}{3} + 3\frac{1}{4} + \left(-7\frac{1}{3}\right)$

22. $10.9 + (-15.6) + 2.1$

June	July	August
$-2\frac{1}{8}$	$1\frac{1}{4}$	$-\frac{9}{16}$

23. NUMBER SENSE When is the sum of two negative mixed numbers an integer?

24. WRITING You are adding two rational numbers with different signs. How can you tell if the sum will be *positive*, *negative*, or *zero*?

25. RESERVOIR The table at the left shows the water level (in inches) of a reservoir for three months compared to the yearly average. Is the water level for the three-month period greater than or less than the yearly average? Explain.

26. BREAK EVEN The table at the right shows the annual profits (in thousands of dollars) of a county fair from 2008 to 2012. What must the 2012 profit be (in hundreds of dollars) to break even over the five-year period?

Year	Profit (thousands of dollars)
2008	2.5
2009	1.75
2010	−3.3
2011	−1.4
2012	?

27. REASONING Is $|a + b| = |a| + |b|$ for all rational numbers a and b? Explain.

28. Repeated Reasoning Evaluate the expression.

$$\frac{19}{20} + \left(\frac{-18}{20}\right) + \frac{17}{20} + \left(\frac{-16}{20}\right) + \cdots + \left(\frac{-4}{20}\right) + \frac{3}{20} + \left(\frac{-2}{20}\right) + \frac{1}{20}$$

Fair Game Review What you learned in previous grades & lessons

Identify the property. Then simplify. *(Section 3.3)*

29. $8 + (-3) + 2 = 8 + 2 + (-3)$

30. $2 \cdot (4.5 \cdot 9) = (2 \cdot 4.5) \cdot 9$

31. $\frac{1}{4} + \left(\frac{3}{4} + \frac{1}{8}\right) = \left(\frac{1}{4} + \frac{3}{4}\right) + \frac{1}{8}$

32. $\frac{3}{7} \cdot \frac{4}{5} \cdot \frac{14}{27} = \frac{3}{7} \cdot \frac{14}{27} \cdot \frac{4}{5}$

33. MULTIPLE CHOICE The regular price of a photo album is \$18. You have a coupon for 15% off. How much is the discount? *(Section 5.6)*

Ⓐ \$2.70 Ⓑ \$3 Ⓒ \$15 Ⓓ \$15.30

12 Study Help

Check It Out
Graphic Organizer
BigIdeasMath.com

You can use a **process diagram** to show the steps involved in a procedure. Here is an example of a process diagram for adding rational numbers.

Adding rational numbers

with the same sign

↓ Add the absolute values of the rational numbers.

↓ Write the sum by using the common sign.

↓ *Example*

−5.5 + (−6.9)

Because the numbers have the same sign, add |−5.5| and |−6.9|.

−5.5 + (−6.9) = −12.4

Use the common sign.

with different signs

↓ Subtract the lesser absolute value from the greater absolute value.

↓ Write the sum by using the sign of the rational number with the greater absolute value.

↓ *Example*

−5.5 + 6.9

Because the numbers have different signs, subtract |−5.5| from |6.9|.

−5.5 + 6.9 = 1.4

Use the sign of 6.9.

On Your Own

Make a process diagram with examples to help you study the topic.

1. writing rational numbers as decimals

After you complete this chapter, make process diagrams with examples for the following topics.

2. subtracting rational numbers
3. multiplying rational numbers
4. dividing rational numbers

"Does this process diagram accurately show how a cat claws furniture?"

Sample Answers

1.

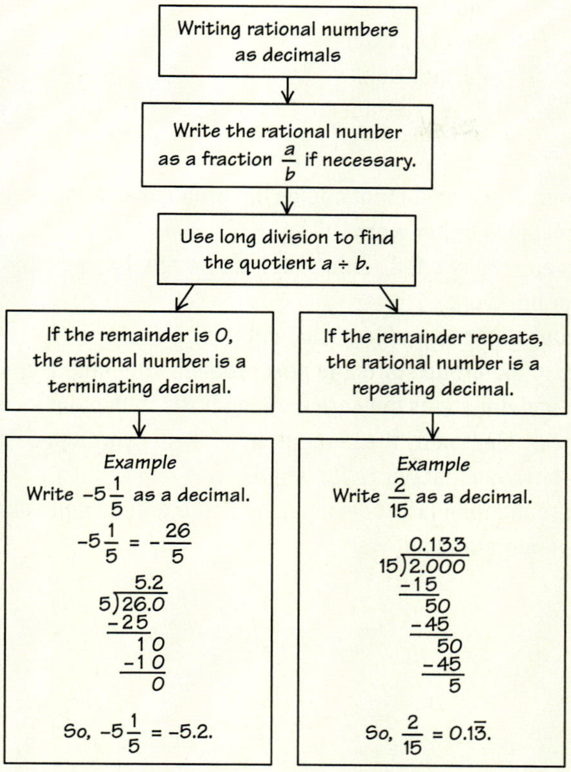

List of Organizers

Available at *BigIdeasMath.com*

Comparison Chart
Concept Circle
Example and Non-Example Chart
Formula Triangle
Four Square
Idea (Definition) and Examples Chart
Information Frame
Information Wheel
Notetaking Organizer
Process Diagram
Summary Triangle
Word Magnet
Y Chart

About this Organizer

A **Process Diagram** can be used to show the steps involved in a procedure. Process diagrams are particularly useful for illustrating procedures with two or more steps, and they can have one or more branches. As shown, students' process diagrams can consist of a single flowchart-type diagram, with example(s) included in the last box to illustrate the steps that precede it. Or, the diagram can have two parallel flowcharts, in which the procedure is stepped out in one chart and an example illustrating each step is shown in the other chart.

Technology for the Teacher

Editable Graphic Organizer

Answers

1. -0.15
2. $-1.8\overline{3}$
3. $-\frac{13}{40}$
4. $-1\frac{7}{25}$
5. $-\frac{1}{3}, -0.2, 0.4, 1.3, \frac{5}{3}$
6. $-\frac{4}{3}, -1.2, -0.8, 0.3, \frac{4}{9}$
7. $-1\frac{7}{40}$
8. $-1\frac{7}{12}$
9. -3.2
10. -6.84
11. $\frac{1}{4}$
12. 1
13. $1\frac{1}{4}$
14. $1\frac{1}{4}$
15. Stock B; Because -3.72 is less than -3.68.
16. $\frac{1}{2}$
17. yes; He gained a total of $54\frac{3}{4}$ yards, which is greater than 50 yards.

Technology for the Teacher

Online Assessment
Assessment Book
ExamView® Assessment Suite

Alternative Quiz Ideas

100% Quiz	Math Log
Error Notebook	Notebook Quiz
Group Quiz	Partner Quiz
Homework Quiz	**Pass the Paper**

Pass the Paper

- Work in groups of four. The first student copies the problem and completes the first step, explaining his or her work.
- The paper is passed and the second student works through the next step, also explaining his or her work.
- This process continues until the problem is completed.
- The second member of the group starts the next problem. Students should be allowed to question and debate as they are working through the quiz.
- Student groups can be selected by the teacher, by students, through a random process, or any way that works for your class.
- The teacher walks around the classroom listening to the groups and asks questions to ensure understanding.

Reteaching and Enrichment Strategies

If students need help. . .	If students got it. . .
Resources by Chapter • Practice A and Practice B • Puzzle Time Lesson Tutorials *BigIdeasMath.com*	Resources by Chapter • Enrichment and Extension • Technology Connection Game Closet at *BigIdeasMath.com* Start the next section

12.1–12.2 Quiz

Check It Out
Progress Check
BigIdeasMath.com

Write the rational number as a decimal. *(Section 12.1)*

1. $-\frac{3}{20}$

2. $-\frac{11}{6}$

Write the decimal as a fraction or a mixed number in simplest form. *(Section 12.1)*

3. -0.325

4. -1.28

Order the numbers from least to greatest. *(Section 12.1)*

5. $-\frac{1}{3}, -0.2, \frac{5}{3}, 0.4, 1.3$

6. $-\frac{4}{3}, -1.2, 0.3, \frac{4}{9}, -0.8$

Add. Write fractions in simplest form. *(Section 12.2)*

7. $-\frac{4}{5} + \left(-\frac{3}{8}\right)$

8. $-\frac{13}{6} + \frac{7}{12}$

9. $-5.8 + 2.6$

10. $-4.28 + (-2.56)$

Evaluate the expression when $x = \frac{3}{4}$ and $y = -\frac{1}{2}$. *(Section 12.2)*

11. $x + y$

12. $2x + y$

13. $x + |y|$

14. $|-x + y|$

15. STOCK The value of Stock A changes −$3.68, and the value of Stock B changes −$3.72. Which stock has the greater loss? Explain. *(Section 12.1)*

16. LEMONADE You drink $\frac{2}{7}$ of a pitcher of lemonade. Your friend drinks $\frac{3}{14}$ of the pitcher. What fraction of the pitcher do you and your friend drink? *(Section 12.2)*

17. FOOTBALL The table shows the statistics of a running back in a football game. Did he gain more than 50 yards total? Explain. *(Section 12.2)*

Quarter	1	2	3	4	Total
Yards	$-8\frac{1}{2}$	23	$42\frac{1}{2}$	$-2\frac{1}{4}$	?

12.3 Subtracting Rational Numbers

Essential Question How can you use what you know about subtracting integers to subtract rational numbers?

1 ACTIVITY: Subtracting Rational Numbers

Work with a partner. Use a number line to find the difference.

a. $-1\frac{1}{2} - \frac{1}{2}$

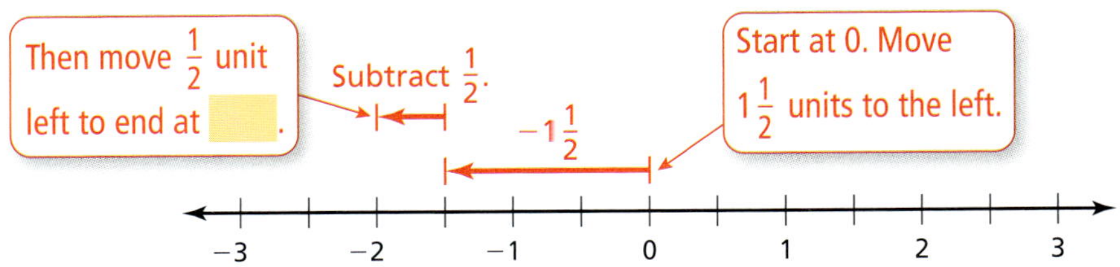

So, $-1\frac{1}{2} - \frac{1}{2} =$ ▭.

b. $\frac{6}{10} - 1\frac{3}{10}$

c. $-1\frac{1}{4} - 1\frac{3}{4}$

d. $-1.9 - 0.8$

e. $0.2 - 0.7$

2 ACTIVITY: Finding Distances on a Number Line

Work with a partner.

a. Plot -3 and 2 on the number line. Then find $-3 - 2$ and $2 - (-3)$. What do you notice about your results?

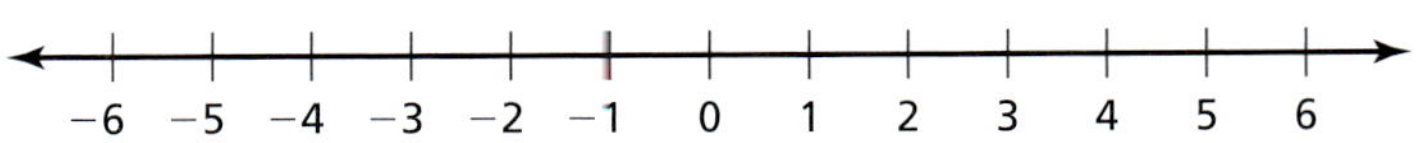

Rational Numbers
In this lesson, you will
- subtract rational numbers.
- solve real-life problems.

b. Plot $\frac{3}{4}$ and 1 on the number line Then find $\frac{3}{4} - 1$ and $1 - \frac{3}{4}$. What do you notice about your results?

c. Choose any two points a and b on a number line. Find the values of $a - b$ and $b - a$. What do the absolute values of these differences represent? Is this true for any pair of rational numbers? Explain.

Laurie's Notes

Introduction

Applying Mathematical Practices

- **Reason Abstractly and Quantitatively:** Mathematically proficient students are able to use a number line to represent the difference of two rational numbers. Representing a difference on a number line means students must attend to the meaning of subtraction and to the meaning of a rational number.

Motivate

- ? "Do you know what an ATM is?" Students should describe features of an automated teller machine.
- Ask students to describe uses and advantages of ATMs. They may have some misconceptions.
- ATMs can be used to deposit money (adding to an account) and to withdraw money (subtracting from an account). Students will investigate transactions in a checkbook.

Activity Notes

Activity 1

- **Reason Abstractly and Quantitatively:** The number line helps students see that the rules for subtracting rational numbers shouldn't be different from the rules for subtracting integers.
- **Teaching Tip:** Suggest to students that they use the tick marks on the number line to help them perform the moves in stages.
- When students have finished, ask volunteers to share their work. Display work at a document camera, if possible.
- Each of these problems involved subtracting a positive number.

Activity 2

- The problems in this activity explore $a - b$ and $b - a$.
- You may need to remind students that when you subtract a positive number n, you move n units to the left. When you subtract a negative number n, you move n units to the right. This is necessary to complete part (a).
- When students have finished the first two parts, ask volunteers to share their work. Display work at a document camera, if possible.
- ? "What did you notice about the results in parts (a) and (b)?" Listen for students saying the answers are opposites. They might also say that subtraction is not commutative.
- Students may need guided questioning for part (c). It is common for students to ask if a and b have to be positive, or if they have to be whole numbers. Repeat that both a and b can be any number on the number line.

What Your Students Will Learn

- Subtract rational numbers with the same sign and with different signs.
- Find the distance between two rational numbers using a number line.

Previous Learning

Students should be comfortable subtracting positive fractions, positive decimals, and integers.

Technology for the Teacher

Dynamic Classroom

Lesson Plans
Complete Materials List

12.3 Record and Practice Journal

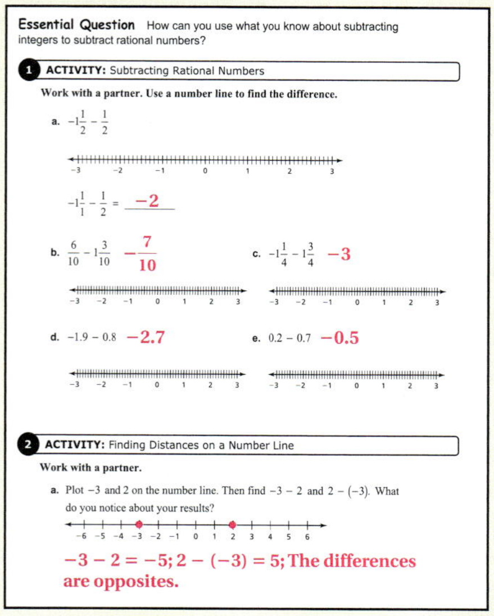

Essential Question How can you use what you know about subtracting integers to subtract rational numbers?

1 ACTIVITY: Subtracting Rational Numbers

Work with a partner. Use a number line to find the difference.

a. $-1\frac{1}{2} - \frac{1}{2}$

$-1\frac{1}{1} - \frac{1}{2} =$ **−2**

b. $\frac{6}{10} - 1\frac{3}{10}$ $-\frac{7}{10}$

c. $-1\frac{1}{4} - 1\frac{3}{4}$ **−3**

d. $-1.9 - 0.8$ **−2.7**

e. $0.2 - 0.7$ **−0.5**

2 ACTIVITY: Finding Distances on a Number Line

Work with a partner.

a. Plot −3 and 2 on the number line. Then find $-3 - 2$ and $2 - (-3)$. What do you notice about your results?

$-3 - 2 = -5$; $2 - (-3) = 5$; The differences are opposites.

English Language Learners

Analyzing Word Problems

Give students a copy of the word problems that are triple-spaced and have wide margins. The additional space gives students plenty of room to add notes. Demonstrate how to underline key words, phrases, and numbers; write down equivalent words in English (or their native language); and draw lines between elements to make the meaning clearer.

12.3 Record and Practice Journal

b. Plot $\frac{3}{4}$ and 1 on the number line. Then find $\frac{3}{4} - 1$ and $1 - \frac{3}{4}$. What do you notice about your results?

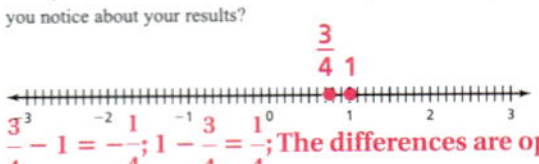

$\frac{3}{4} - 1 = -\frac{1}{4}$; $1 - \frac{3}{4} = \frac{1}{4}$; The differences are opposites.

c. Choose any two points a and b on a number line. Find the values of $a - b$ and $b - a$. What do the absolute values of these differences represent? Is this true for any pair of rational numbers? Explain.

The distance between a and b

3 ACTIVITY: Financial Literacy

Work with a partner. The table shows the balance in a checkbook.

- Deposits and interest are amounts added to the account.
- Amounts shown in parentheses are taken from the account.

Date	Check #	Transaction	Amount	Balance
--	--	Previous Balance	--	100.00
1/02/2013	124	Groceries	(34.57)	65.43
1/07/2013		Check deposit	875.50	940.93
1/11/2013		ATM withdrawal	(40.00)	900.93
1/14/2013	125	Electric company	(78.43)	822.50
1/17/2013		Music store	(10.55)	811.95
1/18/2013	126	Shoes	(47.21)	764.74
1/22/2013		Check deposit	125.00	889.74
1/24/2013		Interest	2.12	891.86
1/25/2013	127	Cell phone	(59.99)	831.87
1/26/2013	128	Clothes	(65.54)	766.33
1/30/2013	129	Cable company	(75.00)	691.33

You can find the balance in the second row two different ways.

$100.00 - 34.57 = 65.43$ Subtract 34.57 from 100.00.

$100.00 + (-34.57) = 65.43$ Add −34.57 to 100.00.

a. Complete the balance column of the table on the previous page.

b. How did you find the balance in the twelfth row?

Sample answer: Subtract 75.00 from the previous balance of 766.33.

c. Use a different way to find the balance in part (b).

Sample answer: Add −75.00 to the previous balance of 766.33.

What Is Your Answer?

4. **IN YOUR OWN WORDS** How can you use what you know about subtracting integers to subtract rational numbers?

To subtract rational numbers, use the same rules for signs used for integers.

5. Give two real-life examples of subtracting rational numbers that are not integers.

Check students' work.

Laurie's Notes

Activity 3

- **Financial Literacy:** Begin with a discussion on how a checkbook and a debit card are used. In order to know your balance at any time, it is necessary to keep a running balance. Checks written must be subtracted from the balance, and deposits are added to the balance. Interest earned is also added to the balance.
- The activity provides additional practice with decimal addition and subtraction. Be sure that students recall the need to align decimal points. Working with partners, students can check their balances after each transaction.
- Discuss part (c). The check written for \$59.99 can be thought of in two ways: balance −\$59.99 or balance + (−\$59.99).
- Talk about the phrase "in the red," which in accounting means a negative balance. The deposits and interest are "in the black" and are positive. They are being added to the balance and the balance grows (or increases). The checks written are "in the red" and are negative. They are being subtracted from the balance and the balance shrinks (or decreases).

What Is Your Answer?

- Ask students to share their real-life examples for Question 5.

Closure

- **Writing:** Explain how $a - b$ and $a + (-b)$ are equivalent. Create an example to further illustrate what you are explaining.

3 ACTIVITY: Financial Literacy

Work with a partner. The table shows the balance in a checkbook.

- Black numbers are amounts added to the account.
- Red numbers are amounts taken from the account.

Date	Check #	Transaction	Amount	Balance
--	--	Previous balance	--	100.00
1/02/2013	124	Groceries	34.57	
1/07/2013		Check deposit	875.50	
1/11/2013		ATM withdrawal	40.00	
1/14/2013	125	Electric company	78.43	
1/17/2013		Music store	10.55	
1/18/2013	126	Shoes	47.21	
1/22/2013		Check deposit	125.00	
1/24/2013		Interest	2.12	
1/25/2013	127	Cell phone	59.99	
1/26/2013	128	Clothes	65.54	
1/30/2013	129	Cable company	75.00	

Math Practice

Interpret Results

What does your answer represent? Does your answer make sense?

You can find the balance in the **second row** two different ways.

$100.00 - 34.57 = 65.43$ Subtract 34.57 from 100.00.

$100.00 + (-34.57) = 65.43$ Add −34.57 to 100.00.

a. Copy the table. Then complete the balance column.

b. How did you find the balance in the **twelfth row**?

c. Use a different way to find the balance in part (b).

What Is Your Answer?

4. **IN YOUR OWN WORDS** How can you use what you know about subtracting integers to subtract rational numbers?

5. Give two real-life examples of subtracting rational numbers that are not integers.

Practice Use what you learned about subtracting rational numbers to complete Exercises 3–5 on page 536.

12.3 Lesson

Check It Out
Lesson Tutorials
BigIdeasMath.com

Key Idea

Subtracting Rational Numbers

Words To subtract rational numbers, use the same rules for signs as you used for integers.

Numbers $\frac{2}{5} - \left(-\frac{1}{5}\right) = \frac{2}{5} + \frac{1}{5} = \frac{2+1}{5} = \frac{3}{5}$

EXAMPLE 1 Subtracting Rational Numbers

Find $-4\frac{1}{7} - \left(-\frac{6}{7}\right)$. **Estimate** $-4 - (-1) = -3$

$-4\frac{1}{7} - \left(-\frac{6}{7}\right) = -4\frac{1}{7} + \frac{6}{7}$ Add the opposite of $-\frac{6}{7}$.

$= -\frac{29}{7} + \frac{6}{7}$ Write the mixed number as an improper fraction.

$= \frac{-29+6}{7}$ Write the sum of the numerators over the common denominator.

$= \frac{-23}{7}$ Add.

$= -3\frac{2}{7}$ Write the improper fraction as a mixed number.

The difference is $-3\frac{2}{7}$. **Reasonable?** $-3\frac{2}{7} \approx -3$ ✓

EXAMPLE 2 Subtracting Rational Numbers

Find 12.8 − 21.6.

$12.8 - 21.6 = 12.8 + (-21.6)$ Add the opposite of 21.6.

$= -8.8$ $|-21.6| > |12.8|$. So, subtract $|12.8|$ from $|-21.6|$.

Use the sign of −21.6.

The difference is −8.8.

On Your Own

Now You're Ready
Exercises 3–11

1. $\frac{1}{3} - \left(-\frac{1}{3}\right)$
2. $-3\frac{1}{3} - \frac{5}{6}$
3. $4\frac{1}{2} - 5\frac{1}{4}$
4. $-8.4 - 6.7$
5. $-20.5 - (-20.5)$
6. $0.41 - (-0.07)$

Laurie's Notes

Introduction

Connect

- **Yesterday:** Students explored how to subtract rational numbers.
- **Today:** Students will formalize the process completed yesterday, and they will subtract rational numbers.

Motivate

- Draw two "dots" on the board (locate them horizontally) and ask students about the distance between them. They may say to measure it with a ruler.
- Draw a line through the points, extending the line beyond the points. Put a tick mark to the left of both dots and label it 0.
- ? "Besides using a ruler, is there another way we could find the distance between the two points?" Listen for scaling the number line and then subtracting the lesser number from the greater number.
- ? "Would this work if 0 is between the two points? if 0 is to the right of both points?" Students will be less certain that subtraction will work. You may need to try simple integer values where students can count.
- Tell students you will return to distance on a number line in the lesson.

Lesson Notes

Key Idea

- Write the Key Idea. Work through the example.
- **Teaching Tip:** Use a different color when you *add the opposite*.

Example 1

- ? "How do you subtract integers?" The statement "add the opposite" should be familiar. Once the problem is written as an addition problem, students should recall the rules for integer addition.
- ? "The problem is now $-4\frac{1}{7} + \frac{6}{7}$. Will the sum be positive or negative?" Negative; the mixed number is negative and has a greater absolute value.
- **FYI:** Some students may choose to solve by first rewriting $-4\frac{1}{7}$ as $-3\frac{8}{7}$.
- **Connection:** The rule for subtracting rational numbers is the same as the rule for subtracting integers. The challenge will be working with fractions!

Example 2

- Write the problem. Students should see that the difference will be negative.
- Rewrite the problem by "adding the opposite."
- Continue to work through the problem as shown.

On Your Own

- **Neighbor Check:** Have students work independently and then have their neighbors check their work. Have students discuss any discrepancies.

Goal Today's lesson is subtracting rational numbers.

Technology for the Teacher

Dynamic Classroom

Lesson Tutorials
Lesson Plans
Answer Presentation Tool

Extra Example 1

Find $-8\frac{2}{3} - 6\frac{1}{6}$. $-14\frac{5}{6}$

Extra Example 2

Find $-3.75 - (-0.96)$. -2.79

On Your Own

1. $\frac{2}{3}$
2. $-4\frac{1}{6}$
3. $-\frac{3}{4}$
4. -15.1
5. 0
6. 0.48

Extra Example 3

Find the distance between the two numbers on the number line.

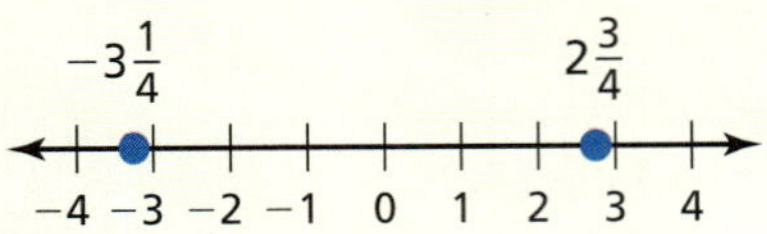

6

Extra Example 4

You have $101.62 in your savings account. You withdraw $45.41 to pay your cell phone bill. Do you have enough money left to buy a video game that costs $49.99?
Yes

On Your Own

7. 7.8

8. no

Differentiated Instruction

Visual

Some students will incorrectly subtract decimals, especially when the second number has more decimal places than the first. Encourage students to use zeros so that the two numbers have the same number of decimal places. For example, 2.35 − 1.457 should be written as

$$\begin{array}{r} 2.350 \\ -\,1.457 \\ \hline 0.893 \end{array} \quad \text{instead of} \quad \begin{array}{r} 2.35 \\ -\,1.457 \\ \hline 0.907 \end{array}. \ ✗$$

Laurie's Notes

Example 3

- **Big Idea:** This problem connects subtraction to finding distances on a number line, an idea that was investigated in the Motivate.
- **Model with Mathematics:** The vertical number line provides a visual model that helps students make a reasonable estimate.
- ? "We want to find the distance between two points. Will the order in which we do the subtraction matter?" Listen for students to state that subtraction is not commutative. However, since we will take the absolute value, you can subtract in either order because the results are opposites.
- Work through the problem, referring to the number line as you work.

Example 4

- To help visualize this problem, fill a glass bowl with water. Float a toy boat in the water so that the distance above the water level is visible.
- ? "If you know the height of the boat above the water and the depth of the boat below the water, how can you find the total height of the boat?" Students will probably say to add the two together. It is also acceptable to subtract the lowest point relative to sea level (a negative number) from the highest point (a positive number).

On Your Own

- **Neighbor Check:** Have students work independently and then have their neighbors check their work. Have students discuss any discrepancies.

Words of Wisdom

- Students often think that 2.1 feet is equivalent to 2 feet, 1 inch. Have students explore which is greater: 2.1 feet or 2 feet, 1 inch.

  $0.1 \text{ ft} \times \frac{12 \text{ in.}}{1 \text{ ft}} = 1.2 \text{ in.}$, so 2.1 feet is greater than 2 feet, 1 inch.

Closure

- Ask students to explain how subtraction of rational numbers is similar to subtraction of integers. *Sample answer:* The sign of the difference is the sign of the number with the greater absolute value.

The distance between any two numbers on a number line is the absolute value of the difference of the numbers.

EXAMPLE 3 Finding Distances Between Numbers on a Number Line

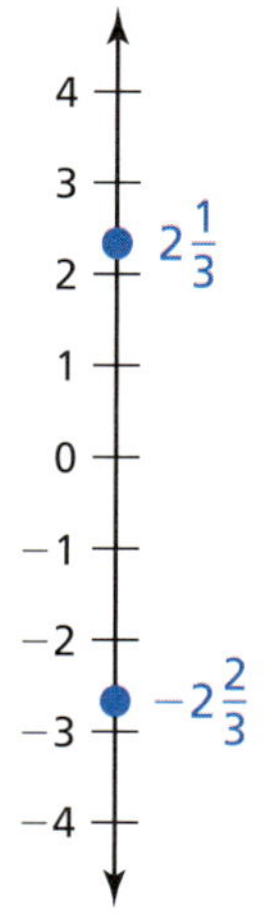

Find the distance between the two numbers on the number line.

To find the distance between the numbers, first find the difference of the numbers.

$$-2\frac{2}{3} - 2\frac{1}{3} = -2\frac{2}{3} + \left(-2\frac{1}{3}\right) \quad \text{Add the opposite of } 2\frac{1}{3}.$$

$$= -\frac{8}{3} + \left(-\frac{7}{3}\right) \quad \text{Write the mixed numbers as improper fractions.}$$

$$= \frac{-15}{3} \quad \text{Add.}$$

$$= -5 \quad \text{Simplify.}$$

∴ Because $|-5| = 5$, the distance between $-2\frac{2}{3}$ and $2\frac{1}{3}$ is 5.

EXAMPLE 4 Real-Life Application

Clearance: 11 ft 8 in.

In the water, the bottom of a boat is 2.1 feet below the surface, and the top of the boat is 8.7 feet above it. Towed on a trailer, the bottom of the boat is 1.3 feet above the ground. Can the boat and trailer pass under the bridge?

Step 1: Find the height h of the boat.

$$h = 8.7 - (-2.1) \quad \text{Subtract the lowest point from the highest point.}$$

$$= 8.7 + 2.1 \quad \text{Add the opposite of } -2.1.$$

$$= 10.8 \quad \text{Add.}$$

Step 2: Find the height t of the boat and trailer.

$$t = 10.8 + 1.3 \quad \text{Add the trailer height to the boat height.}$$

$$= 12.1 \quad \text{Add.}$$

∴ Because 12.1 feet is greater than 11 feet 8 inches, the boat and trailer cannot pass under the bridge.

On Your Own

Now You're Ready
Exercises 13–15

7. Find the distance between −7.5 and −15.3 on a number line.
8. **WHAT IF?** In Example 4, the clearance is 12 feet 1 inch. Can the boat and trailer pass under the bridge?

12.3 Exercises

Check It Out
Help with Homework
BigIdeasMath.com

Vocabulary and Concept Check

1. **WRITING** Explain how to find the difference $-\frac{4}{5} - \frac{3}{5}$.

2. **WHICH ONE DOESN'T BELONG?** Which expression does *not* belong with the other three? Explain your reasoning.

$-\frac{5}{8} - \frac{3}{4}$ $-\frac{3}{4} + \frac{5}{8}$ $-\frac{5}{8} + \left(-\frac{3}{4}\right)$ $-\frac{3}{4} - \frac{5}{8}$

Practice and Problem Solving

Subtract. Write fractions in simplest form.

1 2 3. $\frac{5}{8} - \left(-\frac{7}{8}\right)$ 4. $-1\frac{1}{3} - 1\frac{2}{3}$ 5. $-1 - 2.5$

6. $-5 - \frac{5}{3}$ 7. $-8\frac{3}{8} - 10\frac{1}{6}$ 8. $-\frac{1}{2} - \left(-\frac{5}{9}\right)$

9. $5.5 - 8.1$ 10. $-7.34 - (-5.51)$ 11. $6.673 - (-8.29)$

12. **ERROR ANALYSIS** Describe and correct the error in finding the difference.

✗ $\frac{3}{4} - \frac{9}{2} = \frac{3-9}{4-2} = \frac{-6}{2} = -3$

Find the distance between the two numbers on a number line.

3 13. $-2\frac{1}{2}, -5\frac{3}{4}$ 14. $-2.2, 8.4$ 15. $-7, -3\frac{2}{3}$

16. **SPORTS DRINK** Your sports drink bottle is $\frac{5}{6}$ full After practice, the bottle is $\frac{3}{8}$ full. Write the difference of the amounts after practice and before practice.

17. **SUBMARINE** The figure shows the depths of a submarine.

   a. Find the vertical distance traveled by the submarine.

   b. Find the mean hourly vertical distance traveled by the submarine.

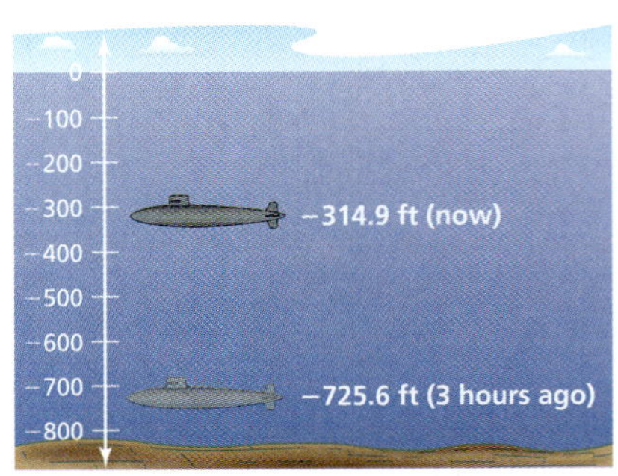

Evaluate.

18. $2\frac{1}{6} - \left(-\frac{8}{3}\right) + \left(-4\frac{7}{9}\right)$ 19. $6.59 + (-7.8) - (-2.41)$ 20. $-\frac{12}{5} + \left|-\frac{13}{6}\right| + \left(-3\frac{2}{3}\right)$

Assignment Guide and Homework Check

Level	Assignment	Homework Check
Advanced	1–5, 8–30 even, 31–35	8, 10, 18, 22, 24

Common Errors

- **Exercises 6–8** Students may try to identify the sign of the answer before finding a common denominator. Remind them that they need to find the common denominator first.
- **Exercises 5, 9–11** Students may forget to line up the decimal points when they subtract decimals. Remind them that the decimal points must be lined up before subtracting. Students may want to use half-inch graph paper to help keep the numbers and decimal points aligned.
- **Exercise 8** Students may not know where to put the negative sign in the fraction. Remind them that the negative can go in the numerator or the denominator (although the numerator is usually best when doing calculations), but not both.

12.3 Record and Practice Journal

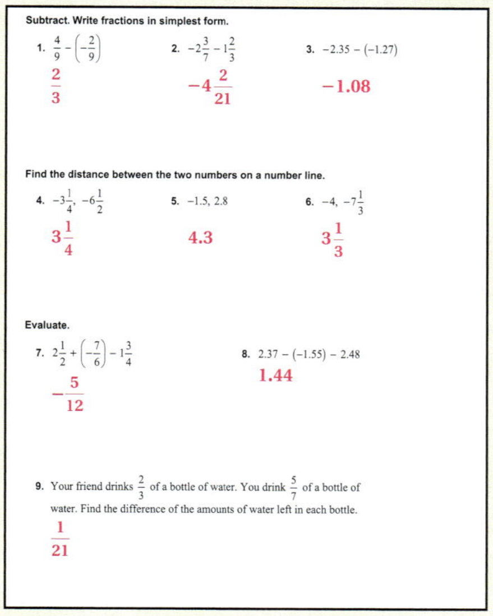
Subtract. Write fractions in simplest form.

1. $\frac{4}{9} - \left(-\frac{2}{9}\right)$ $\frac{2}{3}$
2. $-2\frac{3}{7} - 1\frac{2}{3}$ $-4\frac{2}{21}$
3. $-2.35 - (-1.27)$ -1.08

Find the distance between the two numbers on a number line.

4. $-3\frac{1}{4}, -6\frac{1}{2}$ $3\frac{1}{4}$
5. $-1.5, 2.8$ 4.3
6. $-4, -7\frac{1}{3}$ $3\frac{1}{3}$

Evaluate.

7. $2\frac{1}{2} + \left(-\frac{7}{6}\right) - 1\frac{3}{4}$ $-\frac{5}{12}$
8. $2.37 - (-1.55) - 2.48$ 1.44

9. Your friend drinks $\frac{2}{3}$ of a bottle of water. You drink $\frac{5}{7}$ of a bottle of water. Find the difference of the amounts of water left in each bottle. $\frac{1}{21}$

Vocabulary and Concept Check

1. Instead of subtracting, add the opposite of $\frac{3}{5}$, $-\frac{3}{5}$. Then, add $\left|-\frac{4}{5}\right|$ and $\left|-\frac{3}{5}\right|$, and the sign is negative.
2. $-\frac{3}{4} + \frac{5}{8}$, which equals $-\frac{1}{8}$. All the others equal $-1\frac{3}{8}$.

Practice and Problem Solving

3. $1\frac{1}{2}$
4. -3
5. -3.5
6. $-6\frac{2}{3}$
7. $-18\frac{13}{24}$
8. $\frac{1}{18}$
9. -2.6
10. -1.83
11. 14.963
12. They did not use the least common denominator.

$$\frac{3}{4} - \frac{9}{2} = \frac{3}{4} - \frac{18}{4} = \frac{3-18}{4} = \frac{-15}{4} = -3\frac{3}{4}$$

13. $3\frac{1}{4}$
14. 10.6
15. $3\frac{1}{3}$
16. $\frac{3}{8} - \frac{5}{6} = -\frac{11}{24}$
17. **a.** 410.7 feet

    b. 136.9 feet per hour
18. $\frac{1}{18}$
19. 1.2
20. $-3\frac{9}{10}$

Practice and Problem Solving

21. The difference is an integer when (1) the decimals have the same sign and the digits to the right of the decimal point are the same, or (2) the decimals have different signs and the sum of the decimal parts of the numbers add up to 1.

22. No, the cook needs $\frac{1}{12}$ cup more.

23. $-1\frac{7}{8}$ miles

24–26. See *Taking Math Deeper.*

27. *Sample answer:* $x = -1.8$ and $y = -2.4$; $x = -5.5$ and $y = -6.1$

28. sometimes; It is positive only if the first fraction is greater.

29. always; It is always positive because the first decimal is always greater.

30. $5.24 - (8.85) = -3.61$

Fair Game Review

31. 35.88 **32.** 3

33. $8\frac{2}{3}$ **34.** $2\frac{4}{5}$

35. C

Mini-Assessment

Subtract. Write fractions in simplest form.

1. $\frac{1}{2} - \frac{3}{4}$ $-\frac{1}{4}$
2. $2\frac{2}{5} - \left(-\frac{6}{5}\right)$ $3\frac{3}{5}$
3. $-8.4 - 0.9$ -9.3
4. $-12.55 - (-23.08)$ 10.53
5. The temperature in a town is $-4.7°C$. The temperature decreases $5.4°C$. What is the new temperature? $-10.1°C$

Taking Math Deeper

Exercises 24–26

This problem gives students a chance to find the sum of a long list of signed numbers. To do this efficiently, students can use the Commutative and Associative Properties of Addition.

1 Read and interpret the bar graph.

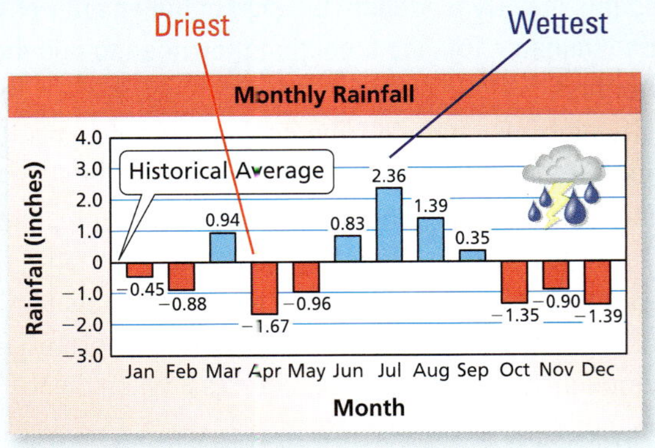

24. Difference $= 2.36 - (-1.67)$
$= 2.36 + 1.67$
$= 4.03$ in.

2 Find the sum of the differences.

Sum of negative numbers		Sum of positive numbers
−0.45		
−0.88		
−1.67		0.94
−0.96		0.83
−1.35		2.36
−0.90		1.39
−1.39		0.35
−7.60		5.87

25. Total sum: $-7.60 + 5.87 = -1.73$ in.

3 Interpret.

26. The total rainfall for the year was 1.73 inches *less* than the historical average.

Reteaching and Enrichment Strategies

If students need help...	If students got it...
Resources by Chapter • Practice A and Practice B • Puzzle Time Record and Practice Journal Practice Differentiating the Lesson Lesson Tutorials Skills Review Handbook	Resources by Chapter • Enrichment and Extension • Technology Connection Start the next section

21. **REASONING** When is the difference of two decimals an integer? Explain.

22. **RECIPE** A cook has $2\frac{2}{3}$ cups of flour. A recipe calls for $2\frac{3}{4}$ cups of flour. Does the cook have enough flour? If not, how much more flour is needed?

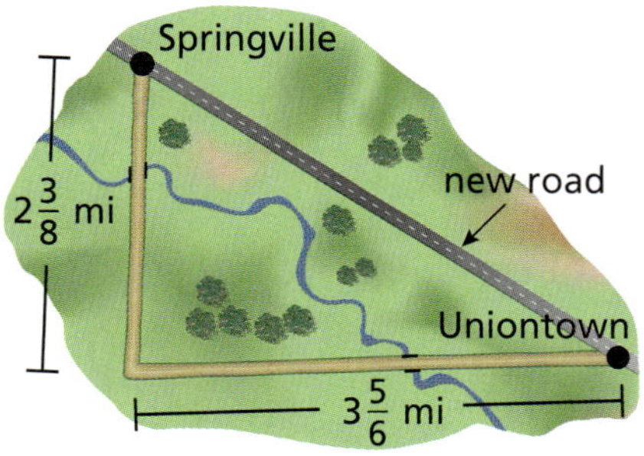

23. **ROADWAY** A new road that connects Uniontown to Springville is $4\frac{1}{3}$ miles long. What is the change in distance when using the new road instead of the dirt roads?

RAINFALL In Exercises 24–26, the bar graph shows the differences in a city's rainfall from the historical average.

24. What is the difference in rainfall between the wettest and the driest months?

25. Find the sum of the differences for the year.

26. What does the sum in Exercise 25 tell you about the rainfall for the year?

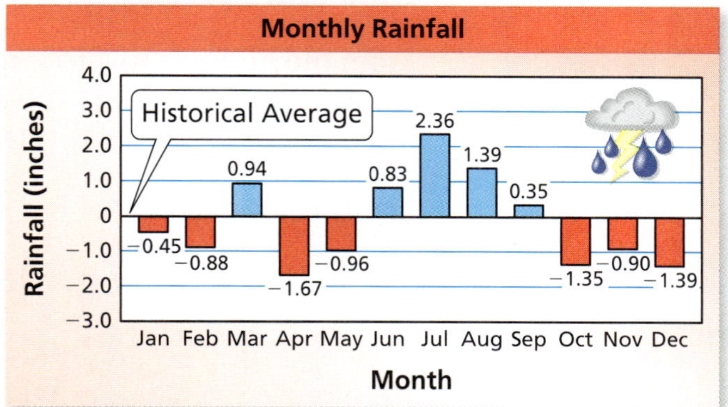

27. **OPEN-ENDED** Write two different pairs of negative decimals, x and y, that make the statement $x - y = 0.6$ true.

REASONING Tell whether the difference between the two numbers is *always, sometimes,* or *never* positive. Explain your reasoning.

28. two negative fractions

29. a positive decimal and a negative decimal

30. **Structure** Fill in the blanks to make the solution correct.

$$5.\square 4 - (\square.8\square) = -3.61$$

Fair Game Review What you learned in previous grades & lessons

Evaluate. *(Section 2.1, Section 2.3, Section 2.5, and Section 2.6)*

31. 5.2×6.9

32. $7.2 \div 2.4$

33. $2\frac{2}{3} \times 3\frac{1}{4}$

34. $9\frac{4}{5} \div 3\frac{1}{2}$

35. **MULTIPLE CHOICE** A sports store has 116 soccer balls. Over 6 months, it sells 8 soccer balls per month. How many soccer balls are in inventory at the end of the 6 months? *(Section 11.3 and Section 11.4)*

Ⓐ −48 Ⓑ 48 Ⓒ 68 Ⓓ 108

12.4 Multiplying and Dividing Rational Numbers

Essential Question Why s the product of two negative rational numbers positive?

In Section 11.4, you used a table to see that the product of two negative integers is a positive integer. In this activity, you will find that same result another way.

1 ACTIVITY: Showing $(-1)(-1) = 1$

Work with a partner. How can you show that $(-1)(-1) = 1$?

From the Additive Inverse Property, you know that $1 + (-1) = 0$. If you can show that $(-1)(-1) + (-1) = 0$ is true, then you have shown that $(-1)(-1) = 1$.

Justify each step.

$$\begin{aligned}(-1)(-1) + (-1) &= (-1)(-1) + 1(-1) \\ &= (-1)[(-1) + 1] \\ &= (-1)0 \\ &= 0\end{aligned}$$

So, $(-1)(-1) = 1$.

2 ACTIVITY: Multiplying by -1

Work with a partner.

a. Graph each number below on three different number lines. Then multiply each number by -1 and graph the product on the appropriate number line.

$2 \qquad 8 \qquad -1$

b. How does multiplying by -1 change the location of the points in part (a)? What is the relationship between the number and the product?

c. Graph each number below on three different number lines. Where do you think the points will be after multiplying by -1? Plot the points. Explain your reasoning.

$\frac{1}{2} \qquad 2.5 \qquad -\frac{5}{2}$

d. What is the relationship between a rational number $-a$ and the product $-1(a)$? Explain your reasoning.

Rational Numbers

In this lesson, you will

- multiply and divide rational numbers.
- solve real-life problems.

Laurie's Notes

Introduction

Applying Mathematical Practices

- **Reason Abstractly and Quantitatively:** Mathematically proficient students are able to follow sequential statements about equivalent expressions. They are able to recognize and state the algebraic properties that support each statement.

Motivate

- Display the table used in Section 11.4 to show $-3(-2) = 6$.
- Discuss this approach, an additive interpretation of multiplication.
- ? "Does this approach make sense for a problem involving fractions, such as $-3\frac{1}{2}\left(-2\frac{3}{4}\right)$?" Students may not be sure how to create a table involving fractions that would show a pattern.
- Explain that they will not use the table approach today. The first three activities use an analytic approach to show that the product of two negatives is positive. This approach requires students to read carefully and recognize the application of the Distributive Property.

Activity Notes

Activity 1

- Activity 1 is an alternate way to show that $(-1)(-1) = 1$.
- **Teaching Tip:** You may want to do a related example before students begin this activity to get them thinking about properties.

$$4 + [2 + (-2)] = 4 + 0 \qquad \text{Additive Inverse Property}$$
$$= 4 \qquad \text{Addition Property of Zero}$$

- **Reason Abstractly and Quantitatively:** Students should read the introduction carefully as it describes the strategy that will be used to show $(-1)(-1) = 1$. When finished, discuss how to show $(-2)(-2) = 4$.

Activity 2

- This activity explores the result of multiplying numbers by -1.
- Part (b) asks two questions. In the first question, students recognize that the point they plot is a reflection (flipped) over 0. The second question is recognizing the relationship between the number and the product (they are opposites).
- **Big Idea:** Multiplying by -1 is the same as taking the opposite of a number. This idea is used in Activity 3.

What Your Students Will Learn

- Multiply and divide rational numbers with the same sign and with different signs.

Previous Learning

Students should be comfortable multiplying and dividing positive fractions, positive decimals, and integers.

Technology for the Teacher

Dynamic Classroom

Lesson Plans
Complete Materials List

12.4 Record and Practice Journal

Essential Question Why is the product of two negative rational numbers positive?

1 ACTIVITY: Showing $(-1)(-1) = 1$

Work with a partner. How can you show that $(-1)(-1) = 1$?

To begin, assume that $(-1)(-1) = 1$ is a true statement. From the Additive Inverse Property, you know that $1 + (-1) = 0$. So, substitute $(-1)(-1)$ for 1 to get $(-1)(-1) + (-1) = 0$. If you can show that $(-1)(-1) + (-1) = 0$ is true, then you have shown that $(-1)(-1) = 1$.

Justify each step.

$(-1)(-1) + (-1) = (-1)(-1) + 1(-1)$ Mult. Prop. of One

$= (-1)[(-1) + 1]$ Dist. Prop.

$= (-1)0$ Add. Inverse Prop.

$= 0$ Mult. Prop. of Zero

$(-1)(-1) =$ 1

2 ACTIVITY: Multiplying by -1

Work with a partner.

a. Graph each number below on three different number lines. Then multiply each number by -1 and graph the product on the appropriate number line.

2 8 -1

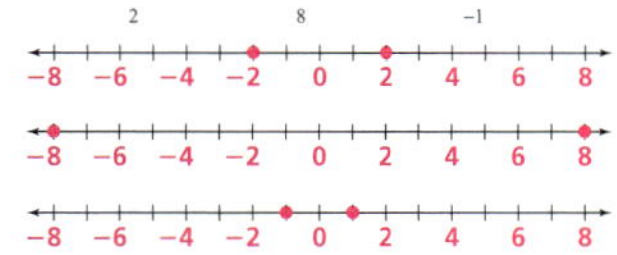

English Language Learners

Simplified Language

Writing stories poses a challenge for English learners. You may want to allow students who struggle with language to outline a story or to create a story using pictures.

12.4 Record and Practice Journal

b. How does multiplying by −1 change the location of the points in part (a)? What is the relationship between the number and the product?

reflects point on the other side of 0; opposites

c. Graph each number below on three different number lines. Where do you think the points will be after multiplying by −1? Plot the points. Explain your reasoning.

$\frac{1}{2}$ 2.5 $-\frac{5}{2}$

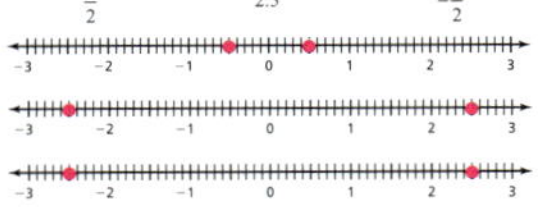

d. What is the relationship between a rational number $-a$ and the product $-1(a)$? Explain your reasoning. They are the same.

3 ACTIVITY: Understanding the product of Rational Numbers

Work with a partner. Let *a* and *b* be positive rational numbers.

a. Because a and b are positive, what do you know about $-a$ and $-b$?

They represent the opposites of a and b and are negative.

b. Justify each step.

$(-a)(-b) = (-1)(a)(-1)(b)$ Mult. by -1 is the same as taking the opposite.

$= (-1)(-1)(a)(b)$ Comm. Prop. of Mult.

$= (1)(a)(b)$ Result of Activity 1

$= ab$ Mult. Prop. of One

c. Because a and b are positive, what do you know about the product ab?

It is positive.

d. What does this tell you about products of rational numbers? Explain.

The products of rational numbers follow the same rules as the products of integers.

4 ACTIVITY: Writing a Story

Work with a partner. Write a story that uses addition, subtraction, multiplication, or division of rational numbers. Check students' work.

- At least one of the numbers in the story has to be negative and *not* an integer.
- Draw pictures to help illustrate what is happening in the story.
- Include the solution of the problem in the story.

If you are having trouble thinking of a story, here are some common uses of negative numbers:

- A profit of −\$15 is a loss of \$15.
- An elevation of −100 feet is a depth of 100 feet below sea level.
- A gain of −5 yards in football is a loss of 5 yards.
- A score of −4 in golf is 4 strokes under par.

What Is Your Answer?

5. **IN YOUR OWN WORDS** Why is the product of two negative rational numbers positive?

Check students' work.

6. **PRECISION** Show that $(-2)(-3) = 6$.

$$(-2)(-3) = (-1)(2)(-1)(3) = (-1)(-1)(2)(3) = 1(2)(3) = 6$$

7. How can you show that the product of a negative rational number and a positive rational number is negative?

$$(-a)(b) = (-1)(a)(b) = -ab$$

Laurie's Notes

Activity 3

- This activity is similar in style to Activity 1.
- **Construct Viable Arguments:** Give students time to work through the steps in this activity. Resist jumping in too soon to give answers. Remind students to read carefully and observe what changes have taken place in each step.
- Discuss the results. If a and b are positive numbers, then $(-a)(-b)$ is the product of two negative numbers. The conclusion is that this product equals ab, which is positive because it is the product of two positive numbers. This won't be surprising to students because of their work with integers. This extends that knowledge to all rational numbers.
- Reasoning abstractly is challenging for many students and this activity may be less convincing to students than observing a pattern with integers. It is important for students to have the opportunity to make sense of an abstract approach. Students will be required to construct these types of proofs on their own in future courses.

Activity 4

- Read through the directions together as a class.
- Have students work in pairs so that brainstorming can occur. Both students should be actively engaged, with one doing the writing while the other draws a diagram to illustrate the problem.
- The four examples showing where negative numbers are commonly used should help students get started.
- Provide at least 20–25 minutes for the brainstorming and writing process. Students' stories should include computations and a final solution.
- **Discuss:** As time allows, have pairs of students share their stories. To help students see when each operation is used, make a table on the board with four columns, one for each operation. Record the context used for each operation.
- **Interdisciplinary:** Some of your language arts colleagues may want to review the students' stories. Speak with them about different possibilities.

What Is Your Answer?

- **Attend to Precision:** Question 6 is connected to Activity 3. Students should recognize and use a similar strategy to the one used in Activity 3.

Closure

- **Writing:** Have students write brief scenarios for all four operations (addition, subtraction, multiplication, and division). Note: The scenarios *do not* need to be connected to one another. Instead of creating a whole story, students just need to write four sentences.

3 ACTIVITY: Understanding the Product of Rational Numbers

Work with a partner. Let *a* and *b* be positive rational numbers.

a. Because a and b are positive, what do you know about $-a$ and $-b$?

b. Justify each step.

$$(-a)(-b) = (-1)(a)(-1)(b)$$
$$= (-1)(-1)(a)(b)$$
$$= (1)(a)(b)$$
$$= ab$$

c. Because a and b are positive, what do you know about the product ab?

d. What does this tell you about products of rational numbers? Explain.

4 ACTIVITY: Writing a Story

Work with a partner. Write a story that uses addition, subtraction, multiplication, or division of rational numbers.

- At least one of the numbers in the story has to be negative and *not* an integer.
- Draw pictures to help illustrate what is happening in the story.
- Include the solution of the problem in the story.

Math Practice

Specify Units

What units are in your story?

If you are having trouble thinking of a story, here are some common uses of negative numbers:

- A profit of $-\$15$ is a loss of $\$15$.
- An elevation of -100 feet is a depth of 100 feet below sea level.
- A gain of -5 yards in football is a loss of 5 yards.
- A score of -4 in golf is 4 strokes under par.

What Is Your Answer?

5. **IN YOUR OWN WORDS** Why is the product of two negative rational numbers positive?
6. **PRECISION** Show that $(-2)(-3) = 6$.
7. How can you show that the product of a negative rational number and a positive rational number is negative?

Practice Use what you learned about multiplying rational numbers to complete Exercises 7–9 on page 542.

12.4 Lesson

Check It Out
Lesson Tutorials
BigIdeasMath.com

Key Idea

Remember

The *reciprocal* of $\frac{a}{b}$ is $\frac{b}{a}$.

Multiplying and Dividing Rational Numbers

Words To multiply or divide rational numbers, use the same rules for signs as you used for integers.

Numbers

$$-\frac{2}{7} \cdot \frac{1}{3} = \frac{-2 \cdot 1}{7 \cdot 3} = \frac{-2}{21} = -\frac{2}{21}$$

$$-\frac{1}{2} \div \frac{4}{9} = \frac{-1}{2} \cdot \frac{9}{4} = \frac{-1 \cdot 9}{2 \cdot 4} = \frac{-9}{8} = -\frac{9}{8}$$

EXAMPLE 1 Dividing Rational Numbers

Find $-5\frac{1}{5} \div 2\frac{1}{3}$.

Estimate $-5 \div 2 = -2\frac{1}{2}$

$-5\frac{1}{5} \div 2\frac{1}{3} = -\frac{26}{5} \div \frac{7}{3}$ Write mixed numbers as improper fractions.

$= \frac{-26}{5} \cdot \frac{3}{7}$ Multiply by the reciprocal of $\frac{7}{3}$.

$= \frac{-26 \cdot 3}{5 \cdot 7}$ Multiply the numerators and the denominators.

$= \frac{-78}{35}$, or $-2\frac{8}{35}$ Simplify.

The quotient is $-2\frac{8}{35}$.

Reasonable? $-2\frac{8}{35} \approx -2\frac{1}{2}$ ✓

EXAMPLE 2 Multiplying Rational Numbers

Find $-2.5 \cdot 3.6$.

$$\begin{array}{r} -2.5 \\ \times\ 3.6 \\ \hline 1\,5\,0 \\ 7\,5\,0 \\ \hline -9.0\,0 \end{array}$$

The decimals have different signs.

The product is negative.

The product is -9.

Laurie's Notes

Introduction

Connect

- **Yesterday:** Students used an analytic approach to show that the product of two negative numbers is positive.
- **Today:** Students will learn the rules for multiplying and dividing rational numbers.

Discuss

- Before beginning the formal lesson, it would be helpful to review rules for multiplying and dividing integers.
  - same signs ⟶ product/quotient is positive
  - different signs ⟶ product/quotient is negative

Lesson Notes

Key Idea

- Write the definition for multiplication and division of rational numbers. Note that the sign of the fraction is written with the numerator when the computation is performed.

Example 1

- **Discuss:** Before starting the first example, take time to discuss estimating products and quotients. This will help students check their answers.
- Work through the problem. Do not skip the initial estimate.
- Remind students that when multiplying or dividing fractions, mixed numbers must be written as improper fractions.
- **Discuss:** There are several important skills involved in this example. Identify each skill with students so that vocabulary is reviewed and each process is made clear.

Example 2

- This example involves multiplying decimals *and* signed numbers.
- Write the example and ask how they might estimate an answer.
- **Extension:** If time permits, repeat this example by converting the decimals to fractions:

$$-2\frac{5}{10} \times 3\frac{6}{10} = -2\frac{1}{2} \times 3\frac{3}{5}$$
$$= \frac{-5}{2} \times \frac{18}{5}$$
$$= \frac{-90}{10}$$
$$= -9$$

Goal Today's lesson is multiplying and dividing rational numbers.

Technology for the Teacher

Dynamic Classroom

Lesson Tutorials
Lesson Plans
Answer Presentation Tool

Extra Example 1

Find $3\frac{1}{4} \div \left(-1\frac{1}{8}\right)$. $-2\frac{8}{9}$

Extra Example 2

Find $-4.8(-5.2)$. 24.96

Extra Example 3

Find $\frac{10}{3} \cdot \left(-3\frac{3}{5}\right) \cdot (-3)$. 36

On Your Own

1. $2\frac{2}{5}$
2. $-\frac{1}{8}$
3. -9.18
4. 3.78
5. $-7\frac{7}{8}$
6. 72

Extra Example 4

Find the mean of -25.63, 37.15, 18.92, and -44.28. -3.46

On Your Own

7. \$58.65

Differentiated Instruction

Inclusion

Remind students that one difference between multiplying decimals and dividing decimals is the placement of the decimal point. In multiplication, the decimal point is placed after the decimals are multiplied. In division, the placement of the decimal point is determined before dividing.

Laurie's Notes

Example 3

- **Look for and Make Use of Structure:** Write the problem. Discuss possible strategies for performing the computation. Students should recognize that algebraic properties can be used to perform the operations in a more efficient manner than how the problem is presented.
- This problem provides a good review of several algebraic properties. The more frequently we refer to these properties by name, the more fluent students become in using them.

On Your Own

- Have three pairs of students choose one question from 1–3 to complete at the board. Have the other students try these problems at their desks. Have the pairs of students explain their work at the board.

? "How can you check that your answer is reasonable?" Use estimation.

- Have three different pairs of students choose one question from 4–6 to complete at the board. Have the other students try these problems at their desks. Have the pairs of students explain their work at the board. Students may need to be reminded that multiplication can be represented using parentheses around one or both of the factors.

Example 4

- **Financial Literacy:** This example uses stock prices to review decimal addition, subtraction, and division. Remind students that the word *mean* is the same as the arithmetic average.
- Explain the stock context and what each column of the table means.

On Your Own

- Predict whether the mean change will be *positive* or *negative*. Explain your reasoning. positive; Students should recognize that the mean of the four stocks is the sum of the change in the first three stocks ($-$\$333.63) and the change in Stock D (\$568.23), divided by four. This will be positive because $-$\$333.63 + \$568.23 is positive.

Closure

- **Exit Ticket:**

  $-2\frac{1}{3} \times 3\frac{2}{3}$ $-8\frac{5}{9}$ $\qquad$ $(-0.5)(-4.2) \div 0.03$ 70

EXAMPLE 3 Multiplying More Than Two Rational Numbers

Find $-\frac{1}{7} \cdot \left[\frac{4}{5} \cdot (-7)\right]$.

You can use properties of multiplication to make the product easier to find.

$$-\frac{1}{7} \cdot \left[\frac{4}{5} \cdot (-7)\right] = -\frac{1}{7} \cdot \left(-7 \cdot \frac{4}{5}\right)$$ Commutative Property of Multiplication

$$= -\frac{1}{7} \cdot (-7) \cdot \frac{4}{5}$$ Associative Property of Multiplication

$$= 1 \cdot \frac{4}{5}$$ Multiplicative Inverse Property

$$= \frac{4}{5}$$ Multiplication Property of One

∴ The product is $\frac{4}{5}$.

On Your Own

Now You're Ready Exercises 10–30

Multiply or divide. Write fractions in simplest form.

1. $-\frac{6}{5} \div \left(-\frac{1}{2}\right)$
2. $\frac{1}{3} \div \left(-2\frac{2}{3}\right)$
3. $1.8(-5.1)$
4. $-6.3(-0.6)$
5. $-\frac{2}{3} \cdot 7\frac{7}{8} \cdot \frac{3}{2}$
6. $-7.2 \cdot 0.1 \cdot (-100)$

EXAMPLE 4 Real-Life Application

An investor owns Stocks A, B, and C. What is the mean change in the value of the stocks?

Account Positions

Stock	Original Value	Current Value	Change
A	600.54	420.15	−180.39
B	391.10	518.38	127.28
C	380.22	99.70	−280.52

$$\text{mean} = \frac{-180.39 + 127.28 + (-280.52)}{3} = \frac{-333.63}{3} = -111.21$$

∴ The mean change in the value of the stocks is −\$111.21.

On Your Own

7. **WHAT IF?** The change in the value of Stock D is \$568.23. What is the mean change in the value of the four stocks?

12.4 Exercises

Check It Out
Help with Homework
BigIdeasMath.com

Vocabulary and Concept Check

1. **WRITING** How is multiplying and dividing rational numbers similar to multiplying and dividing integers?

2. **NUMBER SENSE** Find the reciprocal of $-\frac{2}{5}$.

Tell whether the expression is *positive* or *negative* without evaluating.

3. $-\frac{3}{10} \times \left(-\frac{8}{15}\right)$ 4. $1\frac{1}{2} \div \left(-\frac{1}{4}\right)$ 5. -6.2×8.18 6. $\frac{-8.16}{-2.72}$

Practice and Problem Solving

Multiply.

7. $-1\left(\frac{4}{5}\right)$ 8. $-1\left(-3\frac{1}{2}\right)$ 9. $-0.25(-1)$

Divide. Write fractions in simplest form.

1 10. $-\frac{7}{10} \div \frac{2}{5}$ 11. $\frac{1}{4} \div \left(-\frac{3}{8}\right)$ 12. $-\frac{8}{9} \div \left(-\frac{8}{9}\right)$ 13. $-\frac{1}{5} \div 20$

14. $-2\frac{4}{5} \div (-7)$ 15. $-10\frac{2}{7} \div \left(-4\frac{4}{11}\right)$ 16. $-9 \div 7.2$ 17. $8 \div 2.2$

18. $-3.45 \div (-15)$ 19. $-0.18 \div 0.03$ 20. $3.722 \div (-3.56)$ 21. $12.42 \div (-4.8)$

Multiply. Write fractions in simplest form.

2 3 22. $-\frac{1}{4} \times \left(-\frac{4}{3}\right)$ 23. $\frac{5}{6}\left(-\frac{8}{15}\right)$ 24. $-2\left(-1\frac{1}{4}\right)$

25. $-3\frac{1}{3} \cdot \left(-2\frac{7}{10}\right)$ 26. $0.4 \times (-0.03)$ 27. $-0.05 \times (-0.5)$

28. $-8(0.09)(-0.5)$ 29. $\frac{5}{6} \cdot \left(-4\frac{1}{2}\right) \cdot \left(-2\frac{1}{5}\right)$ 30. $\left(-1\frac{2}{3}\right)^3$

ERROR ANALYSIS **Describe and correct the error.**

31. ✗ $-2.2 \times 3.7 = 8.14$

32. ✗ $-\frac{1}{4} \div \frac{3}{2} = -\frac{4}{1} \times \frac{3}{2} = -\frac{12}{2} = -6$

33. **HOUR HAND** The hour hand of a clock moves $-30°$ every hour. How many degrees does it move in $2\frac{1}{5}$ hours?

34. **SUNFLOWER SEEDS** How many 0.75-pound packages can you make with 6 pounds of sunflower seeds?

Assignment Guide and Homework Check

Level	Assignment	Homework Check
Advanced	1–9, 18–46 even, 47–51	18, 20, 24, 42, 44

Common Errors

- **Exercises 10–15** Students may use the reciprocal of the first fraction instead of the second, or they might forget to write a mixed number as an improper fraction before finding the reciprocal. Review multiplying and dividing fractions and the definition of reciprocal.
- **Exercises 16–21** Students may mix up the dividend and divisor. Remind them that the first number is the dividend and the second is the divisor.
- **Exercises 16–21** Students may forget to shift the decimal point when dividing or they might move the decimal point the wrong number of places. Remind students to use estimation to check their answer and the placement of the decimal.
- **Exercises 35–40** Students may forget to follow the order of operations. Tell them to write parentheses around the multiplication or division parts so that they remember to evaluate them first.

12.4 Record and Practice Journal

Multiply or divide. Write fractions in simplest form.

1. $-\frac{8}{9}\left(-\frac{18}{25}\right)$ **$\frac{16}{25}$**
2. $-4\left(\frac{9}{16}\right)$ **$-2\frac{1}{4}$**
3. $-3\frac{3}{7} \times 2\frac{1}{2}$ **$-8\frac{4}{7}$**
4. $-\frac{2}{3} \div \frac{5}{9}$ **$-1\frac{1}{5}$**
5. $\frac{7}{13} \div (-2)$ **$-\frac{7}{26}$**
6. $-5\frac{5}{8} \div \left(-4\frac{7}{12}\right)$ **$1\frac{5}{22}$**
7. $-1.39 \times (-6.8)$ **9.452**
8. $-10 \div 0.22$ **$-45.\overline{45}$**
9. $-12.166 \div (-1.54)$ **7.9**
10. In a game of tug of war, your team changes $-1\frac{3}{10}$ feet in position every 10 seconds. What is your change in position after 30 seconds?
    $-3\frac{9}{10}$ ft

Vocabulary and Concept Check

1. The same rules for signs of integers are applied to rational numbers.
2. $-\frac{5}{2}$
3. positive
4. negative
5. negative
6. positive

Practice and Problem Solving

7. $-\frac{4}{5}$
8. $3\frac{1}{2}$
9. 0.25
10. $-1\frac{3}{4}$
11. $-\frac{2}{3}$
12. 1
13. $-\frac{1}{100}$
14. $\frac{2}{5}$
15. $2\frac{5}{14}$
16. -1.25
17. $3.\overline{63}$
18. 0.23
19. -6
20. -2.45
21. -2.5875
22. $\frac{1}{3}$
23. $-\frac{4}{9}$
24. $2\frac{1}{2}$ or $\frac{5}{2}$
25. 9
26. -0.012
27. 0.025
28. 0.36
29. $8\frac{1}{4}$ or $\frac{33}{4}$
30. $-4\frac{17}{27}$ or $-\frac{125}{27}$
31. The answer should be negative. $-2.2 \times 3.7 = -8.14$
32. The wrong fraction was inverted.
$$-\frac{1}{4} \div \frac{3}{2} = -\frac{1}{4} \times \frac{2}{3} = -\frac{2}{12} = -\frac{1}{6}$$
33. $-66°$
34. 8 packages

Practice and Problem Solving

35. -19.59 **36.** 1.3

37. -22.667 **38.** $-4\frac{14}{15}$

39. $-5\frac{11}{24}$ **40.** $-1\frac{11}{36}$

41. *Sample answer:* $-\frac{9}{10}, \frac{2}{3}$

42. $191\frac{11}{12}$ yd

43. $3\frac{5}{8}$ gal

44. See *Taking Math Deeper.*

45. -1.28 sec

46. **a.** -0.02 in.

b. See Additional Answers.

Fair Game Review

47. -1.5 **48.** -5.4

49. $4\frac{1}{2}$ **50.** $-8\frac{5}{18}$

51. D

Mini-Assessment

Multiply or divide. Write fractions in simplest form.

1. $-\frac{6}{7}\left(-\frac{5}{2}\right)$ $2\frac{1}{7}$
2. $6\frac{1}{2} \div \left(-2\frac{3}{4}\right)$ $-2\frac{4}{11}$
3. $3.5(-7.65)$ -26.775
4. $-0.25 \div (-0.05)$ 5
5. The cell phone company will add $-\$2.74$ to your next bill for each of the 4 months you were overcharged. How much will be added to your next bill? $-\$10.96$

Taking Math Deeper

Exercise 44

Problems like this one beg for a diagram. It would be easy to misinterpret what "width" is referring to without drawing a diagram and labeling it.

1 Draw a diagram. Label the known and unknown lengths.

2 Find the total width of the boards by multiplying by 15.

$$15\left(9\frac{1}{4}\right) = 15\left(\frac{37}{4}\right) = \frac{555}{4} = 138\frac{3}{4} \text{ in.}$$

3 Subtract the width of the boards from 144 inches.

$$144 - 138\frac{3}{4} = 5\frac{1}{4} \text{ in.}$$

There are 14 spaces, so divide $5\frac{1}{4}$ by 14.

$$5\frac{1}{4} \div 14 = \frac{21}{4} \div 14 = \frac{21}{4} \cdot \frac{1}{14} = \frac{3}{8} \text{ in.}$$

Space ($\frac{3}{8}$ in.)

Reteaching and Enrichment Strategies

If students need help. . .	If students got it. . .
Resources by Chapter • Practice A and Practice B • Puzzle Time Record and Practice Journal Practice Differentiating the Lesson Lesson Tutorials Skills Review Handbook	Resources by Chapter • Enrichment and Extension • Technology Connection Start the next section

Evaluate.

35. $-4.2 + 8.1 \times (-1.9)$

36. $2.85 - 6.2 \div 2^2$

37. $-3.64 \cdot |-5.3| - 1.5^3$

38. $1\frac{5}{9} \div \left(-\frac{2}{3}\right) + \left(-2\frac{3}{5}\right)$

39. $-3\frac{3}{4} \times \frac{5}{6} - 2\frac{1}{3}$

40. $\left(-\frac{2}{3}\right)^2 - \frac{3}{4}\left(2\frac{1}{3}\right)$

41. OPEN-ENDED Write two fractions whose product is $-\frac{3}{5}$.

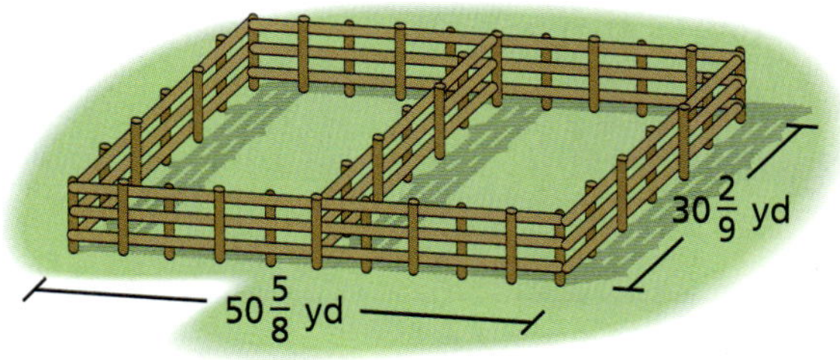

42. FENCING A farmer needs to enclose two adjacent rectangular pastures. How much fencing does the farmer need?

43. GASOLINE A 14.5-gallon gasoline tank is $\frac{3}{4}$ full. How many gallons will it take to fill the tank?

44. PRECISION A section of a boardwalk is made using 15 boards. Each board is $9\frac{1}{4}$ inches wide. The total width of the section is 144 inches. The spacing between each board is equal. What is the width of the spacing between each board?

45. RUNNING The table shows the changes in the times (in seconds) of four teammates. What is the mean change?

Teammate	Change
1	−2.43
2	−1.85
3	0.61
4	−1.45

46. Critical Thinking The daily changes in the barometric pressure for four days are −0.05, 0.09, −0.04, and −0.08 inches.

a. What is the mean change?

b. The mean change after five days is −0.01 inch. What is the change on the fifth day? Explain.

Fair Game Review What you learned in previous grades & lessons

Add or subtract. *(Section 12.2 and Section 12.3)*

47. $-6.2 + 4.7$

48. $-8.1 - (-2.7)$

49. $\frac{9}{5} - \left(-2\frac{7}{10}\right)$

50. $-4\frac{5}{6} + \left(-3\frac{4}{9}\right)$

51. MULTIPLE CHOICE What are the coordinates of the point in Quadrant IV? *(Section 6.5)*

Ⓐ (−4, 1) Ⓑ (−3, −3)

Ⓒ (0, −2) Ⓓ (3, −3)

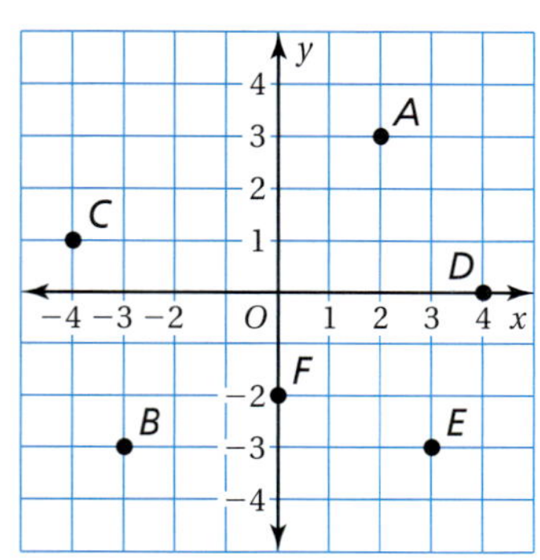

12.3–12.4 Quiz

Check It Out
Progress Check
BigIdeasMath.com

Subtract. Write fractions in simplest form. *(Section 12.3)*

1. $\frac{2}{7} - \left(\frac{6}{7}\right)$

2. $\frac{12}{7} - \left(-\frac{2}{9}\right)$

3. $9.1 - 12.9$

4. $5.647 - (-9.24)$

Find the distance between the two numbers on the number line. *(Section 12.3)*

5.

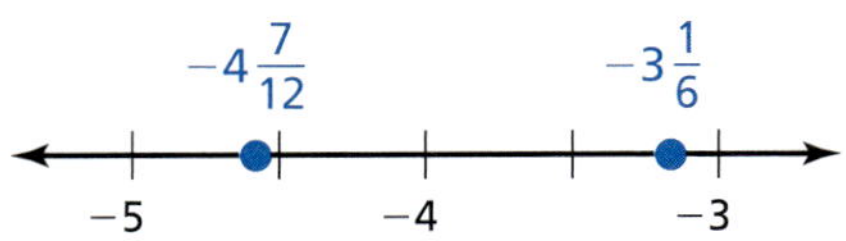

6.

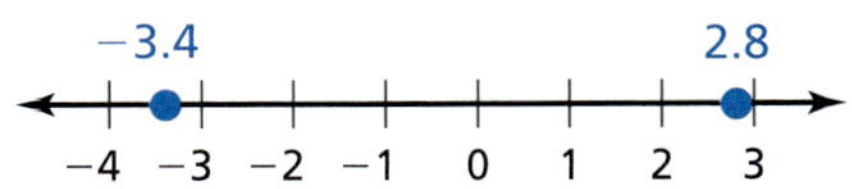

Divide. Write fractions in simplest form. *(Section 12.4)*

7. $\frac{2}{3} \div \left(-\frac{5}{6}\right)$

8. $-8\frac{5}{9} \div \left(-1\frac{4}{7}\right)$

9. $-8.4 \div 2.1$

10. $32.436 \div (-4.24)$

Multiply. Write fractions in simplest form. *(Section 12.4)*

11. $\frac{5}{8} \times \left(-\frac{4}{15}\right)$

12. $-2\frac{3}{8} \times \frac{8}{5}$

13. $-9.4 \times (-4.7)$

14. $-100(-0.6)(0.01)$

15. PARASAILING A parasail is at 200.6 feet above the water. After 5 minutes, the parasail is at 120.8 feet above the water. What is the change in height of the parasail? *(Section 12.3)*

16. TEMPERATURE Use the thermometer shown. How much did the temperature drop from 5:00 P.M. to 10:00 P.M.? *(Section 12.3)*

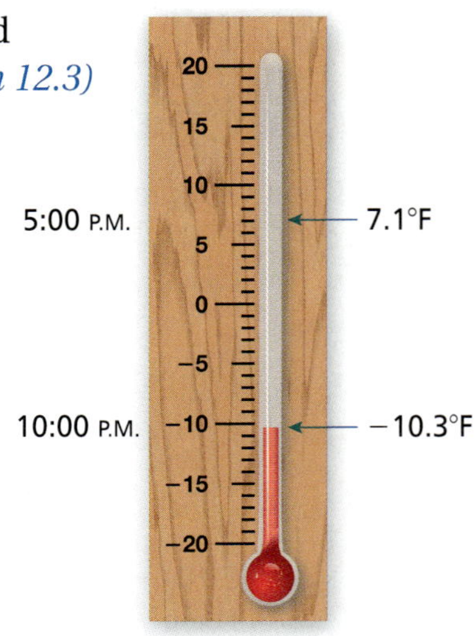

17. LATE FEES You were overcharged \$4.52 on your cell phone bill 3 months in a row. The cell phone company says that it will add −\$4.52 to your next bill for each month you were overcharged. On the next bill, you see an adjustment of −13.28. Is this amount correct? Explain. *(Section 12.4)*

18. CASHEWS How many $1\frac{1}{4}$-pound packages can you make with $7\frac{1}{2}$ pounds of cashews? *(Section 12.4)*

Alternative Assessment Options

Math Chat
Structured Interview
Student Reflective Focus Question
Writing Prompt

Student Reflective Focus Question

Ask students to summarize the rules for adding, subtracting, multiplying, and dividing rational numbers. Be sure that they include examples. Select students at random to present their summaries to the class.

Study Help Sample Answers

Remind students to complete Graphic Organizers for the rest of the chapter.

2.

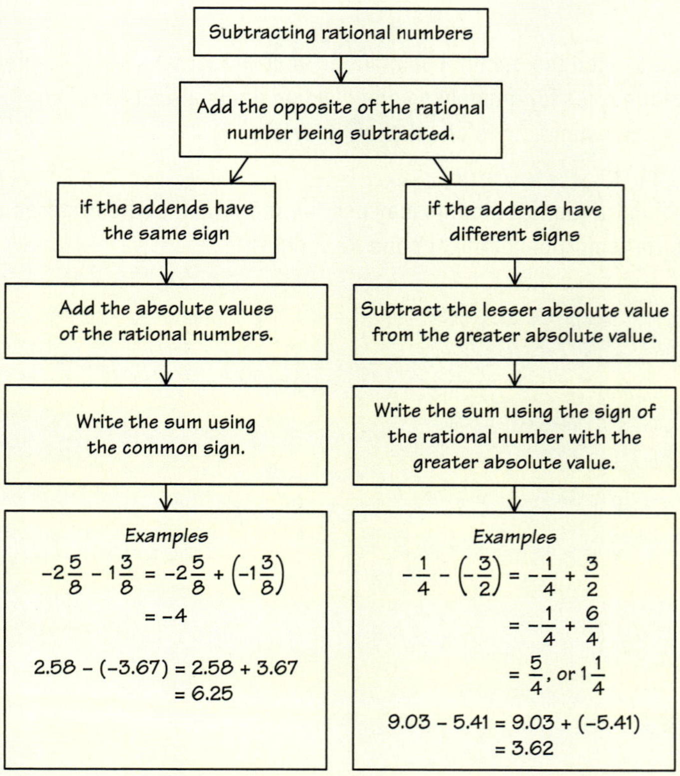

3–4. Available at *BigIdeasMath.com*.

Reteaching and Enrichment Strategies

If students need help. . .	If students got it. . .
Resources by Chapter • Practice A and Practice B • Puzzle Time Lesson Tutorials *BigIdeasMath.com*	Resources by Chapter • Enrichment and Extension • Technology Connection Game Closet at *BigIdeasMath.com* Start the Chapter Review

Answers

1. $-\frac{4}{7}$
2. $1\frac{59}{63}$
3. -3.8
4. 14.887
5. $1\frac{5}{12}$
6. 6.2
7. $-\frac{4}{5}$
8. $5\frac{4}{9}$
9. -4
10. -7.65
11. $-\frac{1}{6}$
12. $-3\frac{4}{5}$
13. 44.18
14. 0.6
15. -79.8 ft
16. 17.4°F
17. No, because $3 \times (-4.52) = -13.56$.
18. 6 packages

Technology for the Teacher

Online Assessment
Assessment Book
ExamView® Assessment Suite

For the Teacher
Additional Review Options

- *BigIdeasMath.com*
- Online Assessment
- Game Closet at *BigIdeasMath.com*
- Vocabulary Help
- Resources by Chapter

Answers

1. $-0.5\overline{3}$
2. 0.625
3. $-2.1\overline{6}$
4. 1.4375
5. $-\frac{3}{5}$
6. $-\frac{7}{20}$
7. $-5\frac{4}{5}$
8. $24\frac{23}{100}$

Review of Common Errors

Exercises 1–4

- Students may forget to carry the negative sign through the division operation.

Exercises 5–8

- Students may use the wrong numerator.

Exercises 9–14

- When adding and subtracting decimals, students may forget to line up the decimal points.

Exercises 9–15

- Students may forget to find a common denominator. Remind students that adding and subtracting fractions always requires a common denominator.

Exercise 16–25

- Students may place the decimal point incorrectly in their answers. Remind students of the rules for multiplying and dividing decimals. Also, remind students to use estimation to check their answers.

Exercises 16–26

- When dividing fractions, students may use the reciprocal of the first fraction instead of the reciprocal of the second fraction.

12 Chapter Review

Check It Out
Vocabulary Help
BigIdeasMath.com

Review Key Vocabulary

rational number, *p. 520*
terminating decimal, *p. 520*
repeating decimal, *p. 520*

Review Examples and Exercises

12.1 Rational Numbers *(pp. 518–523)*

a. Write $4\frac{3}{5}$ as a decimal.

Notice that $4\frac{3}{5} = \frac{23}{5}$.

Divide 23 by 5.

$$\begin{array}{r} 4.6 \\ 5\overline{)23.0} \\ -\,20 \\ \hline 3\,0 \\ -\,3\,0 \\ \hline 0 \end{array}$$

The remainder is 0. So, it is a terminating decimal.

So, $4\frac{3}{5} = 4.6$.

b. Write −0.14 as a fraction in simplest form.

$$-0.14 = -\frac{14}{100}$$

Write the digits after the decimal point in the numerator.

The last digit is in the hundredths place. So, use 100 in the denominator.

$$= -\frac{7}{50}$$

Simplify.

Exercises

Write the rational number as a decimal.

1. $-\frac{8}{15}$
2. $\frac{5}{8}$
3. $-\frac{13}{6}$
4. $1\frac{7}{16}$

Write the decimal as a fraction or a mixed number in simplest form.

5. −0.6
6. −0.35
7. −5.8
8. 24.23

12.2 Adding Rational Numbers (pp. 524–529)

Find $-\frac{7}{2} + \frac{5}{4}$.

$$-\frac{7}{2} + \frac{5}{4} = \frac{-14}{4} + \frac{5}{4}$$ Rewrite using the LCD (least common denominator).

$$= \frac{-14 + 5}{4}$$ Write the sum of the numerators over the common denominator.

$$= \frac{-9}{4}$$ Add.

$$= -2\frac{1}{4}$$ Write the improper fraction as a mixed number.

The sum is $-2\frac{1}{4}$.

Exercises

Add. Write fractions in simplest form.

9. $\frac{9}{10} + \left(-\frac{4}{5}\right)$ **10.** $-4\frac{5}{9} + \frac{8}{9}$ **11.** $-1.6 + (-2.4)$

12.3 Subtracting Rational Numbers (pp. 532–537)

Find $-4\frac{2}{5} - \left(-\frac{3}{5}\right)$.

$$-4\frac{2}{5} - \left(-\frac{3}{5}\right) = -4\frac{2}{5} + \frac{3}{5}$$ Add the opposite of $-\frac{3}{5}$.

$$= -\frac{22}{5} + \frac{3}{5}$$ Write the mixed number as an improper fraction.

$$= \frac{-22 + 3}{5}$$ Write the sum of the numerators over the common denominator.

$$= \frac{-19}{5}, \text{ or } -3\frac{4}{5}$$ Simplify.

The difference is $-3\frac{4}{5}$.

Exercises

Subtract. Write fractions in simplest form.

12. $-\frac{5}{12} - \frac{3}{10}$ **13.** $3\frac{3}{4} - \frac{7}{8}$ **14.** $3.8 - (-7.45)$

15. **TURTLE** A turtle is $20\frac{5}{6}$ inches below the surface of a pond. It dives to a depth of $32\frac{1}{4}$ inches. What is the change in the turtle's position?

Review Game

Rational Numbers

Materials

- questions from the chapter's homework, quizzes, examples, or tests
- 5 index cards, each with a letter in the word HORSE written on it, for each group

Directions

Divide the class into groups. Ask a group one of the questions. If the group is correct, the game continues to the next group. If they answer incorrectly, they receive an H. Each wrong answer in a group will result in that group receiving the next letter in the word HORSE. When a group has all 5 letters, they are out of the game. Choose questions with a wide range of difficulty to control how long the game takes.

Who Wins?

The last group with 4 or fewer letters wins.

For the Student Additional Practice

- Lesson Tutorials
- Multi-Language Glossary
- Self-Grading Progress Check
- *BigIdeasMath.com*
  Dynamic Student Edition
  Student Resources

Answers

9. $\frac{1}{10}$

10. $-3\frac{2}{3}$

11. -4

12. $-\frac{43}{60}$

13. $2\frac{7}{8}$

14. 11.25

15. $-11\frac{5}{12}$ inches

16. $-\frac{3}{4}$

17. $-1\frac{3}{11}$

18. -2

19. 6.16

20. $\frac{28}{81}$

21. $-\frac{16}{45}$

22. 57.23

23. -23.67

24. 5

25. 16

26. -75 ft

My Thoughts on the Chapter

Teacher Tip

Not allowed to write in your teaching edition? Use sticky notes to record your thoughts.

What worked. . .

What did not work. . .

What I would do differently. . .

12.4 Multiplying and Dividing Rational Numbers (pp. 538–543)

a. Find $-4\frac{1}{6} \div 1\frac{1}{3}$.

$$-4\frac{1}{6} \div 1\frac{1}{3} = -\frac{25}{6} \div \frac{4}{3}$$ Write mixed numbers as improper fractions.

$$= \frac{-25}{6} \cdot \frac{3}{4}$$ Multiply by the reciprocal of $\frac{4}{3}$.

$$= \frac{-25 \cdot 3}{6 \cdot 4}$$ Multiply the numerators and the denominators.

$$= \frac{-25}{8}, \text{ or } -3\frac{1}{8}$$ Simplify.

The quotient is $-3\frac{1}{8}$.

b. Find $-1.6 \cdot 2.4$.

$$\begin{array}{r} -1.6 \\ \times\ 2.4 \\ \hline 64 \\ 320 \\ \hline -3.84 \end{array}$$

The decimals have different signs.

The product is negative.

The product is -3.84.

Exercises

Divide. Write fractions in simplest form.

16. $\frac{9}{10} \div \left(-\frac{6}{5}\right)$ **17.** $-\frac{4}{11} \div \frac{2}{7}$ **18.** $6.4 \div (-3.2)$ **19.** $-15.4 \div (-2.5)$

Multiply. Write fractions in simplest form.

20. $-\frac{4}{9}\left(-\frac{7}{9}\right)$ **21.** $\frac{8}{15}\left(-\frac{2}{3}\right)$ **22.** $-5.9(-9.7)$

23. $4.5(-5.26)$ **24.** $-\frac{2}{3} \cdot \left(2\frac{1}{2}\right) \cdot (-3)$ **25.** $-1.6 \cdot (0.5) \cdot (-20)$

26. SUNKEN SHIP The elevation of a sunken ship is -120 feet. Your elevation is $\frac{5}{8}$ of the ship's elevation. What is your elevation?

12 Chapter Test

Check It Out
Test Practice
BigIdeasMath.com

Write the rational number as a decimal.

1. $\frac{7}{40}$ **2.** $-\frac{1}{9}$ **3.** $-\frac{21}{16}$ **4.** $\frac{36}{5}$

Write the decimal as a fraction or a mixed number in simplest form.

5. -0.122 **6.** 0.33 **7.** -4.45 **8.** -7.09

Add or subtract. Write fractions in simplest form.

9. $-\frac{4}{9}+\left(-\frac{23}{18}\right)$ **10.** $\frac{17}{12}-\left(-\frac{1}{8}\right)$ **11.** $9.2+(-2.8)$ **12.** $2.86-12.1$

Multiply or divide. Write fractions in simplest form.

13. $3\frac{9}{10}\times\left(-\frac{8}{3}\right)$ **14.** $-1\frac{5}{6}\div 4\frac{1}{6}$

15. $-4.4\times(-6.02)$ **16.** $-5\div 1.5$

17. $-\frac{3}{5}\cdot\left(2\frac{2}{7}\right)\cdot\left(-3\frac{3}{4}\right)$ **18.** $-6\cdot(-0.05)\cdot(-0.4)$

19. ALMONDS How many 2.25-pound containers can you make with 24.75 pounds of almonds?

20. FISH The elevation of a fish is -27 feet.

a. The fish decreases its elevation by 32 feet, and then increases its elevation by 14 feet. What is its new elevation?

b. Your elevation is $\frac{2}{5}$ of the fish's new elevation. What is your elevation?

21. RAINFALL The table shows the rainfall (in inches) for three months compared to the yearly average. Is the total rainfall for the three-month period greater than or less than the yearly average? Explain.

November	December	January
-0.86	2.56	-1.24

22. BANK ACCOUNTS Bank Account A has \$750.92, and Bank Account B has \$675.44. Account A changes by –\$216.38, and Account B changes by $-$\$168.49. Which account has the greater balance? Explain.

Test Item References

Chapter Test Questions	Section to Review
1–8	12.1
9, 11, 21, 22	12.2
10, 12, 20(a)	12.3
13–19, 20(b)	12.4

Test-Taking Strategies

Remind students to quickly look over the entire test before they start so that they can budget their time. On tests, it is really important for students to **Stop** and **Think.** When students hurry on a test dealing with signed numbers, they often make "sign" errors. Sometimes it helps to represent each problem with a number line to ensure that they are thinking through the process.

Common Errors

- **Exercises 1–4** Students may forget to carry the negative sign through the division operation. Tell them to create a space for the final answer and to write the sign of the number in the space at the beginning.
- **Exercises 9 and 10** Students may forget to find a common denominator. Remind students that adding and subtracting fractions always requires a common denominator.
- **Exercise 14** Students may use the reciprocal of the first fraction instead of the second, or they might forget to write a mixed number as an improper fraction before finding the reciprocal. Review multiplying and dividing fractions and the definition of reciprocal.
- **Exercises 15 and 16** Students may place the decimal point incorrectly in their answers. Remind students of the rules for multiplying and dividing decimals. Also, remind students to use estimation to check their answers.

Reteaching and Enrichment Strategies

If students need help. . .	If students got it. . .
Resources by Chapter • Practice A and Practice B • Puzzle Time Record and Practice Journal Practice Differentiating the Lesson Lesson Tutorials *BigIdeasMath.com* Skills Review Handbook	Resources by Chapter • Enrichment and Extension • Technology Connection Game Closet at *BigIdeasMath.com* Start Cumulative Assessment

Answers

1. 0.175
2. $-0.\overline{1}$
3. -1.3125
4. 7.2
5. $-\frac{61}{500}$
6. $\frac{33}{100}$
7. $-4\frac{9}{20}$
8. $-7\frac{9}{100}$
9. $-1\frac{13}{18}$
10. $1\frac{13}{24}$
11. 6.4
12. -9.24
13. $-10\frac{2}{5}$
14. $-\frac{11}{25}$
15. 26.488
16. $-3.\overline{3}$
17. $5\frac{1}{7}$
18. -0.12
19. 11 containers
20. **a.** -45 feet
    b. -18 feet
21. greater than; The sum of the three months is 0.46.
22. Bank Account A; Bank Account A has $534.54 while Bank Account B only has $506.95.

Technology for the Teacher

Online Assessment
Assessment Book
ExamView® Assessment Suite

Test-Taking Strategies
Available at *BigIdeasMath.com*

After Answering Easy Questions, Relax
Answer Easy Questions First
Estimate the Answer
Read All Choices before Answering
Read Question before Answering
Solve Directly or Eliminate Choices
Solve Problem before Looking at Choices
Use Intelligent Guessing
Work Backwards

About this Strategy
When taking a multiple choice test, be sure to read each question carefully and thoroughly. After reading the question, estimate the answer before trying to solve.

Answers
1. A
2. H
3. −18
4. C
5. I

Technology for the Teacher

Performance Tasks
Online Assessment
Assessment Book
ExamView® Assessment Suite

Item Analysis

1. **A.** Correct answer

   B. The student correctly finds José's height at 5 years old, which was 41 inches, but then reverses the relationship between José and Sean.

   C. When multiplying the rate of growth by the number of elapsed years, the student multiplies only the whole number parts to get $16\frac{3}{4}$.

   D. When multiplying the rate of growth by the number of elapsed years, the student multiplies only the whole number parts to get $16\frac{3}{4}$. After using this to find José's height at 5 years old, the student also reverses the relationship between José and Sean.

2. **F.** The student does not perform the operation correctly.

   G. The student does not perform the operation correctly.

   H. Correct answer

   I. The student does not perform the operation correctly.

3. **Gridded Response:** Correct answer: -18

   Common Error: The student thinks that each number is 2 times the previous number, rather than -2 times the previous number, and gets an answer of 18.

4. **A.** The student thinks that the absolute value of each individual number is negative and finds the sum of -2 and -2.5.

   B. The student correctly simplifies the expression inside the absolute value bars, but then thinks that the absolute value means to take the opposite.

   C. Correct answer

   D. The student takes the absolute value of each individual number and finds the sum of 2 and 2.5.

5. **F.** The student correctly subtracts $\frac{3}{8}$ from $-\frac{7}{4}$, but forgets to find the absolute value.

   G. The student incorrectly finds the sum of the two numbers.

   H. The student incorrectly finds the sum of the two numbers and then finds the absolute value.

   I. Correct answer

12 Cumulative Assessment

Test-Taking Strategy
Estimate the Answer

"Using estimation you can see that there are about 10 tabbies. So about 30 are not tabbies."

1. When José and Sean were each 5 years old, José was $1\frac{1}{2}$ inches taller than Sean. José grew at an average rate of $2\frac{3}{4}$ inches per year from the time that he was 5 years old until the time he was 13 years old. José was 63 inches tall when he was 13 years old. How tall was Sean when he was 5 years old?

A. $39\frac{1}{2}$ in.

B. $42\frac{1}{2}$ in.

C. $44\frac{3}{4}$ in.

D. $47\frac{3}{4}$ in.

2. Which expression represents a positive integer?

F. -6^2

G. $(-3)^3$

H. $(-5)^2$

I. -2^3

3. What is the missing number in the sequence below?

$$\frac{9}{16},\ -\frac{9}{8},\ \frac{9}{4},\ -\frac{9}{2},\ 9,\ \_\_\_\_$$

4. What is the value of the expression below?

$$|-2-(-2.5)|$$

A. -4.5

B. -0.5

C. 0.5

D. 4.5

5. What is the distance between the two numbers on the number line?

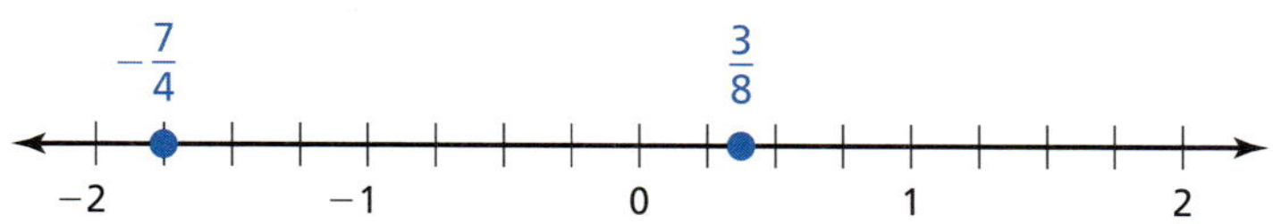

F. $-2\frac{1}{8}$

G. $-1\frac{3}{8}$

H. $1\frac{3}{8}$

I. $2\frac{1}{8}$

6. Sandra was evaluating an expression in the box below.

$$-4\frac{3}{4} \div 2\frac{1}{5} = -\frac{19}{4} \div \frac{11}{5}$$
$$= \frac{-4}{19} \cdot \frac{5}{11}$$
$$= \frac{-4 \cdot 5}{19 \cdot 11}$$
$$= \frac{-20}{209}$$

What should Sandra do to correct the error that she made?

A. Rewrite $-\frac{19}{4}$ as $-\frac{4}{19}$ and multiply by $\frac{11}{5}$.

B. Rewrite $\frac{11}{5}$ as $\frac{5}{11}$ and multiply by $-\frac{19}{4}$.

C. Rewrite $\frac{11}{5}$ as $-\frac{5}{11}$ and multiply by $-\frac{19}{4}$.

D. Rewrite $-4\frac{3}{4}$ as $-\frac{13}{4}$ and multiply by $\frac{5}{11}$.

7. What is the value of the expression below when $q = -2$, $r = -12$, and $s = 8$?

$$\frac{-q^2 - r}{s}$$

F. -2

G. -1

H 1

I 2

8. You are stacking wooden blocks with the dimensions shown below. How many blocks do you need to stack to build a block tower that is $7\frac{1}{2}$ inches tall?

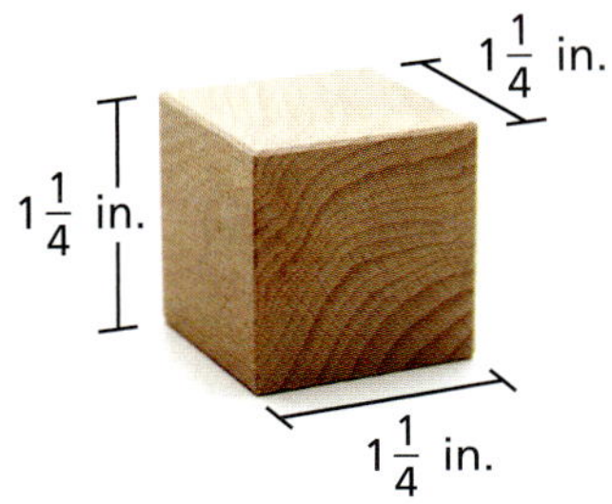

Item Analysis (continued)

6. **A.** The student thinks that you divide two fractions by multiplying the reciprocal of the dividend by the divisor.

   B. Correct answer

   C. The student thinks that you divide two fractions by multiplying the dividend by the opposite of the reciprocal of the divisor.

   D. The student incorrectly rewrites the negative mixed number.

7. **F.** The student subtracts 12 in the numerator.

   G. The student subtracts 12 from 4 in the numerator.

   H. Correct answer

   I. The student thinks that the opposite of the square of -2 is 4.

8. **Gridded Response:** Correct answer: 6

   Common Error: The student divides the integer parts and the fractional parts of the mixed numbers and gets an answer of $7 + 2 = 9$.

Answers

6. B
7. H
8. 6 blocks

Answers

9. B

10. G

11. *Part A* 1.7

Part B 4

Part C $0.9\overline{6}$

Part D 8.7

12. C

Item Analysis (continued)

9. **A.** The student finds $\frac{1}{2}$ of the sum of the base and the height.

B. Correct answer

C. The student finds the sum of the base and the height.

D. The student finds the product of the base and the height.

10. **F.** The student thinks that the cube of -2 is 8.

G. Correct answer

H. The student finds the square of -4 and then thinks that the cube of -2 is 8.

I. The student finds the square of -4.

11. **4 points** The student demonstrates a thorough understanding of interpreting rational numbers on a number line and a thorough conceptual understanding of the four operations using rational numbers. In Part A, the student correctly recognizes that the two greatest values, T and U, have the greatest sum, which is approximately 1.7. In Part B, the student correctly recognizes that the two values that are the farthest apart, U and R, have the greatest difference, which is approximately 4. In Part C, the student correctly recognizes that the two values that have the same sign and also the greatest magnitude, R and S, have the greatest product, which is approximately $0.9\overline{6}$. In Part D, the student correctly recognizes that the two values that have the same sign and also the greatest ratio, R and S, have the greatest quotient, which is approximately 8.7. The student provides clear and complete explanations of the reasoning used.

3 points The student demonstrates an understanding of interpreting rational numbers on a number line and a good conceptual understanding of the four operations using rational numbers, but the student's work and explanations demonstrate an essential but less than thorough understanding.

2 points The student demonstrates an understanding of interpreting rational numbers on a number line and a partial conceptual understanding of the four operations using rational numbers. The student's work and explanations demonstrate a lack of essential understanding.

1 point The student demonstrates a partial understanding of interpreting rational numbers on a number line and a limited conceptual understanding of the four operations using rational numbers. The student's response is incomplete and exhibits many flaws.

0 points The student provided no response, a completely incorrect or incomprehensible response, or a response that demonstrates insufficient understanding of interpreting rational numbers on a number line and an insufficient conceptual understanding of the four operations using rational numbers.

12. **A.** The student thinks that $\frac{-0.4}{-1} = -0.4$ and that $-0.4 + 0.8 = -1.2$.

B. The student thinks that $\frac{-0.4}{-0.2} = -2$.

C. Correct answer

D. The student thinks that $-0.4 + 0.8 = -1.2$.

9. What is the area of a triangle with a base length of $2\frac{1}{2}$ inches and a height of 2 inches?

A. $2\frac{1}{4}$ in.^{2}

B. $2\frac{1}{2}$ in.^{2}

C. $4\frac{1}{2}$ in.^{2}

D. 5 in.^{2}

10. What is the value of the expression below?

$$\frac{-4^2 - (-2)^3}{4}$$

F. -6

G. -2

H. 2

I. 6

11. Four points are graphed on the number line below.

Think Solve Explain

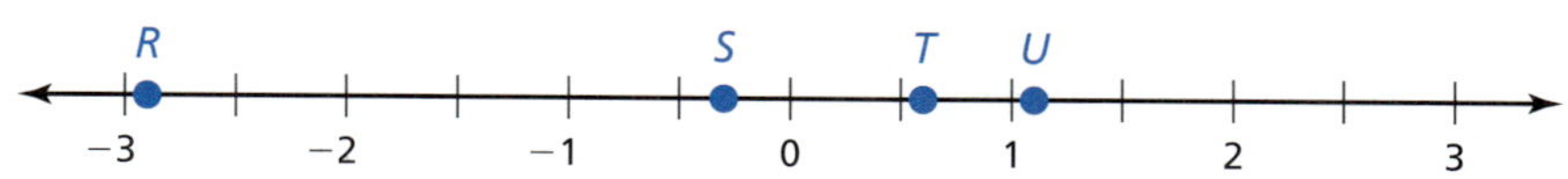

Part A Choose the two points whose values have the greatest sum. Approximate this sum. Explain your reasoning.

Part B Choose the two points whose values have the greatest difference. Approximate this difference. Explain your reasoning.

Part C Choose the two points whose values have the greatest product. Approximate this product. Explain your reasoning.

Part D Choose the two points whose values have the greatest quotient. Approximate this quotient. Explain your reasoning.

12. What number belongs in the box to make the equation true?

$$\frac{-0.4}{\square} + 0.8 = -1.2$$

A. -1

B. -0.2

C. 0.2

D. 1

13 Expressions and Equations

13.1 Algebraic Expressions

13.2 Adding and Subtracting Linear Expressions

13.3 Solving Equations Using Addition or Subtraction

13.4 Solving Equations Using Multiplication or Division

13.5 Solving Two-Step Equations

What Your Students Have Learned

- Use parentheses, brackets, or braces in numerical expressions, and evaluate expressions with these symbols.
- Write and interpret numerical expressions without evaluating them.
- Use and interpret simple equations.
- Write and evaluate numerical expressions involving whole-number exponents.
- Read, write, and evaluate algebraic expressions.
- Apply the properties of operations to generate equivalent expressions.
- Factor out the greatest common factor (GCF) in algebraic and numerical expressions.
- Identify equivalent expressions.
- Determine if a value is a solution of an equation.
- Solve one-step equations.

What Your Students Will Learn

- Add, subtract, factor, and expand linear expressions with rational coefficients.
- Understand that rewriting expressions in different forms can show how the quantities are related.
- Write, graph, and solve one-step equations (includes negative numbers).
- Solve two-step equations.
- Compare algebraic solutions to arithmetic solutions.

Pacing Guide for Chapter 13

Chapter Opener Advanced	1 Day
Section 1 Advanced	1 Day
Section 2 Advanced	2 Days
Study Help/Quiz Advanced	1 Day
Section 3 Advanced	1 Day
Section 4 Advanced	1 Day
Section 5 Advanced	1 Day
Chapter Review/ Chapter Tests Advanced	2 Days
Total Chapter 13 Advanced	10 Days
Year-to-Date Advanced	123 Days

Technology for the Teacher

BigIdeasMath.com
Chapter at a Glance
Complete Materials List
Parent Letters: English and Spanish

What Your Students Have Learned

- Evaluate algebraic expressions by using substitution and the order of operations.
- Write algebraic expressions using key terms and phrases.

Additional Topics for Review

- Powers and Exponents
- Order of Operations
- Greatest Common Factor (GCF)
- The Distributive Property

Try It Yourself

1. $-1\frac{1}{2}$
2. -12
3. 2
4. -14
5. $3q + 5$
6. $n - 9$
7. $6p$
8. $8 \div h$
9. $3t + 4$
10. $7c - 2$

Record and Practice Journal Fair Game Review

1. 6
2. 12
3. 20
4. $\frac{19}{4}$ or $4\frac{3}{4}$
5. 12
6. -1
7. 24 ft^2
8. $8 + y$
9. $p - 6$
10. $7m$
11. $11c - 8$
12. $r - \frac{r}{2}$
13. $9(z + 4)$

Math Background Notes

Vocabulary Review

- Evaluate
- Expression
- Substitute
- Order of Operations

Evaluating Expressions

- Students should know how to evaluate expressions. Because this skill is fairly new to them, you may want to allot slightly more time for practice.
- Remind students that evaluate means to find a value.
- To evaluate an expression, students will need to substitute values into the expression.
- **Teaching Tip:** English Language Learners may find the vocabulary and writing involved in these opening exercises challenging. To ease the transition, help them make analogies. In the same way a sugar substitute replaces sugar, substituting numbers into an expression will replace the variables until you are able to evaluate.
- After they substitute, remind students that they must follow the correct order of operations as they simplify. You may wish to review order of operations prior to completing the examples.

Writing Algebraic Expressions

- Students should know how to write expressions.
- Review key words with students. For example, words such as sum and product alert students to use addition and multiplication.
- **Common Error:** Many students will read Example 3, part (b) and incorrectly write the expression $8 - 3x$. Remind students that they cannot subtract 8 from a number unless they have a number to start with. Always write the number first and then add or subtract as indicated.

Reteaching and Enrichment Strategies

If students need help. . .	If students got it. . .
Record and Practice Journal • Fair Game Review Skills Review Handbook Lesson Tutorials	Game Closet at *BigIdeasMath.com* Start the next section

What You Learned Before

"Hey, Descartes ... *True or False:* The expressions are equivalent."

Evaluating Expressions

Example 1 Evaluate $6x + 2y$ when $x = -3$ and $y = 5$.

$6x + 2y = 6(-3) + 2(5)$ — Substitute -3 for x and 5 for y.

$= -18 + 10$ — Using order of operations, multiply 6 and -3, and 2 and 5.

$= -8$ — Add -18 and 10.

Example 2 Evaluate $6x^2 - 3(y + 2) + 8$ when $x = -2$ and $y = 4$.

$6x^2 - 3(y + 2) + 8 = 6(-2)^2 - 3(4 + 2) + 8$ — Substitute -2 for x and 4 for y.

$= 6(-2)^2 - 3(6) + 8$ — Using order of operations, evaluate within the parentheses.

$= 6(4) - 3(6) + 8$ — Using order of operations, evaluate the exponent.

$= 24 - 18 + 8$ — Using order of operations, multiply 6 and 4, and 3 and 6.

$= 14$ — Subtract 18 from 24. Add the result to 8.

Try It Yourself

Evaluate the expression when $x = -\frac{1}{4}$ and $y = 3$.

1. $2xy$

2. $12x - 3y$

3. $-4x - y + 4$

4. $8x - y^2 - 3$

Writing Algebraic Expressions

Example 3 Write the phrase as an algebraic expression.

a. the sum of twice a number m and four

$2m + 4$

b. eight less than three times a number x

$3x - 8$

Try It Yourself

Write the phrase as an algebraic expression.

5. five more than three times a number q

6. nine less than a number n

7. the product of a number p and six

8. the quotient of eight and a number h

9. four more than three times a number t

10. two less than seven times a number c

13.1 Algebraic Expressions

Essential Question How can you simplify an algebraic expression?

1 ACTIVITY: Simplifying Algebraic Expressions

Work with a partner.

a. Evaluate each algebraic expression when $x = 0$ and when $x = 1$. Use the results to match each expression in the left table with its equivalent expression in the right table.

	Expression	Value When $x = 0$	Value When $x = 1$
A.	$3x + 2 - x + 4$		
B.	$5(x - 3) + 2$		
C.	$x + 3 - (2x + 1)$		
D.	$-4x + 2 - x + 3x$		
E.	$-(1 - x) + 3$		
F.	$2x + x - 3x + 4$		
G.	$4 - 3 + 2(x - 1)$		
H.	$2(1 - x + 4)$		
I.	$5 - (4 - x + 2x)$		
J.	$5x - (2x + 4 - x)$		

	Expression	Value When $x = 0$	Value When $x = 1$
a.	4		
b.	$-x + 1$		
c.	$4x - 4$		
d.	$2x + 6$		
e.	$5x - 13$		
f.	$-2x + 10$		
g.	$x + 2$		
h.	$2x - 1$		
i.	$-2x + 2$		
j.	$-x + 2$		

b. Compare each expression in the left table with its equivalent expression in the right table. In general, how do you think you obtain the equivalent expression in the right column?

Algebraic Expressions
In this lesson, you will
- apply properties of operations to simplify algebraic expressions.
- solve real-life problems.

Laurie's Notes

Introduction

Applying Mathematical Practices

- **Look for and Make Use of Structure:** Mathematically proficient students are able to see an algebraic expression being composed of several terms. In this lesson, students will simplify algebraic expressions by combining like terms.

Motivate

- **Target Math Game Time!** Write the following problem on the board.

$$\square + \square \times \square - \square = \square$$

- **Directions:** Tell students that you are going to randomly generate 4 numbers from -8 to 8, and *as you generate the numbers*, they are to place the numbers in the four boxes to the left of the equal sign. After the fourth number is generated, students should evaluate the expression and write their answer in the red box. The goal is to get as close to the target number of 24 as possible without going over.
- To generate the random numbers, you can use the random number generator on a graphing calculator. Another option is to write the numbers on slips of paper, put the slips of paper into a container, and pull one number at a time. Do not use the same number twice.
- **Extensions:** Play this more than once by changing the target number, changing the range of numbers used, or changing the original expression.
- The goal is to evaluate expressions using the order of operations.

Discuss

- Ask students to recall that an *algebraic expression* is an expression that may contain numbers, operations, and one or more symbols.
- You may wish to do a few examples where you change the sign of the coefficient, such as $4(x + 2)$, $-4(x + 2)$, and $-(x + 2)$.

Activity Notes

Activity 1

- It will be important for students to use the order of operations.
- Encourage students to write the expression, then rewrite it with the value of the variable substituted.
- ? "How did you decide which expressions matched?" The expressions have the same values when $x = 0$ and when $x = 1$.
- ? Discuss answers to parts (a) and (b) and then ask, "What do you think it means to simplify an expression?" Students should say that the expression has fewer terms. They should also say that variable terms have been combined and numeric terms have been combined.
- ? "Have you heard the word *simplify* in mathematics before? Explain." You simplify fractions when you divide out common factors.

What Your Students Will Learn

- Identify terms and like terms of algebraic expressions.
- Apply properties of operations to simplify algebraic expressions.

Previous Learning

Students should know how to evaluate algebraic expressions.

Technology for the Teacher

Dynamic Classroom

Lesson Plans
Complete Materials List

13.1 Record and Practice Journal

Essential Question How can you simplify an algebraic expression?

1 ACTIVITY: Simplifying Algebraic Expressions

Work with a partner.

a. Evaluate each algebraic expression when $x = 0$ and when $x = 1$. Use the results to match each expression in the left table with its equivalent expression in the right table.

		Expression	Value When $x = 0$	Value When $x = 1$
d	A.	$3x + 2 - x + 4$	6	8
e	B.	$5(x - 3) + 2$	−13	−8
j	C.	$x + 3 - (2x + 1)$	2	1
i	D.	$-4x + 2 - x + 3x$	2	0
g	E.	$-(1 - x) + 3$	2	3
a	F.	$2x + x - 3x + 4$	4	4
h	G.	$4 - 3 + 2(x - 1)$	−1	1
f	H.	$2(1 - x + 4)$	10	8
b	I.	$5 - (4 - x + 2x)$	1	0
c	J.	$5x - (2x + 4 - x)$	−4	0

	Expression	Value When $x = 0$	Value When $x = 1$
a.	4	4	4
b.	$-x + 1$	1	0
c.	$4x - 4$	−4	0
d.	$2x + 6$	6	8
e.	$5x - 13$	−13	−8
f.	$-2x + 10$	10	8
g.	$x + 2$	2	3
h.	$2x - 1$	−1	1
i.	$-2x + 2$	2	0
j.	$-x + 2$	2	1

b. Compare each expression in the left table with its equivalent expression in the right table. In general, how do you think you obtain the equivalent expression in the right table?
Combine like terms

Differentiated Instruction

Inclusion

When simplifying expressions with parentheses such as $3(x + 2) - 5$, students should use the Distributive Property first. This follows the order of operations.

13.1 Record and Practice Journal

2 ACTIVITY: Writing a Math Lesson

Work with a partner. Use your results from Activity 1 to write a lesson on simplifying an algebraic expression.

Check students' work.

Simplifying an Algebraic Expression

Key Idea Use the following steps to simplify an algebraic expression.

1.

2.

3.

Describe steps you can use to simplify an expression.

Examples ← Write 3 examples. Use expressions from Activity 1.

a.

b.

c.

Exercises ← Write 3 exercises. Use expressions different from the ones in Activity 1.

Simplify the expression.

1.

2.

3.

What Is Your Answer?

3. **IN YOUR OWN WORDS** How can you simplify an algebraic expression? Give an example that demonstrates your procedure.

Sample answer: You can simplify an algebraic expression by: (1) using the Distributive Property to eliminate parentheses, (2) use the Commutative and Associative Properties of Addition to get the variable terms together and the terms without variables together, (3) combine the variable terms and combine the terms without variables.

4. **REASONING** Why would you want to simplify an algebraic expression? Discuss several reasons.

Check students' work.

Laurie's Notes

Activity 2

- This activity will give you insight into how students think about simplifying algebraic expressions. The previous activity contained only like variable terms, so students will likely not consider expressions with unlike variable terms such as $3x + 2y - 4x + 5$.
- **Construct Viable Arguments and Critique the Reasoning of Others:** When partners have finished, have them exchange their lessons with another pair of students. Students should read the lessons they receive and write critiques to the authors of the lessons. They should provide feedback on what is clear in the lesson and what can be made more clear.
- Have students complete the exercises of the lesson they received.

What Is Your Answer?

- Students should be able to answer Question 3 independently.
- Discuss student answers for Question 4.

Closure

- **Exit Ticket:** Which of the following expressions would simplify to $3x + 4$? If the expression does not simplify to $3x + 4$, what does it simplify to?

A. $-6 + 3x + 2$ $3x - 4$

B. $3(x + 4)$ $3x + 12$

C. $2x + 8 + x - 2$ $3x + 6$

D. $-5x + 4 + 8x$ $3x + 4$

2 ACTIVITY: Writing a Math Lesson

Work with a partner. Use your results from Activity 1 to write a lesson on simplifying an algebraic expression.

Math Practice

Communicate Precisely

What can you do to make sure that you are communicating exactly what is needed in the Key Idea?

Simplifying an Algebraic Expression

Key Idea Use the following steps to simplify an algebraic expression.

1.
2.
3.

Describe steps you can use to simplify an expression.

Examples

a.

b.

c.

Write 3 examples. Use expressions from Activity 1.

Exercises

Simplify the expression.

1.
2.
3.

Write 3 exercises. Use expressions different from the ones in Activity 1.

What Is Your Answer?

3. **IN YOUR OWN WORDS** How can you simplify an algebraic expression? Give an example that demonstrates your procedure.

4. **REASONING** Why would you want to simplify an algebraic expression? Discuss several reasons.

Practice Use what you learned about simplifying algebraic expressions to complete Exercises 12–14 on page 558.

13.1 Lesson

Check It Out
Lesson Tutorials
BigIdeasMath.com

Key Vocabulary
like terms, *p. 556*
simplest form, *p. 556*

Parts of an algebraic expression are called *terms*. **Like terms** are terms that have the same variables raised to the same exponents. Constant terms are also like terms. To identify terms and like terms in an expression, first write the expression as a sum of its terms.

EXAMPLE 1 Identifying Terms and Like Terms

Identify the terms and like terms in each expression.

a. $9x - 2 + 7 - x$

Rewrite as a sum of terms.

$$9x + (-2) + 7 + (-x)$$

Terms: $9x$, -2, 7, $-x$

Like terms: $9x$ and $-x$, -2 and 7

b. $z^2 + 5z - 3z^2 + z$

Rewrite as a sum of terms.

$$z^2 + 5z + (-3z^2) + z$$

Terms: z^2, $5z$, $-3z^2$, z

Like terms: z^2 and $-3z^2$, $5z$ and z

An algebraic expression is in **simplest form** when it has no like terms and no parentheses. To *combine* like terms that have variables, use the Distributive Property to add or subtract the coefficients.

EXAMPLE 2 Simplifying an Algebraic Expression

Study Tip
To subtract a variable term, add the term with the opposite coefficient.

Simplify $\frac{3}{4}y + 12 - \frac{1}{2}y - 6$.

$$\frac{3}{4}y + 12 - \frac{1}{2}y - 6 = \frac{3}{4}y + 12 + \left(-\frac{1}{2}y\right) + (-6)$$ Rewrite as a sum.

$$= \frac{3}{4}y + \left(-\frac{1}{2}y\right) + 12 + (-6)$$ Commutative Property of Addition

$$= \left[\frac{3}{4} + \left(-\frac{1}{2}\right)\right]y + 12 + (-6)$$ Distributive Property

$$= \frac{1}{4}y + 6$$ Combine like terms.

On Your Own

Now You're Ready
Exercises 5–10 and 12–17

Identify the terms and like terms in the expression.

1. $y + 10 - \frac{3}{2}y$
2. $2r^2 + 7r - r^2 - 9$
3. $7 + 4p - 5 + p + 2q$

Simplify the expression.

4. $14 - 3z + 8 + z$
5. $2.5x + 4.3x - 5$
6. $\frac{3}{8}b - \frac{3}{4}b$

Multi-Language Glossary at BigIdeasMath.com

Laurie's Notes

Introduction

Connect

- **Yesterday:** Students explored simplifying algebraic expressions by evaluating two expressions for more than one value of the variable.
- **Today:** Students will simplify algebraic expressions by combining like terms.

Motivate

? Ask students if they have ever heard the phrase "you can't add apples and oranges."

- Some students may have heard this phrase before. Ask them what they think the phrase means. They might talk about needing a common denominator.

Words of Wisdom

- Students often have difficulty with simplifying algebraic expressions. They must be comfortable with integer operations and be able to apply the Commutative and Distributive Properties. For example: $5x + 7 - 3x$ can be rewritten as $5x + 7 + (-3x) = 5x + (-3x) + 7$.

Lesson Notes

Discuss

- **Attend to Precision:** Like terms are also referred to as *similar terms*. Be sure to note that in the definition of like terms, the variables are raised to the same exponents.

Example 1

- Terms are separated by addition. The expression $9x - 2 + 7 - x$ can be written as $9x + (-2) + 7 + (-x)$, so it has four terms. This form will help students simplify because they can see the sign associated with each term.
- **Common Error:** When identifying and writing the terms, make sure students include the sign of the term.
- **Attend to Precision:** Make sure students understand that the coefficient of $-x$ is -1. Similarly, the exponent of the variable in the terms $5z$ and z is 1.

Example 2

? "What do you call the number that is multiplied by the variable?" coefficient

- Discuss what it means to write an algebraic expression in simplest form.
- Ask students to identify the coefficient of each term. Identify the constant terms.
- Remind students about the Commutative and Distributive Properties.
- Have students show the step that uses the Distributive Property until they become proficient.

On Your Own

- Check that students have not forgotten to include the sign of the term.

Goal Today's lesson is to simplify algebraic expressions.

Technology for the Teacher

Dynamic Classroom

Lesson Tutorials
Lesson Plans
Answer Presentation Tool

Extra Example 1

Identify the terms and like terms in each expression.

a. $3y - 2 - 4y + 6$
Terms: $3y$, -2, $-4y$, 6
Like terms: $3y$ and $-4y$, -2 and 6

b. $w + 5w^2 + 2w^2 - 7w$
Terms: w, $5w^2$, $2w^2$, $-7w$
Like terms: w and $-7w$, $5w^2$ and $2w^2$

Extra Example 2

Simplify each expression.

a. $8u + 5u - 7u$ $6u$

b. $6d - 5 - 4d + 6$ $2d + 1$

On Your Own

1. Terms: y, 10, $-\frac{3}{2}y$
   Like terms: y and $-\frac{3}{2}y$
2. Terms: $2r^2$, $7r$, $-r^2$, -9
   Like terms: $2r^2$ and $-r^2$
3. Terms: 7, $4p$, -5, p, $2q$
   Like terms: 7 and -5, $4p$ and p
4. $-2z + 22$
5. $6.8x - 5$
6. $-\frac{3}{8}b$

English Language Learners

Vocabulary

English learners will benefit from understanding that a term is a number, a variable, or the product of a number and a variable. Like terms are terms that have identical variable parts.

3 and 16 are like terms because they contain no variable.

$4x$ and $7x$ are like terms because they have the same variable x.

$5a$ and $5b$ are not like terms because they have different variables.

Extra Example 3

Simplify $12g + 4 - 5g$. $7g + 4$

On Your Own

7. $3q - 1$
8. $5g - 8$
9. $-3x + 8$

Extra Example 4

Each person in Example 4 buys a ticket, a small drink, and a small popcorn. Write an expression in simplest form that represents the amount of money the group spends at the movies. $12.25x$

On Your Own

10. $14x$;

$$7.50x + 3.50x + 3x$$
$$= (7.50 + 3.50 + 3)x$$
$$= 14x$$

Laurie's Notes

Example 3

- Students have not had a lot of practice with a fractional factor in the Distributive Property.
- **Teaching Tip:** Use arrows to show the $-\frac{1}{2}$ being distributed over the $6n$ and the 4.

$$-\frac{1}{2}(6n + 4)$$

On Your Own

- Students should write the original expression, followed by each step in the simplifying process.
- **Common Error:** In Questions 7, 8, and 9, students often forget to distribute the constant over *both* of the terms inside the parentheses.
- **Neighbor Check:** Have students work independently and then have their neighbors check their work. Have students discuss any discrepancies.

Example 4

- Discuss the information provided in the side column. You might ask how these prices compare to the prices at a local movie theater.
- Work through the example with students. Students are writing an algebraic expression, so it is important to identify what the variable represents in the problem. Do not skip this step!
- When you finish the problem, ask students what the total cost would be for a group of 4 people. $\$14.25 \times 4 = \$57$

On Your Own

- Ask a student to share his or her answer.

Closure

- Simplify the following algebraic expressions.
  - **a.** $5x - 8 + 2x^2 + 7x$ $2x^2 + 12x - 8$
  - **b.** $4n + 6(n - 4)$ $10n - 24$

EXAMPLE 3 Simplifying an Algebraic Expression

Simplify $-\frac{1}{2}(6n + 4) + 2n$.

$$-\frac{1}{2}(6n + 4) + 2n = -\frac{1}{2}(6n) + \left(-\frac{1}{2}\right)(4) + 2n \quad \text{Distributive Property}$$

$$= -3n + (-2) + 2n \quad \text{Multiply.}$$

$$= -3n + 2n + (-2) \quad \text{Commutative Property of Addition}$$

$$= (-3 + 2)n + (-2) \quad \text{Distributive Property}$$

$$= -n - 2 \quad \text{Simplify.}$$

On Your Own

Now You're Ready
Exercises 18–20

Simplify the expression.

7. $3(q + 1) - 4$
8. $-2(g + 4) + 7g$
9. $7 - 4\left(\frac{3}{4}x - \frac{1}{4}\right)$

EXAMPLE 4 Real-Life Application

Each person in a group buys a ticket, a medium drink, and a large popcorn. Write an expression in simplest form that represents the amount of money the group spends at the movies. Interpret the expression.

Words Each ticket is $7.50, each medium drink is $2.75, and each large popcorn is $4.

Variable The same number of each item is purchased. So, x can represent the number of tickets, the number of medium drinks, and the number of large popcorns.

Expression $7.50x \quad + \quad 2.75x \quad + \quad 4x$

$$7.50x + 2.75x + 4x = (7.50 + 2.75 + 4)x \quad \text{Distributive Property}$$

$$= 14.25x \quad \text{Add coefficients.}$$

∴ The expression $14.25x$ indicates that the total cost per person is $14.25.

On Your Own

10. **WHAT IF?** Each person buys a ticket, a large drink, and a small popcorn. How does the expression change? Explain.

13.1 Exercises

Check It Out
Help with Homework
BigIdeasMath.com

Vocabulary and Concept Check

1. **WRITING** Explain how to identify the terms of $3y - 4 - 5y$.
2. **WRITING** Describe how to combine like terms in the expression $3n + 4n - 2$.
3. **VOCABULARY** Is the expression $3x + 2x - 4$ in simplest form? Explain.
4. **REASONING** Which algebraic expression is in simplest form? Explain.

$5x - 4 + 6y$	$4x + 8 - x$
$3(7 + y)$	$12n - n$

Practice and Problem Solving

Identify the terms and like terms in the expression.

1 5. $t + 8 + 3t$
6. $3z + 4 + 2 + 4z$
7. $2n - n - 4 + 7n$
8. $-x - 9x^2 + 12x^2 + 7$
9. $1.4y + 5 - 4.2 - 5y^2 + z$
10. $\frac{1}{2}s - 4 + \frac{3}{4}s + \frac{1}{8} - s^3$

11. **ERROR ANALYSIS** Describe and correct the error in identifying the like terms in the expression.

$3x - 5 + 2x^2 + 9x = 3x + 2x^2 + 9x - 5$

Like Terms: $3x$, $2x^2$, and $9x$

Simplify the expression.

2 12. $12g + 9g$
13. $11x + 9 - 7$
14. $8s - 11s + 6$
15. $4.2v - 5 - 6.5v$
16. $8 + 4a + 6.2 - 9a$
17. $\frac{2}{5}y - 4 + 7 - \frac{9}{10}y$

3 18. $4(b - 6) + 19$
19. $4p - 5(p + 6)$
20. $-\frac{2}{3}(12c - 9) + 14c$

21. **HIKING** On a hike, each hiker carries the items shown. Write an expression in simplest form that represents the weight carried by x hikers. Interpret the expression.

Assignment Guide and Homework Check

Level	Assignment	Homework Check
Advanced	1–4, 6–10 even, 11–14, 16–28 even, 30–32	8, 10, 18, 20, 26

Common Errors

- **Exercises 5–10** When identifying and writing terms, make sure students include the sign of the term. Students may find it helpful to write the original problem using addition. For example, $3n - 10 + 2n = 3n + (-10) + 2n$.
- **Exercises 5–10** Students may confuse like variables with like terms. Remind them that the same variables must be raised to the same exponents for terms to be like terms. The terms $3x$ and $7x^2$ are not like terms because one has an exponent of 1 and the other has an exponent of 2.
- **Exercises 13–20** The subtraction operation can confuse students. It is not obvious to them why it is okay to rewrite $6t - 24 + 3t$ as $6t + 3t - 24$. Tell students to write the original problem using addition, and then use the Commutative Property.

$$6t - 24 + 3t = 6t + (-24) + 3t = 6t + 3t + (-24)$$
$$= 6t + 3t - 24 = 9t - 24$$

- **Exercises 18–20** Students often forget to distribute the constant over *both* of the terms inside the parentheses. Remind them of the Distributive Property, $a(b + c) = ab + ac$.

13.1 Record and Practice Journal

Identify the terms and like terms in the expression.

1. $3x + 4 - 7x - 6$
   Terms: $3x, 4, -7x, -6$
   Like terms: $3x$ and $-7x$, 4 and -6

2. $-9 + 2.5y - 0.7y + 1 + 6.4y^2$
   Terms: $-9, 2.5y, -0.7y, 1, 6.4y^2$
   Like terms: $2.5y$ and $-0.7y$, -9 and 1

Simplify the expression.

3. $5a - 2a + 9$
   $3a + 9$

4. $m - \frac{1}{6} - 4m + \frac{5}{6}$
   $-3m + \frac{2}{3}$

5. $2.3w - 7 + 8.1 - 3w$
   $-0.7w + 1.1$

6. $7(d - 1) + 2$
   $7d - 5$

7. $13g + 2(4k - g)$
   $11g + 8k$

8. $20(p + 2) + 16(-3 - p)$
   $4p - 8$

9. Write an expression in simplest form that represents the cost for shampooing and cutting w women's hair and m men's hair.

	Women	Men
Cut	\$15	\$7
Shampoo	\$5	\$2

$20w + 9m$

Vocabulary and Concept Check

1. Terms of an expression are separated by addition. Rewrite the expression as $3y + (-4) + (-5y)$. The terms in the expression are $3y$, -4, and $-5y$.
2. Use the Distributive Property to add the coefficients of the like terms $3n$ and $4n$.
   $$3n + 4n - 2 = (3 + 4)n - 2$$
   $$= 7n - 2$$
3. no; The like terms $3x$ and $2x$ should be combined.
   $$3x + 2x - 4 = (3 + 2)x - 4$$
   $$= 5x - 4$$
4. $5x - 4 + 6y$; There are no like terms or parentheses in the expression.

Practice and Problem Solving

5. Terms: t, 8, $3t$;
   Like terms: t and $3t$
6. Terms: $3z$, 4, 2, $4z$;
   Like terms: $3z$ and $4z$, 4 and 2
7. Terms: $2n$, $-n$, -4, $7n$;
   Like terms: $2n$, $-n$, and $7n$
8. Terms: $-x$, $-9x^2$, $12x^2$, 7;
   Like terms: $-9x^2$ and $12x^2$
9. Terms: $1.4y$, 5, -4.2, $-5y^2$, z;
   Like terms: 5 and -4.2
10. Terms: $\frac{1}{2}s$, -4, $\frac{3}{4}s$, $\frac{1}{8}$, $-s^3$;
    Like terms: $\frac{1}{2}s$ and $\frac{3}{4}s$,
    -4 and $\frac{1}{8}$
11. $2x^2$ is not a like term, because x is squared. The like terms are $3x$ and $9x$.
12. $21g$
13. $11x + 2$
14. $-3s + 6$
15. $-2.3v - 5$
16. $14.2 - 5a$
17. $3 - \frac{1}{2}y$

Practice and Problem Solving

18. $4b - 5$ **19.** $-p - 30$

20. $6c + 6$

21. $10.2x$; The weight carried by each hiker is 10.2 pounds.

22. -9; -9; *Sample answer:* Simplifying the expression first is easier because you only have to substitute once instead of substituting three times.

23. yes; Both expressions simplify to $11x^2 + 3y$.

24–28. See Additional Answers.

29. See *Taking Math Deeper*.

Fair Game Review

30. 14.5 in., 14.8 in., 15 in., 15.3 in., 15.8 in.

31. 0.52 m, 0.545 m, 0.55 m, 0.6 m, 0.65 m

32. C

Mini-Assessment

Identify the terms and like terms in the expression.

1. $4r + 2 - 6 + 3r$

Terms: $4r, 2, -6, 3r$
Like terms: $4r$ and $3r$, 2 and -6

2. $5h^2 - 3h^2 - 4h + 3h + 7$

Terms: $5h^2, -3h^2, -4h, 3h, 7$
Like terms: $5h^2$ and $-3h^2$, $-4h$ and $3h$

Simplify the expression.

3. $6m + 7 - 3m - 1$ $3m + 6$

4. $3(5b + 2) - 4$ $15b + 2$

5. Write an expression in simplest form that represents the perimeter of the polygon. $(3x + 9)$ m

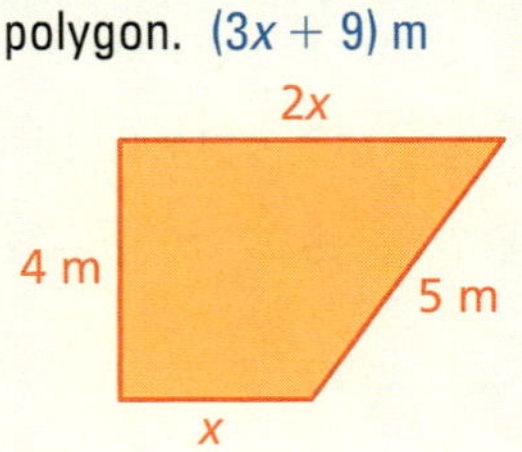

Taking Math Deeper

Exercise 29

In this problem, students have to realize that when you subtract the two red strips, you have subtracted their intersection twice.

1 Area of Gold = Total Area − Vertical Strip − Horizontal Strip + Intersection

$$= 12(20) - 12x - 20x + x^2$$

a. $$= 240 - 32x + x^2$$

Notice that the intersection of the two red strips is subtracted twice, so it must be added back into the expression once.

2 When $x = 3$, the area of the gold foil is

$$\text{Area} = 240 - 32x + x^2$$
$$= 240 - 32(3) + 3^2$$
$$= 240 - 96 + 9$$

b. $$= 153 \text{ in.}^2$$

3 **c.** This pattern is used as the flag of England.

England

Historical Greece

Georgia

Note: The flag for the United Kingdom (England, Scotland, Wales, and Northern Ireland) is different and has a criss-cross pattern.

Reteaching and Enrichment Strategies

If students need help...	If students got it...
Resources by Chapter • Practice A and Practice B • Puzzle Time Record and Practice Journal Practice Differentiating the Lesson Lesson Tutorials Skills Review Handbook	Resources by Chapter • Enrichment and Extension • Technology Connection Start the next section

22. **STRUCTURE** Evaluate the expression $-8x + 5 - 2x - 4 + 5x$ when $x = 2$ before and after simplifying. Which method do you prefer? Explain.

23. **REASONING** Are the expressions $8x^2 + 3(x^2 + y)$ and $7x^2 + 7y + 4x^2 - 4y$ equivalent? Explain your reasoning.

24. **CRITICAL THINKING** Which solution shows a correct way of simplifying $6 - 4(2 - 5x)$? Explain the errors made in the other solutions.

Ⓐ $6 - 4(2 - 5x) = 6 - 4(-3x) = 6 + 12x$

Ⓑ $6 - 4(2 - 5x) = 6 - 8 + 20x = -2 + 20x$

Ⓒ $6 - 4(2 - 5x) = 2(2 - 5x) = 4 - 10x$

Ⓓ $6 - 4(2 - 5x) = 6 - 8 - 20x = -2 - 20x$

25. **BANNER** Write an expression in simplest form that represents the area of the banner.

	Car	Truck
Wash	\$8	\$10
Wax	\$12	\$15

26. **CAR WASH** Write an expression in simplest form that represents the earnings for washing and waxing x cars and y trucks.

MODELING Draw a diagram that shows how the expression can represent the area of a figure. Then simplify the expression.

27. $5(2 + x + 3)$

28. $(4 + 1)(x + 2x)$

29. Critical Thinking You apply gold foil to a piece of red poster board to make the design shown.

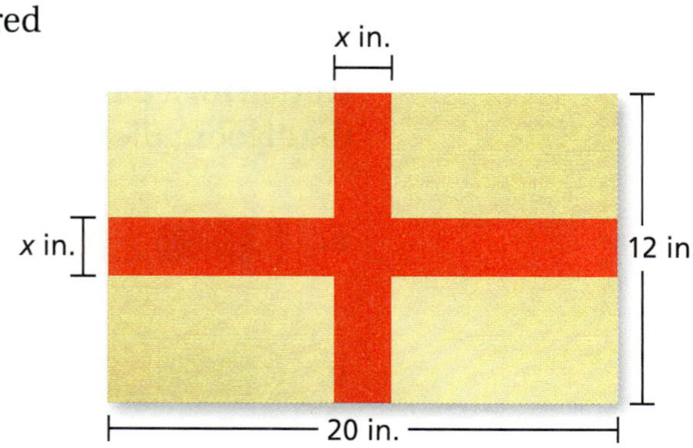

a. Write an expression in simplest form that represents the area of the gold foil.

b. Find the area of the gold foil when $x = 3$.

c. The pattern at the right is called "St. George's Cross." Find a country that uses this pattern as its flag.

Fair Game Review What you learned in previous grades & lessons

Order the lengths from least to greatest. *(Skills Review Handbook)*

30. 15 in., 14.8 in., 15.8 in., 14.5 in., 15.3 in.

31. 0.65 m, 0.6 m, 0.52 m, 0.55 m, 0.545 m

32. **MULTIPLE CHOICE** A bird's nest is 12 feet above the ground. A mole's den is 12 inches below the ground. What is the difference in height of these two positions? *(Section 11.3)*

Ⓐ 24 in. Ⓑ 11 ft Ⓒ 13 ft Ⓓ 24 ft

13.2 Adding and Subtracting Linear Expressions

Essential Question How can you use algebra tiles to add or subtract algebraic expressions?

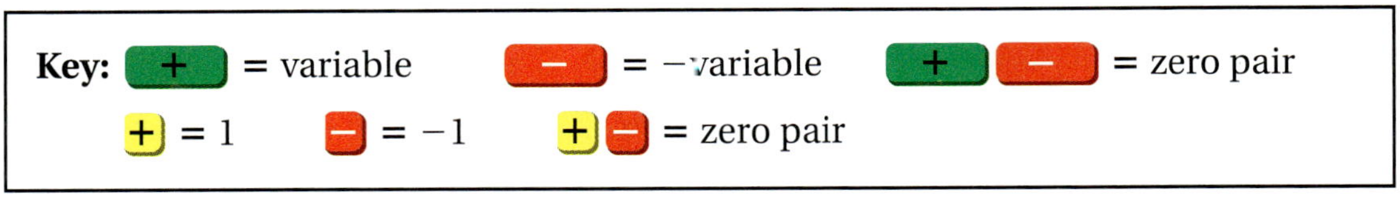

1 ACTIVITY: Writing Algebraic Expressions

Work with a partner. Write an algebraic expression shown by the algebra tiles.

a. 

b. 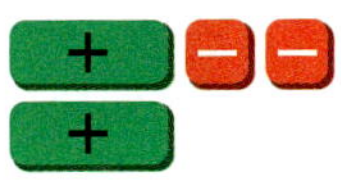

c. 

d. 

2 ACTIVITY: Adding Algebraic Expressions

Work with a partner. Write the sum of two algebraic expressions modeled by the algebra tiles. Then use the algebra tiles to simplify the expression.

a. 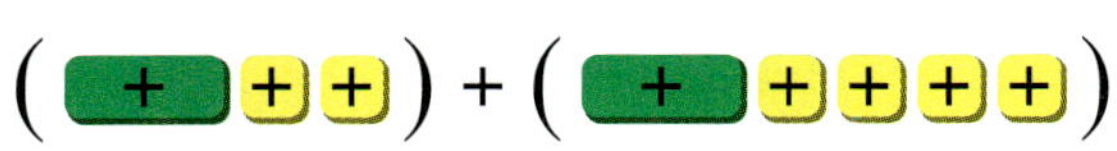

b. 

c. 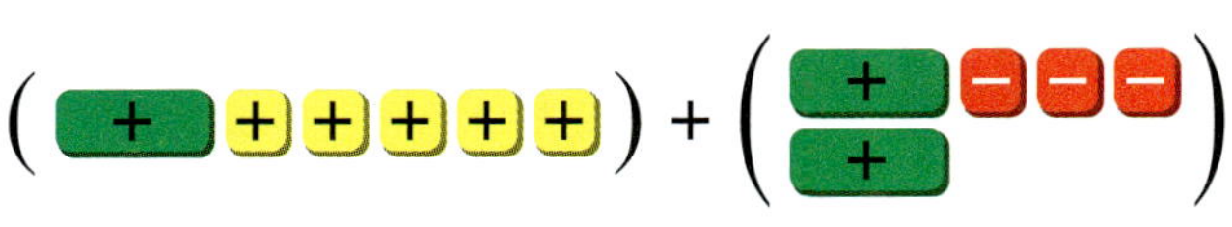

d. 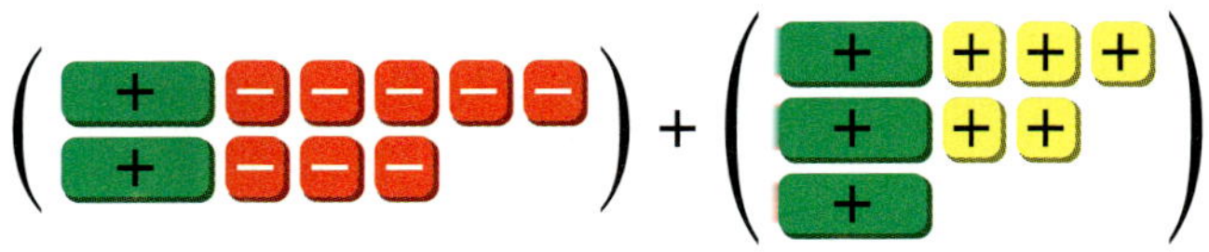

Linear Expressions

In this lesson, you will

- apply properties of operations to add and subtract linear expressions.
- solve real-life problems.

Laurie's Notes

Introduction

Applying Mathematical Practices

- **Reason Abstractly and Quantitatively:** Algebra tiles help students make sense of algebraic expressions by modeling them and finding sums and differences. Algebra tiles are a concrete representation, deepening student understanding of the meaning of each expression.

Discuss

- **FYI:** Show students a collection of yellow integer-tiles and one green variable-tile. Define the yellow integer-tile as having dimensions 1 by 1 with an area of 1 square unit and the variable-tile as having dimensions 1 by x with an area of x square units. The Record and Practice Journal has algebra tiles and they are also available commercially. Be sure to point out to students that the variable-tile is NOT an integral length, meaning you should not be able to *measure* the length of the variable-tile by lining up yellow integer-tiles. The length of the tile is a variable—x!
- Display a collection of tiles, say 1 variable-tile, 3 yellow integer-tiles ($+3$) and 2 red integer-tiles (-2). Say, "These algebra tiles represent an algebraic expression and just as you simplify algebraic expressions, you are going to simplify expressions modeled by the algebra tiles."

Activity Notes

Activity 1

- **Management Tip:** Distribute a set of algebra tiles to each pair of students. Presort them in baggies for easy distribution and collection.
- Even though the collection of tiles is shown, encourage students to make the collection shown with their own algebra tiles.
- Remind students that any letter can be used to represent a variable.
- Some students may write expressions that represent each algebra tile such as "$x + x - 1 - 1$" for part (b). Ask them "Is your expression in simplest form? If not, how can you write it in simplest form?"
- Ask for volunteers to share their results.
- Students may write $3 + x$ for part (a). Explain that it is more common to state the x-term first, as $x + 3$. The Commutative Property of Addition assures that $3 + x$ and $x + 3$ are equivalent.

Activity 2

- Have each partner represent one of the expressions using their tiles. To add, have them combine their tiles together in the common work space. Then simplify by removing any zero pairs.
- Note that all expressions in Activity 2 have positive coefficients.
- Ask for volunteers to explain how they used the tiles to simplify.
- **Reason Abstractly and Quantitatively:** Handling the tiles helps students understand that $x + x = 2x$ and not x^2. Students who have worked with algebra tiles should not make that mistake.

What Your Students Will Learn

- Use a vertical or horizontal method and apply properties of operations to add or subtract linear expressions.

Previous Learning

Students should know how to simplify algebraic expressions.

Technology for the Teacher

Dynamic Classroom

Lesson Plans
Complete Materials List

13.2 Record and Practice Journal

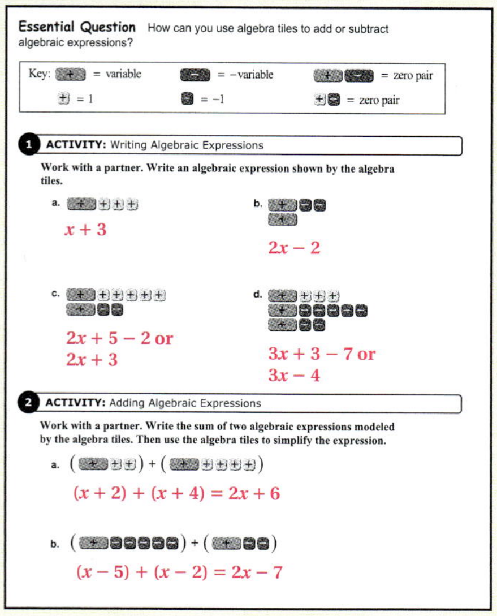

Essential Question How can you use algebra tiles to add or subtract algebraic expressions?

Key: [variable tile] = variable; [negative variable tile] = −variable; [variable tile][negative variable tile] = zero pair; [+] = 1; [−] = −1; [+][−] = zero pair

1 ACTIVITY: Writing Algebraic Expressions

Work with a partner. Write an algebraic expression shown by the algebra tiles.

a. $x + 3$

b. $2x - 2$

c. $2x + 5 - 2$ or $2x + 3$

d. $3x + 3 - 7$ or $3x - 4$

2 ACTIVITY: Adding Algebraic Expressions

Work with a partner. Write the sum of two algebraic expressions modeled by the algebra tiles. Then use the algebra tiles to simplify the expression.

a. $(x + 2) + (x + 4) = 2x + 6$

b. $(x - 5) + (x - 2) = 2x - 7$

Differentiated Instruction

Auditory

Have students verbally describe the difference of two algebraic expressions, such as

$2x - (x + 1)$.

This expression is read as "two x minus the quantity x plus one." Remind students that "the quantity x plus one" means that x and 1 are grouped together.

13.2 Record and Practice Journal

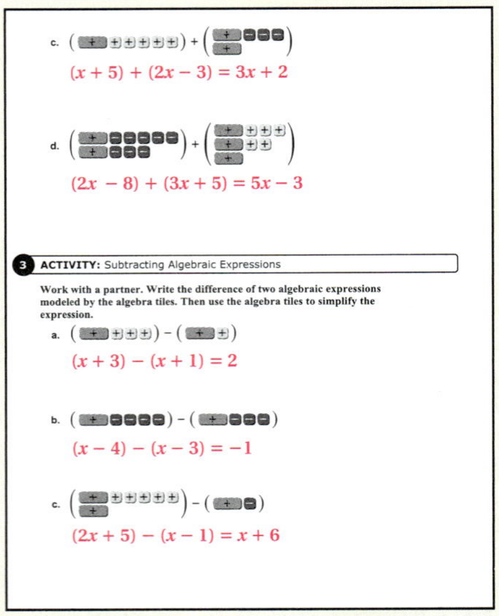

c. $(x + 5) + (2x - 3) = 3x + 2$

d. $(2x - 8) + (3x + 5) = 5x - 3$

3 ACTIVITY: Subtracting Algebraic Expressions

Work with a partner. Write the difference of two algebraic expressions modeled by the algebra tiles. Then use the algebra tiles to simplify the expression.

a. $(x + 3) - (x + 1) = 2$

b. $(x - 4) - (x - 3) = -1$

c. $(2x + 5) - (x - 1) = x + 6$

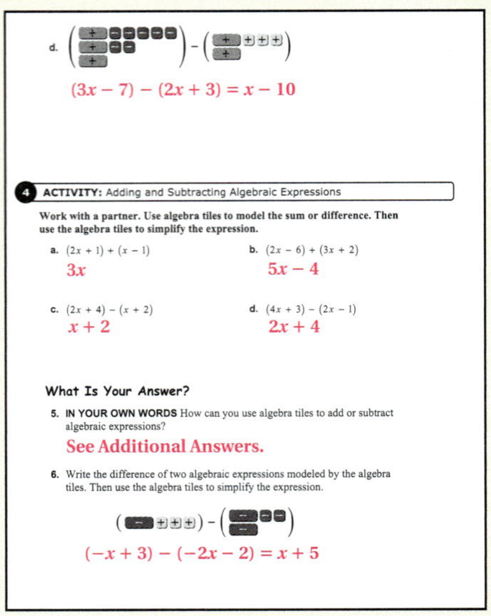

d. $(3x - 7) - (2x + 3) = x - 10$

4 ACTIVITY: Adding and Subtracting Algebraic Expressions

Work with a partner. Use algebra tiles to model the sum or difference. Then use the algebra tiles to simplify the expression.

a. $(2x + 1) + (x - 1)$ $3x$

b. $(2x - 6) + (3x + 2)$ $5x - 4$

c. $(2x + 4) - (x + 2)$ $x + 2$

d. $(4x + 3) - (2x - 1)$ $2x + 4$

What Is Your Answer?

5. IN YOUR OWN WORDS How can you use algebra tiles to add or subtract algebraic expressions?
See Additional Answers.

6. Write the difference of two algebraic expressions modeled by the algebra tiles. Then use the algebra tiles to simplify the expression.

$(-x + 3) - (-2x - 2) = x + 5$

Laurie's Notes

Activity 3

- Review the meaning of subtraction and how it is performed using integer tiles. Begin by representing the first expression on the work space. Students find the difference by removing the algebra tiles in the second expression from the algebra tiles in the first expression.
- For parts (c) and (d), students will need to add zero pairs in order to subtract the second expression.
- Ask for volunteers to share their results. This should include modeling at least one of the problems at the document camera or overhead projector so that the language of the subtraction is heard.
- **Reason Abstractly and Quantitatively:** Using the algebra tiles, it should be clear that $3x - 2x = 1x$. Without the algebra tiles, students may incorrectly reason that $3x - 2x = 1$. They can lose track of what the expressions $3x$ and $2x$ represent and see them only as symbols, not quantities.

Activity 4

- Now students start with the expressions and create a model using algebra tiles. To model $x - 1$ in part (a), students can think of the equivalent expression $x + (-1)$.
- If time permits, have students model these problems so that classmates hear the language associated with performing these operations.
- Students may need help with part (d). Model $4x + 3$. When students go to remove $2x - 1$, remind them that $2x - 1$ can be written as $2x + (-1)$.

What Is Your Answer?

- In Question 6, students need to add zero pairs to the first expression in order to "take away" the algebra tiles in the second expression. This is also the first problem that uses the negative variable-tile.

Closure

- Use algebra tiles to create the following model and simplify it.

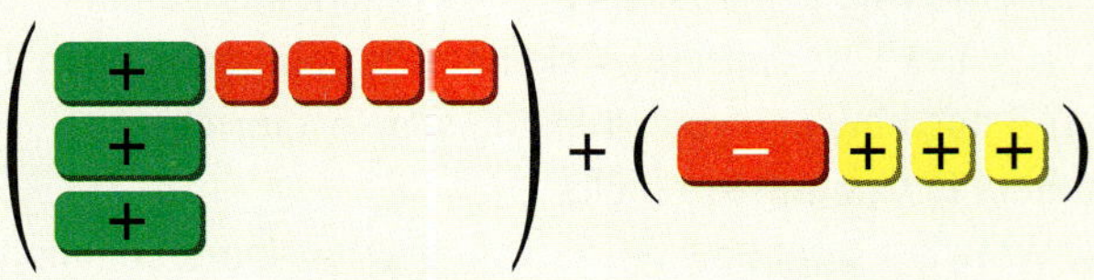

$2x - 1$

3 ACTIVITY: Subtracting Algebraic Expressions

Math Practice

Use Expressions

What do the tiles represent? How does this help you write an expression?

Work with a partner. Write the difference of two algebraic expressions modeled by the algebra tiles. Then use the algebra tiles to simplify the expression.

a. 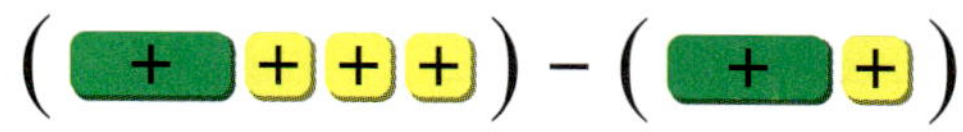

b. 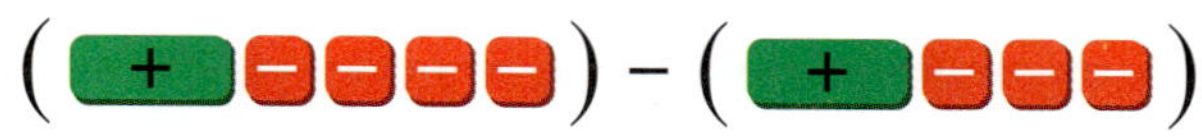

c. 

d. 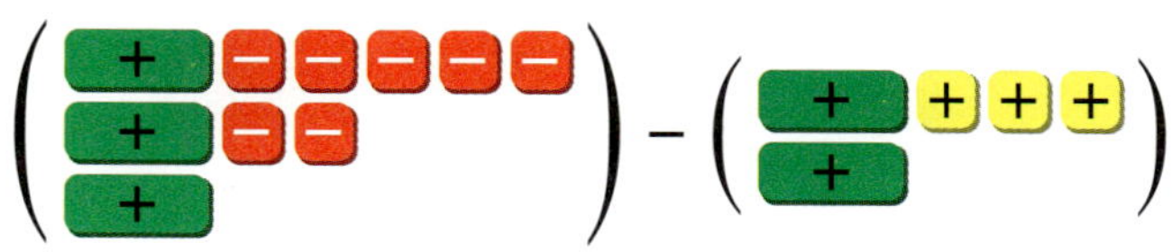

4 ACTIVITY: Adding and Subtracting Algebraic Expressions

Work with a partner. Use algebra tiles to model the sum or difference. Then use the algebra tiles to simplify the expression.

a. $(2x + 1) + (x - 1)$

b. $(2x - 6) + (3x + 2)$

c. $(2x + 4) - (x + 2)$

d. $(4x + 3) - (2x - 1)$

What Is Your Answer?

5. **IN YOUR OWN WORDS** How can you use algebra tiles to add or subtract algebraic expressions?

6. Write the difference of two algebraic expressions modeled by the algebra tiles. Then use the algebra tiles to simplify the expression.

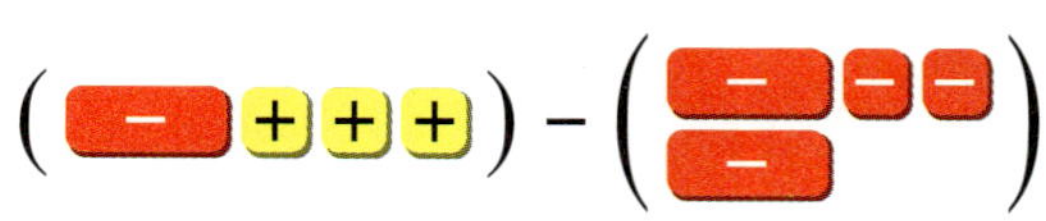

Practice Use what you learned about adding and subtracting algebraic expressions to complete Exercises 6 and 7 on page 564.

13.2 Lesson

Check It Out
Lesson Tutorials
BigIdeasMath.com

Key Vocabulary
linear expression, *p. 562*

A **linear expression** is an algebraic expression in which the exponent of the variable is 1.

Linear Expressions	$-4x$	$3x + 5$	$5 - \frac{1}{6}x$
Nonlinear Expressions	x^2	$-7x^3 + x$	$x^5 + 1$

You can use a vertical or a horizontal method to add linear expressions.

EXAMPLE 1 Adding Linear Expressions

Find each sum.

a. $(x - 2) + (3x + 8)$

Vertical method: Align like terms vertically and add.

$$\begin{array}{r} x - 2 \\ + \; 3x + 8 \\ \hline 4x + 6 \end{array}$$

b. $(-4y + 3) + (11y - 5)$

Horizontal method: Use properties of operations to group like terms and simplify.

$$(-4y + 3) + (11y - 5) = -4y + 3 + 11y - 5$$ Rewrite the sum.

$$= -4y + 11y + 3 - 5$$ Commutative Property of Addition

$$= (-4y + 11y) + (3 - 5)$$ Group like terms.

$$= 7y - 2$$ Combine like terms.

EXAMPLE 2 Adding Linear Expressions

Find $2(-7.5z + 3) + (5z - 2)$.

$$2(-7.5z + 3) + (5z - 2) = -15z + 6 + 5z - 2$$ Distributive Property

$$= -15z + 5z + 6 - 2$$ Commutative Property of Addition

$$= -10z + 4$$ Combine like terms.

On Your Own

Now You're Ready
Exercises 8–16

Find the sum.

1. $(x + 3) + (2x - 1)$
2. $(-8z + 4) + (8z - 7)$
3. $(4 - n) + 2(-5n + 3)$
4. $\frac{1}{2}(w - 6) + \frac{1}{4}(w + 12)$

Multi-Language Glossary at BigIdeasMath.com

Laurie's Notes

Introduction

Connect

- **Yesterday:** Students used algebra tiles to develop an understanding of how to add and subtract algebraic expressions.
- **Today:** Students will use a horizontal or vertical format to add and subtract linear expressions.

Motivate

- Draw a vertical line on the middle of your board. On the left write "These are" and on the right write "These are not." On each side write examples of expressions that are linear (left side) and are not linear (right side).
- Explain that you are not giving names to either side yet. You are just trying to have them be good detectives in thinking about what characteristics they see.

? "Can you give other examples of what you think would be on the left or right?"

? "What feature(s) distinguish the expressions that **are**, from the expressions that **are not**?" Listen for a reference to exponents (right side) and the lack of exponents (left side).

Discuss

- Tell students that the expressions on the left are examples of **linear expressions**, which are algebraic expressions in which the exponent of the variable is 1. This is not a precise definition because $\frac{1}{x}$ has a variable with an exponent of 1 and it is not a linear expression. However, this description is appropriate for this grade level.

Lesson Notes

Example 1

- **Connection:** When you add (or subtract) whole numbers, you use the place values of the numbers. The same is true when you add (or subtract) decimals—lining up the decimal points assures that this happens. Lining up place values is similar to lining up like terms. Make this connection for students as you begin to work these problems.
- Using the vertical method, students should see the connection to adding two whole numbers.
- **Teaching Tip:** Before adding, rewrite $x - 2$ as $x + (-2)$.
- The Commutative Property of Addition is used to change the order of the terms so that like terms are adjacent to one another.

Example 2

- **Attend to Precision:** Ask for a volunteer to read the problem. Listen for, "Two times the quantity negative seven point five *z* plus 3, plus the quantity five *z* minus two." Students should be able to read this.

? "What is the first step?" Simplify $2(-7.5z + 3)$.

Goal Today's lesson is to add and subtract **linear expressions**.

Technology for the Teacher

Dynamic Classroom

Lesson Tutorials
Lesson Plans
Answer Presentation Tool

Extra Example 1

Find each sum.

a. $(-2x + 2) + (4x - 7)$ $2x - 5$

b. $(7y - 5) + (3y + 8)$ $10y + 3$

Extra Example 2

Find $(7w - 6) + 5(-2.4w + 1)$
$-5w - 1$

On Your Own

1. $3x + 2$
2. -3
3. $-11n + 10$
4. $\frac{3}{4}w$

English Language Learners

Vocabulary

The word *variable* is often used in algebra and is represented by a letter. The letter stands for a number that changes, or *varies*. Students find it helpful when the letter is meaningful to the problem. For instance, use the letter *t* to represent a unit of time or use the letter *d* to represents dollars.

Extra Example 3

Find each difference.

a. $(-3x + 7) - (4x - 8)$ $-7x + 15$

b. $-3(2y - 9) - (5y + 4)$ $-11y + 23$

Extra Example 4

The original price of a coffee table is *d* dollars. You use a coupon and buy the table for $(d - 4)$ dollars. You paint the table and sell it for $(3d + 1)$ dollars. Write an expression that represents your earnings from buying and selling the coffee table. Interpret the expression.

$(3d + 1) - (d - 4)$; You earn $(2d + 5)$ dollars.

On Your Own

5. $2m - 15$

6. $-8c - 25$

7. $4

Laurie's Notes

Example 3

? "How do you think you subtract linear expressions?" Subtract like terms.

? Write part (a) and ask, "Can you subtract the quantity $(-x + 6)$ by removing the parentheses? No, you must subtract each term in the linear expression. So, you add the opposite.

- My experience is that students make more errors when subtracting linear expressions using the vertical method unless they take the time to rewrite the problem as shown where *adding the opposite* is obvious. As stated in the Study Tip, to find the opposite of a linear expression you can multiply the expression by -1.
- **Reason Abstractly and Quantitatively:** It may be helpful to rewrite $(5x + 6) - (-x + 6)$ as $(5x + 6) + [-(-x + 6)]$ and then $(5x + 6) + (-1)(-x + 6)$. This is the Multiplication Property of -1.
- Write part (b). This may be easier for students to understand than part (a) because the constant 2 is written in the problem, whereas the constant 1 in part (a) is not written. When students rewrite the problem as *add the opposite*, they can see that -2 needs to be distributed.
- **Look for and Make Use of Structure:** Using the Commutative Property to rewrite $7y + 5 - 8y + 6$ as $7y - 8y + 5 + 6$ is not obvious to all students. Take time to probe for understanding. Subtracting $8y$ is the same as adding the opposite of $8y$. You may need to work through these extra steps so students make sense of how the order of the terms can be changed.

? "Do you prefer the vertical or horizontal method? Why?" Answers will vary.

Example 4

- Have a quick discussion about how to calculate the earnings when buying something and reselling it.
- Ask for a volunteer to read the problem. Remind students that the variable *d* is unknown, and you are not writing an expression for the selling price.

? "What is the value of the coupon? Explain." $2, because you purchase the hat for $(d - 2)$ dollars.

- Write the verbal model and substitute the linear expressions.

? "This is a subtraction problem. What is our next step?" Add the opposite.

? "If you pay $(d - 2)$ dollars for an item and earn $(d - 2)$ dollars back, what does this mean?" If students are having difficulty interpreting this problem, substitute a value for *d*, such as $20, then explain.

- Students can verify that the selling price of $(2d - 4)$ dollars is twice that of the purchase price $(d - 2)$ by multiplying by 2. You could decide to have a quick review of factoring by factoring 2 out of the selling price, or wait and do this as an introduction to 13.2 Extension.

Closure

- **Exit Ticket:** Find the sum or difference.
  $2(3x - 4) + (2x - 5)$ $8x - 13$ $2(3x - 4) - (2x - 5)$ $4x - 3$

To subtract one linear expression from another, add the opposite of each term in the expression. You can use a vertical or a horizontal method.

EXAMPLE 3 Subtracting Linear Expressions

Study Tip

To find the opposite of a linear expression, you can multiply the expression by -1.

Find each difference.

a. $(5x + 6) - (-x + 6)$

b. $(7y + 5) - 2(4y - 3)$

a. Vertical method: Align like terms vertically and subtract.

$$\begin{array}{r} (5x + 6) \\ -\,(-x + 6) \\ \hline \end{array} \quad \xrightarrow{\text{Add the opposite.}} \quad \begin{array}{r} 5x + 6 \\ +\quad x - 6 \\ \hline 6x \end{array}$$

b. Horizontal method: Use properties of operations to group like terms and simplify.

$(7y + 5) - 2(4y - 3) = 7y + 5 - 8y + 6$ — Distributive Property

$= 7y - 8y + 5 + 6$ — Commutative Property of Addition

$= (7y - 8y) + (5 + 6)$ — Group like terms.

$= -y + 11$ — Combine like terms.

EXAMPLE 4 Real-Life Application

The original price of a cowboy hat is d dollars. You use a coupon and buy the hat for $(d - 2)$ dollars. You decorate the hat and sell it for $(2d - 4)$ dollars. Write an expression that represents your earnings from buying and selling the hat. Interpret the expression.

earnings = selling price − purchase price — Use a model.

$= (2d - 4) - (d - 2)$ — Write the difference.

$= (2d - 4) + (-d + 2)$ — Add the opposite.

$= 2d - d - 4 + 2$ — Group like terms.

$= d - 2$ — Combine like terms.

You earn $(d - 2)$ dollars. You also paid $(d - 2)$ dollars, so you doubled your money by selling the hat for twice as much as you paid for it.

On Your Own

Now You're Ready
Exercises 19–24

Find the difference.

5. $(m - 3) - (-m + 12)$

6. $-2(c + 2.5) - 5(1.2c + 4)$

7. **WHAT IF?** In Example 4, you sell the hat for $(d + 2)$ dollars. How much do you earn from buying and selling the hat?

13.2 Exercises

Check It Out
Help with Homework
BigIdeasMath.com

Vocabulary and Concept Check

VOCABULARY **Determine whether the algebraic expression is a linear expression. Explain.**

1. $x^2 + x + 1$
2. $-2x - 8$
3. $x - x^4$
4. **WRITING** Describe two methods for adding or subtracting linear expressions.
5. **DIFFERENT WORDS, SAME QUESTION** Which is different? Find "both" answers.

Subtract x from $3x - 1$.	Find $3x - 1$ decreased by x.
What is x more than $3x - 1$?	What is the difference of $3x - 1$ and x?

Practice and Problem Solving

Write the sum or difference of two algebraic expressions modeled by the algebra tiles. Then use the algebra tiles to simplify the expression.

6. 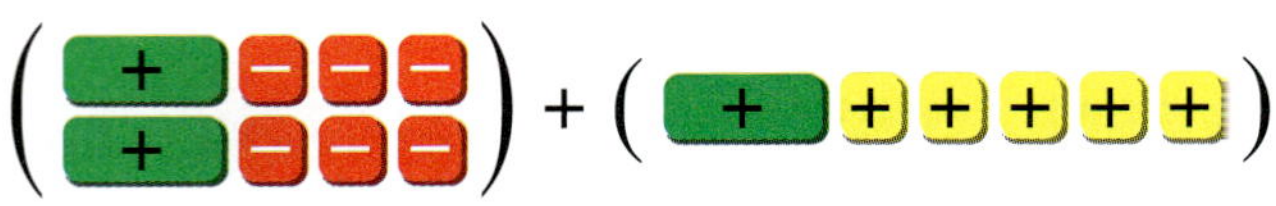

7. 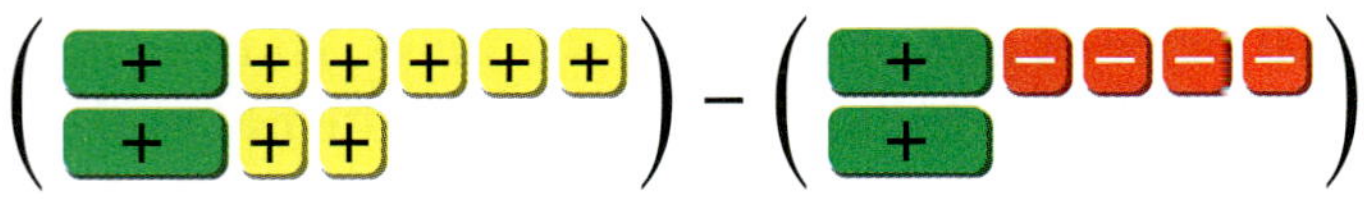

Find the sum.

1 2 8. $(n + 8) + (n - 12)$

9. $(7 - b) + (3b + 2)$

10. $(2w - 9) + (-4w - 5)$

11. $(2x - 6) + 4(x - 3)$

12. $5(-3.4k - 7) + (3k + 21)$

13. $(1 - 5q) + 2(2.5q + 8)$

14. $3(2 - 0.9h) + (-1.3h - 4)$

15. $\frac{1}{3}(9 - 6m) + \frac{1}{4}(12m - 8)$

16. $-\frac{1}{2}(7z + 4) + \frac{1}{5}(5z - 15)$

17. **BANKING** You start a new job. After w weeks, you have $(10w + 120)$ dollars in your savings account and $(45w + 25)$ dollars in your checking account. Write an expression that represents the total in both accounts.

18. **FIREFLIES** While catching fireflies, you and a friend decide to have a competition. After m minutes, you have $(3m + 13)$ fireflies and your friend has $(4m + 6)$ fireflies.
    a. Write an expression that represents the number of fireflies you and your friend caught together.
    b. The competition ends after 5 minutes. Who has more fireflies?

Assignment Guide and Homework Check

Level	Assignment	Homework Check
Advanced	1–7, 10–24 even, 25, 26–30 even, 32–35	10, 16, 18, 24, 28

Common Errors

- **Exercise 30** Students may count the corner tiles twice. Remind them that the corner tiles are the end of one length and the beginning of another, and should not be counted twice.
- **Exercise 31** Students may try to make distance negative. Remind them that distance is always positive.

13.2 Record and Practice Journal

Find the sum or difference.

1. $(x - 2) + (x + 6)$

   $2x + 4$

2. $(2n - 4) - (4n - 3)$

   $-2n - 1$

3. $2(-3y - 1) + (2y + 7)$

   $-4y + 5$

4. $(1 - 3k) - 4(2 + 2.5k)$

   $-13k - 7$

5. $(6g - 9) + \frac{1}{3}(15 - 9g)$

   $3g - 4$

6. $\frac{1}{2}(2r + 4) - \frac{1}{4}(16 - 8r)$

   $3r - 2$

7. You earn $(4x + 12)$ points after completing x levels of a video game and then lose $(2x - 5)$ points. Write an expression that represents the total number of points you have now.

   $2x + 17$

Vocabulary and Concept Check

1. not linear; An exponent of a variable is not equal to 1.
2. linear; The exponent of the variable is equal to 1.
3. not linear; An exponent of a variable is not equal to 1.
4. Vertical method: Align like terms vertically and add or subtract the opposite. Horizontal method: Group like terms using properties of operations and simplify.
5. What is x more than $3x - 1$?; $4x - 1$; $2x - 1$

Practice and Problem Solving

6. *Sample answer:* $(2x - 6) + (x + 5) = 3x - 1$
7. *Sample answer:* $(2x + 7) - (2x - 4) = 11$
8. $2n - 4$
9. $2b + 9$
10. $-2w - 14$
11. $6x - 18$
12. $-14k - 14$
13. 17
14. $-4h + 2$
15. $m + 1$
16. $-2\frac{1}{2}z - 5$
17. $55w + 145$
18. **a.** $7m + 19$

    b. you

Practice and Problem Solving

19. $-3g - 4$
20. $9d + 3$
21. $-12y + 20$
22. $14n - 29$
23. $-2c$
24. $x + 10\frac{1}{2}$
25. See Additional Answers.
26. a. 7 fireflies per minute
    b. 19 fireflies
27. no; If the variable terms are opposites, the sum is a numerical expression.
28. $8n$
29. $0.25x + 0.15$
30. See *Taking Math Deeper.*
31. $|x - 3|$, or equivalently $|-x + 3|$; 0; 6

Fair Game Review

32. $-\frac{7}{15}$
33. $\frac{2}{5}$
34. $2\frac{2}{15}$
35. D

Mini-Assessment

Find the sum or difference.

1. $(5m + 3) + (-8m + 8)$ $-3m + 11$
2. $(4 - x) + (2x + 5)$ $x + 9$
3. $(8x - 3) - (2x + 6)$ $6x - 9$
4. $(2 - 7y) - 3(y - 9)$ $-10y + 29$
5. A rectangle has side lengths $(x + 5)$ meters and $(2x - 1)$ meters. Write an expression in simplest form that represents the perimeter of the rectangle.
   $6x + 8$ meters

Taking Math Deeper

Exercise 30

It is easy to count tiles more than once in this problem and then write an incorrect expression. You can avoid this pitfall by drawing a diagram.

1. Let w represent the width of the room in feet. The expression $(w + 10)$ represents the length of the room. Draw the room using 1-foot-by-1-foot tiles.

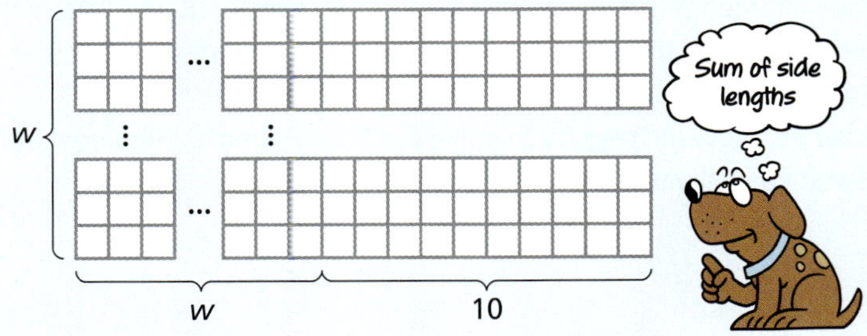

2. Using the diagram, you can see that if you find the sum $w + w + (w + 10) + (w + 10)$, then you will count each corner tile twice. So, you must subtract 4 from this sum.

$$w + w + (w + 10) + (w + 10) - 4$$
$$= w + w + w + w + 10 + 10 - 4$$
$$= 4w + 16$$

3. Another way to find the sum is to keep track when you are counting each corner tile. Starting at the bottom and adding side lengths counterclockwise, you can write

$$w + 10 + (w - 1) + (w + 9) + (w - 2)$$
$$= w + w + w + w + 10 - 1 + 9 - 2$$
$$= 4w + 16.$$

So, an expression for the number of tiles along the outside of the room is $4w + 16$.

Project

Research the costs of at least 3 different types of floor tiles. Choose a reasonable value for the width and find how much more it would cost to tile the room with the most expensive tile than with the least expensive tile.

Reteaching and Enrichment Strategies

If students need help...	If students got it...
Resources by Chapter • Practice A and Practice B • Puzzle Time Record and Practice Journal Practice Differentiating the Lesson Lesson Tutorials Skills Review Handbook	Resources by Chapter • Enrichment and Extension • Technology Connection Start the next section

Find the difference.

3 **19.** $(-2g + 7) - (g + 11)$ **20.** $(6d + 5) - (2 - 3d)$ **21.** $(4 - 5y) - 2(3.5y - 8)$

22. $(2n - 9) - 5(-2.4n + 4)$ **23.** $\frac{1}{8}(-8c + 16) - \frac{1}{3}(6 + 3c)$ **24.** $\frac{3}{4}(3x + 6) - \frac{1}{4}(5x - 24)$

25. ERROR ANALYSIS Describe and correct the error in finding the difference.

✗
$$\begin{aligned}(4m + 9) - 3(2m - 5) &= 4m + 9 - 6m - 15\\ &= 4m - 6m + 9 - 15\\ &= -2m - 6\end{aligned}$$

26. STRUCTURE Refer to the expressions in Exercise 18.

a. How many fireflies are caught each minute during the competition?

b. How many fireflies are caught before the competition starts?

27. LOGIC Your friend says the sum of two linear expressions is always a linear expression. Is your friend correct? Explain.

28. GEOMETRY The expression $17n + 11$ represents the perimeter (in feet) of the triangle. Write an expression that represents the measure of the third side.

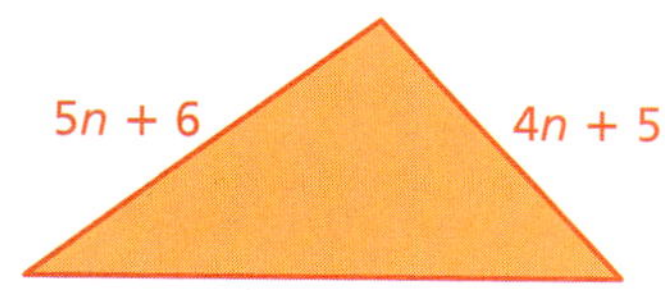

29. TAXI Taxi Express charges $2.60 plus $3.65 per mile, and Cab Cruiser charges $2.75 plus $3.90 per mile. Write an expression that represents how much more Cab Cruiser charges than Taxi Express.

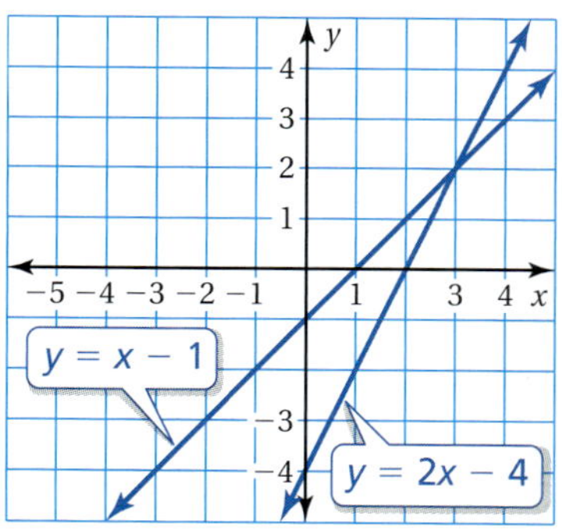

30. MODELING A rectangular room is 10 feet longer than it is wide. One-foot-by-one-foot tiles cover the entire floor. Write an expression that represents the number of tiles along the outside of the room.

31. **Reasoning** Write an expression in simplest form that represents the vertical distance between the two lines shown. What is the distance when $x = 3$? when $x = -3$?

Fair Game Review What you learned in previous grades & lessons

Evaluate the expression when $x = -\frac{4}{5}$ and $y = \frac{1}{3}$. *(Section 12.2)*

32. $x + y$ **33.** $2x + 6y$ **34.** $-x + 4y$

35. MULTIPLE CHOICE What is the surface area of a cube that has a side length of 5 feet? *(Section 8.2)*

Ⓐ 25 ft^2 Ⓑ 75 ft^2 Ⓒ 125 ft^2 Ⓓ 150 ft^2

Extension 13.2 Factoring Expressions

Check It Out
Lesson Tutorials
BigIdeasMath.com

Key Vocabulary
factoring an expression, *p. 566*

When **factoring an expression**, you write the expression as a product of factors. You can use the Distributive Property to factor expressions.

EXAMPLE 1 Factoring Out the GCF

Factor $24x - 18$ using the GCF.

Find the GCF of $24x$ and 18 by writing their prime factorizations.

$$24x = 2 \cdot 2 \cdot 2 \cdot 3 \cdot x$$
$$18 = 2 \cdot 3 \cdot 3$$

Circle the common prime factors.

So, the GCF of $24x$ and 18 is $2 \cdot 3 = 6$. Use the GCF to factor the expression.

$$24x - 18 = 6(4x) - 6(3) \qquad \text{Rewrite using GCF.}$$
$$= 6(4x - 3) \qquad \text{Distributive Property}$$

So, $24x - 18 = 6(4x - 3)$.

You can also use the Distributive Property to factor out any rational number from an expression.

EXAMPLE 2 Factoring Out a Fraction

Factor $\frac{1}{2}$ out of $\frac{1}{2}x + \frac{3}{2}$.

Write each term as a product of $\frac{1}{2}$ and another factor.

$$\frac{1}{2}x = \frac{1}{2} \cdot x \qquad \text{Think: } \frac{1}{2}x \text{ is } \frac{1}{2} \text{ times what?}$$
$$\frac{3}{2} = \frac{1}{2} \cdot 3 \qquad \text{Think: } \frac{3}{2} \text{ is } \frac{1}{2} \text{ times what?}$$

Use the Distributive Property to factor out $\frac{1}{2}$.

$$\frac{1}{2}x + \frac{3}{2} = \frac{1}{2} \cdot x + \frac{1}{2} \cdot 3 \qquad \text{Rewrite the expression.}$$
$$= \frac{1}{2}(x + 3) \qquad \text{Distributive Property}$$

So, $\frac{1}{2}x + \frac{3}{2} = \frac{1}{2}(x + 3)$.

Linear Expressions
In this extension, you will
- factor linear expressions.

Multi-Language Glossary at BigIdeasMath.com

Laurie's Notes

Introduction

Connect

- **Yesterday:** Students found the sums and differences of linear expressions.
- **Today:** Students will factor linear expressions.

Discuss

- In today's lesson students will be factoring expressions. The connection you want them to make is that the Distributive Property states there is an equality between two expressions: $a(b + c) = ab + ac$. Students should be comfortable *distributing the factor of a*. Here they will *factor out the a*.
- Students often view the equal sign in the Distributive Property as an arrow pointing to the right. Remind them that the arrow also points to the left!

Lesson Notes

Example 1

? "Do $24x$ and 18 have any common factors?" 2, 3, and 6

- Now students may find it odd that you did not just ask about 24 and 18. Instead you included the variable factor. Explain that an algebraic term has factors, just like numbers.
- Write the prime factorizations of $24x$ and 18. Students will be comfortable seeing 6 as the GCF of $24x$ and 18. The variable x has not changed that.
- Say, "You want to rewrite $24x - 18$ as a product." Get students started by writing $24x - 18 = \_\_\_\_(\_\_\_\_ - \_\_\_\_)$

? "How can you use the Distributive Property?" Listen for students to mention that the GCF is the factor you want to remove.

- **Look for and Make Use of Structure:** When you finish, be sure that students recognize that this is the Distributive Property, with the arrow pointing left!

$6(4x - 3) = 24x - 18$ Distributive Property ⟶ expanding

$24x - 18 = 6(4x - 3)$ Distributive Property ⟵ factoring

Example 2

- **Look for and Make Use of Structure:** Writing each term as a product helps students see the common factor. This is particularly true with fractions.

? "How can you write $\frac{1}{2}x$ as a product?" *Sample answer:* $\frac{1}{2} \cdot x$

? "How can you write $\frac{3}{2}$ as a product?" *Sample answer:* $\frac{1}{2} \cdot 3$

- Because there is a common factor of $\frac{1}{2}$ in each term, you can factor it out.

$\frac{1}{2}(x + 3) = \frac{1}{2}x + \frac{3}{2}$ Distributive Property ⟶ expanding

$\frac{1}{2}x + \frac{3}{2} = \frac{1}{2}(x + 3)$ Distributive Property ⟵ factoring

What Your Students Will Learn

- Factor linear expressions by factoring out the Greatest Common Factor (GCF) or the coefficient of the variable.

Goal Today's lesson is factoring linear expressions.

Technology for the Teacher

Dynamic Classroom

Lesson Tutorials
Lesson Plans
Answer Presentation Tool

Extra Example 1

Factor $6x - 27$ using the GCF.
$3(2x - 9)$

Extra Example 2

Factor $\frac{1}{5}$ out of $\frac{1}{5}x - \frac{4}{5}$.
$\frac{1}{5}(x - 4)$

Record and Practice Journal Extension 13.2 Practice

1. $7(1 + 4)$
2. $25(1 + 2)$
3. $7(b - 1)$
4. $8(a - 2)$
5. $4(2x + 3)$
6. $12(y + 2t)$
7. $10(w + 5z)$
8. $2(5v + 6u)$
9. $3(3a + 5b)$
10. $\frac{1}{2}(a - 1)$
11. $\frac{1}{4}(d - 3)$
12. $\frac{5}{6}\left(s + \frac{4}{5}\right)$
13. $\frac{3}{10}\left(y - \frac{4}{3}\right)$
14. $1.1(x + 9)$
15. $3.4(c + 3)$
16. $-2(3x - 5)$
17. $-\frac{1}{3}\left(-y + \frac{9}{2}\right)$
18. $(2x + 3)$ ft

Laurie's Notes

Example 3

- This example looks at factoring out a negative number. If students are comfortable with the two directions of the Distributive Property identified in the last two examples, then they should be able to discuss the role of the negative factor in this example.
- As a prompt, write $-4p + 10 = -2(\_\_\_\_ + \_\_\_\_)$
- ? "-2 times what is $-4p$?" $2p$
- ? "-2 times what is 10?" -5
- Fill in the blanks, $-4p + 10 = -2(2p + (-5))$, which can be written as $-2(2p - 5)$.

Closure

- Match the algebraic expression on the left with its factored form on the right.

1. $12x + 6$ C	A. $-6(2x - 1)$
2. $12x - 6$ B	B. $6(2x - 1)$
3. $-12x - 6$ D	C. $6(2x + 1)$
4. $-12x + 6$ A	D. $-6(2x + 1)$

Extra Example 3

Factor -4 out of $-12r - 20$

$-4(3r + 5)$

Practice

1. $3(3 + 7)$
2. $16(2 - 3)$
3. $2(4x + 1)$
4. $3(y - 8)$
5. $4(5z - 2)$
6. $5(3w + 13)$
7. $4(9a + 4b)$
8. $7(3m - 7n)$
9. $\frac{1}{3}(b - 1)$
10. $\frac{3}{8}(d + 2)$
11. $2.2(x + 2)$
12. $4\left(h - \frac{3}{4}\right)$
13. $-\frac{1}{2}(x - 12)$
14. $-\frac{1}{4}(2x + 5y)$
15. $(3x - 8)$ ft
16. See Additional Answers.
17. *Sample answer:* $2x - 1$ and x, $2x$ and $x - 1$

Mini-Assessment

1. Factor $21w + 28$ using the GCF.
   $7(3w + 4)$
2. Factor $\frac{1}{3}$ out of $\frac{1}{3}p - \frac{2}{3}$.
   $\frac{1}{3}(p - 2)$
3. Factor -6 out of $-12y - 42$.
   $-6(2y + 7)$

EXAMPLE 3 Factoring Out a Negative Number

Factor -2 out of $-4p + 10$.

> **Math Practice**
> **View as Components**
> How does rewriting each term as a product help you see the common factor?

Write each term as a product of -2 and another factor.

$-4p = -2 \cdot 2p$ — Think: $-4p$ is -2 times what?

$10 = -2 \cdot (-5)$ — Think: 10 is -2 times what?

Use the Distributive Property to factor out -2.

$-4p + 10 = -2 \cdot 2p + (-2) \cdot (-5)$ — Rewrite the expression.

$= -2[2p + (-5)]$ — Distributive Property

$= -2(2p - 5)$ — Simplify.

So, $-4p + 10 = -2(2p - 5)$.

Practice

Factor the expression using the GCF.

1. $9 + 21$
2. $32 - 48$
3. $8x + 2$
4. $3y - 24$
5. $20z - 8$
6. $15w + 65$
7. $36a + 16b$
8. $21m - 49n$

Factor out the coefficient of the variable.

9. $\frac{1}{3}b - \frac{1}{3}$
10. $\frac{3}{8}d + \frac{3}{4}$
11. $2.2x + 4.4$
12. $4h - 3$

13. Factor $-\frac{1}{2}$ out of $-\frac{1}{2}x + 6$.

14. Factor $-\frac{1}{4}$ out of $-\frac{1}{2}x - \frac{5}{4}y$.

15. **WRESTLING** A square wrestling mat has a perimeter of $(12x - 32)$ feet. Write an expression that represents the side length of the mat (in feet).

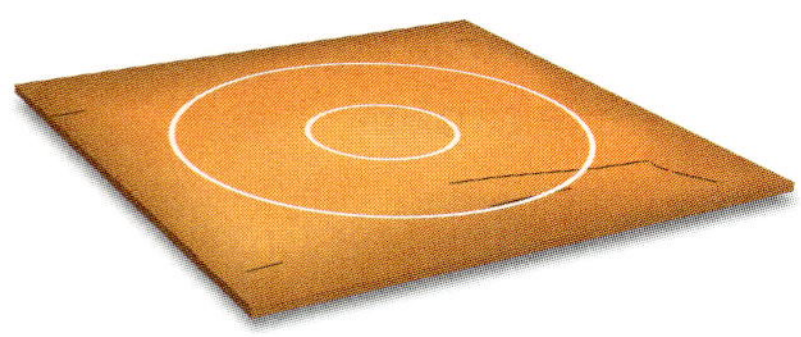

16. **MAKING A DIAGRAM** A table is 6 feet long and 3 feet wide. You extend the table by inserting two identical table *leaves*. The longest side length of each rectangular leaf is 3 feet. The extended table is rectangular with an area of $(18 + 6x)$ square feet.
    a. Make a diagram of the table and leaves.
    b. Write an expression that represents the length of the extended table. What does x represent?

17. **STRUCTURE** The area of the trapezoid is $\left(\frac{3}{4}x - \frac{1}{4}\right)$ square centimeters. Write two different pairs of expressions that represent possible lengths of the bases.

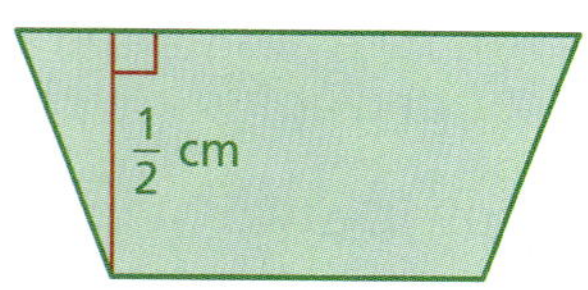

13 Study Help

Check It Out
Graphic Organizer
BigIdeasMath.com

You can use a **four square** to organize information about a topic. Each of the four squares can be a category, such as *definition, vocabulary, example, non-example, words, algebra, table, numbers, visual, graph,* or *equation*. Here is an example of a four square for like terms.

Like Terms

Definition	Examples
Terms that have the same variables raised to the same exponents	2 and −3, 3x and −7x, x^2 and $6x^2$
Words	**Non-Examples**
To *combine* like terms that have variables, use the Distributive Property to add or subtract the coefficients.	y and 4, 3x and −4y, $6x^2$ and 2x

On Your Own

Make four squares to help you study these topics.

1. simplest form
2. linear expression
3. factoring expressions

After you complete this chapter, make four squares for the following topics.

4. equivalent equations
5. solving equations using addition or subtraction
6. solving equations using multiplication or division
7. solving two-step equations

"My four square shows that my new red skateboard is faster than my old blue skateboard."

Sample Answers

1.

Definition	**Words**
An algebraic expression is in *simplest form* when it has: 1. no like terms and 2. no parentheses.	To write an algebraic expression in simplest form: Step 1: Rewrite as a sum. Step 2: Use the Distributive Property on parentheses, if necessary. Step 3: Rearrange terms. Step 4: Combine like terms.

Simplest form

Example	**Example**
$5x^2 + 6x - 3x^2 + 8 - x$ $= 5x^2 + 6x + (-3x^2) + 8 + (-1x)$ $= 5x^2 + (-3x^2) + 6x + (-1x) + 8$ $= [5 + (-3)]x^2 + [6 + (-1)]x + 8$ $= 2x^2 + 5x + 8$	$9 - 3\left(\frac{2}{3}m - \frac{1}{3}\right) + 3m$ $= 9 + (-3)\left(\frac{2}{3}m + \left(-\frac{1}{3}\right)\right) + 3m$ $= 9 + (-3)\left(\frac{2}{3}m\right) + (-3)\left(-\frac{1}{3}\right) + 3m$ $= 9 + (-2m) + 1 + 3m$ $= (-2m) + 3m + 9 + 1$ $= (-2 + 3)m + (9 + 1)$ $= m + 10$

2.

Definition	**Examples**
An algebraic expression in which the exponent of the variable is 1	$-7x,\ 2x + 3,\ 8 - \frac{1}{4}x$ **Non-examples:** $x^3,\ -5x^2 + x,\ x^7 - 9$

Linear expression

Example	**Example**
Adding linear expressions: $(7 - w) + 3(-2w + 4)$ $= 7 + (-1w) + 3(-2w) + 3(4)$ $= 7 + (-1w) + (-6w) + 12$ $= (-1w) + (-6w) + 7 + 12$ $= -7w + 19$	Subtracting linear expressions: $(4y + 7) - (y - 8)$ $4y + 7 \Rightarrow 4y + 7$ $-(1y - 8) \qquad + (-1y) + 8$ $3y + 15$

3.

Words	**Example**
Write the expression as a product of factors. You can use the Distributive Property.	Factor $12a - 30$ using the GCF. $12a = ② \cdot 2 \cdot ③ \cdot a$ $30 = ② \cdot ③ \cdot 5$ $\text{GCF} = 2 \cdot 3 = 6$ $12a - 30 = 6(2a) - 6(5)$ $= 6(2a - 5)$

Factoring expressions

Example	**Example**
Factor $\frac{1}{4}$ out of $\frac{1}{4}r + \frac{3}{4}$. $\frac{1}{4}r = \frac{1}{4} \cdot r$ $\frac{3}{4} = \frac{1}{4} \cdot 3$ $\frac{1}{4}r + \frac{3}{4} = \frac{1}{4} \cdot r + \frac{1}{4} \cdot 3$ $= \frac{1}{4}(r + 3)$	Factor -7 out of $-21p + 28$. $-21p = -7 \cdot 3p$ $28 = -7 \cdot (-4)$ $-21p + 28$ $= -7(3p) + (-7)(-4)$ $= -7(3p - 4)$

List of Organizers

Available at *BigIdeasMath.com*

Comparison Chart
Concept Circle
Example and Non-Example Chart
Formula Triangle
Four Square
Idea (Definition) and Examples Chart
Information Frame
Information Wheel
Notetaking Organizer
Process Diagram
Summary Triangle
Word Magnet
Y Chart

About this Organizer

A **Four Square** can be used to organize information about a topic. Students write the topic in the "bubble" in the middle of the four square. Then students write concepts related to the topic in the four squares surrounding the bubble. Any concept related to the topic can be used. Encourage students to include concepts that will help them learn the topic. Students can place their four squares on note cards to use as a quick study reference.

Technology for the Teacher

Editable Graphic Organizer

Answers

1. Terms: $11x$, $2x$; Like terms: $11x$ and $2x$
2. Terms: $9x$, $-5x$ Like terms: $9x$ and $-5x$
3. Terms: $21x$, 6, $-x$, -5; Like terms: $21x$ and $-x$; 6 and -5
4. Terms: $8x$, 14, $-3x$, 1; Like terms: $8x$ and $-3x$; 14 and 1
5. $8x$
6. $-7 + 7x$
7. $2x + 6$
8. $5x + 4$
9. $4s + 4$
10. $12t - 1$
11. $-13k + 8$
12. $\frac{7}{12}q$
13. $3n - 10$
14. $9h + 2$
15. $5(c - 3)$
16. $\frac{2}{9}(j + 3)$
17. $2.4(n + 4)$
18. $-6(z - 2)$
19. $32.67x$
20. $3n + 1$; The total number of apples you and your friend picked is one more than 3 full baskets.
21. $8w$

Technology for the Teacher

Online Assessment
Assessment Book
ExamView® Assessment Suite

Alternative Quiz Ideas

100% Quiz	Math Log
Error Notebook	Notebook Quiz
Group Quiz	Partner Quiz
Homework Quiz	Pass the Paper

Group Quiz

Students work in groups. Give each group a large index card. Each group writes five questions that they feel evaluate the material they have been studying. On a separate piece of paper, students solve the problems. When they are finished, they exchange cards with another group. The new groups work through the questions on the card.

Reteaching and Enrichment Strategies

If students need help. . .	If students got it. . .
Resources by Chapter • Practice A and Practice B • Puzzle Time Lesson Tutorials *BigIdeasMath.com*	Resources by Chapter • Enrichment and Extension • Technology Connection Game Closet at *BigIdeasMath.com* Start the next section

13.1–13.2 Quiz

Check It Out
Progress Check
BigIdeasMath.com

Identify the terms and like terms in the expression. *(Section 13.1)*

1. $11x + 2x$

2. $9x - 5x$

3. $21x + 6 - x - 5$

4. $8x + 14 - 3x + 1$

Simplify the expression. *(Section 13.1)*

5. $2(3x + x)$

6. $-7 + 3x + 4x$

7. $2x + 4 - 3x + 2 + 3x$

8. $7x + 6 + 3x - 2 - 5x$

Find the sum or difference. *(Section 13.2)*

9. $(s + 12) + (3s - 8)$

10. $(9t + 5) + (3t - 6)$

11. $(2 - k) + 3(-4k + 2)$

12. $\frac{1}{4}(q - 12) + \frac{1}{3}(q + 9)$

13. $(n - 8) - (-2n + 2)$

14. $-3(h - 4) - 2(-6h + 5)$

Factor out the coefficient of the variable. *(Section 13.2)*

15. $5c - 15$

16. $\frac{2}{9}j + \frac{2}{3}$

17. $2.4n + 9.6$

18. $-6z + 12$

19. **PAINTING** You buy the same number of brushes, rollers, and paint cans. Write an expression in simplest form that represents the total amount of money you spend for painting supplies. *(Section 13.1)*

20. **APPLES** A basket holds n apples. You pick $2n - 3$ apples, and your friend picks $n + 4$ apples. Write an expression that represents the number of apples you and your friend picked. Interpret the expression. *(Section 13.2)*

21. **EXERCISE** Write an expression in simplest form for the perimeter of the exercise mat. *(Section 13.1)*

13.3 Solving Equations Using Addition or Subtraction

Essential Question How can you use algebra tiles to solve addition or subtraction equations?

1 ACTIVITY: Solving Equations

Work with a partner. Use algebra tiles to model and solve the equation.

a. $x - 3 = -4$

Model the equation $x - 3 = -4$.

To get the variable tile by itself, remove the ____ tiles on the left side by adding ____ ____ tiles to each side.

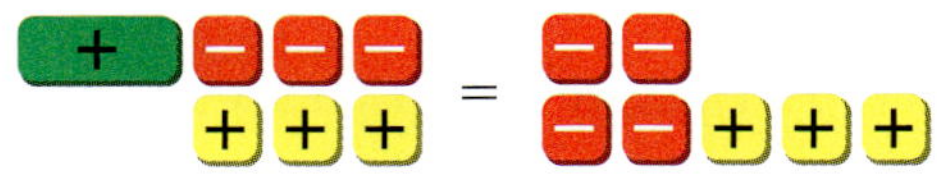

How many *zero pairs* can you remove from each side? ____ Circle them.

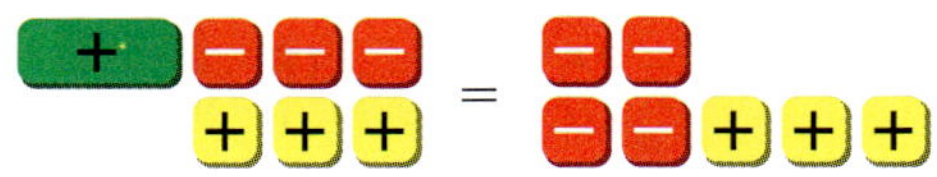

The remaining tile shows the value of x.

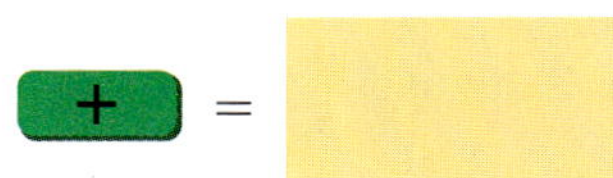

So, $x =$ ____.

b. $z - 6 = 2$ **c.** $p - 7 = -3$ **d.** $-15 = t - 5$

2 ACTIVITY: Solving Equations

Work with a partner. Use algebra tiles to model and solve the equation.

a. $-5 = n + 2$

Model the equation $-5 = n + 2$.

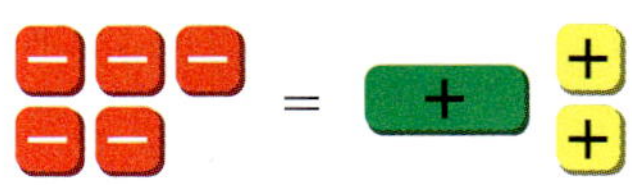

Remove the ____ tiles on the right side by adding ____ ____ tiles to each side.

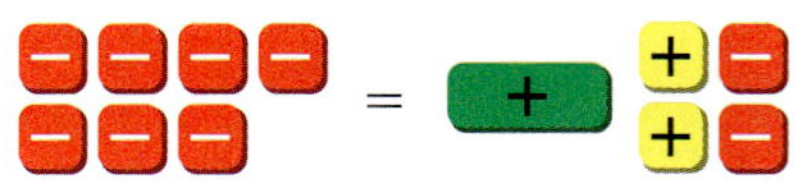

How many *zero pairs* can you remove from the right side? ____ Circle them.

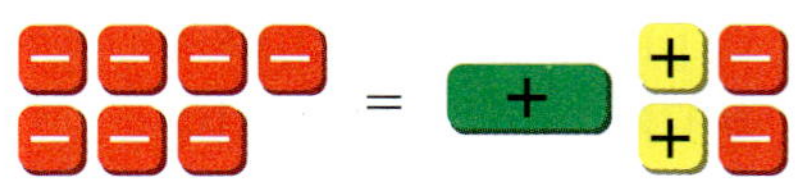

The remaining tiles show the value of n.

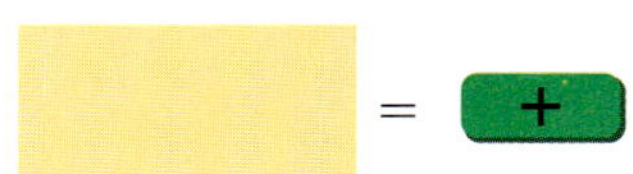

So, $n =$ ____.

b. $y + 10 = -5$ **c.** $7 + b = -1$ **d.** $8 = 12 + z$

Solving Equations

In this lesson, you will

- write simple equations.
- solve equations using addition or subtraction.
- solve real-life problems.

Laurie's Notes

Introduction

Applying Mathematical Practices

- **Model with Mathematics:** Algebra tiles can help students make sense of equations. Algebra tiles are a concrete representation, deepening student understanding of what it means to solve an equation.

Motivate

- Show students a collection of algebra tiles and ask them what the collection represents.
- ? "Can the collection be simplified? (Can you remove zero pairs?)"
- ? "What is the expression represented by the collection?"
- **Model:** As a class, model the equations $x + 3 = 7$ and $x + 2 = 5$ using algebra tiles. These do not require a zero pair to solve and will help remind students how to solve equations using algebra tiles.

Activity Notes

Activity 1

- ? There are two points to make at the beginning. First, ask students what it means to solve an equation. Second, mention that students need to think of $x - 3$ as $x + (-3)$ when using algebra tiles. (You can only *add* a positive tile or a negative tile.) to find the value of the variable that makes the equation true
- ? "To get the variable tile by itself, what do you have to do to both sides of the equation?" Students may mention removing 3 red tiles from each side or adding 3 yellow tiles to each side.
- **Reason Abstractly and Quantitatively:** Subtracting -3 is equivalent to adding 3. In the activity, the approach is to add 3 to each side. Removing the red tiles is intuitive to students when the symbolic representation is introduced, but it is adding 3 (inverse operations) that will make sense. Mathematically proficient students recognize the equivalence.
- After adding 3 to each side, the green variable-tile is equal to -1.
- Students may say they can use mental math to solve. It is the tactile experience of adding and removing tiles that you want them to experience.

Activity 2

- **Representation:** While the equations $-5 = n + 2$ and $n + 2 = -5$ are the same to mathematics teachers, students may see these as very different equations. Students even see $2 + n = -5$ as a different equation. Take time to discuss the equivalence of all three equations.
- Ask students why $x = 4$ is equivalent to $4 = x$. Student conceptions and misconceptions show up when the original equation is modified.
- ? When students finish part (a) ask, "What did you do to both sides of the equation in order to solve?" Add two red tiles.
- ? **Reason Abstractly and Quantitatively:** "Adding -2 is equivalent to what?" Subtracting 2

What Your Students Will Learn

- Write simple equations using key terms and phrases.
- Solve equations using the Addition Property of Equality or the Subtraction Property of Equality.

Previous Learning

Students have solved equations with whole numbers.

Technology for the Teacher

Dynamic Classroom

Lesson Plans
Complete Materials List

13.3 Record and Practice Journal

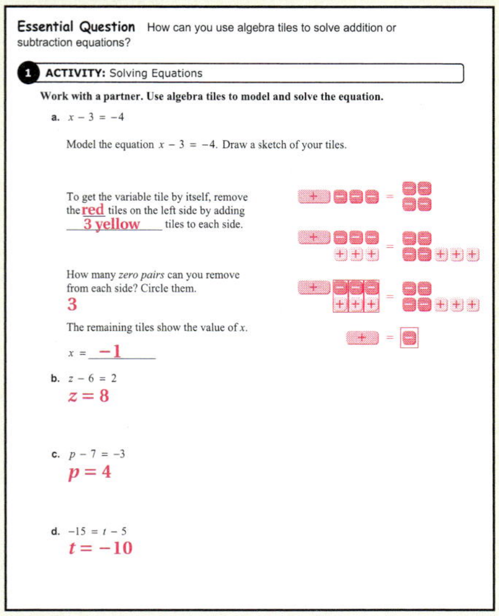

Essential Question How can you use algebra tiles to solve addition or subtraction equations?

1 ACTIVITY: Solving Equations

Work with a partner. Use algebra tiles to model and solve the equation.

a. $x - 3 = -4$

Model the equation $x - 3 = -4$. Draw a sketch of your tiles.

To get the variable tile by itself, remove the red tiles on the left side by adding 3 yellow tiles to each side.

How many *zero pairs* can you remove from each side? Circle them.
3

The remaining tiles show the value of x.

$x =$ -1

b. $z - 6 = 2$
$z = 8$

c. $p - 7 = -3$
$p = 4$

d. $-15 = t - 5$
$t = -10$

Differentiated Instruction

Kinesthetic

When working out solutions, ask two students to assist you at the board or overhead. Assign one student to the left side of the equation and the other student to the right side. Each student is responsible for performing the operations on his/her side of the equation. Emphasize that in order for both sides of the equation to remain equal, both students must perform the same operation at the same time to solve the equation.

13.3 Record and Practice Journal

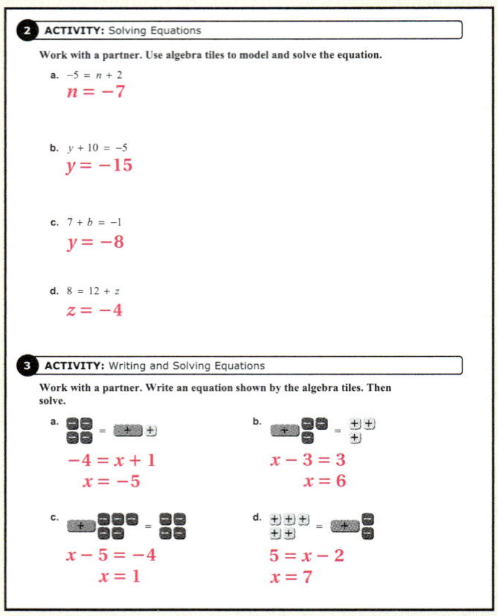

2 ACTIVITY: Solving Equations

Work with a partner. Use algebra tiles to model and solve the equation.

a. $-5 = n + 2$
$n = -7$

b. $y + 10 = -5$
$y = -15$

c. $7 + b = -1$
$y = -8$

d. $8 = 12 + z$
$z = -4$

3 ACTIVITY: Writing and Solving Equations

Work with a partner. Write an equation shown by the algebra tiles. Then solve.

a. $-4 = x + 1$
$x = -5$

b. $x - 3 = 3$
$x = 6$

c. $x - 5 = -4$
$x = 1$

d. $5 = x - 2$
$x = 7$

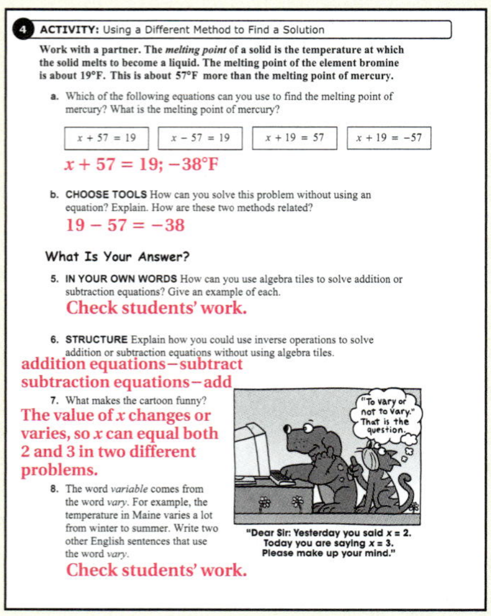

4 ACTIVITY: Using a Different Method to Find a Solution

Work with a partner. The *melting point* of a solid is the temperature at which the solid melts to become a liquid. The melting point of the element bromine is about 19°F. This is about 57°F more than the melting point of mercury.

a. Which of the following equations can you use to find the melting point of mercury? What is the melting point of mercury?

$x + 57 = 19$ | $x - 57 = 19$ | $x + 19 = 57$ | $x + 19 = -57$

$x + 57 = 19$; -38°F

b. **CHOOSE TOOLS** How can you solve this problem without using an equation? Explain. How are these two methods related?
$19 - 57 = -38$

What Is Your Answer?

5. **IN YOUR OWN WORDS** How can you use algebra tiles to solve addition or subtraction equations? Give an example of each.
Check students' work.

6. **STRUCTURE** Explain how you could use inverse operations to solve addition or subtraction equations without using algebra tiles.
addition equations—subtract
subtraction equations—add

7. What makes the cartoon funny?
The value of x changes or varies, so x can equal both 2 and 3 in two different problems.

"Dear Sir: Yesterday you said $x = 2$. Today you are saying $x = 3$. Please make up your mind."

8. The word *variable* comes from the word *vary*. For example, the temperature in Maine varies a lot from winter to summer. Write two other English sentences that use the word *vary*.
Check students' work.

Laurie's Notes

Activity 3

- **Model with Mathematics:** Students are asked to write "an" equation, not "the" equation. This is because there are many correct answers. Students can use any variable and different forms of the equation. For example, part (b) could be written as $n - 3 = 3$ or $n + (-3) = 3$.
- **?** Ask students to share the different equations they wrote.
- **Extension:** Share common tasks that *undo* one another. Examples: tying and untying your shoes, filling and emptying a glass, and opening and closing a door

Activity 4

- At this point, students should be able to figure out how to solve an equation like this. If they have difficulty, they can sketch algebra tiles on a piece of paper and find the solution, despite the large numbers.
- It is important to compare algebraic and arithmetic solutions. Explain to students that setting up equations is important since the equations will get more and more complex. Solving arithmetically may seem easy now to students.

What Is Your Answer?

- **Think-Pair-Share:** Students should read each question independently and then work in pairs to answer the questions. When they have answered the questions, the pair should compare their answers with another group and discuss any discrepancies.

Closure

- Translate the following model into symbols and explain in words how it could be solved.

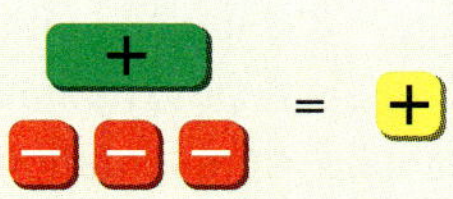

Sample answer: $x - 3 = 1$; Add three yellow tiles to each side. The result will be $x = 4$.

Math Practice

Interpret Results

How can you add tiles to make zero pairs? Explain how this helps you solve the equation.

3 ACTIVITY: Writing and Solving Equations

Work with a partner. Write an equation shown by the algebra tiles. Then solve.

a. 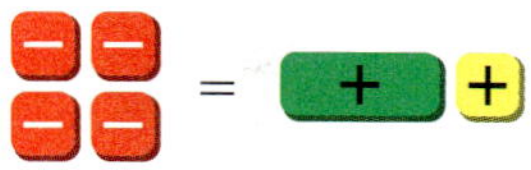

b. 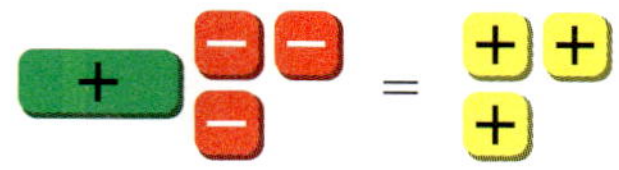

c. 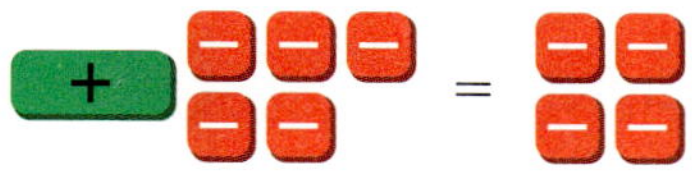

d. 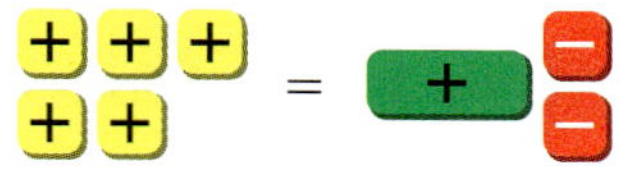

4 ACTIVITY: Using a Different Method to Find a Solution

Work with a partner. The *melting point* of a solid is the temperature at which the solid melts to become a liquid. The melting point of the element bromine is about 19°F. This is about 57°F more than the melting point of mercury.

a. Which of the following equations can you use to find the melting point of mercury? What is the melting point of mercury?

$x + 57 = 19$ $\quad$ $x - 57 = 19$ $\quad$ $x + 19 = 57$ $\quad$ $x + 19 = -57$

b. **CHOOSE TOOLS** How can you solve this problem without using an equation? Explain. How are these two methods related?

What Is Your Answer?

5. **IN YOUR OWN WORDS** How can you use algebra tiles to solve addition or subtraction equations? Give an example of each.

6. **STRUCTURE** Explain how you could use inverse operations to solve addition or subtraction equations without using algebra tiles.

7. What makes the cartoon funny?

"Dear Sir: Yesterday you said $x = 2$. Today you are saying $x = 3$. Please make up your mind."

8. The word *variable* comes from the word *vary*. For example, the temperature in Maine varies a lot from winter to summer.

   Write two other English sentences that use the word *vary*.

Practice Use what you learned about solving addition or subtraction equations to complete Exercises 5–8 on page 574.

13.3 Lesson

Check It Out
Lesson Tutorials
BigIdeasMath.com

Key Vocabulary
equivalent equations, *p. 572*

Two equations are **equivalent equations** if they have the same solutions. The Addition and Subtraction Properties of Equality can be used to write equivalent equations.

Key Ideas

Addition Property of Equality

Words Adding the same number to each side of an equation produces an equivalent equation.

Algebra If $a = b$, then $a + c = b + c$.

Subtraction Property of Equality

Words Subtracting the same number from each side of an equation produces an equivalent equation.

Algebra If $a = b$, then $a - c = b - c$.

Remember

Addition and subtraction are inverse operations.

EXAMPLE 1 Solving Equations

a. Solve $x - 5 = -1$.

$x - 5 = -1$ — Write the equation.

Undo the subtraction. → $\underline{+5 \quad +5}$ — Addition Property of Equality

$x = 4$ — Simplify.

The solution is $x = 4$.

Check

$x - 5 = -1$

$4 - 5 \stackrel{?}{=} -1$

$-1 = -1$ ✓

b. Solve $z + \frac{3}{2} = \frac{1}{2}$.

$z + \frac{3}{2} = \frac{1}{2}$ — Write the equation.

Undo the addition. → $\underline{-\frac{3}{2} \quad -\frac{3}{2}}$ — Subtraction Property of Equality

$z = -1$ — Simplify.

The solution is $z = -1$.

Check

$z + \frac{3}{2} = \frac{1}{2}$

$-1 + \frac{3}{2} \stackrel{?}{=} \frac{1}{2}$

$\frac{1}{2} = \frac{1}{2}$ ✓

On Your Own

Now You're Ready
Exercises 5–20

Solve the equation. Check your solution.

1. $p - 5 = -2$
2. $w + 13.2 = 10.4$
3. $x - \frac{5}{6} = -\frac{1}{6}$

Multi-Language Glossary at BigIdeasMath.com

Laurie's Notes

Introduction

Connect

- **Yesterday:** Students used algebra tiles to model solving equations.
- **Today:** Students will formalize the process using the Addition and Subtraction Properties of Equality.

Motivate

- Have two students stand at the front of the room and write an "=" on the board between them. Hand each the same number of items (i.e., pencils, paper clips, etc.). The students should verify that they have the same number of items. Then give each two more of the same item. Verify that the number of items they have is equal. Finally, take four of the items from each student. Verify that they have the same amount.
- **Discuss:** This is the essence of the two properties used today—as long as each side of the equation has the same amount added to it or subtracted from it, the two sides of the equation are still equal.

Lesson Notes

Key Ideas

- Discuss how the activity in the introduction modeled the two properties.

? "What are inverse operations?" Inverse operations undo one another.

- Ask students to give examples of inverse operations. They may say addition and subtraction or multiplication and division. They may even offer actions such as opening and closing a door.

Example 1

- Work through each part as a class. Notice that a vertical format is used. Use color to show the quantity being added to or subtracted from each side.

? **Discuss:** The equations in parts (a) and (b) have the variable on the left.

  - "Would part (a) have the same solution if it was written as $-1 = x - 5$?" yes
  - "Would part (b) have the same solution if it was written as $\frac{1}{2} = z + \frac{3}{2}$?" yes

On Your Own

- After students have completed Example 1, they should be able to do these questions independently.

Words of Wisdom

- **Struggling Students:** If students have difficulty with the *On Your Own* questions, assess whether it is algebraic (how to solve equations) or computational (how to add or subtract rational numbers). Use this information to guide your instruction. Provide colored pencils so students can record the quantity being added to or subtracted from each side.
- Encourage students to be neat and to keep their equal signs lined up.

Goal Today's lesson is solving equations using addition or subtraction.

Technology for the Teacher

Dynamic Classroom

Lesson Tutorials
Lesson Plans
Answer Presentation Tool

Extra Example 1

a. Solve $t + 6 = -5$. -11

b. Solve $y - \frac{4}{5} = -\frac{2}{5}$. $\frac{2}{5}$

On Your Own

1. $p = 3$
2. $w = -2.8$
3. $x = \frac{2}{3}$

Laurie's Notes

Extra Example 2

You spent \$7.25 this week. This is \$3.65 less than you spent last week. Write and solve an equation to find the amount s you spent last week.
$s - 3.65 = 7.25$, \$10.90

On Your Own

4. $P - 145.25 = 120.50$

Extra Example 3

You have -1 point after Level 2 of a video game. Your score is 24 points less than your friend's score. Write and solve an equation to find your friend's score after Level 2. $-1 = f - 24$, 23 points

On Your Own

5. 15 points

English Language Learners

Vocabulary

In this section, students learn to use *inverse* (or *opposite*) operations to solve equations. Students use addition to solve a subtraction equation and use subtraction to solve an addition equation. Review these pairs of words that are essential to understanding mathematics. Give students one word of a pair and ask them to provide the opposite.

odd, even	positive, negative
add, subtract	sum, difference
multiply, divide	product, quotient
plus, minus	

Example 2

- **Financial Literacy:** Discuss the word *profit* and how it is computed: income − expenses = profit.
- The second sentence contains key information. When translated into symbols, students can tell that "this profit" refers to "the profit this week."
- The color-coding in this text is very helpful in assisting students as they translate from words to symbols. Students may not recognize that "is" translates to "equals,' so give a quick example. (Evan is $5\frac{1}{2}$ feet tall means the same as $E = 5.5$.)

On Your Own

- **Neighbor Check:** Have students work independently and then have their neighbors check their work. Have students discuss any discrepancies.

Example 3

- This example includes a line graph as a way to present information about the problem. Take time to have students *read and interpret* the information in the line graph.

? Here are some questions to ask about the graph.

  - "What information is displayed on each axis of the line graph?" The horizontal axis shows the level of a video game, and the vertical axis shows the number of points scored.
  - "Were the scores ever tied?" Yes, at the very start and at some point in Level 3.
  - "Who was ahead after Level 2?" your friend
  - "What does '33 points' on the line graph mean?" It is the difference of your score and your friend's score after Level 4.
  - "Describe each player's performance from start to finish." *Sample answer:* Your friend did better than you at the beginning, but after Level 2 your score increased and your friend's score decreased. You ended up with 33 more points than your friend.

- **Model with Mathematics:** Take time to discuss the verbal model and how it translates information from the line graph. Mathematically proficient students are able to identify important quantities in a graph and make use of them to solve problems.

On Your Own

- Encourage students to write the key words and phrases using colored pencils and then translate the words to symbols.

Closure

- **Exit Ticket:**
  $p - 3.5 = -1.3$ 2.2 $-4.2 + m = 8.6$ 12.8

EXAMPLE 2 Writing an Equation

A company has a profit of \$750 this week. This profit is \$900 more than the profit P last week. Which equation can be used to find P?

(A) $750 = 900 - P$ (B) $750 = P + 900$

(C) $900 = P - 750$ (D) $900 = P + 750$

Words	The profit this week	is	\$900	more than	the profit last week.
Equation	750	=	P	+	900

The equation is $750 = P + 900$. The correct answer is (B).

On Your Own

Now You're Ready Exercises 22–25

4. A company has a profit of \$120.50 today. This profit is \$145.25 less than the profit P yesterday. Write an equation that can be used to find P.

EXAMPLE 3 Real-Life Application

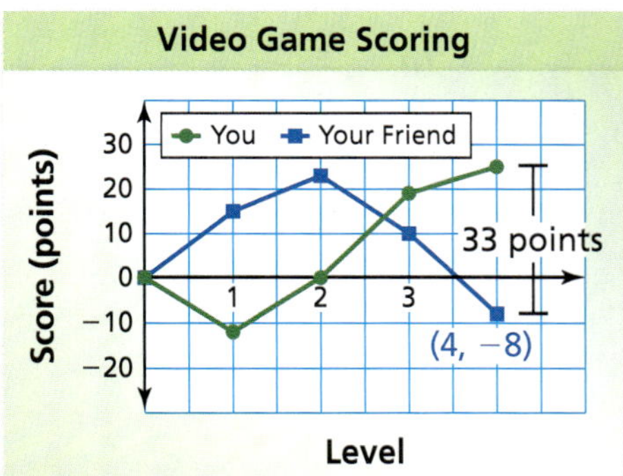

The line graph shows the scoring while you and your friend played a video game. Write and solve an equation to find your score after Level 4.

You can determine the following from the graph.

Words	Your friend's score	is	33 points	less than	your score.
Variable	Let s be your score after Level 4.				
Equation	−8	=	s	−	33

$-8 = s - 33$ Write equation.

$\underline{+33} \quad \underline{+33}$ Addition Property of Equality

$25 = s$ Simplify.

Your score after Level 4 is 25 points.

Reasonable? From the graph, your score after Level 4 is between 20 points and 30 points. So, 25 points is a reasonable answer.

On Your Own

5. **WHAT IF?** You have −12 points after Level 1. Your score is 27 points less than your friend's score. What is your friend's score?

13.3 Exercises

Check It Out
Help with Homework
BigIdeasMath.com

Vocabulary and Concept Check

1. **VOCABULARY** What property would you use to solve $m + 6 = -4$?
2. **VOCABULARY** Name two inverse operations.
3. **WRITING** Are the equations $m + 3 = -5$ and $m = -2$ equivalent? Explain.
4. **WHICH ONE DOESN'T BELONG?** Which equation does *not* belong with the other three? Explain your reasoning.

$x + 3 = -1$ | $x + 1 = -5$ | $x - 2 = -6$ | $x - 9 = -13$

Practice and Problem Solving

Solve the equation. Check your solution.

1 5. $a - 6 = 13$
6. $-3 = z - 8$
7. $-14 = k + 6$
8. $x + 4 = -14$
9. $c - 7.6 = -4$
10. $-10.1 = w + 5.3$
11. $\frac{1}{2} = q + \frac{2}{3}$
12. $p - 3\frac{1}{6} = -2\frac{1}{2}$
13. $g - 9 = -19$
14. $-9.3 = d - 3.4$
15. $4.58 + y = 2.5$
16. $x - 5.2 = -18.73$
17. $q + \frac{5}{9} = \frac{1}{6}$
18. $-2\frac{1}{4} = r - \frac{4}{5}$
19. $w + 3\frac{3}{8} = 1\frac{5}{6}$
20. $4\frac{2}{5} + k = -3\frac{2}{11}$

21. **ERROR ANALYSIS** Describe and correct the error in finding the solution.

$$\begin{array}{rcr} x + 8 & = & 10 \\ +8 & & +8 \\ \hline x & = & 18 \end{array}$$

Write the word sentence as an equation. Then solve.

2 22. 4 less than a number n is -15.
23. 10 more than a number c is 3.
24. The sum of a number y and -3 is -8.
25. The difference between a number p and 6 is -14.

In Exercises 26–28, write an equation. Then solve.

26. **DRY ICE** The temperature of dry ice is -109.3°F. This is 184.9°F less than the outside temperature. What is the outside temperature?
27. **PROFIT** A company makes a profit of \$1.38 million. This is \$2.54 million more than last year. What was the profit last year?
28. **HELICOPTER** The difference in elevation of a helicopter and a submarine is $18\frac{1}{2}$ meters. The elevation of the submarine is $-7\frac{3}{4}$ meters. What is the elevation of the helicopter?

Assignment Guide and Homework Check

Level	Assignment	Homework Check
Advanced	1–8, 14–20 even, 21, 22–40 even, 41–45	18, 30, 34, 36

Common Errors

- **Exercises 5–20** Students may use the same operation in solving for x instead of the inverse operation. Demonstrate that this will not work to simplify the equation. Students most likely ignored the side with the variable when they made this mistake. Remind them to check their answers in the original equation.
- **Exercises 5–20** Students may add or subtract the number on the side of the equation without the variable. For example, they might write $-14 + 14 = k + 6 + 14$ instead of $-14 - 6 = x + 6 - 6$. Remind students that they are trying to get the variable by itself, so they have to start with the side that the variable is on and use the inverse of that operation.
- **Exercises 29–31** Students may try to use inverse operations to combine like terms. Remind them that inverse operations are used on both sides of the equation.

13.3 Record and Practice Journal

Solve the equation. Check your solution.

1. $y + 12 = -26$ **−38**
2. $15 + c = -12$ **−27**
3. $-16 = d + 21$ **−37**
4. $n + 12.8 = -0.3$ **−13.1**
5. $1\frac{1}{8} = g - 4\frac{2}{5}$ $5\frac{21}{40}$
6. $-5.47 + k = -14.19$ **−8.72**

Write the word sentence as an equation. Then solve.

7. 42 less than x is -50. $x - 42 = -50;\ -8$
8. 32 is the sum of a number z and 9. $32 = z + 9;\ 23$
9. A clothing company makes a profit of $2.3 million. This is $4.1 million more than last year. What was the profit last year? **−$1.8 million**
10. A drop on a wooden roller coaster is $-98\frac{1}{2}$ feet. A drop on a steel roller coaster is $100\frac{1}{4}$ feet lower than the drop on the wooden roller coaster. What is the drop on the steel roller coaster? $-198\frac{3}{4}$ **ft**

Vocabulary and Concept Check

1. Subtraction Property of Equality
2. *Sample answer:* addition and subtraction
3. No, $m = -8$ not -2 in the first equation.
4. The equation $x + 1 = -5$ does not belong because its solution is $x = -6$ and the solution of the other equations is $x = -4$.

Practice and Problem Solving

5. $a = 19$
6. $z = 5$
7. $k = -20$
8. $x = -18$
9. $c = 3.6$
10. $w = -15.4$
11. $q = -\frac{1}{6}$
12. $p = \frac{2}{3}$
13. $g = -10$
14. $d = -5.9$
15. $y = -2.08$
16. $x = -13.53$
17. $q = -\frac{7}{18}$
18. $r = -1\frac{9}{20}$
19. $w = -1\frac{13}{24}$
20. $k = -7\frac{32}{55}$
21. The 8 should have been subtracted rather than added.

$$\begin{aligned} x + 8 &= 10 \\ -8 &\quad -8 \\ x &= 2 \end{aligned}$$

22. $n - 4 = -15;\ n = -11$
23. $c + 10 = 3;\ c = -7$
24. $y + (-3) = -8;\ y = -5$
25. $p - 6 = -14;\ p = -8$
26. $t - 184.9 = -109.3;\ 75.6°\text{F}$
27. $p + 2.54 = 1.38;$ $-\$1.16$ million
28. $h - \left(-7\frac{3}{4}\right) = 18\frac{1}{2};\ 10\frac{3}{4}$ m

Practice and Problem Solving

29. $x + 8 = 12$; 4 cm

30. $x + 20.4 = 24.2$; 3.8 in.

31. $x + 22.7 = 34.6$; 11.9 ft

32. $305 = h + 153$; 152 ft

33. See *Taking Math Deeper.*

34. $d + 24\frac{1}{3} = 65\frac{3}{5}$; $41\frac{4}{15}$ km

35. $m + 30.3 + 40.8 = 180$; 108.9°

36. $p + 63.43 + 87.15 + 81.96 = 311.62$; more than 79.08

37. -9

38. $2, -2$

39. $6, -6$

40. $13, -13$

Fair Game Review

41. -56 **42.** -72

43. -9 **44.** -6.5

45. B

Mini-Assessment

Solve the equation.

1. $x + 3.6 = -4.75$ $x = -8.35$
2. $-15.8 = y - 24.3$ $y = 8.5$
3. $t - 2\frac{2}{3} = -\frac{5}{2}$ $t = \frac{1}{6}$
4. $-\frac{5}{6} = z + \frac{1}{8}$ $z = -\frac{23}{24}$
5. You withdrew \$47.25 from your checking account. Now your balance is $-\$23.75$. Write and solve an equation to find the amount of money in your account before you withdrew the money. $x - 47.25 = -23.75$; \$23.50

Taking Math Deeper

Exercise 33

It's surprising how difficult this problem can be for students. There are two reasons for this. One is that you are not given the location of 0 on the vertical number line. The second is that the information is not given in the order it is used.

1. Draw a vertical number line. Locate the jumping platform at 0.

2. Draw the first jump. Draw the second jump so that the first jump is higher.

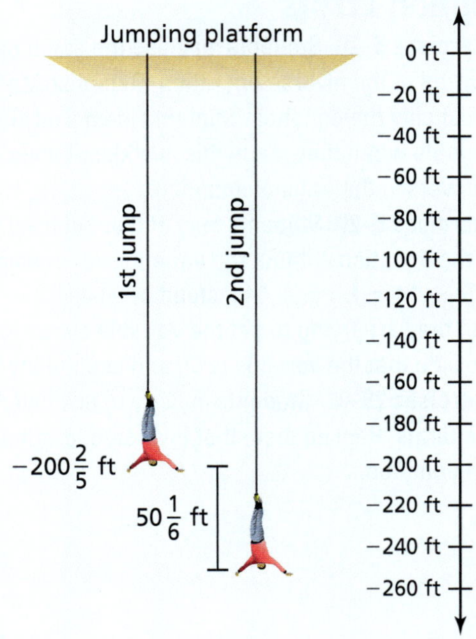

3. Subtract to find the height of the second jump.

$$-200\frac{2}{5} - 50\frac{1}{6} = -250 - \frac{2}{5} - \frac{1}{6}$$
$$= -250 - \frac{12}{30} - \frac{5}{30}$$
$$= -250\frac{17}{30} \text{ ft}$$

Project

Research bungee jumping. What safety requirements are necessary for a bungee jumping business?

Reteaching and Enrichment Strategies

If students need help. . .	If students got it. . .
Resources by Chapter • Practice A and Practice B • Puzzle Time Record and Practice Journal Practice Differentiating the Lesson Lesson Tutorials Skills Review Handbook	Resources by Chapter • Enrichment and Extension • Technology Connection Start the next section

GEOMETRY **Write and solve an equation to find the unknown side length.**

29. Perimeter = 12 cm

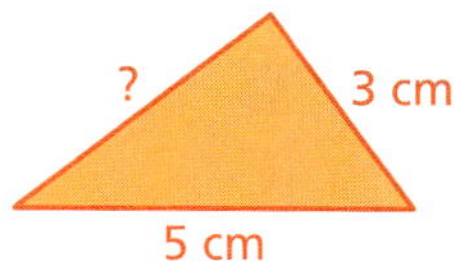

30. Perimeter = 24.2 in.

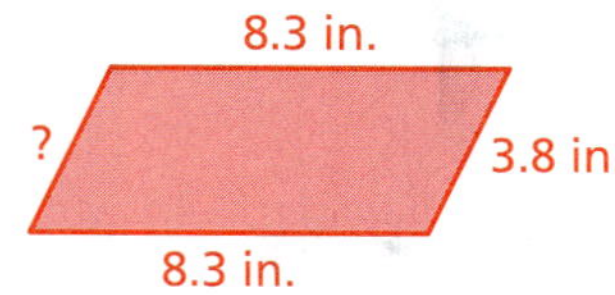

31. Perimeter = 34.6 ft

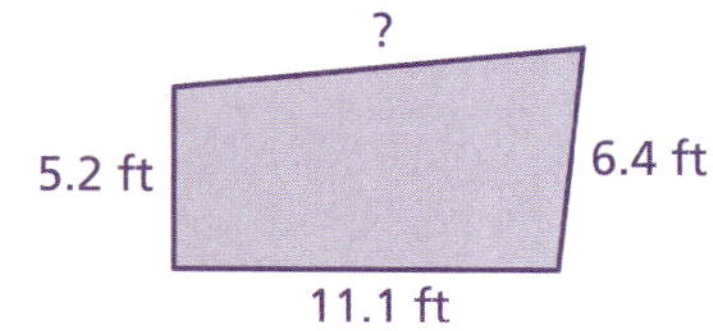

In Exercises 32–36, write an equation. Then solve.

32. **STATUE OF LIBERTY** The total height of the Statue of Liberty and its pedestal is 153 feet more than the height of the statue. What is the height of the statue?

33. **BUNGEE JUMPING** Your first jump is $50\frac{1}{6}$ feet higher than your second jump. Your first jump reaches $-200\frac{2}{5}$ feet. What is the height of your second jump?

34. **TRAVEL** Boatesville is $65\frac{3}{5}$ kilometers from Stanton. A bus traveling from Stanton is $24\frac{1}{3}$ kilometers from Boatesville. How far has the bus traveled?

35. **GEOMETRY** The sum of the measures of the angles of a triangle equals 180°. What is the measure of the missing angle?

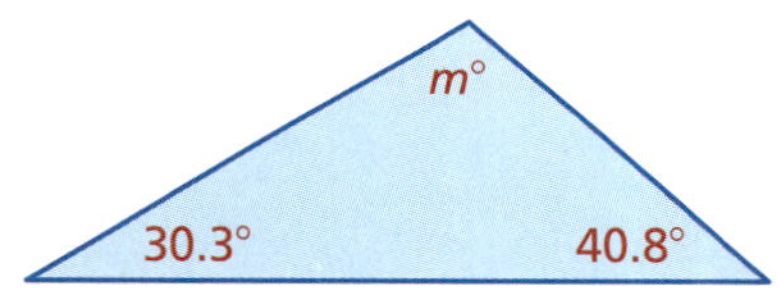

36. **SKATEBOARDING** The table shows your scores in a skateboarding competition. The leader has 311.62 points. What score do you need in the fourth round to win?

Round	1	2	3	4
Points	63.43	87.15	81.96	?

37. **CRITICAL THINKING** Find the value of $2x - 1$ when $x + 6 = 2$.

Critical Thinking **Find the values of x.**

38. $|x| = 2$

39. $|x| - 2 = 4$

40. $|x| + 5 = 18$

Fair Game Review What you learned in previous grades & lessons

Multiply or divide. *(Section 11.4 and Section 11.5)*

41. -7×8

42. $6 \times (-12)$

43. $18 \div (-2)$

44. $-26 \div 4$

45. **MULTIPLE CHOICE** A class of 144 students voted for a class president. Three-fourths of the students voted for you. Of the students who voted for you, $\frac{5}{9}$ are female. How many female students voted for you? *(Section 12.4)*

Ⓐ 50 Ⓑ 60 Ⓒ 80 Ⓓ 108

13.4 Solving Equations Using Multiplication or Division

Essential Question How can you use multiplication or division to solve equations?

1 ACTIVITY: Using Division to Solve Equations

Work with a partner. Use algebra tiles to model and solve the equation.

a. $3x = -12$

Model the equation $3x = -12$.

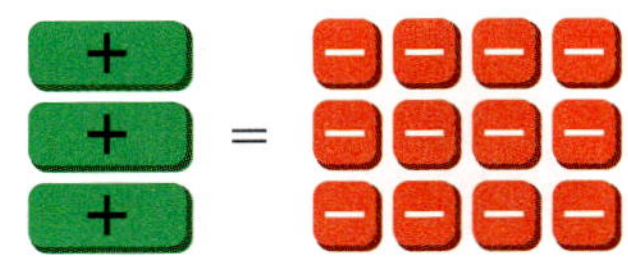

Your goal is to get one variable tile by itself. Because there are ____ variable tiles, divide the ____ tiles into ____ equal groups. Circle the groups.

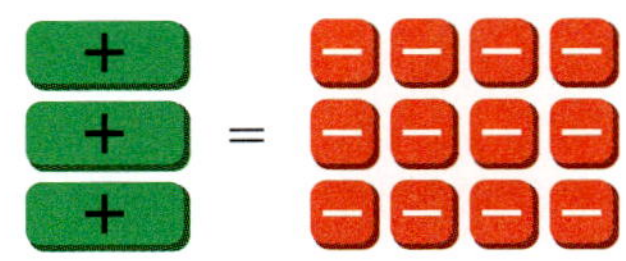

Keep one of the groups. This shows the value of x.

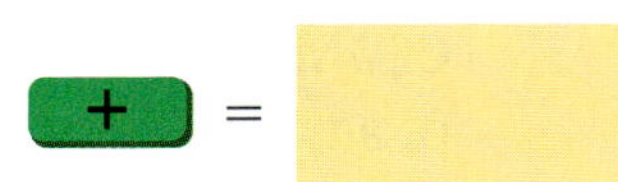

So, $x =$ ____.

b. $2k = -8$

c. $-15 = 3t$

d. $-20 = 5m$

e. $4h = -16$

2 ACTIVITY: Writing and Solving Equations

Work with a partner. Write an equation shown by the algebra tiles. Then solve.

a.

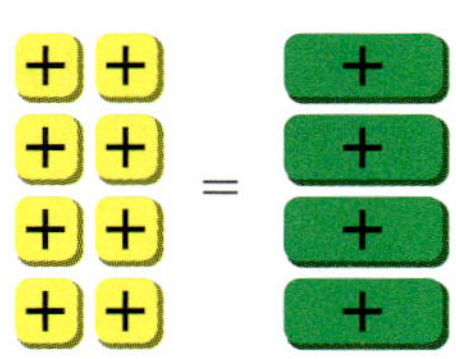

b.

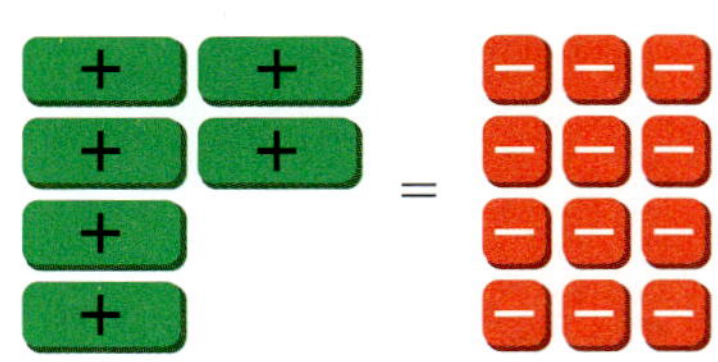

c.

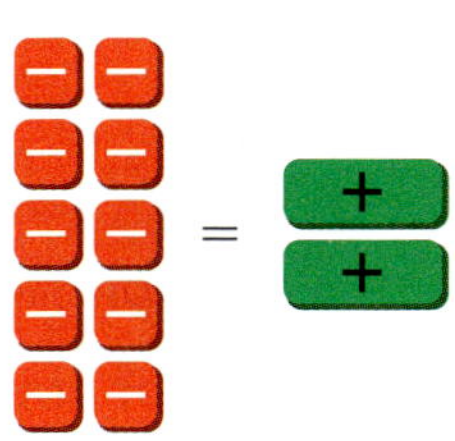

d.

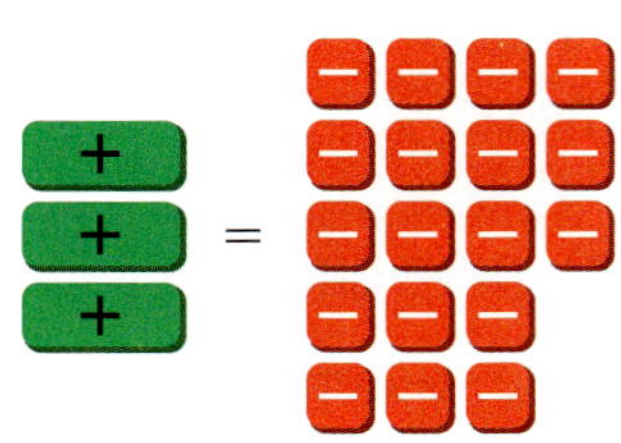

Solving Equations

In this lesson, you will

- solve equations using multiplication or division.
- solve real-life problems.

Laurie's Notes

Introduction

Applying Mathematical Practices

- **Model with Mathematics:** Algebra tiles can help students make sense of equations. Algebra tiles are a concrete representation, deepening student understanding of what it means to solve an equation.

Motivate

- **Model:** Display two green variable-tiles and four yellow integer-tiles to the class.

? "If two green tiles equal four yellow tiles, what does one green tile equal?" two yellow tiles

? "How did you decide that one green tile equals two yellow tiles?" Divide each side into groups. The number of groups is the number of variable-tiles.

Activity Notes

Activity 1

- Model the first equation as students model the equation at their desks. Write the corresponding algebraic equation represented by the tiles with the first and last step. Encourage students to do the same.

? **Discuss:** Remind students that the goal is to find the value of just one green variable-tile. "If three green tiles equal 12 red tiles, what is the value of each green tile? How did you find your answer?" 4; To get one green tile, you need three groups. So, divide the 12 red tiles into three equal groups.

- Remind students that variables can be on either side of the equation. If students are more comfortable with variables on the left, they can write part (c) as $3t = -15$.

Activity 2

- **Think-Pair-Share:** After students work on the problems in pairs, ask for volunteers to work the problems for the class. Listen for how students describe the solutions.

? "Why is it difficult to model the equation $\frac{1}{3}x = 6$ with algebra tiles?"

You can't show $\frac{1}{3}$ of a green variable-tile, but you can talk about the meaning. If $\frac{1}{3}$ of a green variable-tile is 6, then $\frac{2}{3}$ would be 12, and $\frac{3}{3}$ (or a whole green tile) would be 18.

What Your Students Will Learn

- Solve equations using the Multiplication Property of Equality or the Division Property of Equality.

Previous Learning

Students used algebra tiles to model solving equations involving addition and subtraction.

Technology for the Teacher

Dynamic Classroom

Lesson Plans
Complete Materials List

13.4 Record and Practice Journal

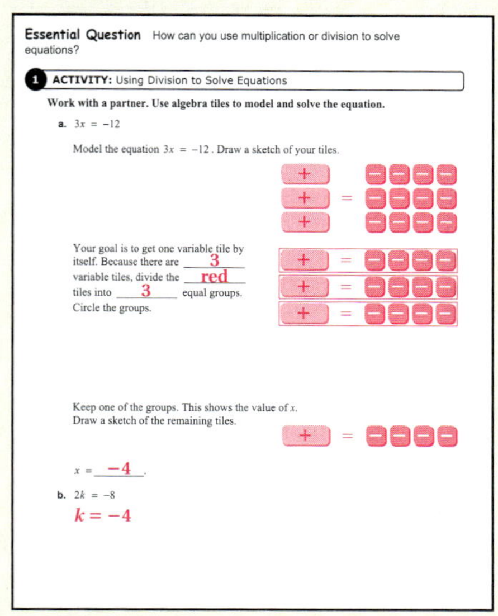

Essential Question How can you use multiplication or division to solve equations?

1 ACTIVITY: Using Division to Solve Equations

Work with a partner. Use algebra tiles to model and solve the equation.

a. $3x = -12$

Model the equation $3x = -12$. Draw a sketch of your tiles.

Your goal is to get one variable tile by itself. Because there are 3 variable tiles, divide the red tiles into 3 equal groups. Circle the groups.

Keep one of the groups. This shows the value of x. Draw a sketch of the remaining tiles.

$x = -4$.

b. $2k = -8$

$k = -4$

Differentiated Instruction

Visual

To model a division equation, such as $\frac{d}{4} = -3$, use the variable tile to represent the fractional part of a variable.

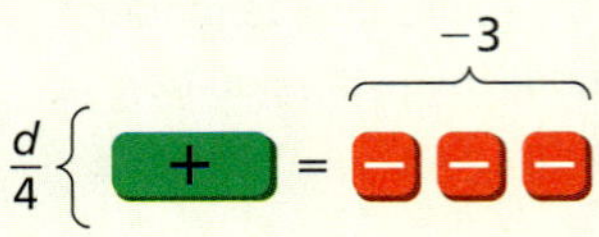

Then, model the solution.

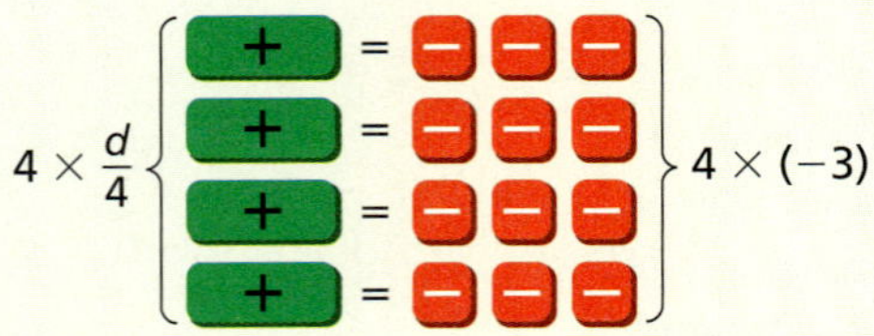

So, $d = -12$.

13.4 Record and Practice Journal

c. $-15 = 3t$
$t = -5$

d. $-20 = 5m$
$m = -4$

e. $4h = -16$
$h = -4$

2 ACTIVITY: Writing and Solving Equations

Work with a partner. Write an equation shown by the algebra tiles. Then solve.

a. 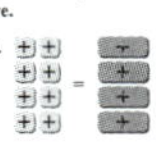
$8 = 4x$
$x = 2$

b. 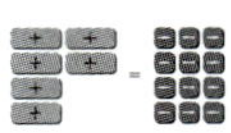
$6x = -12$
$x = -2$

c. 
$-10 = 2x$
$x = -5$

d. 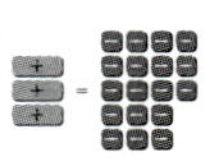
$3x = -18$
$x = -6$

3 ACTIVITY: Using a Different Method to Find a Solution

Work with a partner. Choose the equation you can use to solve each problem. Solve the equation. Then explain how to solve the problem without using an equation. How are the two methods related?

a. For the final part of a race, a handcyclist travels 32 feet each second across a distance of 400 feet. How many seconds does it take for the handcyclist to travel the last 400 feet of the race?

$32x = 400$	$400x = 32$	$\frac{x}{32} = 400$	$\frac{x}{400} = 32$

$32x = 400$; 12.5 sec

b. The melting point of the element radon is about −96°F. The melting point of nitrogen is about 3.6 times the melting point of radon. What is the melting point of nitrogen?

$3.6x = -96$	$x + 96 = 3.6$	$\frac{x}{3.6} = -96$	$-96x = 3.6$

$\frac{x}{3.6} = -96$; −345.6°F

c. This year, a hardware store has a profit of −$6.0 million. This profit is $\frac{3}{4}$ of last year's profit. What is last year's profit?

$\frac{x}{-6} = \frac{3}{4}$	$-6x = \frac{3}{4}$	$\frac{3}{4} + x = -6$	$\frac{3}{4}x = -6$

$\frac{3}{4}x = -6$; −$8 million

What Is Your Answer?

4. IN YOUR OWN WORDS How can you use multiplication or division to solve equations? Give an example of each.
Sample answer: Use inverse operations.

Laurie's Notes

Activity 3

- Students often comment that they solve the problem by just doing a computation. Students might tell you, "I just know what to do" and say having to write and solve an equation makes it harder. You want students to practice the process of identifying what the unknown is, how to represent the unknown with a variable, and then use inverse operations to solve.
- **Make Sense of Problems and Persevere in Solving Them:** Having four possible equations requires students to make sense of each equation and to determine which equation represents the problem. Students need to understand the arithmetic solution and look at the various steps needed to solve algebraically in order to see the relationship.
- It is important to compare algebraic and arithmetic solutions. Explain to students that setting up equations is important since the equations will get more and more complex. Solving arithmetically may seem easy now to students.

What Is Your Answer?

- **Neighbor Check:** Have students work independently and then have their neighbors check their work. Have students discuss any discrepancies.

Closure

- **Exit Ticket:** Solve $\frac{x}{2} = -14$ and $2x = -14$. -28; -7

3 ACTIVITY: Using a Different Method to Find a Solution

Math Practice

Analyze Givens

How can you use the given information to decide which equation represents the situation?

Work with a partner. Choose the equation you can use to solve each problem. Solve the equation. Then explain how to solve the problem without using an equation. How are the two methods related?

a. For the final part of a race, a handcyclist travels 32 feet each second across a distance of 400 feet. How many seconds does it take for the handcyclist to travel the last 400 feet of the race?

$32x = 400$	$400x = 32$
$\frac{x}{32} = 400$	$\frac{x}{400} = 32$

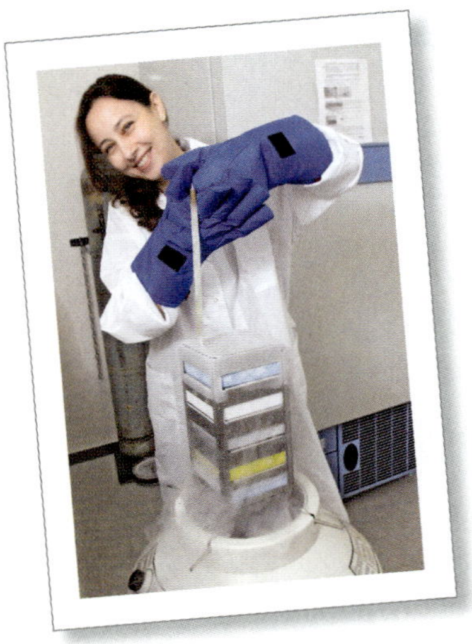

b. The melting point of the element radon is about −96°F. The melting point of nitrogen is about 3.6 times the melting point of radon. What is the melting point of nitrogen?

$3.6x = -96$	$x + 96 = 3.6$
$\frac{x}{3.6} = -96$	$-96x = 3.6$

c. This year, a hardware store has a profit of −$6.0 million. This profit is $\frac{3}{4}$ of last year's profit. What is last year's profit?

$\frac{x}{-6} = \frac{3}{4}$	$-6x = \frac{3}{4}$
$\frac{3}{4} + x = -6$	$\frac{3}{4}x = -6$

What Is Your Answer?

4. **IN YOUR OWN WORDS** How can you use multiplication or division to solve equations? Give an example of each.

Practice Use what you learned about solving equations to complete Exercises 7–10 on page 580.

13.4 Lesson

Check It Out
Lesson Tutorials
BigIdeasMath.com

Key Ideas

Multiplication Property of Equality

Words Multiplying each side of an equation by the same number produces an equivalent equation.

Algebra If $a = b$, then $a \cdot c = b \cdot c$.

Division Property of Equality

Words Dividing each side of an equation by the same number produces an equivalent equation.

Algebra If $a = b$, then $a \div c = b \div c$, $c \neq 0$.

Remember

Multiplication and division are inverse operations.

EXAMPLE 1 Solving Equations

a. Solve $\frac{x}{3} = -6$.

$\frac{x}{3} = -6$ Write the equation.

Undo the division. → $3 \cdot \frac{x}{3} = 3 \cdot (-6)$ Multiplication Property of Equality

$x = -18$ Simplify.

The solution is $x = -18$.

Check

$\frac{x}{3} = -6$

$\frac{-18}{3} \stackrel{?}{=} -6$

$-6 = -6$ ✓

b. Solve $18 = -4y$.

$18 = -4y$ Write the equation.

Undo the multiplication. → $\frac{18}{-4} = \frac{-4y}{-4}$ Division Property of Equality

$-4.5 = y$ Simplify.

The solution is $y = -4.5$.

Check

$18 = -4y$

$18 \stackrel{?}{=} -4(-4.5)$

$18 = 18$ ✓

On Your Own

Now You're Ready
Exercises 7–18

Solve the equation. Check your solution.

1. $\frac{x}{5} = -2$ **2.** $-a = -24$ **3.** $3 = -1.5n$

Laurie's Notes

Introduction

Connect

- **Yesterday:** Students used algebra tiles to model solving equations.
- **Today:** Students will formalize the process of solving equations using the Multiplication and Division Properties of Equality.

Motivate

- Have two students stand at the front of the room. Hand a third student an odd number of index cards without telling the student how many cards he or she has been given.
- Ask the student with the index cards to share them equally between the two students. The student may pause when he or she realizes that there is an odd number of cards. Give the student time to realize that the remaining card needs to be divided into two pieces and that each student will receive one-half of a card.
- Ask the students holding the index cards to verify that they have the same number of cards.

Lesson Notes

Key Ideas

- Write the Properties of Equality on the board.
- ? If you started the class with the index card activity, then ask students which property was modeled in the opening activity. Division Property of Equality
- Remind students of how multiplication and division are represented with a variable, such as $4x$ and $\frac{x}{4}$.

Example 1

- Work through each problem. If possible, use colors to show the multiplication or division on each side of the equation.
- **FYI:** Note that the -6 is written in parentheses in the solution of part (a). When you do this step in class, you may want to write both numbers in parentheses $(3)(-6)$ to avoid students thinking that the multiplication dot is a decimal point.
- ? "Could the problem be represented as $(-6) \cdot 3$ instead of $3 \cdot (-6)$? Why or why not?" yes; This is an example of the Commutative Property of Multiplication.

On Your Own

- If students have difficulty as they work these problems, assess whether it is algebraic (how to solve equations) or computational (how to multiply or divide rational numbers). Use this information to guide your instruction.
- You may want to provide colored pencils to students so that they can highlight the quantity being multiplied or divided on each side.
- Encourage students to be neat and to keep the equal signs lined up.

Goal Today's lesson is solving equations using multiplication or division.

Technology for the Teacher

Dynamic Classroom

Lesson Tutorials
Lesson Plans
Answer Presentation Tool

Extra Example 1

a. Solve $\frac{c}{8} = -7$. -56

b. Solve $-5p = -32$. 6.4

On Your Own

1. $x = -10$
2. $a = 24$
3. $n = -2$

Extra Example 2

Solve $-\frac{5}{9}m = 25$. -45

On Your Own

4. $x = -21$
5. $b = -3\frac{1}{8}$
6. $h = -24$

Extra Example 3

The record low temperature in Nevada is −50°F. The record low temperature in Montana is 1.4 times the record low temperature in Nevada. What is the record low temperature in Montana? −70°F

On Your Own

7. −80°F

English Language Learners

Graphic Organizer
When solving a one-step equation, students must remember to isolate the variable. Encourage students to make a table in their notebooks that will help them remember which operation to use to solve a one-step equation.

Operation on Variable	Operation to Solve	Example
Addition	Subtraction	$a + 3 = -5$
Subtraction	Addition	$b - 4 = 2$
Multiplication	Division	$c \cdot (-2) = 7$
Division	Multiplication	$\frac{d}{-4} = -8$

Laurie's Notes

Example 2

- Explain that $\frac{x}{3}$ and $\frac{1}{3}x$ are equivalent. Discuss how to multiply a fraction and a whole number: $\frac{1}{3}x = \frac{1}{3} \cdot \frac{x}{1} = \frac{x}{3}$. Repeat to show that $\frac{4}{5}x = \frac{4x}{5}$.
- ? "What is x being multiplied by?" $-\frac{4}{5}$
- ? "Can you divide both sides by $-\frac{4}{5}$?" yes
- **Look for and Make Use of Structure:** Dividing by a fraction is equivalent to multiplying by its reciprocal.
- Students may need a quick review of multiplying fractions.
- **FYI:** You may want to emphasize that you are dividing each side by $-\frac{4}{5}$. This will emphasize the connection to multiplying by the reciprocal $-\frac{5}{4}$, and that both of these processes are equivalent.

Words of Wisdom

- When checking a solution, read it out loud. It is helpful for students to hear (as well as to see) what they are reading.

On Your Own

- **Think-Pair-Share:** Students should read each question independently and then work in pairs to answer the questions. When they have answered the questions, the pair should compare their answers with another group and discuss any discrepancies.

Example 3

- Encourage students to look at the artwork next to the problem. The first sentence contains key information that is translated into the equation.
- **Model with Mathematics:** The color-coding in the text is very helpful in assisting students as they translate from words to symbols. You may want to use color-coding when you do other examples.

On Your Own

- **Think-Pair-Share:** Students should read each question independently and then work in pairs to answer the questions. When they have answered the questions, the pair should compare their answers with another group and discuss any discrepancies.

Closure

- **Writing:** The variable in a one-step equation is being multiplied by $-\frac{3}{4}$. Describe how to solve the equation for x. You divide both sides of the equation by $-\frac{3}{4}$, which is the same as multiplying by the reciprocal. So, you multiply both sides of the equation by $-\frac{4}{3}$ and simplify.

Assignment Guide and Homework Check

Level	Assignment	Homework Check
Advanced	1–10, 16–22 even, 23, 30–40 even, 39, 41–45	20, 30, 36, 39

Common Errors

- **Exercises 7–18** When the variable is multiplied by a negative number, students may not remember to keep the negative with the number and will really solve for $-x$ instead of x. Do an example of one of these problems on the board. Solve for $-x$ and ask students if x is by itself. If they do not realize it, remind them that there is a -1 in front of the variable and that they must divide by -1 to "get the variable by itself."
- **Exercises 19–22** Students may not understand why they should multiply by the reciprocal and may try to divide by the reciprocal. Ask students how they would solve the problem without using the reciprocal (divide by the fractional coefficient). Then ask how to divide a number by a fraction (multiply by the reciprocal). It is a short cut to multiply by the reciprocal from the beginning.
- **Exercises 19–22** If students have a difficult time grasping the idea of multiplying by the reciprocal, have them write out each step instead of just multiplying by the reciprocal. Remind students to check their answers.

13.4 Record and Practice Journal

Solve the equation. Check your solution.

1. $\frac{d}{5} = -6$ — -30
2. $8x = -6$ — $-\frac{3}{4}$
3. $-15 = \frac{z}{-2}$ — 30
4. $3.2n = -0.8$ — -0.25
5. $-\frac{3}{10}h = 15$ — -50
6. $-1.1k = -1.21$ — 1.1

Write the word sentence as an equation. Then solve.

7. A number divided by -8 is 7. — $\frac{x}{-8} = 7;\ -56$
8. The product of -12 and a number is 60. — $-12x = 60;\ -5$
9. You earn \$0.85 for every cup of hot chocolate you sell. How many cups do you need to sell to earn \$55.25? — 65 cups

Vocabulary and Concept Check

1. Multiplication is the inverse operation of division, so it can undo division.
2. *Sample answer:* opening and closing a door; waking up and going to sleep
3. dividing by 5
4. adding 12
5. multiplying by -8
6. subtracting -6

Practice and Problem Solving

7. $h = 5$
8. $t = 9$
9. $n = -14$
10. $k = -27$
11. $m = -2$
12. $t = -4$
13. $x = -8$
14. $b = 40$
15. $p = -8$
16. $d = -9$
17. $n = 8$
18. $p = 24$
19. $g = -16$
20. $c = -20$
21. $f = 6\frac{3}{4}$
22. $y = -16\frac{1}{4}$
23. They should divide by -4.2.

$$-4.2x = 21$$
$$\frac{-4.2x}{-4.2} = \frac{21}{-4.2}$$
$$x = -5$$

24. $\frac{x}{-9} = -16;\ x = 144$
25. $\frac{2}{5}x = \frac{3}{20};\ x = \frac{3}{8}$
26. $15x = -75;\ x = -5$
27. $\frac{x}{-1.5} = 21;\ x = -31.5$
28. $0.75n = 36$; 48 newspapers
29. $\frac{x}{30} = 12\frac{3}{5}$; 378 ft

Practice and Problem Solving

30–33. Sample answers are given.

30. a. $3x = -9$ **b.** $\frac{x}{2} = -1.5$

31. a. $-2x = 4.4$ **b.** $\frac{x}{1.1} = -2$

32. a. $5x = -\frac{5}{2}$ **b.** $\frac{x}{2} = -\frac{1}{4}$

33. a. $4x = -5$ **b.** $\frac{x}{5} = -\frac{1}{4}$

34. All of them except "multiply each side by $-\frac{2}{3}$."

35. $-1.26n = -10.08$; 8 days

36. $\frac{3}{4}s = 1464$; 1952 students

37. -50 ft

38. See *Taking Math Deeper*.

39. $-5, 5$

40. $1\frac{3}{5}$ days

Fair Game Review

41. -7 **42.** -9

43. 12 **44.** -9

45. B

Mini-Assessment

Solve the equation.

1. $7x = -84$ $x = -12$
2. $-0.3y = 2.4$ $y = -8$
3. $\frac{1}{2}m = -\frac{6}{7}$ $m = -1\frac{5}{7}$
4. $4\frac{1}{2} = -\frac{8}{9}k$ $k = -5\frac{1}{16}$
5. A stock has a return of $-\$1.40$ per day. Write and solve an equation to find the number of days until the total return is $-\$12.60$.
   $-1.4x = -12.6$; 9 days

Taking Math Deeper

Exercise 38

This problem is a good example of the type of question that can occur on a standardized test. For this problem, remind students to *read the question carefully.*

1 Organize the given information.

	Store A	Store B
Price	\$150.60	x

\$150.60 is $\frac{5}{6}x$.

2 Write and solve an equation for x.

$150.60 = \frac{5}{6}x$ Write the equation.

$6(150.60) = 5x$ Multiply each side by 6.

$\frac{6(150.60)}{5} = x$ Divide each side by 5.

$\$180.72 = x$ Simplify.

The bike costs \$180.72 at Store B.

3 How much do you save?

$180.72 - 150.60 = \$30.12$

You save \$30.12 at Store A.

Project

Find the prices of bikes in at least 2 different stores in your area. How much can you save by making your purchase at the store with the lowest price? Do you think it pays to comparison shop? Why or why not?

Reteaching and Enrichment Strategies

If students need help. . .	If students got it. . .
Resources by Chapter • Practice A and Practice B • Puzzle Time Record and Practice Journal Practice Differentiating the Lesson Lesson Tutorials Skills Review Handbook	Resources by Chapter • Enrichment and Extension • Technology Connection Start the next section

OPEN-ENDED **(a) Write a multiplication equation that has the given solution. (b) Write a division equation that has the same solution.**

30. -3 **31.** -2.2 **32.** $-\frac{1}{2}$ **33.** $-1\frac{1}{4}$

34. **REASONING** Which of the methods can you use to solve $-\frac{2}{3}c = 16$?

Multiply each side by $-\frac{2}{3}$.	Multiply each side by $-\frac{3}{2}$.
Divide each side by $-\frac{2}{3}$.	Multiply each side by 3, then divide each side by -2.

35. **STOCK** A stock has a return of $-\$1.26$ per day. Write and solve an equation to find the number of days until the total return is $-\$10.08$.

36. **ELECTION** In a school election, $\frac{3}{4}$ of the students vote. There are 1464 ballots. Write and solve an equation to find the number of students.

37. **OCEANOGRAPHY** Aquarius is an underwater ocean laboratory located in the Florida Keys National Marine Sanctuary. Solve the equation $\frac{31}{25}x = -62$ to find the value of x.

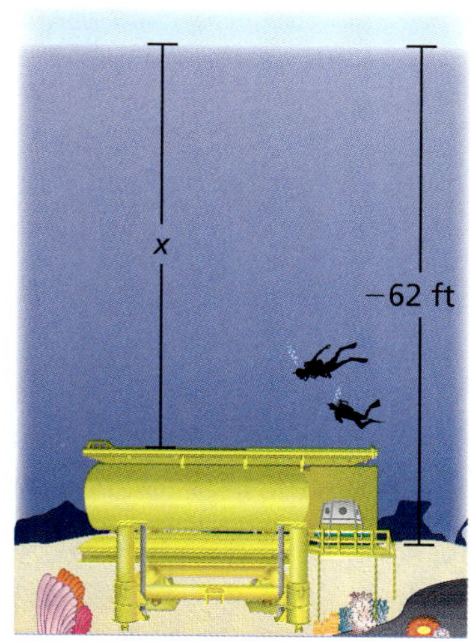

38. **SHOPPING** The price of a bike at Store A is $\frac{5}{6}$ the price at Store B. The price at Store A is \$150.60. Write and solve an equation to find how much you save by buying the bike at Store A.

39. **CRITICAL THINKING** Solve $-2|m| = -10$.

40. **Number Sense** In four days, your family drives $\frac{5}{7}$ of a trip. Your rate of travel is the same throughout the trip. The total trip is 1250 miles. In how many more days will you reach your destination?

Fair Game Review What you learned in previous grades & lessons

Subtract. *(Section 11.3)*

41. $5 - 12$ **42.** $-7 - 2$ **43.** $4 - (-8)$ **44.** $-14 - (-5)$

45. **MULTIPLE CHOICE** Of the 120 apartments in a building, 75 have been scheduled to receive new carpet. What fraction of the apartments have not been scheduled to receive new carpet? *(Skills Review Handbook)*

(A) $\frac{1}{4}$ (B) $\frac{3}{8}$ (C) $\frac{5}{8}$ (D) $\frac{3}{4}$

13.5 Solving Two-Step Equations

Essential Question How can you use algebra tiles to solve a two-step equation?

1 ACTIVITY: Solving a Two-Step Equation

Work with a partner. Use algebra tiles to model and solve $2x - 3 = -5$.

Model the equation $2x - 3 = -5$.

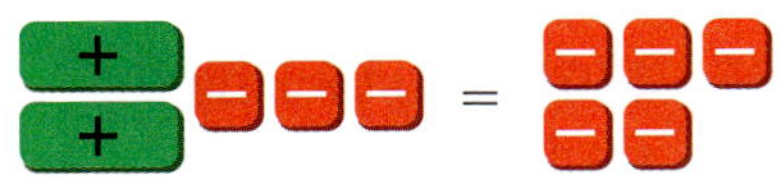

Remove the ____ red tiles on the left side by adding ____ yellow tiles to each side.

How many *zero pairs* can you remove from each side? ____
Circle them.

Because there are ____ green tiles, divide the red tiles into ____ equal groups. Circle the groups.

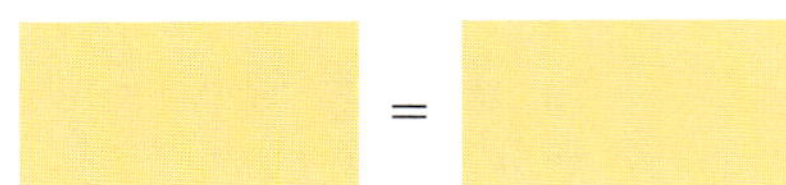

Keep one of the groups. This shows the value of x.

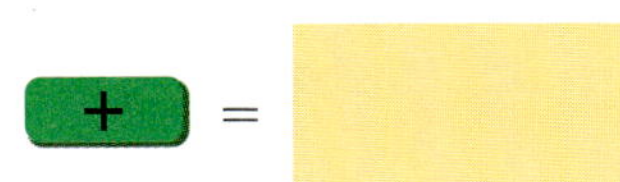

So, $x =$ ____.

2 ACTIVITY: The Math behind the Tiles

Work with a partner. Solve $2x - 3 = -5$ without using algebra tiles. Complete each step. Then answer the questions.

Use the steps in Activity 1 as a guide.

$2x - 3 = -5$	Write the equation.
$2x - 3 + \_\_\_\_ = -5 + \_\_\_\_$	Add ____ to each side.
$2x = \_\_\_\_$	Simplify.
$\dfrac{2x}{\_\_\_\_} = \dfrac{\_\_\_\_}{\_\_\_\_}$	Divide each side by ____.
$x = \_\_\_\_$	Simplify.

So, $x =$ ____.

a. Which step is first, adding 3 to each side or dividing each side by 2?

b. How are the above steps related to the steps in Activity 1?

Solving Equations
In this lesson, you will
- solve two-step equations.
- solve real-life problems.

Laurie's Notes

Introduction

Applying Mathematical Practice

- **Model with Mathematics:** Algebra tiles can help students make sense of equations. Algebra tiles are a concrete representation, deepening student understanding of what it means to solve an equation.

Motivate

- Write the number four on a slip of paper and put it in an envelope. Seal the envelope. Write the number 15 on the outside of the envelope.
- Hold the envelope up to your forehead and say, "I'm thinking of a number. When I double the number and add 7, I get an answer of 15." Show that the number 15 is written on the outside of the envelope.

? "What number did I start with?" 4

? "Can anyone explain how they know what number I was thinking of?" Listen for students to "undo" your process by working backwards: subtract 7 from 15 (to get 8) and then divide by 2 (to get 4).

- Open the envelope and reveal the four on your slip of paper.

? "Why didn't you divide by 2 first and then subtract 7?" You need to do the steps in the reverse order to undo the calculations.

- Explain that today you will investigate how to solve equations with two operations, like the number puzzle. These are called two-step equations.

Activity Notes

Activity 1

- Have students write the corresponding algebraic equations that result with each step to connect the model to the algebraic representation.
- **Discuss:** When students have finished, summarize by saying that the goal is to find the value of just one green variable-tile, so it should seem reasonable to "get rid of" the red integer-tiles on the left-hand side. This is referred to as *isolating the variable.*

? "How did you get the green variable-tiles by themselves?" Add three yellow integer-tiles to each side or remove three red integer-tiles from each side.

- Take time to look at each method discussed in the question above. Adding three yellow tiles is represented by $2x - 3 + 3 = -5 + 3$. Taking three red tiles away from each side results in $2x = -2$.
- **Reason Abstractly and Quantitatively:** Mathematically proficient students recognize the equivalence of adding 3 and subtracting -3.

Activity 2

? After you complete Activity 2, ask, "What happens if you divide each side by 2 first? Will you get the same answer?"

$\frac{2x-3}{2} = -\frac{5}{2}$ simplifies to $x - \frac{3}{2} = -\frac{5}{2}$. This introduces fractions into the problem, but you will still get the same answer.

? "Which method do you prefer?" Most students will want to avoid fractions.

What Your Students Will Learn

- Solve two-step equations by applying properties of equality.
- Combine like terms before solving algebraic equations.

Previous Learning

Students solved one-step equations.

Technology for the Teacher

Dynamic Classroom

Lesson Plans

Complete Materials List

13.5 Record and Practice Journal

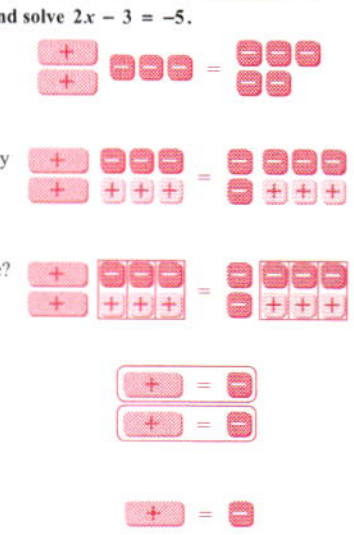

Essential Question How can you use algebra tiles to solve a two-step equation?

1 ACTIVITY: Solving a Two-Step Equation

Work with a partner. Use algebra tiles to model and solve $2x - 3 = -5$.

Model the equation $2x - 3 = -5$. Draw a sketch of your tiles.

Remove the __3__ red tiles on the left side by adding __3__ yellow tiles to each side.

How many *zero pairs* can you remove from each side? Circle them. 3

Because there are __2__ green tiles, divide the red tiles into __2__ equal groups. Circle the groups.

Keep one of the groups. This shows the value of x. Draw a sketch of the remaining tiles.

$x =$ __-1__.

2 ACTIVITY: The Math behind the Tiles

Work with a partner. Solve $2x - 3 = -5$ without using algebra tiles. Complete each step. Then answer the questions.

a. Which step is first, adding 3 to each side or dividing each side by 2?
adding 3 to each side

b. How are the above steps related to the steps in Activity 1?
They are the same.

Differentiated Instruction

Auditory

Write the equation $5h + 6 = 1$ on the board or overhead. Ask students to tell you the steps needed to solve the equation. Repeat the steps out loud as you solve the problem. Ask students for another way to solve the problem. Which way is more efficient? Why?

13.5 Record and Practice Journal

3 ACTIVITY: Solving Equations Using Algebra Tiles

Work with a partner.

- **Write an equation shown by the algebra tiles.**
- **Use algebra tiles to model and solve the equation.**
- **Check your answer by solving the equation without using algebra tiles.**

a. 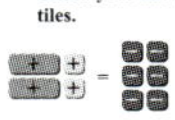

$2x + 2 = -6$
$x = -4$

b. 

$-13 = 3x - 4$
$x = -3$

4 ACTIVITY: Working Backwards

Work with a partner.

a. Your friend pauses a video game to get a drink. You continue the game. You double the score by saving a princess. Then you lose 75 points because you do not collect the treasure. You finish the game with –25 points. How many points did you have when you started?

One way to solve the problem is to work backwards. To do this, start with the end result and retrace the events.

You started the game with ___25___ points.

b. You triple your account balance by making a deposit. Then you withdraw \$127.32 to buy groceries. Your account is now overdrawn by \$10.56. By working backwards, find your account balance before you made the deposit.

\$38.92

What Is Your Answer?

5. **IN YOUR OWN WORDS** How can you use algebra tiles to solve a two-step equation?

See Additional Answers.

6. When solving the equation $4x + 1 = -11$, what is the first step?

Subtract 1 from each side.

7. **REPEATED REASONING** Solve the equation $2x - 75 = -25$. How do your steps compare with the strategy of working backwards in Activity 4?

$x = 25$; The steps are the same.

Laurie's Notes

Activity 3

- Ask for volunteers to share their work with the class.
- **Attend to Precision:** Listen to the language that students use when they explain their solutions. If they say, "I'll put two red tiles on each side," ask them to express their steps mathematically. They *should* say, "Add 2 red tiles to each side."
- **Reason Abstractly and Quantitatively:** You want students to be able to connect their manipulation of the tiles with operations they'll record symbolically.
- Students may use different methods to solve part (b). Some students may add four yellow tiles to each side (add the opposite), and others may remove four red tiles from each side (subtract −4).
- After one method of solving an equation is described, be sure to ask if anyone approached the problem in another way so students don't think their method is wrong.

Activity 4

- If you did not do the activity in the introduction, you may want to before this activity. Both use the strategy *Working Backwards*.
- Have students make up their own number puzzles and have their partners work backwards to guess their number.

What Is Your Answer?

- Some students may not have the appropriate language to describe the process yet. Have them focus on creating examples.
- For Question 7, listen for the idea that it is the last operation performed that is "undone" first in the solving process.

Closure

- Match the equation with the first step used in solving the equation.

Equation	First Step in Solving
1) $4x - 3 = -7$	A) Divide by -3.
2) $5 = -2x + 4$	B) Subtract 2.
3) $-3x + 2 = 6$	C) Multiply by $\frac{1}{3}$.
4) $-4 = 3x - 2$	D) Add 3.
	E) Add 2.
	F) Subtract 4.

Answers: 1D, 2F, 3B, 4E

3 ACTIVITY: Solving Equations Using Algebra Tiles

Work with a partner.

- **Write an equation shown by the algebra tiles.**
- **Use algebra tiles to model and solve the equation.**
- **Check your answer by solving the equation without using algebra tiles.**

a. 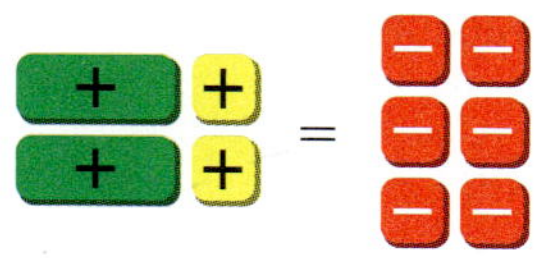

b. 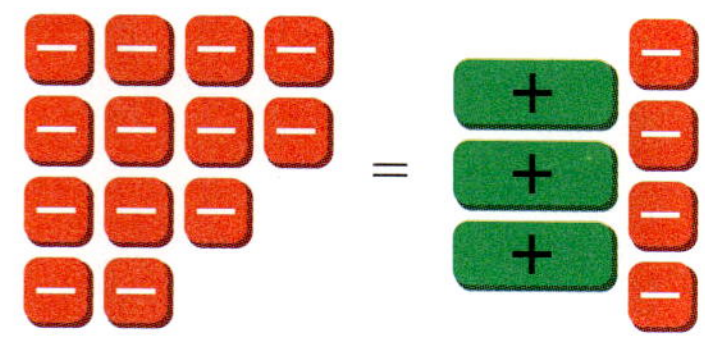

4 ACTIVITY: Working Backwards

Math Practice

Maintain Oversight

How does working backwards help you decide which operation to do first? Explain.

Work with a partner.

a. **Sample:** Your friend pauses a video game to get a drink. You continue the game. You double the score by saving a princess. Then you lose 75 points because you do not collect the treasure. You finish the game with -25 points. How many points did you have when you started?

One way to solve the problem is to work backwards. To do this, start with the end result and retrace the events.

You have -25 points at the end of the game.	$\mathbf{-25}$
You lost 75 points for not collecting the treasure, so add 75 to -25.	$\mathbf{-25 + 75 = 50}$
You doubled your score for saving the princess, so find half of 50.	$\mathbf{50 \div 2 = 25}$

So, you started the game with 25 points.

b. You triple your account balance by making a deposit. Then you withdraw \$127.32 to buy groceries. Your account is now overdrawn by \$10.56. By working backwards, find your account balance before you made the deposit.

What Is Your Answer?

5. **IN YOUR OWN WORDS** How can you use algebra tiles to solve a two-step equation?

6. When solving the equation $4x + 1 = -11$, what is the first step?

7. **REPEATED REASONING** Solve the equation $2x - 75 = -25$. How do your steps compare with the strategy of working backwards in Activity 4?

Practice Use what you learned about solving two-step equations to complete Exercises 6–11 on page 586.

13.5 Lesson

Check It Out
Lesson Tutorials
BigIdeasMath.com

EXAMPLE 1 Solving a Two-Step Equation

Solve $-3x + 5 = 2$. Check your solution.

	$-3x + 5 = 2$	Write the equation.
Undo the addition.	$\quad -5 \quad -5$	Subtraction Property of Equality
	$-3x = -3$	Simplify.
Undo the multiplication.	$\dfrac{-3x}{-3} = \dfrac{-3}{-3}$	Division Property of Equality
	$x = 1$	Simplify.

Check

$-3x + 5 = 2$

$-3(1) + 5 \stackrel{?}{=} 2$

$-3 + 5 \stackrel{?}{=} 2$

$2 = 2$ ✓

The solution is $x = 1$.

On Your Own

Now You're Ready
Exercises 6–17

Solve the equation. Check your solution.

1. $2x + 12 = 4$ **2.** $-5c + 9 = -16$ **3.** $3(x - 4) = 9$

EXAMPLE 2 Solving a Two-Step Equation

Solve $\frac{x}{8} - \frac{1}{2} = -\frac{7}{2}$. Check your solution.

$\dfrac{x}{8} - \dfrac{1}{2} = -\dfrac{7}{2}$	Write the equation.
$\quad +\dfrac{1}{2} \quad +\dfrac{1}{2}$	Addition Property of Equality
$\dfrac{x}{8} = -3$	Simplify.
$8 \cdot \dfrac{x}{8} = 8 \cdot (-3)$	Multiplication Property of Equality
$x = -24$	Simplify.

Check

$\frac{x}{8} - \frac{1}{2} = -\frac{7}{2}$

$\frac{-24}{8} - \frac{1}{2} \stackrel{?}{=} -\frac{7}{2}$

$-3 - \frac{1}{2} \stackrel{?}{=} -\frac{7}{2}$

$-\frac{7}{2} = -\frac{7}{2}$ ✓

Study Tip

You can simplify the equation in Example 2 before solving. Multiply each side by the LCD of the fractions, 8.

$\frac{x}{8} - \frac{1}{2} = -\frac{7}{2}$

$x - 4 = -28$

$x = -24$

The solution is $x = -24$.

On Your Own

Now You're Ready
Exercises 20–25

Solve the equation. Check your solution.

4. $\frac{m}{2} + 6 = 10$ **5.** $-\frac{z}{3} + 5 = 9$ **6.** $\frac{2}{5} + 4a = -\frac{6}{5}$

Laurie's Notes

Introduction

Connect

- **Yesterday:** Students used algebra tiles to model solving two-step equations.
- **Today:** Students will solve equations by undoing the operations in the reverse order of how the expression would have been evaluated.

Motivate

? "Four friends each purchase a large beverage and share a $9 pizza. The total bill before tax is $16. What is the cost of each beverage?" $1.75

- Ask students to explain how they solved this problem. Listen for students to mention subtracting the cost of the pizza from the total before dividing by four.

Lesson Notes

Example 1

? **Vocabulary Review:** "In the expression $-3x + 5$, what is -3 called?" the coefficient

- Work through the example. Before doing each step, ask students what the next step should be.
- Take the time to check the solution so that students see this as important.

On Your Own

- Students may be uncertain of how to solve Question 3 because of the parentheses. Remind students about the Distributive Property.
- Review the methods students use to solve each problem. In Question 3, for example, some students may distribute the three and some may realize they can divide both sides of the equation by three.
- **Challenge:** Ask students to describe two methods for solving Question 1. Some may notice that each number in the equation can be divided by 2.

Example 2

? "If you knew the value of *x*, how would you evaluate the expression $\frac{x}{8} - \frac{1}{2}$?" *Sample answer:* Divide the number by 8 and subtract $\frac{1}{2}$; Some students might say that you need to find a common denominator and then subtract the fractions.

? "What is the first step to solve this equation?" Add $\frac{1}{2}$ to each side.

? "What is the second step to solve this equation?" Multiply each side by 8.

On Your Own

- **Think-Pair-Share:** Students should read each question independently and then work in pairs to answer the questions. When they have answered the questions, the pair should compare their answers with another group and discuss any discrepancies.

Goal Today's lesson is solving two-step equations.

Technology for the Teacher

Dynamic Classroom

Lesson Tutorials
Lesson Plans
Answer Presentation Tool

Extra Example 1

Solve $4t - 7 = -15$. Check your solution. -2

On Your Own

1. $x = -4$
2. $c = 5$
3. $x = 7$

Extra Example 2

Solve $\frac{n}{9} + \frac{2}{3} = -\frac{2}{3}$. Check your solution. -12

On Your Own

4. $m = 8$
5. $z = -12$
6. $a = -\frac{2}{5}$

Laurie's Notes

Extra Example 3

Solve $12.5 = 0.3m - 2.8m$. -5

Extra Example 4

A taxi charges \$2.50 plus \$2 for every mile traveled. Find the number of miles traveled for a fare of \$10.50. 4 miles

On Your Own

7. $y = 8$
8. $x = -5$
9. $m = 10$
10. 9.5 ft

English Language Learners

Verbal Clues

English learners should become familiar with words and phrases that give clues to the types of operations required. Word problems calling for two-step equations almost always contain words such as *per*, *each*, and *every*. These are clues to quantities that will appear in the equation, usually as the coefficient of the variable. Point out these words in the exercises and identify the terms associated with them in the equations used to solve the problems.

Example 3

- This problem requires students to *combine like terms* as the first step.
- ? "What do we call $3y$ and $-8y$?" They are like terms.

Example 4

- Note that the unknown value in this problem is the starting height. Roller coasters do not begin on the ground!
- **Reason Abstractly and Quantitatively:** The table helps students develop the ability to translate from words to symbols.

On Your Own

- **Think-Pair-Share**: Students should read each question independently and then work in pairs to answer the questions. When they have answered the questions, the pair should compare their answers with another group and discuss any discrepancies.

Closure

- What does it mean to *isolate the variable term*? Use inverse operations to get the variable by itself.
- What are like or similar terms? Give examples. Terms that can be combined are like terms. Some examples are $3x$ and $-2x$, $5a$ and $-a$, and 5 and -8.
- Explain how the solutions of the two equations are similar.

  $4x - 5 = 7 \qquad \frac{4}{3}x - \frac{5}{3} = \frac{7}{3}$

  You can multiply each term of the second equation by 3 and then the two equations will be the same. The solution of the two equations is the same.

EXAMPLE 3 Combining Like Terms Before Solving

Solve $3y - 8y = 25$.

$3y - 8y = 25$ Write the equation.

$-5y = 25$ Combine like terms.

$y = -5$ Divide each side by -5.

∴ The solution is $y = -5$.

EXAMPLE 4 Real-Life Application

The height at the top of a roller coaster hill is 10 times the height h of the starting point. The height decreases 100 feet from the top to the bottom of the hill. The height at the bottom of the hill is -10 feet. Find h.

Location	Verbal Description	Expression
Start	The height at the start is h.	h
Top of hill	The height at the top of the hill is 10 times the starting height h.	$10h$
Bottom of hill	The height decreases by 100 feet. So, subtract 100.	$10h - 100$

The height at the bottom of the hill is -10 feet. Solve $10h - 100 = -10$ to find h.

$10h - 100 = -10$ Write equation.

$10h = 90$ Add 100 to each side.

$h = 9$ Divide each side by 10.

∴ So, the height at the start is 9 feet.

On Your Own

Now You're Ready
Exercises 29–34

Solve the equation. Check your solution.

7. $4 - 2y + 3 = -9$ **8.** $7x - 10x = 15$ **9.** $-8 = 1.3m - 2.1m$

10. **WHAT IF?** In Example 4, the height at the bottom of the hill is -5 feet. Find the height h.

13.5 Exercises

Check It Out
Help with Homework
BigIdeasMath.com

Vocabulary and Concept Check

1. **WRITING** How do you solve two-step equations?

Match the equation with the first step to solve it.

2. $4 + 4n = -12$
3. $4n = -12$
4. $\frac{n}{4} = -12$
5. $\frac{n}{4} - 4 = -12$

A. Add 4.
B. Subtract 4.
C. Multiply by 4.
D. Divide by 4.

Practice and Problem Solving

Solve the equation. Check your solution.

1 6. $2v + 7 = 3$
7. $4b + 3 = -9$
8. $17 = 5k - 2$
9. $-6t - 7 = 17$
10. $8n + 16.2 = 1.6$
11. $-5g + 2.3 = -18.8$
12. $2t - 5 = -10$
13. $-4p + 9 = -5$
14. $11 = -5x - 2$
15. $4 + 2.2h = -3.7$
16. $-4.8f + 6.4 = -8.48$
17. $7.3y - 5.18 = -51.9$

ERROR ANALYSIS **Describe and correct the error in finding the solution.**

18. ✗

$$-6 + 2x = -10$$
$$-6 + \frac{2x}{2} = -\frac{10}{2}$$
$$-6 + x = -5$$
$$x = 1$$

19. ✗

$$-3x + 2 = -7$$
$$-3x = -9$$
$$-\frac{3x}{3} = \frac{-9}{3}$$
$$x = -3$$

Solve the equation. Check your solution.

2 20. $\frac{3}{5}g - \frac{1}{3} = -\frac{10}{3}$
21. $\frac{a}{4} - \frac{5}{6} = -\frac{1}{2}$
22. $-\frac{1}{3} + 2z = -\frac{5}{6}$
23. $2 - \frac{b}{3} = -\frac{5}{2}$
24. $-\frac{2}{3}x + \frac{3}{7} = \frac{1}{2}$
25. $-\frac{9}{4}v + \frac{4}{5} = \frac{7}{8}$

In Exercises 26–28, write an equation. Then solve.

26. **WEATHER** Starting at 1:00 P.M., the temperature changes -4 degrees per hour. How long will it take to reach $-1°$?

Temperature at 1:00 P.M.

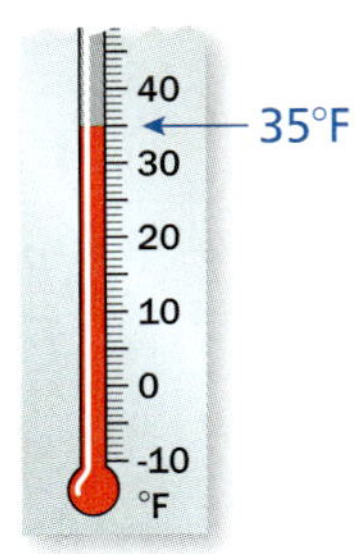

27. **BOWLING** It costs \$2.50 to rent bowling shoes. Each game costs \$2.25. You have \$9.25. How many games can you bowl?

28. **CELL PHONES** A cell phone company charges a monthly fee plus \$0.25 for each text message. The monthly fee is \$30.00 and you owe \$59.50. How many text messages did you have?

Assignment Guide and Homework Check

Level	Assignment	Homework Check
Advanced	1–11, 18–40 even, 41–46	24, 28, 30, 36, 40

Common Errors

- **Exercises 6–17** Students may divide the coefficient first instead of adding or subtracting first. Tell them that while this is a valid method, they must remember to divide each part of the equation by the coefficient.
- **Exercises 20–25** Students may immediately multiply each term by one of the denominators without thinking if it will help them solve for the variable. Ask them to check if all the denominators would be eliminated.
- **Exercises 32–34** Students may try to add or subtract without distributing. Remind them that when parentheses are present, they either need to use the Distributive Property or they need to undo the multiplication first. All of the exercises can be solved using either method.

13.5 Record and Practice Journal

Solve the equation. Check your solution.

1. $3a - 5 = -14$ **-3**
2. $10 = -2c + 22$ **6**
3. $18 = -5b - 17$ **-7**
4. $-12 = -8z + 12$ **3**
5. $1.3n - 0.03 = -9$ **-6.9**
6. $-\frac{5}{11}h + \frac{7}{9} = \frac{2}{9}$ **$1\frac{2}{9}$**
7. The length of a rectangle is 3 meters less than twice its width.
   a. Write an equation to find the length of the rectangle.
   $\ell = 2w - 3$
   b. The length of the rectangle is 11 meters. What is the width of the rectangle?
   7 meters

Vocabulary and Concept Check

1. Eliminate the constants on the side with the variable. Then solve for the variable using either division or multiplication.

2. B
3. D
4. C
5. A

Practice and Problem Solving

6. $v = -2$
7. $b = -3$
8. $k = 3\frac{4}{5}$
9. $t = -4$
10. $n = -1.825$
11. $g = 4.22$
12. $t = -2\frac{1}{2}$
13. $p = 3\frac{1}{2}$
14. $x = -2\frac{3}{5}$
15. $h = -3.5$
16. $f = 3.1$
17. $y = -6.4$
18. The steps are out of order.

$$-6 + 2x = -10$$
$$2x = -4$$
$$\frac{2x}{2} = \frac{-4}{2}$$
$$x = -2$$

19. Each side should be divided by -3, not 3.

$$-3x + 2 = -7$$
$$-3x = -9$$
$$\frac{-3x}{-3} = \frac{-9}{-3}$$
$$x = 3$$

20. $g = -5$
21. $a = 1\frac{1}{3}$
22. $z = -\frac{1}{4}$
23. $b = 13\frac{1}{2}$
24. $x = -\frac{3}{28}$
25. $v = -\frac{1}{30}$
26. $-4x + 35 = -1$; 9 hours (10:00 P.M.)

Practice and Problem Solving

27. $2.5 + 2.25x = 9.25$; 3 games

28. $30 + 0.25x = 59.5$;
118 text messages

29. $v = -5$ **30.** $t = -13$

31. $d = -12$ **32.** $x = -1$

33. $m = -9$ **34.** $y = -4$

35. *Sample answer:* You travel halfway up a ladder. Then you climb down two feet and are 8 feet above the ground. How long is the ladder? $x = 20$

36. $12\frac{3}{4}$ ft **37.** the initial fee

38. the coldest surface temperature on the moon

39. See *Taking Math Deeper*.

40. -21 ft

41. decrease the length by 10 cm; $2(25 + x) + 2(12) = 54$

Fair Game Review

42. -34.72 **43.** $-6\frac{2}{3}$

44. $-3\frac{1}{8}$ **45.** 6.2

46. C

Mini-Assessment

Solve the equation.

1. $4x + 16.4 = -3.6$ $x = -5$
2. $-8.46 = -2.1n - 2.16$ $n = 3$
3. $-\frac{4}{5} + \frac{1}{2}m = -\frac{1}{5}$ $m = 1\frac{1}{5}$
4. $-\frac{5}{9} = \frac{2}{3}\ell - \frac{1}{3}$ $\ell = -\frac{1}{3}$
5. A gym charges $8.75 for each swimming class and a one-time registration fee of $12.50. A student paid a total of $56.25. Write and solve an equation to find the number of swimming classes the student took.
$8.75x + 12.5 = 56.25$; 5 classes

Taking Math Deeper

Exercise 39

This problem asks students to answer a question *with* and *without* using algebra.

1 Summarize the given information.

1. You caught x insects on Saturday.
2. 5 of the insects escaped.
3. The remaining insects form 3 groups of 9 each.

2 **a.** Work backwards.

3. There are $3(9) = 27$ insects remaining.
2. Add the 5 that escaped.
1. You caught $27 + 5$, or 32, insects on Saturday.

3 **b.** Write and solve an equation.

1. x insects on Saturday
2. $(x - 5)$ are remaining.
3. $\frac{(x-5)}{3} = 9$

$\frac{x-5}{3} = 9$ Write the equation.

$x - 5 = 27$ Multiply each side by 3.

$x = 32$ Add 5 to each side.

You caught 32 insects on Saturday.

Reteaching and Enrichment Strategies

If students need help. . .	If students got it. . .
Resources by Chapter • Practice A and Practice B • Puzzle Time Record and Practice Journal Practice Differentiating the Lesson Lesson Tutorials Skills Review Handbook	Resources by Chapter • Enrichment and Extension • Technology Connection Start the next section

Solve the equation. Check your solution.

3 **29.** $3v - 9v = 30$ **30.** $12t - 8t = -52$ **31.** $-8d - 5d + 7d = 72$

32. $6(x - 2) = -18$ **33.** $-4(m + 3) = 24$ **34.** $-8(y + 9) = -40$

35. **WRITING** Write a real-world problem that can be modeled by $\frac{1}{2}x - 2 = 8$. Then solve the equation.

36. **GEOMETRY** The perimeter of the parallelogram is 102 feet. Find m.

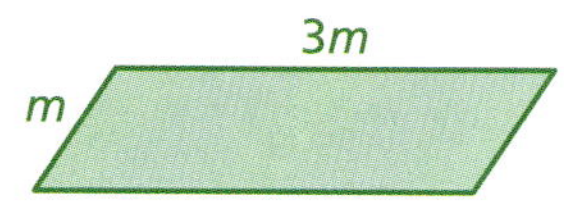

REASONING **Exercises 37 and 38 are missing information. Tell what information you need to solve the problem.**

37. **TAXI** A taxi service charges an initial fee plus \$1.80 per mile. How far can you travel for \$12?

38. **EARTH** The coldest surface temperature on the Moon is 57 degrees colder than twice the coldest surface temperature on Earth. What is the coldest surface temperature on Earth?

39. **PROBLEM SOLVING** On Saturday, you catch insects for your science class. Five of the insects escape. The remaining insects are divided into three groups to share in class. Each group has nine insects. How many insects did you catch on Saturday?

a. Solve the problem by working backwards.

b. Solve the equation $\frac{x - 5}{3} = 9$. How does the answer compare with the answer to part (a)?

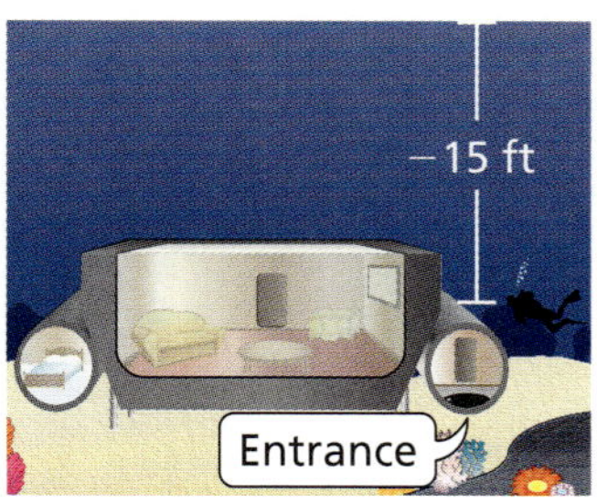

40. **UNDERWATER HOTEL** You must scuba dive to the entrance of your room at Jules' Undersea Lodge in Key Largo, Florida. The diver is 1 foot deeper than $\frac{2}{3}$ of the elevation of the entrance. What is the elevation of the entrance?

41. Geometry How much should you change the length of the rectangle so that the perimeter is 54 centimeters? Write an equation that shows how you found your answer.

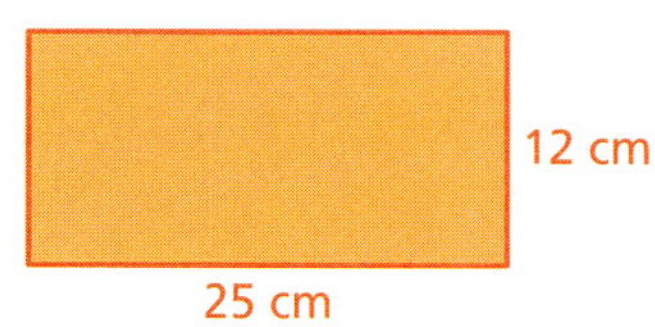

Fair Game Review What you learned in previous grades & lessons

Multiply or divide. *(Section 12.4)*

42. -6.2×5.6 **43.** $\frac{8}{3} \times \left(-2\frac{1}{2}\right)$ **44.** $\frac{5}{2} \div \left(-\frac{4}{5}\right)$ **45.** $-18.6 \div (-3)$

46. **MULTIPLE CHOICE** Which fraction is *not* equivalent to 0.75? *(Skills Review Handbook)*

Ⓐ $\frac{15}{20}$ Ⓑ $\frac{9}{12}$ Ⓒ $\frac{6}{9}$ Ⓓ $\frac{3}{4}$

13.3–13.5 Quiz

Check It Out
Progress Check
BigIdeasMath.com

Solve the equation. Check your solution. *(Section 13.3, Section 13.4, and Section 13.5)*

1. $-6.5 + x = -4.12$

2. $4\frac{1}{2} + p = -5\frac{3}{4}$

3. $-\frac{b}{7} = 4$

4. $-2w + 3.7 = -0.5$

Write the word sentence as an equation. Then solve. *(Section 13.3 and Section 13.4)*

5. The difference between a number b and 7.4 is -6.3.

6. $5\frac{2}{5}$ more than a number a is $7\frac{1}{2}$.

7. A number x multiplied by $\frac{3}{8}$ is $-\frac{15}{32}$.

8. The quotient of two times a number k and -2.6 is 12.

Write and solve an equation to find the value of x. *(Section 13.3 and Section 13.5)*

9. Perimeter = 26

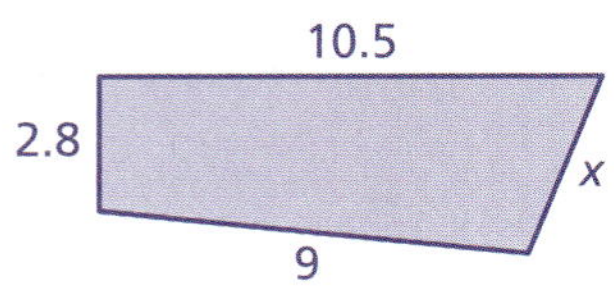

10. Perimeter = 23.59

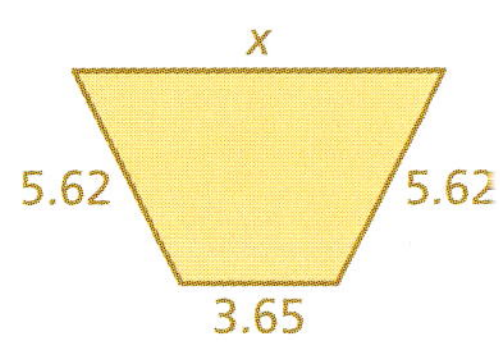

11. Perimeter = 33

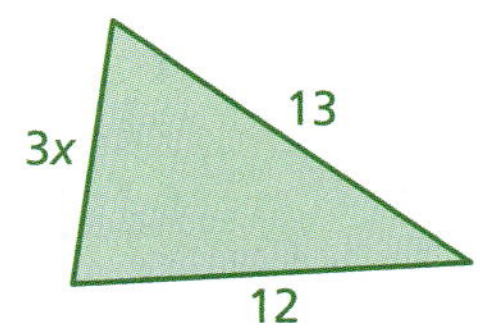

12. BANKING You withdraw \$29.79 from your bank account. Now your balance is $-\$20.51$. Write and solve an equation to find the amount of money in your bank account before you withdrew the money. *(Section 13.3)*

13. WATER LEVEL During a drought, the water level of a lake changes $-3\frac{1}{5}$ feet per day. Write and solve an equation to find how long it takes for the water level to change -16 feet. *(Section 13.4)*

14. BASKETBALL A basketball game has four quarters. The length of a game is 32 minutes. You play the entire game except for $4\frac{1}{2}$ minutes. Write and solve an equation to find the mean time you play per quarter. *(Section 13.5)*

15. SCRAPBOOKING The mat needs to be cut to have a 0.5-inch border on all four sides. *(Section 13.5)*

a. How much should you cut from the left and right sides?

b. How much should you cut from the top and bottom?

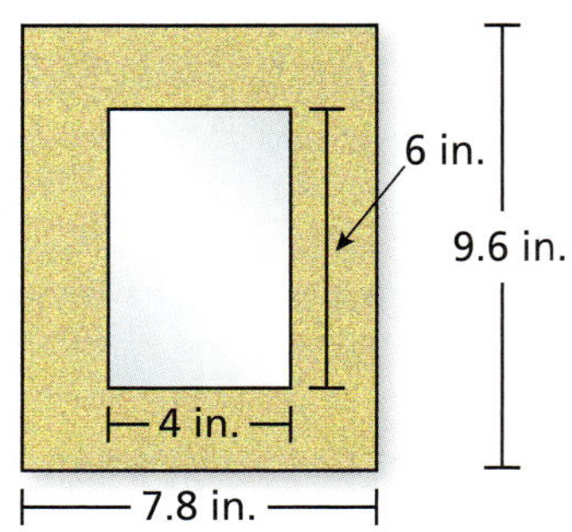

Alternative Assessment Options

Math Chat	Student Reflective Focus Question
Structured Interview	Writing Prompt

Structured Interview

Interviews can occur formally or informally. Ask a student to perform a task and explain it, describing his or her thought process throughout the task. Probe the student for more information. Do not ask leading questions. Keep a rubric or notes.

Teacher Prompts	Student Answers	Teacher Notes
Tell me a story about buying flowers. Include Cost of a vase: \$8.50 Cost per rose: \$2.25 Total amount spent: \$26.50 Number of roses bought: ?	I bought flowers for my aunt for her birthday. I bought a vase that costs \$8.50 and roses that cost \$2.25 each. I spent a total of \$26.50. So, I bought 8 roses.	Student can write and solve a two-step equation.
Add to your story using this phrase. \$50 bill and receive ? back	I gave the florist a \$50 bill and received \$23.50 back in change.	Student can write and solve an equation involving addition or subtraction.

Study Help Sample Answers

Remind students to complete Graphic Organizers for the rest of the chapter.

4.

Words	Algebra
Two equations are equivalent equations if they have the same solutions. You can use the Addition, Subtraction, Multiplication, and Division Properties of Equality to write equivalent equations.	$a = b$ and $a + c = b + c$ $a = b$ and $a - c = b - c$ $a = b$ and $a \cdot c = b \cdot c$ $a = b$ and $\frac{a}{c} = \frac{b}{c}$, $c \neq 0$

Equivalent equations

Examples	Non-Examples
$x - 7 = 2$ and $x - 7 + 7 = 2 + 7$ $2d + 5 = -7$ and $2d + 5 - 5 = -7 - 5$ $24 = \frac{y}{-4}$ and $-4 \cdot 24 = -4 \cdot \frac{y}{-4}$ $3c = -12$ and $\frac{3c}{3} = \frac{-12}{3}$	$x + 7 = 2$ and $x + 7 - 7 = 2 + 7$ $3c = -4$ and $\frac{3c}{3} = 3 \cdot (-4)$ $7 = m + 3$ and $7 - 7 = m + 3 - 3$ $3x + 7 = 3$ and $3x = 7 + 3$

5–7. Available at *BigIdeasMath.com.*

Reteaching and Enrichment Strategies

If students need help. . .	If students got it. . .
Resources by Chapter • Practice A and Practice B • Puzzle Time Lesson Tutorials *BigIdeasMath.com*	Resources by Chapter • Enrichment and Extension • Technology Connection Game Closet at *BigIdeasMath.com* Start the Chapter Review

Answers

1. $x = 2.38$
2. $p = -10\frac{1}{4}$
3. $b = -28$
4. $w = 2.1$
5. $b - 7.4 = -6.8$; $b = 0.6$
6. $5\frac{2}{5} + a = 7\frac{1}{2}$; $a = 2\frac{1}{10}$
7. $\frac{3}{8}x = -\frac{15}{32}$; $x = -1\frac{1}{4}$
8. $\frac{2k}{-2.6} = 12$; $k = -15.6$
9. $x + 22.3 = 26$; $x = 3.7$
10. $x + 14.89 = 23.59$; $x = 8.7$
11. $3x + 25 = 33$; $x = 2\frac{2}{3}$
12. $x - 29.79 = -20.51$; \$9.28
13. $-3\frac{1}{5}x = -16$; 5 days
14. $4x + 4\frac{1}{2} = 32$; $6\frac{7}{8}$ minutes
15. **a.** 1.4 in.

    b. 1.3 in.

Technology for the Teacher

Online Assessment
Assessment Book
ExamView® Assessment Suite

For the Teacher
Additional Review Options

- *BigIdeasMath.com*
- Online Assessment
- Game Closet at *BigIdeasMath.com*
- Vocabulary Help
- Resources by Chapter

Answers

1. Terms: z, 8, $-4z$; Like terms: z and $-4z$
2. Terms: $3n$, 7, $-n$, -3; Like terms: $3n$ and $-n$, 7 and -3
3. Terms: $10x^2$, $-y$, 12, $-3x^2$; Like terms: $10x^2$ and $-3x^2$
4. $-4h$
5. $3.5r - 7$
6. $\frac{9}{20}x + 12$
7. $3q + 21$
8. $2m - 18$
9. $1.5n - 3.2$

Review of Common Errors

Exercises 1–3

- When identifying and writing terms, make sure students include the sign of the term. They may find it helpful to write the original problem using addition.
- Students may confuse like variables with like terms. Remind them that the same variables must be raised to the same exponents for terms to be like terms.

Exercises 4–9

- The subtraction operation can confuse students. Tell students to write the original problem using addition and then to use the Commutative Property.

Exercises 7–9

- Students often forget to distribute the constant over *both* of the terms inside the parentheses. Remind them of the Distributive Property, $a(b + c) = ab + ac$.

Exercises 16–23

- Students may use the same operation in solving for x instead of the inverse operation.

Exercises 24–32

- When the variable is multiplied by a negative number, students forget to keep the negative with the number and solve for $-x$ instead of x.

Exercises 33–37

- Students might divide the coefficient first instead of adding or subtracting.

13 Chapter Review

Check It Out Vocabulary Help BigIdeasMath.com

Review Key Vocabulary

like terms, *p. 556*
simplest form, *p. 556*
linear expression, *p. 562*
factoring an expression, *p. 566*
equivalent equations, *p. 572*

Review Examples and Exercises

13.1 Algebraic Expressions *(pp. 554–559)*

a. Identify the terms and like terms in the expression $6y + 9 + 3y - 7$.

Rewrite as a sum of terms.

$$\underbrace{6y} + \underbrace{9} + \underbrace{3y} + \underbrace{(-7)}$$

Terms: $6y$, 9, $3y$, -7

Like terms: $6y$ and $3y$, 9 and -7

b. Simplify $\frac{2}{3}y + 14 - \frac{1}{6}y - 8$.

$$\frac{2}{3}y + 14 - \frac{1}{6}y - 8 = \frac{2}{3}y + 14 + \left(-\frac{1}{6}y\right) + (-8)$$ Rewrite as a sum.

$$= \frac{2}{3}y + \left(-\frac{1}{6}y\right) + 14 + (-8)$$ Commutative Property of Addition

$$= \left[\frac{2}{3} + \left(-\frac{1}{6}\right)\right]y + 14 + (-8)$$ Distributive Property

$$= \frac{1}{2}y + 6$$ Combine like terms.

Exercises

Identify the terms and like terms in the expression.

1. $z + 8 - 4z$
2. $3n + 7 - n - 3$
3. $10x^2 - y + 12 - 3x^2$

Simplify the expression.

4. $4h - 8h$
5. $6.4r - 7 - 2.9r$
6. $\frac{3}{5}x + 19 - \frac{3}{20}x - 7$
7. $3(2 + q) + 15$
8. $\frac{1}{8}(16m - 8) - 17$
9. $-1.5(4 - n) + 2.8$

13.2 Adding and Subtracting Linear Expressions (pp. 560–567)

a. Find (5z + 4) + (3z − 6).

$$\begin{array}{r} 5z + 4 \\ + \; 3z - 6 \\ \hline 8z - 2 \end{array}$$

Align like terms vertically and add.

b. Factor $\frac{1}{4}$ out of $\frac{1}{4}x - \frac{3}{4}$.

Write each term as a product of $\frac{1}{4}$ and another factor.

$$\frac{1}{4}x = \frac{1}{4} \cdot x \qquad -\frac{3}{4} = \frac{1}{4} \cdot (-3)$$

Use the Distributive Property to factor out $\frac{1}{4}$.

$$\frac{1}{4}x - \frac{3}{4} = \frac{1}{4} \cdot x + \frac{1}{4} \cdot (-3) = \frac{1}{4}(x - 3)$$

So, $\frac{1}{4}x - \frac{3}{4} = \frac{1}{4}(x - 3)$.

Exercises

Find the sum or difference.

10. $(c - 4) + (3c + 9)$

11. $\frac{2}{5}(d - 10) - \frac{2}{3}(d + 6)$

Factor out the coefficient of the variable.

12. $2b + 8$

13. $\frac{1}{4}y + \frac{3}{8}$

14. $1.7j - 3.4$

15. $-5p + 20$

13.3 Solving Equations Using Addition or Subtraction (pp. 570–575)

Solve $x - 9 = -6$.

$x - 9 = -6$ Write the equation.

Undo the subtraction. → $\underline{+9} \quad \underline{+9}$ Addition Property of Equality

$x = 3$ Simplify.

Check

$x - 9 = -6$

$3 - 9 \stackrel{?}{=} -6$

$-6 = -6$ ✓

Exercises

Solve the equation. Check your solution.

16. $p - 3 = -4$

17. $6 + q = 1$

18. $-2 + j = -22$

19. $b - 19 = -11$

20. $n + \frac{3}{4} = \frac{1}{4}$

21. $v - \frac{5}{6} = -\frac{7}{8}$

22. $t - 3.7 = 1.2$

23. $\ell + 15.2 = -4.5$

Review Game

Expressions and Equations

Materials

- color-coded spinner with operation symbols
- 20 cards numbered -10 through 10, excluding 0
- 2 pencils
- paper

Players: 2

Directions

One player draws three cards and writes each number in one of the shaded boxes in the equation below.

$\blacksquare\, x \ \square\ \blacksquare = \blacksquare$

The player then spins the spinner. The spinner is used to determine the operation to be used. The player writes the operation in the unshaded box above. For example, if a player draws -10, -3, and 2, and spins *red* $+$, the equation may be $-3x + 2 = -10$.

Each player solves the equation, exchanges his or her work with the other player, and checks the other player's work. Each player receives 1 point for the round for a correct solution. Each round can result in a total of 0, 1, or 2 points received between the two players. Players take turns drawing/spinning and writing the equations. Students can play several rounds or for a predetermined amount of time.

Who Wins?

The player with the most points wins.

For the Student
Additional Practice

- Lesson Tutorials
- Multi-Language Glossary
- Self-Grading Progress Check
- *BigIdeasMath.com*
  Dynamic Student Edition
  Student Resources

Answers

10. $4c + 5$ **11.** $-\frac{4}{15}d - 8$

12. $2(b + 4)$ **13.** $\frac{1}{4}\left(y + \frac{3}{2}\right)$

14. $1.7(j - 2)$ **15.** $-5(p - 4)$

16. $p = -1$ **17.** $q = -5$

18. $j = -20$ **19.** $b = 8$

20. $n = -\frac{1}{2}$ **21.** $v = -\frac{1}{24}$

22. $t = 4.9$ **23.** $\ell = -19.7$

24. $x = -24$ **25.** $y = -49$

26. $z = 3$ **27.** $w = 50$

28. $x = -2$ **29.** $y = -5$

30. $z = 6$ **31.** $w = -0.5$

32. $-16°F$ **33.** $c = 7$

34. $w = -\frac{8}{9}$ **35.** $w = -12$

36. $x = -3.5$ **37.** 11 years

My Thoughts on the Chapter

Teacher Tip

Not allowed to write in your teaching edition? Use sticky notes to record your thoughts.

What worked. . .

What did not work. . .

What I would do differently. . .

13.4 Solving Equations Using Multiplication or Division (pp. 576–581)

Solve $\frac{x}{5} = -7$.

	$\frac{x}{5} = -7$	Write the equation.
Undo the division. →	$5 \cdot \frac{x}{5} = 5 \cdot (-7)$	Multiplication Property of Equality
	$x = -35$	Simplify.

Check

$\frac{x}{5} = -7$

$\frac{-35}{5} \stackrel{?}{=} -7$

$-7 = -7$ ✓

Exercises

Solve the equation. Check your solution.

24. $\frac{x}{3} = -8$ **25.** $-7 = \frac{y}{7}$ **26.** $-\frac{z}{4} = -\frac{3}{4}$ **27.** $-\frac{w}{20} = -2.5$

28. $4x = -8$ **29.** $-10 = 2y$ **30.** $-5.4z = -32.4$ **31.** $-6.8w = 3.4$

32. TEMPERATURE The mean temperature change is -3.2°F per day for 5 days. What is the total change over the 5-day period?

13.5 Solving Two-Step Equations (pp. 582–587)

Solve $-6y + 7 = -5$. Check your solution.

$-6y + 7 = -5$	Write the equation.
$\underline{-7} \quad \underline{-7}$	Subtraction Property of Equality
$-6y = -12$	Simplify.
$\frac{-6y}{-6} = \frac{-12}{-6}$	Division Property of Equality
$y = 2$	Simplify.

∴ The solution is $y = 2$.

Check

$-6y + 7 = -5$

$-6(2) + 7 \stackrel{?}{=} -5$

$-12 + 7 \stackrel{?}{=} -5$

$-5 = -5$ ✓

Exercises

Solve the equation. Check your solution.

33. $-2c + 6 = -8$ **34.** $3(3w - 4) = -20$

35. $\frac{w}{6} + \frac{5}{8} = -1\frac{3}{8}$ **36.** $-3x - 4.6 = 5.9$

37. EROSION The floor of a canyon has an elevation of -14.5 feet. Erosion causes the elevation to change by -1.5 feet per year. How many years will it take for the canyon floor to have an elevation of -31 feet?

13 Chapter Test

Check It Out
Test Practice
BigIdeasMath.com

Simplify the expression.

1. $8x - 5 + 2x$

2. $2.5w - 3y + 4w$

3. $3(5 - 2n) + 9n$

4. $\frac{5}{7}x - 15 - \frac{9}{14}x - 9$

Find the sum or difference.

5. $(3j + 11) + (8j - 7)$

6. $\frac{3}{4}(8p + 12) + \frac{3}{8}(16p - 8)$

7. $(2r - 13) - (-6r + 4)$

8. $-2.5(2s - 5) - 3(4.5s - 5.2)$

Factor out the coefficient of the variable.

9. $3n - 24$

10. $\frac{1}{2}q + \frac{5}{2}$

Solve the equation. Check your solution.

11. $7x = -3$

12. $2(x + 1) = -2$

13. $\frac{2}{9}g = -8$

14. $z + 14.5 = 5.4$

15. $-14 = 6c$

16. $\frac{2}{7}k - \frac{3}{8} = -\frac{19}{8}$

17. **HAIR SALON** Write an expression in simplest form that represents the income from w women and m men getting a haircut and a shampoo.

	Women	Men
Haircut	\$45	\$15
Shampoo	\$12	\$7

18. **RECORD** A runner is compared with the world record holder during a race. A negative number means the runner is ahead of the time of the world record holder. A positive number means that the runner is behind the time of the world record holder. The table shows the time difference between the runner and the world record holder for each lap. What time difference does the runner need for the fourth lap to match the world record?

Lap	Time Difference
1	−1.23
2	0.45
3	0.18
4	?

19. **GYMNASTICS** You lose 0.3 point for stepping out of bounds during a floor routine. Your final score is 9.124. Write and solve an equation to find your score before the penalty.

20. **PERIMETER** The perimeter of the triangle is 45. Find the value of x.

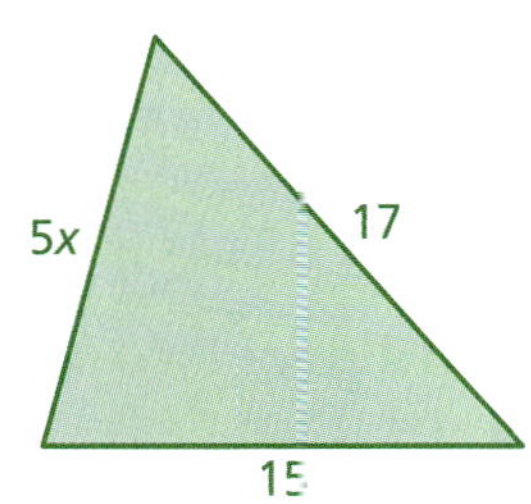

Test Item References

Chapter Test Questions	Section to Review
1–4, 17	13.1
5–10	13.2
11–16, 19	13.3
11–16	13.4
11–16, 18, 20	13.5

Test-Taking Strategies

Remind students to quickly look over the entire test before they start so that they can budget their time. On tests, it is really important for students to **Stop** and **Think.** When students hurry on a test dealing with signed numbers, they often make "sign" errors. There are equations on the test, so remind students to always check their solutions.

Common Errors

- **Exercises 1–8** The subtraction operation can confuse students. Tell students to write the original problem using addition and then to use the Commutative Property.
- **Exercises 11–16** Students may use the same operations, instead of inverse operations, when solving for the variable. Demonstrate to students that this will not give the correct solution. Also, students may use the properties of equality improperly by adding, subtracting, multiplying, or dividing on one side of the equation only, or by using inverse operations on opposite sides of the equation. Remind students of the properties and to check their answers in the original equation.

Reteaching and Enrichment Strategies

If students need help. . .	If students got it. . .
Resources by Chapter • Practice A and Practice B • Puzzle Time Record and Practice Journal Practice Differentiating the Lesson Lesson Tutorials *BigIdeasMath.com* Skills Review Handbook	Resources by Chapter • Enrichment and Extension • Technology Connection Game Closet at *BigIdeasMath.com* Start Cumulative Assessment

Answers

1. $10x - 5$
2. $6.5w - 3y$
3. $15 + 3n$
4. $\frac{1}{14}x + 6$
5. $11j + 4$
6. $12p + 6$
7. $8r - 17$
8. $-18.5s + 28.1$
9. $3(n - 8)$
10. $\frac{1}{2}(q + 5)$
11. $x = -\frac{3}{7}$
12. $x = -2$
13. $g = -36$
14. $z = -9.1$
15. $c = -2\frac{1}{3}$
16. $k = -7$
17. $57w + 22m$
18. 0.6
19. $x - 0.3 = 9.124; 9.424$
20. $2\frac{3}{5}$

Technology for the Teacher

Online Assessment
Assessment Book
ExamView® Assessment Suite

Test-Taking Strategies

Available at *BigIdeasMath.com*

After Answering Easy Questions, Relax

Answer Easy Questions First

Estimate the Answer

Read All Choices before Answering

Read Question before Answering

Solve Directly or Eliminate Choices

Solve Problem before Looking at Choices

Use Intelligent Guessing

Work Backwards

About this Strategy

When taking a multiple choice test, be sure to read each question carefully and thoroughly. After skimming the test and answering the easy questions, stop for a few seconds, take a deep breath, and relax. Work through the remaining questions carefully, using your knowledge and test-taking strategies. Remember, you already completed many of the questions on the test!

Answers

1. B
2. -9
3. G
4. B

Technology for the Teacher

Performance Tasks

Online Assessment

Assessment Book

ExamView® Assessment Suite

Item Analysis

1. **A.** A student multiplies instead of divides.
   B. Correct answer
   C. The student exchanges the dividend and divisor.
   D. The student sets the equation equal to the wrong integer.

2. **Gridded Response:** Correct answer: -9

   Common Error: The student incorrectly substitutes and evaluates $\frac{(-6)(0) - 0^2}{4}$ to get 0 as an answer.

3. **F.** The student adds -38 and -14.
   G. Correct answer
   H. The student adds 38 and -14.
   I. The student adds 38 and 14.

4. **A.** The student finds the temperature of the first thermometer.
   B. Correct answer
   C. The student finds the median of the temperatures.
   D. The student finds the mean of 8°, 10°, and 12°.

13 Cumulative Assessment

Test-Taking Strategy
After Answering Easy Questions, Relax

"After answering the easy questions, relax and try the harder ones. For this, 2x = 12, so x = 6 hyenas."

1. Which equation represents the word sentence shown below?

> The quotient of a number b and 0.3 equals negative 10.

A. $0.3b = 10$

B. $\frac{b}{0.3} = -10$

C. $\frac{0.3}{b} = -10$

D. $\frac{b}{0.3} = 10$

2. What is the value of the expression below when $c = 0$ and $d = -6$?

$$\frac{cd - d^2}{4}$$

3. What is the value of the expression below?

$$-38 - (-14)$$

F. -52

G. -24

H. 24

I. 52

4. The daily low temperatures last week are shown below.

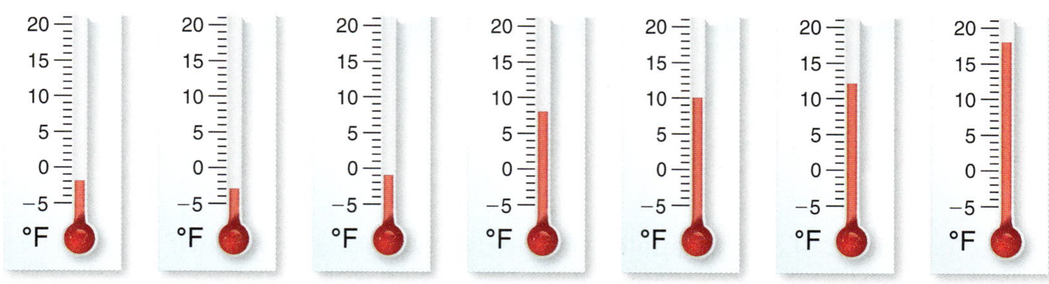

What is the mean low temperature of last week?

A. -2°F

B. 6°F

C. 8°F

D. 10°F

5. Which equation is equivalent to the equation shown below?

$$-\frac{3}{4}x + \frac{1}{8} = -\frac{3}{8}$$

F. $-\frac{3}{4}x = -\frac{3}{8} - \frac{1}{8}$

G. $-\frac{3}{4}x = -\frac{3}{8} + \frac{1}{8}$

H. $x + \frac{1}{8} = -\frac{3}{8} \cdot \left(-\frac{4}{3}\right)$

I. $x + \frac{1}{8} = -\frac{3}{8} \cdot \left(-\frac{3}{4}\right)$

6. What is the value of the expression below?

$$-0.28 \div (-0.07)$$

7. Karina was solving the equation in the box below.

$$-96 = -6(x - 15)$$
$$-96 = -6x - 90$$
$$-96 + 90 = -6x - 90 + 90$$
$$-6 = -6x$$
$$\frac{-6}{-6} = \frac{-6x}{-6}$$
$$1 = x$$

What should Karina do to correct the error that she made?

A. First add 6 to both sides of the equation.

B. First subtract x from both sides of the equation.

C. Distribute the -6 to get $6x - 90$.

D. Distribute the -6 to get $-6x + 90$.

8. The perimeter of the rectangle is 400 inches. What is the value of j? (All measurements are in inches.)

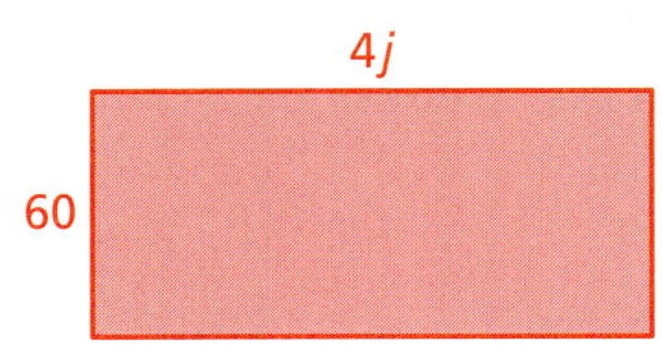

F. 35

G. 85

H. 140

I. 200

Item Analysis (continued)

5. **F.** Correct answer

   G. The student adds $\frac{1}{8}$ to both sides instead of subtracting.

   H. The student creates an equation that is not equivalent to the given equation, because if the student multiplies both sides of the equation by $-\frac{4}{3}$, this would require distributing the multiplication on the left side of the equation, thereby also multiplying $\frac{1}{8}$ by $-\frac{4}{3}$.

   I. The student creates an equation that is not equivalent to the given equation, because if the student multiplies both sides of the equation by $-\frac{3}{4}$, this would require distributing the multiplication on the left side of the equation, thereby also multiplying $\frac{1}{8}$ by $-\frac{3}{4}$. Furthermore, multiplying $-\frac{3}{4}x$ by $-\frac{3}{4}$ does not result in a product of x.

6. **Gridded Response:** Correct answer: 4

   Common Error: The student thinks that the quotient of two negative numbers is a negative number, or the student misplaces the decimal point in the answer.

7. **A.** The student thinks that the inverse operation of multiplying by -6 is adding 6.

   B. The student disregards that the -6 first must be distributed and violates the order of operations.

   C. The student does not distribute the negative sign to the two terms inside the parentheses.

   D. Correct answer

8. **F.** Correct answer

   G. The student solves $60 + 4j = 400$.

   H. The student finds the length, $4j = 140$ inches.

   I. The student adds the length and the width.

Answers

5. F
6. 4
7. D
8. F

Answers

9. B

10. H

11. *Part A*

$116 = 43.50 + 7.25w$

$w = 10$

10 weeks

Part B

$116 - 20 = 43.50 + 8.75w$

$w = 6$

4 weeks sooner because in Part A it takes 10 weeks and in Part B it takes 6 weeks. The difference is $10 - 6 = 4$ weeks.

Item Analysis (continued)

9. **A.** The student incorrectly performs the order of operations.

B. Correct answer

C. The student incorrectly performs the order of operations.

D. The student incorrectly performs the order of operations.

10. **F.** $-2\frac{1}{4} - \left(8\frac{3}{8}\right)$

G. $-(2 + 8)\frac{1 + 3}{4 + 8}$

H. Correct answer

I. $(8 - 2)\frac{3 - 1}{8 - 4}$

11. **4 points** The student demonstrates a thorough understanding of writing and solving two-step equations and of interpreting the results. In Part A, the student correctly writes an equation such as $116 = 43.50 + 7.25w$ and solves to get $w = 10$. The student shows appropriate work and states that it would take 10 weeks to save the money before you can purchase the bicycle. In Part B, the student correctly adjusts the equation written in Part A to get an equation such as $116 - 20 = 43.50 + 8.75w$ and solves to get $w = 6$. The student shows appropriate work and states you could have purchased the bicycle 4 weeks sooner because $10 - 6 = 4$.

3 points The student demonstrates a good understanding of writing and solving two-step equations, and interpreting the results, but the student's work and explanations demonstrate an essential but less than thorough understanding.

2 points The student demonstrates a partial understanding of writing and solving two-step equations and of interpreting the results. The student's work and explanations demonstrate a lack of essential understanding.

1 point The student demonstrates a limited understanding of writing and solving two-step equations and of interpreting the results. The student's response is incomplete and exhibits many flaws.

0 points The student provided no response, a completely incorrect or incomprehensible response, or a response that demonstrates insufficient understanding of writing and solving two-step equations, and interpreting the results.

9. Jacob was evaluating the expression below when $x = -2$ and $y = 4$.

$$3 + x^2 \div y$$

His work is in the box below.

$$\begin{aligned} 3 + x^2 \div y &= 3 + (-2^2) \div 4 \\ &= 3 - 4 \div 4 \\ &= 3 - 1 \\ &= 3 \end{aligned}$$

What should Jacob do to correct the error that he made?

A. Divide 3 by 4 before subtracting.

B. Square -2, then divide.

C. Square then divide.

D. Subtract 4 from 3 before dividing.

10. Which number is equivalent to the expression shown below?

$$-2\frac{1}{4} - \left(-8\frac{3}{8}\right)$$

F. $-10\frac{5}{8}$

G. $-10\frac{1}{3}$

H. $6\frac{1}{8}$

I. $6\frac{1}{2}$

11. You want to buy the bicycle. You already have \$43.50 saved and plan to save an additional \$7.25 every week.

Think Solve Explain

Part A Write and solve an equation to find the number of weeks you need to save before you can purchase the bicycle.

Part B How much sooner could you purchase the bicycle if you had a coupon for \$20 off and saved \$8.75 every week? Explain your reasoning.

14 Ratios and Proportions

What Your Students Have Learned

- Generate numerical patterns given rules, identify the relationship, and form ordered pairs.
- Plot points in the first quadrant of the coordinate plane.
- Convert standard measurement units within a measurement system.
- Graph ordered pairs in all four quadrants of the coordinate plane.
- Understand ratios and describe ratio relationships.
- Compare ratios using tables.
- Use ratio reasoning to convert measurement units.
- Understand rates and unit rates.

What Your Students Will Learn

- Find unit rates associated with ratios of fractions, areas, and other quantities in like or different units.
- Decide whether two quantities are proportional using ratio tables and graphs.
- Use graphs to determine whether two ratios form a proportion.
- Identify the constant of proportionality (unit rate) in tables, graphs, equations, diagrams, and verbal descriptions.
- Represent proportional relationships with equations.
- Explain what a point (x, y) means on a proportional graph in context, particularly $(0, 0)$ and $(1, r)$, where r is the unit rate.
- Use proportionality to solve problems.

Pacing Guide for Chapter 14

Chapter Opener Advanced	1 Day
Section 1 Advanced	1 Day
Section 2 Advanced	2 Days
Section 3 Advanced	1 Day
Study Help/Quiz Advanced	1 Day
Section 4 Advanced	1 Day
Section 5 Advanced	1 Day
Section 6 Advanced	1 Day
Chapter Review/ Chapter Tests Advanced	2 Days
Total Chapter 14 Advanced	11 Days
Year-to-Date Advanced	134 Days

Technology for the Teacher

BigIdeasMath.com
Chapter at a Glance
Complete Materials List
Parent Letters: English and Spanish

What Your Students Have Learned

- Simplify fractions by using factors.
- Identify equivalent fractions.
- Solve one-step equations.

Additional Topics for Review

- Identifying Patterns in Tables
- Operations on Fractions and Decimals
- Coordinate Plane
- Graphing Ordered Pairs
- Equations in Two Variables
- Ratios
- Ratio Tables
- Rates
- Unit Rates
- Converting Measures

Try It Yourself

1.	$\frac{1}{12}$	**2.**	$\frac{1}{3}$
3.	$\frac{3}{4}$	**4.**	$\frac{2}{3}$
5.	no	**6.**	no
7.	no	**8.**	yes
9.	-15	**10.**	3
11.	$\frac{9}{2}$	**12.**	-28

Record and Practice Journal Fair Game Review

1.	$\frac{1}{6}$	**2.**	$\frac{2}{3}$
3.	$\frac{1}{5}$	**4.**	$\frac{1}{2}$
5.	$\frac{4}{9}$	**6.**	$\frac{4}{5}$

7–20. See Additional Answers.

Math Background Notes

Vocabulary Review

- Greatest Common Factor
- Equivalent Fractions
- Equation
- Inverse Operations
- Properties of Equality

Simplifying Fractions

- Students should know how to simplify fractions.
- Some students may have learned simplifying fractions as reducing fractions.
- Remind students that you must divide the numerator and the denominator by the same factor. This is equivalent to dividing by one which does not change the value of the fraction but does change the form.

Identifying Equivalent Fractions

- Students should know how to identify equivalent fractions.
- Encourage students to simplify the fraction with the greater numbers. If the fraction simplifies to the second fraction, students can conclude the fractions are equivalent.
- **Teaching Tip:** Some students may have difficulty simplifying fractions. This makes the search for equivalent fractions difficult. Encourage these students to start with the fraction with lesser numbers and multiply the numerator and denominator by the same factor to see if they can produce the second fraction.

Solving Equations

- Students should know how to solve equations.
- Remind students that whatever is performed on one side of the equation must be performed on the other side.
- Remind students that variables can appear on either side of the equal sign.
- **Common Error:** Students may do the same operation on both sides instead of the inverse operation. Remind them that to get the variable alone, they need to use the inverse operation.

Reteaching and Enrichment Strategies

If students need help. . .	If students got it. . .
Record and Practice Journal • Fair Game Review Skills Review Handbook Lesson Tutorials	Game Closet at *BigIdeasMath.com* Start the next section

What You Learned Before

Simplifying Fractions

Example 1 Simplify $\frac{4}{8}$.

$$\frac{4 \div 4}{8 \div 4} = \frac{1}{2}$$

Example 2 Simplify $\frac{10}{15}$.

$$\frac{10 \div 5}{15 \div 5} = \frac{2}{3}$$

Identifying Equivalent Fractions

Example 3 Is $\frac{1}{4}$ equivalent to $\frac{13}{52}$?

$$\frac{13 \div 13}{52 \div 13} = \frac{1}{4}$$

$\frac{1}{4}$ is equivalent to $\frac{13}{52}$.

Example 4 Is $\frac{30}{54}$ equivalent to $\frac{5}{8}$?

$$\frac{30 \div 6}{54 \div 6} = \frac{5}{9}$$

$\frac{30}{54}$ is *not* equivalent to $\frac{5}{8}$.

Solving Equations

Example 5 Solve $12x = 168$.

$12x = 168$ Write the equation.

$\frac{12x}{12} = \frac{168}{12}$ Division Property of Equality

$x = 14$ Simplify.

Check

$12x = 168$

$12(14) \stackrel{?}{=} 168$

$168 = 168$ ✓

Try It Yourself

Simplify.

1. $\frac{12}{144}$
2. $\frac{15}{45}$
3. $\frac{75}{100}$
4. $\frac{16}{24}$

Are the fractions equivalent? Explain.

5. $\frac{15}{60} \stackrel{?}{=} \frac{3}{4}$
6. $\frac{2}{5} \stackrel{?}{=} \frac{24}{144}$
7. $\frac{15}{20} \stackrel{?}{=} \frac{3}{5}$
8. $\frac{2}{8} \stackrel{?}{=} \frac{16}{64}$

Solve the equation. Check your solution.

9. $\frac{y}{-5} = 3$
10. $0.6 = 0.2a$
11. $-2w = -9$
12. $\frac{1}{7}n = -4$

14.1 Ratios and Rates

Essential Question How do rates help you describe real-life problems?

The Meaning of a Word ● Rate

When you rent snorkel gear at the beach, you should pay attention to the rental **rate**. The rental rate is in dollars per hour.

1 ACTIVITY: Finding Reasonable Rates

Work with a partner.

a. Match each description with a verbal rate.

b. Match each verbal rate with a numerical rate.

c. Give a reasonable numerical rate for each description. Then give an unreasonable rate.

Description	*Verbal Rate*	*Numerical Rate*
Your running rate in a 100-meter dash	Dollars per year	$\frac{\square\text{ in.}}{\text{yr}}$
The fertilization rate for an apple orchard	Inches per year	$\frac{\square\text{ lb}}{\text{acre}}$
The average pay rate for a professional athlete	Meters per second	$\frac{\$\square}{\text{yr}}$
The average rainfall rate in a rain forest	Pounds per acre	$\frac{\square\text{ m}}{\text{sec}}$

Ratios and Rates

In this lesson, you will

- find ratios, rates, and unit rates.
- find ratios and rates involving ratios of fractions.

2 ACTIVITY: Simplifying Expressions That Contain Fractions

Work with a partner. Describe a situation where the given expression may apply. Show how you can rewrite each expression as a division problem. Then simplify and interpret your result.

a. $\dfrac{\frac{1}{2}\text{ c}}{4\text{ fl oz}}$ **b.** $\dfrac{2\text{ in.}}{\frac{3}{4}\text{ sec}}$ **c.** $\dfrac{\frac{3}{8}\text{ c sugar}}{\frac{3}{5}\text{ c flour}}$ **d.** $\dfrac{\frac{5}{6}\text{ gal}}{\frac{2}{3}\text{ sec}}$

Laurie's Notes

Introduction

Applying Mathematical Practices

- **Look for and Make Use of Structure:** In working with ratios and rates, students will make connections to their work with fractions.

Motivate

- **Model:** In an area visible to students, set a wind-up toy in motion. If a toy is not available, (quietly) ask a student to walk across the room at a constant speed.
- ? "How fast is the toy or student moving?" There will be no exact answer.
- ? "How do you measure the *rate* that the toy or student is moving? Which two pieces of information do you need to find the *rate*?"
- Provide measuring tape and a stopwatch. Ask volunteers to compute the rate.
- Discuss why you might use a convenient unit of time (i.e., 5 seconds) versus trying to use a convenient unit of distance (i.e., 10 feet).
- Write the information measured on the board in words, and also as a numerical rate.

Activity Notes

Meaning of the Word

- Begin with a general discussion of rates that students should understand and ask for reasonable values for each: speed limit (65 miles per hour), heart rate (70 beats per minute), and gas mileage (35 miles per gallon).

Activity 1

- Share and discuss sample answers when students have finished.
- ? "How would you describe what a rate is to someone who doesn't know?" Listen for a comparison of two quantities where the units are different. The word "per" is used when comparing the two quantities.
- *Note:* All of the rates in this activity are **unit rates**. The denominator references a single unit (i.e., inches per one month, miles per one hour).
- Share with students that rates do not have to be unit rates. Example: $1.89 per 12 ounces and 100 shared minutes per 4 people.

Activity 2

- Before students begin Activity 2, explain that sometimes a rate can have fractions in the numerator, denominator, or both. The point of this activity is to simplify complex fractions. Students should not feel overwhelmed by the sight of a complex fraction.
- **Look for and Make Use of Structure:** The units may prevent some students from *seeing* the division. If students are stuck, then ask how the problem would look without units.
- Point out how the expression in part (c) is different from the others because the units (cups) are the same in the numerator and the denominator.
- After they have finished, have some students share their answers.

What Your Students Will Learn

- Find ratios, rates, and unit rates using division.
- Find ratios and rates involving ratios of fractions using graphs and ratio tables.

Previous Learning

Students have used reasoning about multiplication and division to solve ratio and rate problems. Knowing measurement abbreviations is helpful.

Technology for the Teacher

Dynamic Classroom

Lesson Plans
Complete Materials List

14.1 Record and Practice Journal

Essential Question How do rates help you describe real-life problems?

1 ACTIVITY: Finding Reasonable Rates

Work with a partner.

a. Match each description with a verbal rate.

b. Match each verbal rate with a numerical rate.

c. Give a reasonable numerical rate for each description. Then give an unreasonable rate.

See Additional Answers.

Description	*Verbal Rate*	*Numerical Rate*
Your running rate in a 100-meter dash	Dollars per year	$= \frac{\square \text{ in.}}{\text{yr}}$
The fertilization rate for an apple orchard	Inches per year	$= \frac{\square \text{ lb}}{\text{acre}}$
The average pay rate for a professional athlete	Meters per second	$= \frac{\$\square}{\text{yr}}$
The average rainfall rate in a rainforest	Pounds per acre	$= \frac{\square \text{ m}}{\text{sec}}$

2 ACTIVITY: Simplifying Expressions That Contain Fractions

Work with a partner. Describe a situation where the given expression may apply. Show how you can rewrite each expression as a division problem. Then simplify and interpret your result.

a. $\frac{\frac{1}{2}\text{ c}}{4\text{ fl oz}}$ $\frac{1}{8}$ c/fl oz

b. $\frac{2\text{ in.}}{\frac{3}{4}\text{ sec}}$ $\frac{8}{3}$, or $2\frac{2}{3}$ in./sec

English Language Learners

Have students add a glossary to their math notebooks. Key vocabulary words should be added as they are introduced. Illustrations next to the vocabulary words will help in understanding and reinforcing the concept.

14.1 Record and Practice Journal

c. $\dfrac{\frac{3}{8}\text{ c sugar}}{\frac{3}{5}\text{ c flour}}$ $\frac{5}{8}$ cup sugar for every cup flour

d. $\dfrac{\frac{5}{6}\text{ gal}}{\frac{2}{3}\text{ sec}}$ $\frac{5}{4}$, or $1\frac{1}{4}$ gal/sec

3 ACTIVITY: Using Ratio Tables to Find Equivalent Rates

Work with a partner. A communications satellite in orbit travels about 18 miles every 4 seconds.

a. Identify the rate in this problem. $\dfrac{18\text{ mi}}{4\text{ sec}}$

b. Recall that you can use *ratio tables* to find and organize equivalent ratios and rates. Complete the ratio table below.

Time (seconds)	4	8	12	16	20
Distance (miles)	18	36	54	72	90

c. How can you use a ratio table to find the speed of the satellite in miles per minute? miles per hour?

270 mi/min and 16,200 mi/h

d. How far does the satellite travel in 1 second? Solve this problem (1) by using a ratio table and (2) by evaluating a quotient.

4.5 miles

e. How far does the satellite travel in $\frac{1}{2}$ second? Explain your steps.

2.25 miles

4 ACTIVITY: Unit Analysis

Work with a partner. Describe a situation where the product may apply. Then find each product and list the units.

a. $10\text{ gal} \times \dfrac{22\text{ mi}}{\text{gal}}$ 220 mi

b. $\dfrac{7}{2}\text{ lb} \times \dfrac{\$3}{\frac{1}{2}\text{lb}}$ \$21

c. $\dfrac{1}{2}\text{ sec} \times \dfrac{30\text{ ft}^2}{\text{sec}}$ 15 ft^2

What Is Your Answer?

5. **IN YOUR OWN WORDS** How do rates help you describe real-life problems? Give two examples.

Sample answer: Describe how fast or slow something is happening.

6. To estimate the annual salary for a given hourly pay rate, multiply by 2 and insert "000" at the end.

Sample: \$10 per hour is about \$20,000 per year.

a. Explain why this works. Assume the person is working 40 hours a week.

40 hours per week ≈ 2000 hours per year

b. Estimate the annual salary for an hourly pay rate of \$8 per hour.

\$16,000 per year

c. You earn \$1 million per month. What is your annual salary?

\$12 million per year

"We had someone apply for the job. He says he would like \$1 million a month, but will settle for \$8 an hour."

d. Why is the cartoon funny?

\$8 an hour is much less than \$1 million per month.

Laurie's Notes

Activity 3

- Students use a ratio table in part (b). They should have studied ratio tables in Section 5.2. If students are not familiar with ratio tables, then they may need extra help with this activity.
- In my experience, students are more successful correctly answering the types of questions in this activity when they organize their work in a ratio table *and* when they label the units. The ratio table is an organizing structure and the units guide the computations.
- Take time to discuss each part of the activity.

Activity 4

- **Look for and Make Use of Structure:** Explain that some real-life problems involve the product of an amount and a rate. The structure you want students to observe is the product of a whole number and a fraction. There is a numerical component and a units component.
- Unit analysis, which again students should have learned in Section 5.7, is also known as *dimensional analysis*.
- Students may have difficulty thinking of a context for part (c).

What Is Your Answer?

- The estimation rule in Question 6 is very interesting.
- Students will need to try a few examples to see why it works. One example to try: the 40-hour work week.

Closure

- Describe three common rates and give a numerical example of each. Listen to be sure that students are comparing two units using the word "per."

3 ACTIVITY: Using Ratio Tables to Find Equivalent Rates

Work with a partner. A communications satellite in orbit travels about 18 miles every 4 seconds.

a. Identify the rate in this problem.

b. Recall that you can use *ratio tables* to find and organize equivalent ratios and rates. Complete the ratio table below.

Time (seconds)	4	8	12	16	20
Distance (miles)					

c. How can you use a ratio table to find the speed of the satellite in miles per minute? miles per hour?

d. How far does the satellite travel in 1 second? Solve this problem (1) by using a ratio table and (2) by evaluating a quotient.

e. How far does the satellite travel in $\frac{1}{2}$ second? Explain your steps.

4 ACTIVITY: Unit Analysis

Math Practice

View as Components

What is the product of the numbers? What is the product of the units? Explain.

Work with a partner. Describe a situation where the product may apply. Then find each product and list the units.

a. $10 \text{ gal} \times \frac{22 \text{ mi}}{\text{gal}}$ **b.** $\frac{7}{2} \text{ lb} \times \frac{\$3}{\frac{1}{2} \text{ lb}}$ **c.** $\frac{1}{2} \text{ sec} \times \frac{30 \text{ ft}^2}{\text{sec}}$

What Is Your Answer?

5. **IN YOUR OWN WORDS** How do rates help you describe real-life problems? Give two examples.

6. To estimate the annual salary for a given hourly pay rate, multiply by 2 and insert "000" at the end.

Sample: \$10 per hour is about \$20,000 per year.

a. Explain why this works. Assume the person is working 40 hours a week.

b. Estimate the annual salary for an hourly pay rate of \$8 per hour.

c. You earn \$1 million per month. What is your annual salary?

d. Why is the cartoon funny?

"We had someone apply for the job. He says he would like \$1 million a month, but will settle for \$8 an hour."

Practice Use what you discovered about ratios and rates to complete Exercises 7–10 on page 603.

14.1 Lesson

Check It Out
Lesson Tutorials
BigIdeasMath.com

Key Vocabulary
ratio, *p. 600*
rate, *p. 600*
unit rate, *p. 600*
complex fraction, *p. 601*

A **ratio** is a comparison of two quantities using division.

$\frac{3}{4}$, 3 to 4, 3 : 4

A **rate** is a ratio of two quantities with different units.

$\frac{60 \text{ miles}}{2 \text{ hours}}$

A rate with a denominator of 1 is called a **unit rate**.

$\frac{30 \text{ miles}}{1 \text{ hour}}$

EXAMPLE 1 Finding Ratios and Rates

There are 45 males and 60 females in a subway car. The subway car travels 2.5 miles in 5 minutes.

a. Find the ratio of males to females.

$$\frac{\text{males}}{\text{females}} = \frac{45}{60} = \frac{3}{4}$$

∴ The ratio of males to females is $\frac{3}{4}$.

b. Find the speed of the subway car.

$$2.5 \text{ miles in 5 minutes} = \frac{2.5 \text{ mi}}{5 \text{ min}} = \frac{2.5 \text{ mi} \div 5}{5 \text{ min} \div 5} = \frac{0.5 \text{ mi}}{1 \text{ min}}$$

∴ The speed is 0.5 mile per minute.

EXAMPLE 2 Finding a Rate from a Ratio Table

The ratio table shows the costs for different amounts of artificial turf. Find the unit rate in dollars per square foot.

× 4 × 4 × 4

Amount (square feet)	25	100	400	1600
Cost (dollars)	100	400	1600	6400

× 4 × 4 × 4

Use a ratio from the table to find the unit rate.

$$\frac{\text{cost}}{\text{amount}} = \frac{\$100}{25 \text{ ft}^2}$$ Use the first ratio in the table.

$$= \frac{\$4}{1 \text{ ft}^2}$$ Simplify.

∴ So, the unit rate is \$4 per square foot.

Remember
The abbreviation ft^2 means *square feet*.

Multi-Language Glossary at BigIdeasMath.com

Laurie's Notes

Introduction

Connect

- **Yesterday:** Students explored rates.
- **Today:** Students determine rates from words, tables, and graphs.

Motivate

- ? "A pitcher for a baseball team is able to throw a fastball approximately 132 feet in 1 second. How fast would this be in miles per hour?" 90 mi/h
- Explain that in this lesson they will be working with equivalent rates written in different units.

Lesson Notes

Discuss

- Remind students of the definitions for *ratio, rate,* and *unit rate.*
- ? Ask students to give their own examples of each.

Example 1

- **Common Error:** Be sure students read the question carefully. Order matters when writing a ratio. Writing $\frac{60}{45}$ would be incorrect.
- In part (b), 2.5 miles per 5 minutes is a rate. 0.5 mile per 1 minute is a unit rate.
- ? **Extension:** "How would you change 2.5 miles per 5 minutes into a unit rate in miles per hour?" To answer this question, use dimensional analysis introduced in the investigation.

  $\frac{2.5 \text{ miles}}{5 \text{ minutes}} \times \frac{60 \text{ minutes}}{1 \text{ hour}} = 30$ miles per hour

  - Besides dividing out a common unit of minutes, divide out a common factor of 5.

Example 2

- Read the problem and write the ratio table. Ask students for assistance in filling in the table.
- ? Ask questions to review ratio tables. "Instead of multiplying each column by 4, could you multiply each column by 3?" yes "Could you add the first two columns and say that 125 ft^2 costs \$500?" yes
- ? "What is the goal when finding a unit rate?" finding a rate with a denominator of 1
- **Look for and Make Use of Structure:** Explain that any equivalent ratio can be used to find the **unit rate**. Each simplifies to 4 : 1.
- **Common Error:** Students may try to find a unit rate using $\frac{25}{100}$ instead of $\frac{100}{25}$ because of the way the ratio table is set up. The problem asks for the unit rate in *dollars per square foot.* Ask students, "What is more useful to know—how many square feet of sod you can purchase for \$1 or the cost of 1 square foot of sod?"

Goal Today's lesson is determining **ratios** and **rates**.

Technology for the Teacher

Dynamic Classroom

Lesson Tutorials
Lesson Plans
Answer Presentation Tool

Extra Example 1

a. There are 12 dogs and 15 cats at the pet store. Find the ratio of cats to dogs. $\frac{5}{4}$

b. You bicycle 30 blocks in 20 minutes. Find your speed. 1.5 blocks per minute

Extra Example 2

The table shows the amount of money you can raise by dancing for a charity. Find your unit rate in dollars per hour.

Time (hours)	6	12	18	24
Money (dollars)	\$90	\$180	\$270	\$360

\$15 per hour

On Your Own

1. $\frac{4}{3}$ **2.** $\frac{4}{7}$

3. 4.8 mi per sec

Extra Example 3

The graph shows the distance that you walk. Find your rate in feet per second.

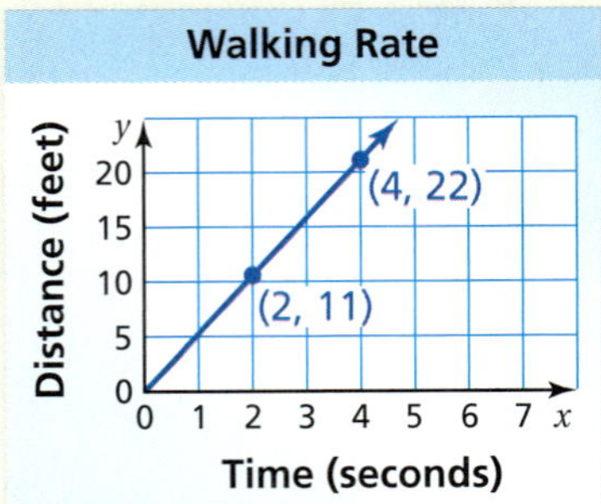

5.5 feet per second

On Your Own

4. No; the unit rate is still $\frac{1}{2}$ mile per minute because the rates represented by points on the graph are equivalent.

Differentiated Instruction

Visual

Students may see little difference between fractions and ratios. Fractions are one type of ratio, part-to-whole. Ratios include part-to-part and whole-to-whole.

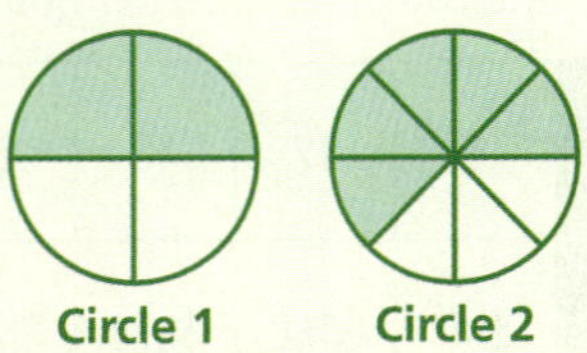

Part-to-whole of Circle 1:

$$\frac{\text{unshaded parts}}{\text{whole number of parts}} = \frac{2}{4}$$

Part-to-part of Circle 2:

$$\frac{\text{shaded parts}}{\text{unshaded parts}} = \frac{5}{3}$$

Whole-to-whole:

$$\frac{\text{parts of Circle 1}}{\text{parts of Circle 2}} = \frac{4}{8}$$

Laurie's Notes

On Your Own

? "How do you calculate a speed?" distance ÷ time

? "Speed is a rate. Why?" You are comparing two quantities with different units.

- Students should work with partners on these three problems.
- **Extension:** Have students estimate the speed in miles per hour.

$$\frac{14.4\text{ miles}}{3\text{ seconds}} \times \frac{3600\text{ seconds}}{1\text{ hour}} = \frac{17{,}280\text{ miles}}{\text{hour}}$$

Write

- Define a *complex fraction* and give examples. Including an example with units would be helpful, such as $\frac{30\text{ meters}}{\frac{1}{2}\text{ minute}}$.

Example 3

? Ask general questions about subways—have you ridden on one, how fast do you think they move, how much does it cost to ride on one, and so on. Some students may have no context for these questions.

- Refer to the graph and ask students to identify the quantities on each axis.

? "What information do you need to find the speed of the subway car?" how much time it takes to travel a given distance

- Choose and interpret a point on the line. For instance, the point $\left(\frac{1}{2}, \frac{1}{4}\right)$ indicates that the subway car travels $\frac{1}{4}$ mile in $\frac{1}{2}$ minute.
- **Reason Abstractly and Quantitatively** and **Look for and Make Use of Structure:** At this point, students may very quickly observe that this is equivalent to $\frac{1}{2}$ mile in 1 minute. It is important for students to see the rate written as a complex fraction. The process shown would be the same for a subway car that travels, for instance, $\frac{4}{7}$ mile in $\frac{5}{6}$ minutes.
- Finish working the problem as shown.

On Your Own

- **Construct Viable Arguments and Critique the Reasoning of Others:** There are several ways in which students may explain their reasoning. Take time to hear a variety of approaches.

On Your Own

Now You're Ready
Exercises 11–24

1. In Example 1, find the ratio of females to males.
2. In Example 1, find the ratio of females to total passengers.
3. The ratio table shows the distance that the *International Space Station* travels while orbiting Earth. Find the speed in miles per second.

Time (seconds)	3	6	9	12
Distance (miles)	14.4	28.8	43.2	57.6

A **complex fraction** has at least one fraction in the numerator, denominator, or both. You may need to simplify complex fractions when finding ratios and rates.

EXAMPLE 3 Finding a Rate from a Graph

The graph shows the speed of a subway car. Find the speed in miles per minute. Compare the speed to the speed of the subway car in Example 1.

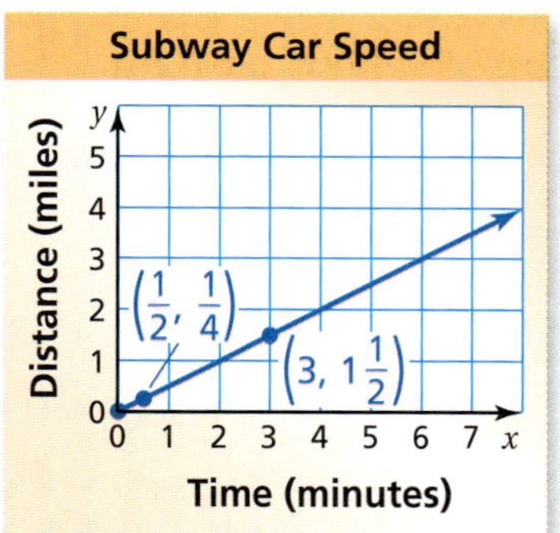

Step 1: Choose and interpret a point on the line.

The point $\left(\frac{1}{2}, \frac{1}{4}\right)$ indicates that the subway car travels $\frac{1}{4}$ mile in $\frac{1}{2}$ minute.

Step 2: Find the speed.

$$\frac{\text{distance traveled}}{\text{elapsed time}} = \frac{\frac{1}{4}}{\frac{1}{2}}$$ ← miles, ← minutes

$$= \frac{1}{4} \div \frac{1}{2}$$ Rewrite the quotient.

$$= \frac{1}{4} \cdot 2 = \frac{1}{2}$$ Simplify.

∴ The speed of the subway car is $\frac{1}{2}$ mile per minute.

Because $\frac{1}{2}$ mile per minute = 0.5 mile per minute, the speeds of the two subway cars are the same.

On Your Own

Now You're Ready
Exercise 28

4. You use the point $\left(3, 1\frac{1}{2}\right)$ to find the speed of the subway car. Does your answer change? Explain your reasoning.

EXAMPLE 4 Solving a Ratio Problem

You mix $\frac{1}{2}$ cup of yellow paint for every $\frac{3}{4}$ cup of blue paint to make 15 cups of green paint. How much yellow paint and blue paint do you use?

> **Math Practice**
> **Analyze Givens**
> What information is given in the problem? How does this help you know that the ratio table needs a "total" column? Explain.

Method 1: The ratio of yellow paint to blue paint is $\frac{1}{2}$ to $\frac{3}{4}$. Use a ratio table to find an equivalent ratio in which the total amount of yellow paint and blue paint is 15 cups.

Yellow (cups)	Blue (cups)	Total (cups)
$\frac{1}{2}$	$\frac{3}{4}$	$\frac{1}{2} + \frac{3}{4} = \frac{5}{4}$
2	3	5
6	9	15

(× 4 from row 1 to row 2; × 3 from row 2 to row 3)

So, you use 6 cups of yellow paint and 9 cups of blue paint.

Method 2: Use the fraction of the green paint that is made from yellow paint and the fraction of the green paint that is made from blue paint. You use $\frac{1}{2}$ cup of yellow paint for every $\frac{3}{4}$ cup of blue paint, so the fraction of the green paint that is made from yellow paint is

$$\frac{\frac{1}{2}}{\frac{1}{2}+\frac{3}{4}} = \frac{\frac{1}{2}}{\frac{5}{4}} = \frac{1}{2} \bullet \frac{4}{5} = \frac{2}{5}.$$

(yellow → numerator; green → denominator)

Similarly, the fraction of the green paint that is made from blue paint is

$$\frac{\frac{3}{4}}{\frac{1}{2}+\frac{3}{4}} = \frac{\frac{3}{4}}{\frac{5}{4}} = \frac{3}{4} \bullet \frac{4}{5} = \frac{3}{5}.$$

(blue → numerator; green → denominator)

So, you use $\frac{2}{5} \bullet 15 = 6$ cups of yellow paint and $\frac{3}{5} \bullet 15 = 9$ cups of blue paint.

On Your Own

Now You're Ready
Exercises 33 and 34

5. How much yellow paint and blue paint do you use to make 20 cups of green paint?

Laurie's Notes

Example 4

? "What primary colors do you mix to get green?" yellow and blue

? "When you mix $\frac{1}{2}$ cup of yellow paint with $\frac{1}{2}$ cup of blue paint, what do you get?" 1 cup of green paint

- Ask a volunteer to read the problem.

? "How much green paint do you want to make?" 15 cups

- Point out that the ratio table shows the total number of cups of green paint for each combination of yellow paint and blue paint listed.
- Supply the thinking behind the multiplications used in the ratio table: multiply by 4 first so that there is a whole number of cups of green paint, then multiply by 3 so that there are 15 cups of green paint as specified in the problem.
- Encourage students to think about different ways of finding the solution using a ratio table. Have them try something to see whether the results give insight to the next step. For instance, if students double the original amounts of yellow and blue paint, then there would be 1 cup of yellow, $1\frac{1}{2}$ cups of blue, and $2\frac{1}{2}$ cups total. These amounts could be doubled and then tripled, giving the results shown in the table.
- **Model with Mathematics:** Method 2 shows a different approach. You could model it with a tape diagram as follows.

$$\frac{1}{2}+\frac{3}{4}=\frac{2}{4}+\frac{3}{4}=\frac{5}{4}$$

- Every $\frac{5}{4}$ cups of green paint are made up of $\frac{2}{4}$ cup of yellow paint and $\frac{3}{4}$ cup of blue paint. In other words, every 5 parts of green paint are made up of 2 parts yellow and 3 parts blue. So, $\frac{2}{5}$ of the green paint is made up of yellow paint and $\frac{3}{5}$ is made up of blue paint. Point out that these fractions have a sum of 1, as should be expected.
- Find $\frac{2}{5}$ of 15 and $\frac{3}{5}$ of 15 to answer the question.
- **Connection:** After you finish going over Method 2, go back to the ratio table used in Method 1 and show that $\frac{2}{5}$ of each total listed is made up of yellow paint and $\frac{3}{5}$ of each total listed is made up of blue paint.

On Your Own

- If time permits, then ask students to solve the problem using both methods shown in Example 4.

Closure

- **Exit Ticket:** Write 4.8 meters per 3 seconds as a unit rate. 1.6 meters per second

Extra Example 4

You mix $\frac{1}{2}$ cup of red paint for every $\frac{1}{4}$ cup of blue paint to make 12 cups of purple paint. How much red paint and blue paint do you use? 8 cups of red paint and 4 cups of blue paint

On Your Own

5. 8 cups of yellow paint and 12 cups of blue paint

Vocabulary and Concept Check

1. It has a denominator of 1.
2. Unit rates are easier to compare.
3. *Sample answer:* A basketball player runs 10 feet down the court in 2 seconds.
4. \$15 per gal
5. \$0.10 per fl oz
6. \$0.20 per egg

Practice and Problem Solving

7. \$72
8. \$28
9. 870 MB
10. 57 mi
11. $\frac{5}{9}$
12. $\frac{9}{4}$
13. $\frac{7}{3}$
14. $\frac{17}{3}$
15. $\frac{4}{3}$
16. $\frac{14}{27}$
17. 60 mi/h
18. 32 mi per gal
19. \$2.40 per lb
20. \$0.80 per can
21. 54 words per min
22. 8.7 m per h
23. 4.5 servings per package
24. 3.6 ft per yr
25. 4.8 MB per min

Assignment Guide and Homework Check

Level	Assignment	Homework Check
Advanced	1–10, 18–38 even, 40–43	18, 22, 30, 32, 38

Common Errors

- **Exercises 11–16** Students may put the wrong number in the numerator. Remind them that the first number or object is the numerator and the second is the denominator.
- **Exercises 17–22** Students may find the unit rate but forget to include the units. Remind them that the units are necessary for understanding a unit rate, or any rate.
- **Exercises 23 and 24** When finding the rate from the table, students may put the wrong unit in the numerator. Tell them that the unit in the second row of the table will be the unit in the numerator of the rate.

14.1 Record and Practice Journal

Write the ratio as a fraction in simplest form.

1. 8 to 14
   $\frac{4}{7}$
2. 36 even : 12 odd
   $\frac{3}{1}$
3. 42 vanilla to 48 chocolate
   $\frac{7}{8}$

Find the unit rate.

4. \$2.50 for 5 ounces
   \$0.50 per ounce
5. 15 degrees in 2 hours
   7.5 degrees per hour
6. 183 miles in 3 hours
   61 miles per hour

Use the ratio table to find the unit rate with the specified units.

7. pounds per box

Boxes	0	1	2	3
Pounds	0	30	60	90

30 pounds per box

8. cost per notebook

Notebooks	0	5	10	15
Cost (dollars)	0	9.45	18.90	28.35

\$1.89 per notebook

9. You create 15 centerpieces for a party in 5 hours.
   a. What is the unit rate?
   3 centerpieces per hour
   b. How long will it take you to make 42 centerpieces?
   14 hours

14.1 Exercises

Check It Out
Help with Homework
BigIdeasMath.com

Vocabulary and Concept Check

1. **VOCABULARY** How can you tell when a rate is a unit rate?
2. **WRITING** Why do you think rates are usually written as unit rates?
3. **OPEN-ENDED** Write a real-life rate that applies to you.

Estimate the unit rate.

4. \$74.75

5. \$1.19

6. \$2.35

Practice and Problem Solving

Find the product. List the units.

7. $8\text{ h} \times \frac{\$9}{\text{h}}$
8. $8\text{ lb} \times \frac{\$3.50}{\text{lb}}$
9. $\frac{29}{2}\text{ sec} \times \frac{60\text{ MB}}{\text{sec}}$
10. $\frac{3}{4}\text{ h} \times \frac{19\text{ mi}}{\frac{1}{4}\text{ h}}$

Write the ratio as a fraction in simplest form.

1 11. 25 to 45
12. 63 : 28
13. 35 girls : 15 boys
14. 51 correct : 9 incorrect
15. 16 dogs to 12 cats
16. $2\frac{1}{3}$ feet : $4\frac{1}{2}$ feet

Find the unit rate.

17. 180 miles in 3 hours
18. 256 miles per 8 gallons
19. \$9.60 for 4 pounds
20. \$4.80 for 6 cans
21. 297 words in 5.5 minutes
22. $21\frac{3}{4}$ meters in $2\frac{1}{2}$ hours

Use the ratio table to find the unit rate with the specified units.

2 23. servings per package

Packages	3	6	9	12
Servings	13.5	27	40.5	54

24. feet per year

Years	2	6	10	14
Feet	7.2	21.6	36	50.4

25. **DOWNLOAD** At 1:00 P.M., you have 24 megabytes of a movie. At 1:15 P.M., you have 96 megabytes. What is the download rate in megabytes per minute?

26. **POPULATION** In 2007, the U.S. population was 302 million people. In 2012, it was 314 million. What was the rate of population change per year?

27. **PAINTING** A painter can paint 350 square feet in 1.25 hours. What is the painting rate in square feet per hour?

3 **28.** **TICKETS** The graph shows the cost of buying tickets to a concert.

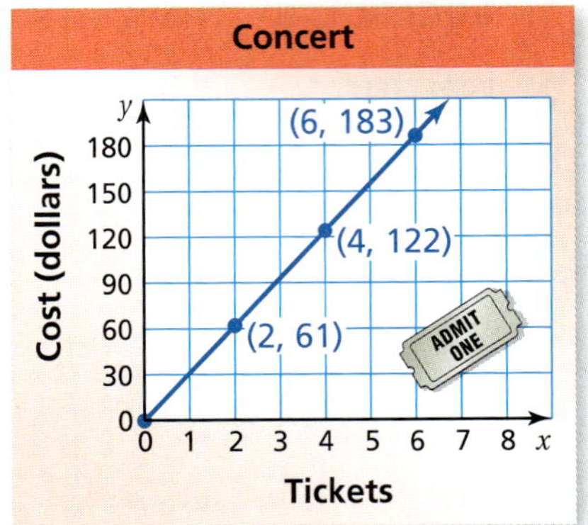

a. What does the point (4, 122) represent?

b. What is the unit rate?

c. What is the cost of buying 10 tickets?

29. **CRITICAL THINKING** Are the two statements equivalent? Explain your reasoning.

- The ratio of boys to girls is 2 to 3.
- The ratio of girls to boys is 3 to 2.

30. **TENNIS** A sports store sells three different packs of tennis balls. Which pack is the best buy? Explain.

31. **FLOORING** It costs \$68 for 16 square feet of flooring. How much does it cost for 12 square feet of flooring?

32. **OIL SPILL** An oil spill spreads 25 square meters every $\frac{1}{6}$ hour. How much area does the oil spill cover after 2 hours?

4 **33.** **JUICE** You mix $\frac{1}{4}$ cup of juice concentrate for every 2 cups of water to make 18 cups of juice. How much juice concentrate and water do you use?

34. **LANDSCAPING** A supplier sells $2\frac{1}{4}$ pounds of mulch for every $1\frac{1}{3}$ pounds of gravel. The supplier sells 172 pounds of mulch and gravel combined. How many pounds of each item does the supplier sell?

35. **HEART RATE** Your friend's heart beats 18 times in 15 seconds when at rest. While running, your friend's heart beats 25 times in 10 seconds.

a. Find the heart rate in beats per minute at rest and while running.

b. How many more times does your friend's heart beat in 3 minutes while running than while at rest?

Common Errors

- **Exercise 30** Students may find the reciprocal of the unit rate and come up with the incorrect conclusion. Remind them that the unit rate represents the rate for one unit, so they should divide by the number of units.
- **Exercises 32–34** Students may have difficulty simplifying the complex fractions that result In the solutions of these exercises. A quick review of complex fractions may be helpful. Be sure to point out that a fraction bar means division, so they can rewrite the complex fraction as *numerator* ÷ *denominator*.

Practice and Problem Solving

26. 2.4 million people per year

27. 280 square feet per hour

28. **a.** It costs \$122 for 4 tickets.

b. \$30.50 per ticket

c. \$305

29. no; Although the relative number of boys and girls are the same, the two ratios are inverses.

30. The 9-pack is the best buy at \$2.55 per container.

31. \$51

32. 300 square meters

33. 2 cups of juice concentrate, 16 cups of water

34. 108 pounds of mulch, 64 pounds of gravel

35. **a.** rest: 72 beats per minute running: 150 beats per minute

b. 234 beats

Differentiated Instruction

Auditory

Ask students to recall when they have heard or used expressions with the word *per*, such as dollars per gallon. Make a list of the expressions. Explain that these are all rates and have a denominator of 1. They are called *unit rates.* Ask students how they would find the unit rate from a rate such as \$42.50 for 10 gallons of gasoline. Conversely, ask how they would use a unit rate of 25 miles per gallon to determine the number of miles traveled using 10 gallons. Motivate the students by posting the list and adding to it as you work through the chapter.

Practice and Problem Solving

36. **a.** whole milk

b. orange juice

37. See *Taking Math Deeper.*

38. **a.** 16 cups of red paint, 10 cups of blue paint

b. $3\frac{1}{5}$ cups of red paint, 2 cups of blue paint, $\frac{4}{5}$ cup of white paint

39. **a.** you; $\frac{1}{3}$ mile per hour faster

b. $3\frac{1}{2}$ hours

c. you; $1\frac{1}{6}$ miles

Fair Game Review

40. > **41.** <

42. = **43.** B

Mini-Assessment

Write the ratio as a fraction in simplest form.

1. 30 to 50 $\frac{3}{5}$ **2.** 3 : 12 $\frac{1}{4}$

Find the unit rate.

3. 165 miles in 3 hours 55 mi/h

4. $9.60 for 8 cans $1.20 per can

5. The graph shows the cost of buying movie tickets.

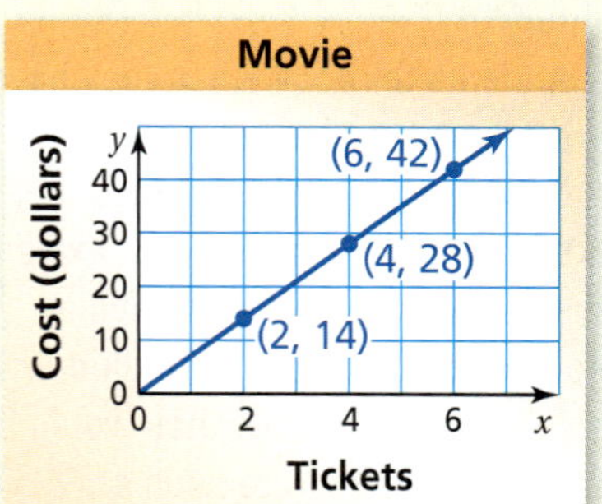

a. What does the point (4, 28) represent? $28 for 4 tickets

b. What is the unit rate? $7 per ticket

c. What is the cost of buying 9 tickets? $63

Taking Math Deeper

Exercise 37

This is a nice real-life problem that deals with rates. If students search the Internet for "fire hydrant colors" they can get the information about the rates in gallons per minute (GPM).

1 **a.** Perform an Internet search and make a table from the results.

Blue	1500 + GPM	Very good
Green	1000–1499 GPM	Good
Yellow	500–999 GPM	Adequate
Red	Below 500 GPM	Inadequate

2 A number line helps display the information.

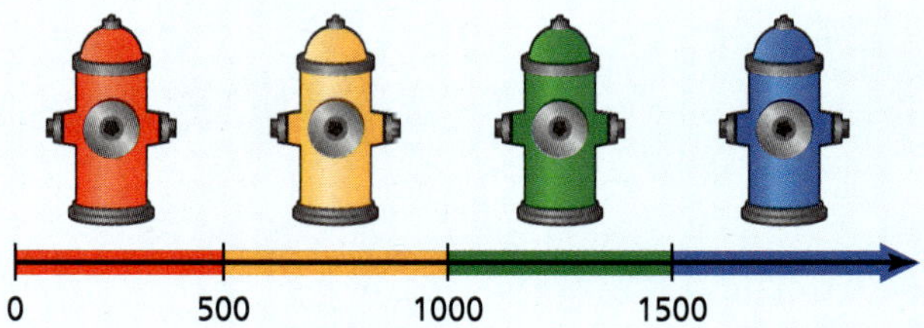

3 **b.** Knowing the rate at which the water comes out of the fire hydrant is critical. If a firefighter pumps water out at too high a rate, the system of water pipes in the ground could be stressed and burst.

Reteaching and Enrichment Strategies

If students need help. . .	If students got it. . .
Resources by Chapter • Practice A and Practice B • Puzzle Time Record and Practice Journal Practice Differentiating the Lesson Lesson Tutorials Skills Review Handbook	Resources by Chapter • Enrichment and Extension • Technology Connection Start the next section

36. PRECISION The table shows nutritional information for three beverages.

Beverage	Serving Size	Calories	Sodium
Whole milk	1 c	146	98 mg
Orange juice	1 pt	210	10 mg
Apple juice	24 fl oz	351	21 mg

a. Which has the most calories per fluid ounce?

b. Which has the least sodium per fluid ounce?

37. RESEARCH Fire hydrants are painted one of four different colors to indicate the rate at which water comes from the hydrant.

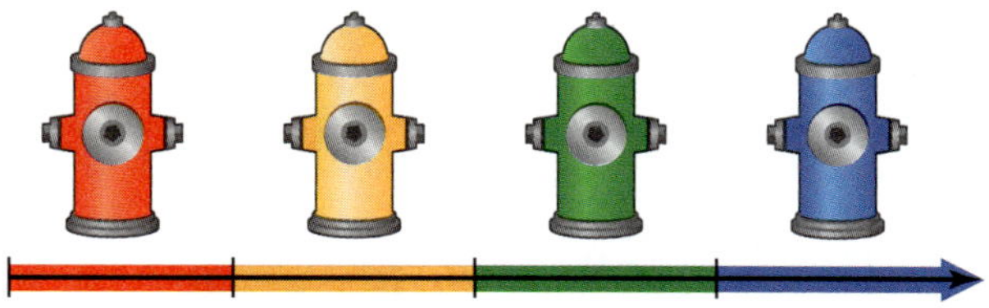

a. Use the Internet to find the ranges of the rates for each color.

b. Research why a firefighter needs to know the rate at which water comes out of a hydrant.

38. PAINT You mix $\frac{2}{5}$ cup of red paint for every $\frac{1}{4}$ cup of blue paint to make $1\frac{5}{8}$ gallons of purple paint.

a. How much red paint and blue paint do you use?

b. You decide that you want to make a lighter purple paint. You make the new mixture by adding $\frac{1}{10}$ cup of white paint for every $\frac{2}{5}$ cup of red paint and $\frac{1}{4}$ cup of blue paint. How much red paint, blue paint, and white paint do you use to make $\frac{3}{8}$ gallon of lighter purple paint?

39. Critical Thinking You and a friend start hiking toward each other from opposite ends of a 17.5-mile hiking trail. You hike $\frac{2}{3}$ mile every $\frac{1}{4}$ hour. Your friend hikes $2\frac{1}{3}$ miles per hour.

a. Who hikes faster? How much faster?

b. After how many hours do you meet?

c. When you meet, who hiked farther? How much farther?

Fair Game Review What you learned in previous grades & lessons

Copy and complete the statement using <, >, or =. *(Section 12.1)*

40. $\frac{9}{2} \square \frac{8}{3}$

41. $-\frac{8}{15} \square \frac{10}{18}$

42. $\frac{-6}{24} \square \frac{-2}{8}$

43. MULTIPLE CHOICE Which fraction is greater than $-\frac{2}{3}$ and less than $-\frac{1}{2}$? *(Section 12.1)*

Ⓐ $-\frac{3}{4}$ Ⓑ $-\frac{7}{12}$ Ⓒ $-\frac{5}{12}$ Ⓓ $-\frac{3}{8}$

14.2 Proportions

Essential Question How can proportions help you decide when things are "fair"?

The Meaning of a Word ● Proportional

When you work toward a goal, your success is usually **proportional** to the amount of work you put in.

An equation stating that two ratios are equal is a **proportion**.

1 ACTIVITY: Determining Proportions

Work with a partner. Tell whether the two ratios are equivalent. If they are not equivalent, change the next day to make the ratios equivalent. Explain your reasoning.

a. On the first day, you pay $5 for 2 boxes of popcorn. The next day, you pay $7.50 for 3 boxes.

First Day Next Day

$$\frac{\$5.00}{2 \text{ boxes}} \stackrel{?}{=} \frac{\$7.50}{3 \text{ boxes}}$$

b. On the first day, it takes you $3\frac{1}{2}$ hours to drive 175 miles. The next day, it takes you 5 hours to drive 200 miles.

First Day Next Day

$$\frac{3\frac{1}{2} \text{ h}}{175 \text{ mi}} \stackrel{?}{=} \frac{5 \text{ h}}{200 \text{ mi}}$$

Proportions
In this lesson, you will
- use equivalent ratios to determine whether two ratios form a proportion.
- use the Cross Products Property to determine whether two ratios form a proportion.

c. On the first day, you walk 4 miles and burn 300 calories. The next day, you walk $3\frac{1}{3}$ miles and burn 250 calories.

First Day Next Day

$$\frac{4 \text{ mi}}{300 \text{ cal}} \stackrel{?}{=} \frac{3\frac{1}{3} \text{ mi}}{250 \text{ cal}}$$

d. On the first day, you paint 150 square feet in $2\frac{1}{2}$ hours. The next day, you paint 200 square feet in 4 hours.

First Day Next Day

$$\frac{150 \text{ ft}^2}{2\frac{1}{2} \text{ h}} \stackrel{?}{=} \frac{200 \text{ ft}^2}{4 \text{ h}}$$

Laurie's Notes

Introduction

Applying Mathematical Practices

- **Construct Viable Arguments and Critique the Reasoning of Others:** To develop an understanding of proportions, ask students to explain their reasoning. A proportion is an equation stating that two ratios are equivalent. Explanations offered by students need to be connected to this definition.

Motivate

- Ask for two volunteers. Hand student A 8 square tiles and student B 4 square tiles.
- Make up a story as to why student A starts off with more than student B.
- ? "What is the ratio of student A's tiles to student B's tiles?" 2 : 1
- ? "What is the ratio of student B's tiles to student A's tiles?" 1 : 2
- ? "If I give student A 2 more tiles, how many should I give student B so that they still have the same ratio? Explain your reasoning." 1; Student A has twice as many as student B, so you need to give him/her twice as many tiles each time.
- Hand each student 2 more tiles.
- ? Is the ratio of student A's tiles to student B's tiles still 2 : 1? Explain." No, the ratio is $10 : 6 = 5 : 3 \neq 2 : 1$.
- Ask additional questions if time permits.

Activity Notes

Meaning of the Word

- Explain the meaning of the word proportional. Reference the activity used to motivate today's lesson for examples.

Activity 1

- Note the use of color to distinguish the ratios and help students focus on the writing of the proportions. The ratios on the left contain information from the first day, and the ratios on the right contain information from the second day.
- **Management Tip:** For this activity, students will work in pairs. To allow the activity to run smoothly and to save time in class, plan a partner for each student before class.
- Discuss student explanations.
- **Construct Viable Arguments and Critique the Reasoning of Others:** Encourage students to share their strategies. How did they decide whether the ratios were equal or not? There will be different strategies, and it is important to hear a variety.

What Your Students Will Learn

- Use equivalent ratios to determine whether two ratios form a proportion or whether two quantities are proportional.
- Use the Cross Products Property to identify proportional realtionships.

Previous Learning

Students have written and simplified ratios.

Technology for the Teacher

Dynamic Classroom

Lesson Plans
Complete Materials List

14.2 Record and Practice Journal

Essential Question How can proportions help you decide when things are "fair"?

1 ACTIVITY: Determining Proportions

Work with a partner. Tell whether the two ratios are equivalent. If they are not equivalent, change the next day to make the ratios equivalent. Explain your reasoning.

a. On the first day, you pay $5 for 2 boxes of popcorn. The next day, you pay $7.50 for 3 boxes.

First Day Next Day

$\frac{\$5.00}{2 \text{ boxes}} \stackrel{?}{=} \frac{\$7.50}{3 \text{ boxes}}$

equivalent

b. On the first day, it takes you $3\frac{1}{2}$ hours to drive 175 miles. The next day, it takes you 5 hours to drive 200 miles.

First Day Next Day

$\frac{3\frac{1}{2}\text{ h}}{175\text{ mi}} \stackrel{?}{=} \frac{5\text{ h}}{200\text{ mi}}$

not equivalent; 250 miles in 5 hours

c. On the first day, you walk 4 miles and burn 300 calories. The next day, you walk $3\frac{1}{3}$ miles and burn 250 calories.

First Day Next Day

$\frac{4\text{ mi}}{300\text{ cal}} \stackrel{?}{=} \frac{3\frac{1}{3}\text{ mi}}{250\text{ cal}}$

equivalent

d. On the first day, you paint 150 square feet in $2\frac{1}{2}$ hours. The next day, you paint 200 square feet in 4 hours.

First Day Next Day

$\frac{150\text{ ft}^2}{2\frac{1}{2}\text{ h}} \stackrel{?}{=} \frac{200\text{ ft}^2}{4\text{ h}}$

not equivalent; 240 square feet in 4 hours

English Language Learners

Word Problems

Most word problems follow a standard format that allows English learners to recognize key words that are integral to writing a mathematical statement of the problem. Most numbers given in a word problem are used. Analyzing the units in the mathematical statement and determining the units of the solution give students confidence that they are on the right path for solving the problem.

14.2 Record and Practice Journal

2 ACTIVITY: Checking a Proportion

Work with a partner.

a. It is said that "one year in a dog's life is equivalent to seven years in a human's life." Explain why Newton thinks he has a score of 105 points. Did he solve the proportion correctly?

$\frac{1 \text{ year}}{7 \text{ years}} \stackrel{?}{=} \frac{15 \text{ points}}{105 \text{ points}}$

yes

"I got 15 on my online test. That's 105 in dog points! Isn't that an A+?"

b. If Newton thinks his score is 98 points, how many points does he actually have? Explain your reasoning. 14

$\frac{1 \text{ year}}{7 \text{ years}} = \frac{14 \text{ points}}{98 \text{ points}}$

3 ACTIVITY: Determining Fairness

Work with a partner. Write a ratio for each sentence. Compare the ratios. If they are equal, then the answer is "It is fair." If they are not equal, then the answer is "It is not fair." Explain your reasoning.

a. You pay $184 for 2 tickets to a concert. & I pay $266 for 3 tickets to the same concert. ➡ **Is this fair?**

It is not fair.

b. You get 75 points for answering 15 questions correctly. & I get 70 points for answering 14 questions correctly. ➡ **Is this fair?**

It is fair.

c. You trade 24 football cards for 15 baseball cards. & I trade 20 football cards for 32 baseball cards. ➡ **Is this fair?**

It is not fair.

What Is Your Answer?

4. Find a recipe for something you like to eat. Then show how two of the ingredient amounts are proportional when you double or triple the recipe.

Check students' work.

5. **IN YOUR OWN WORDS** How can proportions help you decide when things are "fair"? Give an example.

When 2 ratios are equivalent, things are fair.

Laurie's Notes

Activity 2

- Help students think through this problem.
- The ratio often heard is 1 dog year : 7 human years. Years may be in the ratio of 1 : 7, but there is no reason why test scores need to be.
- The proportion that is being used is $\frac{1 \text{ dog year}}{7 \text{ human years}} = \frac{15 \text{ test points}}{105 \text{ dog points}}$.
- While the numeric proportion is a true proportion, the units of $\frac{1}{7} = \frac{15}{105}$ make no sense.

Activity 3

- Students like to think about fairness when deciding if two ratios are equal. When they finish, have students share their reasoning.
- One common strategy is to compare unit rates.

Words of Wisdom

- All of the work today is to build an understanding of what it means for two ratios to be equal and to develop different strategies for deciding if the ratios are equal.
- **Construct Viable Arguments and Critique the Reasoning of Others:** Student discussion can often be quite interesting as they explain their reasoning. It is important to focus on both the computations and the information that students use to explain their reasoning.

What Is Your Answer?

- Question 4 can be assigned for homework.

Closure

- **Writing Prompt:** One way to decide if 6 hours for $4.80 is the same rate as 10 hours for $8 is . . . *Sample answer:* to write each comparison as a ratio and then find the unit rates.

$\frac{\$4.80}{6 \text{ hours}} = \$0.80 \text{ per hour}$

$\frac{\$8.00}{10 \text{ hours}} = \$0.80 \text{ per hour}$

2 ACTIVITY: Checking a Proportion

Work with a partner.

a. It is said that "one year in a dog's life is equivalent to seven years in a human's life." Explain why Newton thinks he has a score of 105 points. Did he solve the proportion correctly?

$$\frac{1 \text{ year}}{7 \text{ years}} \stackrel{?}{=} \frac{15 \text{ points}}{105 \text{ points}}$$

b. If Newton thinks his score is 98 points, how many points does he actually have? Explain your reasoning.

"I got 15 on my online test. That's 105 in dog points! Isn't that an A+?"

3 ACTIVITY: Determining Fairness

Math Practice

Justify Conclusions

What information can you use to justify your conclusion?

Work with a partner. Write a ratio for each sentence. Compare the ratios. If they are equal, then the answer is "It is fair." If they are not equal, then the answer is "It is not fair." Explain your reasoning.

a. You pay $184 for 2 tickets to a concert. & I pay $266 for 3 tickets to the same concert. → **Is this fair?**

b. You get 75 points for answering 15 questions correctly. & I get 70 points for answering 14 questions correctly. → **Is this fair?**

c. You trade 24 football cards for 15 baseball cards. & I trade 20 football cards for 32 baseball cards. → **Is this fair?**

What Is Your Answer?

4. Find a recipe for something you like to eat. Then show how two of the ingredient amounts are proportional when you double or triple the recipe.

5. **IN YOUR OWN WORDS** How can proportions help you decide when things are "fair"? Give an example.

Practice Use what you discovered about proportions to complete Exercises 15–20 on page 610.

14.2 Lesson

Check It Out
Lesson Tutorials
BigIdeasMath.com

Key Vocabulary

proportion, *p. 608*
proportional, *p. 608*
cross products, *p. 609*

Key Idea

Proportions

Words A **proportion** is an equation stating that two ratios are equivalent. Two quantities that form a proportion are **proportional**.

Numbers $\frac{2}{3} = \frac{4}{6}$ The proportion is read "2 is to 3 as 4 is to 6."

EXAMPLE 1 Determining Whether Ratios Form a Proportion

Tell whether $\frac{6}{4}$ and $\frac{8}{12}$ form a proportion.

Compare the ratios in simplest form.

$$\frac{6}{4} = \frac{6 \div 2}{4 \div 2} = \frac{3}{2}$$

$$\frac{8}{12} = \frac{8 \div 4}{12 \div 4} = \frac{2}{3}$$

The ratios are *not* equivalent.

So, $\frac{6}{4}$ and $\frac{8}{12}$ do *not* form a proportion.

EXAMPLE 2 Determining Whether Two Quantities Are Proportional

Tell whether *x* and *y* are proportional.

x	y
$\frac{1}{2}$	3
1	6
$\frac{3}{2}$	9
2	12

Compare each ratio *x* to *y* in simplest form.

$$\frac{\frac{1}{2}}{3} = \frac{1}{6} \qquad \frac{1}{6} \qquad \frac{\frac{3}{2}}{9} = \frac{1}{6} \qquad \frac{2}{12} = \frac{1}{6}$$

The ratios are equivalent.

So, *x* and *y* are proportional.

Reading

Two quantities that are proportional are in a *proportional relationship.*

On Your Own

Now You're Ready
Exercises 5–14

Tell whether the ratios form a proportion.

1. $\frac{1}{2}, \frac{5}{10}$
2. $\frac{4}{6}, \frac{18}{24}$
3. $\frac{10}{3}, \frac{5}{6}$
4. $\frac{25}{20}, \frac{15}{12}$
5. Tell whether *x* and *y* are proportional.

Birdhouses Built, *x*	1	2	4	6
Nails Used, *y*	12	24	48	72

Multi-Language Glossary at BigIdeasMath.com

Laurie's Notes

Introduction

Connect

- **Yesterday:** Students explored pairs of rates and decided if they were equivalent, or fair.
- **Today:** Students will use multiplication and division, and the Cross Products Property to decide if two ratios are equal.

Motivate

- Draw the following on the board: $\frac{\square}{\square} = \frac{\square}{\square}$
- Ask students to use the numbers 2, 3, 4, and 6 placing one number in each square to make two ratios. They should list all combinations that are different $\left(\text{i.e., } \frac{2}{4} = \frac{3}{6} \text{ is not different from } \frac{3}{6} = \frac{2}{4}\right)$.

? "How did you decide where to place the numbers?" Listen for ideas related to fractions (reducing, using the Cross Products Property, etc.).

- Record student solutions to reference later.

Lesson Notes

Key Idea

- Write the definition of proportion on the board.
- **FYI:** Without units associated with the numeric values, students think of proportions as fractions.
- If students are comfortable with writing equivalent fractions and simplifying fractions, they will generally have a good sense about working with proportions.

Example 1

- The strategy is to write the ratios in simplest form.

? "What is the relationship between $\frac{2}{3}$ and $\frac{3}{2}$?" They are reciprocals.

Example 2

- Copy the table of values. Students may see x and y as ordered pairs or see the table as a vertical ratio table.

? "How can you decide whether x and y are proportional?" Determine whether each ratio $x : y$ is equivalent

On Your Own

- **Construct Viable Arguments and Critique the Reasoning of Others:** Students may come up with different strategies for answering Question 5. Ask them to explain their reasoning.

Goal Today's lesson is comparing ratios using **proportions** and the **Cross Products** Property.

Technology for the Teacher

Dynamic Classroom

Lesson Tutorials
Lesson Plans
Answer Presentation Tool

Extra Example 1

Tell whether $\frac{10}{18}$ and $\frac{45}{81}$ form a proportion.
yes

Extra Example 2

Tell whether x and y are proportional.

x	y
$\frac{1}{2}$	4
1	8
$\frac{3}{2}$	12
2	16

yes

On Your Own

1. yes
2. no
3. no
4. yes
5. yes

Laurie's Notes

Key Ideas

- Write the Key Ideas on the board.
- ? "Why can't *b* or *d* equal zero?" or "Which number should *b* or *d* not be equal to? Why not?" Zero, because it would lead to division by zero.
- Use the Cross Products Property to verify that each of the solutions written at the beginning of class is a proportion.
- Discuss the Study Tip. It shows why the Cross Products Property is true.

Example 3

- Ask the students to read the problem.
- Work through each method of the solution.
- ? **Connection:** When you have finished each method, tie this lesson to equivalent ratios by asking, "If you are swimming at a constant rate and you swam 4 laps in 2.4 minutes, how long should it take you to complete 16 laps?" 4 times as long, or 9.6 min
- ? "What does this mean in the context of the problem?" You slowed down after 4 laps.

Closure

- Write an example of two ratios that are equal. Explain how you know they are equal.
- Write an example of two ratios that are not equal. Explain how you know they are not equal.

Extra Example 3

You run the first 3 laps around the gym in 1.5 minutes. You complete 24 laps in 12 minutes. Is the number of laps proportional to your time? The number of laps is proportional to the time.

On Your Own

6. yes

Differentiated Instruction

Visual

As suggested in the Study Tip, the Cross Products Property is an application of the Multiplication Property of Equality. For those students having trouble following the steps shown in the Study Tip, show the following more detailed steps on the board or overhead.

$\frac{a}{b} = \frac{c}{d}$	Proportion
$b \cdot d \cdot \frac{a}{b} = b \cdot d \cdot \frac{c}{d}$	Multiply each side by $b \cdot d$.
$\cancel{b}d \cdot \frac{a}{\cancel{b}} = b\cancel{d} \cdot \frac{c}{\cancel{d}}$	Divide out common factors.
$d \cdot a = b \cdot c$	Simplify.
$ad = bc$	Commutative Property of Multiplication

Key Ideas

Cross Products

In the proportion $\frac{a}{b} = \frac{c}{d}$, the products $a \cdot d$ and $b \cdot c$ are called **cross products**.

Cross Products Property

Words The cross products of a proportion are equal.

Numbers

$$\frac{2}{3} = \frac{4}{6}$$

$2 \cdot 6 = 3 \cdot 4$

Algebra

$$\frac{a}{b} = \frac{c}{d}$$

$ad = bc$, where $b \neq 0$ and $d \neq 0$

Study Tip

You can use the Multiplication Property of Equality to show that the cross products are equal.

$$\frac{a}{b} = \frac{c}{d}$$

$$bd \cdot \frac{a}{b} = bd \cdot \frac{c}{d}$$

$$ad = bc$$

EXAMPLE 3 Identifying Proportional Relationships

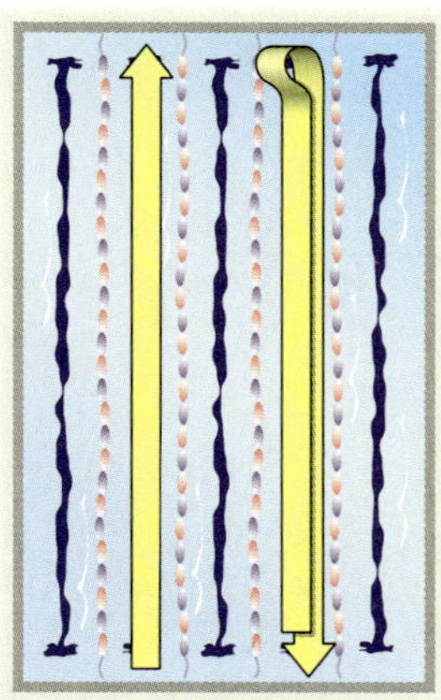

1 length 1 lap

You swim your first 4 laps in 2.4 minutes. You complete 16 laps in 12 minutes. Is the number of laps proportional to your time?

Method 1: Compare unit rates.

$$\frac{2.4 \text{ min}}{4 \text{ laps}} = \frac{0.6 \text{ min}}{1 \text{ lap}} \quad (\div 4)$$

$$\frac{12 \text{ min}}{16 \text{ laps}} = \frac{0.75 \text{ min}}{1 \text{ lap}} \quad (\div 16)$$

The unit rates are *not* equivalent.

So, the number of laps is *not* proportional to the time.

Method 2: Use the Cross Products Property.

$$\frac{2.4 \text{ min}}{4 \text{ laps}} \stackrel{?}{=} \frac{12 \text{ min}}{16 \text{ laps}}$$ Test to see if the rates are equivalent.

$$2.4 \cdot 16 \stackrel{?}{=} 4 \cdot 12$$ Find the cross products.

$$38.4 \neq 48$$ The cross products are *not* equal.

So, the number of laps is *not* proportional to the time.

On Your Own

Now You're Ready Exercises 15–20

6. You read the first 20 pages of a book in 25 minutes. You read 36 pages in 45 minutes. Is the number of pages read proportional to your time?

14.2 Exercises

Check It Out
Help with Homework
BigIdeasMath.com

Vocabulary and Concept Check

1. **VOCABULARY** What does it mean for two ratios to form a proportion?
2. **VOCABULARY** What are two ways you can tell that two ratios form a proportion?
3. **OPEN-ENDED** Write two ratios that are equivalent to $\frac{3}{5}$.
4. **WHICH ONE DOESN'T BELONG?** Which ratio does *not* belong with the other three? Explain your reasoning.

$\frac{4}{10}$ $\frac{2}{5}$ $\frac{3}{5}$ $\frac{6}{15}$

Practice and Problem Solving

Tell whether the ratios form a proportion.

1 5. $\frac{1}{3}, \frac{7}{21}$
6. $\frac{1}{5}, \frac{6}{30}$
7. $\frac{3}{4}, \frac{24}{18}$
8. $\frac{2}{5}, \frac{40}{16}$
9. $\frac{48}{9}, \frac{16}{3}$
10. $\frac{18}{27}, \frac{33}{44}$
11. $\frac{7}{2}, \frac{16}{6}$
12. $\frac{12}{10}, \frac{14}{12}$

Tell whether *x* and *y* are proportional.

2 13.

x	1	2	3	4
y	7	8	9	10

14.

x	2	4	6	8
y	5	10	15	20

Tell whether the two rates form a proportion.

3 15. 7 inches in 9 hours; 42 inches in 54 hours
16. 12 players from 21 teams; 15 players from 24 teams
17. 440 calories in 4 servings; 300 calories in 3 servings
18. 120 units made in 5 days; 88 units made in 4 days
19. 66 wins in 82 games; 99 wins in 123 games
20. 68 hits in 172 at bats; 43 hits in 123 at bats

21. **FITNESS** You can do 90 sit-ups in 2 minutes. Your friend can do 135 sit-ups in 3 minutes. Do these rates form a proportion? Explain.

22. **HEART RATES** Find the heart rates of you and your friend. Do these rates form a proportion? Explain.

	Heartbeats	Seconds
You	22	20
Friend	18	15

Assignment Guide and Homework Check

Level	Assignment	Homework Check
Advanced	1–4, 6–14 even, 15–20, 22–32 even, 33–37	14, 22, 26, 28, 30

Common Errors

- **Exercises 5–12** Students may have difficulty understanding why you can write the ratio in simplest form. Tell students to compare the ratio to a fraction. Simplifying ratios is the same as writing equivalent fractions.
- **Exercises 15–20** Students may mix up the rates and incorrectly find that they are not proportional. For example, they might write $\frac{7\text{ inches}}{9\text{ hours}} \stackrel{?}{=} \frac{54\text{ hours}}{42\text{ inches}}$. Remind students about writing a rate and help them to identify which unit goes in the numerator.

14.2 Record and Practice Journal

Tell whether the ratios form a proportion.

1. $\frac{1}{5}, \frac{5}{15}$ **no**
2. $\frac{2}{3}, \frac{12}{18}$ **yes**
3. $\frac{15}{2}, \frac{4}{30}$ **no**
4. $\frac{56}{21}, \frac{8}{3}$ **yes**
5. $\frac{5}{8}, \frac{62.5}{100}$ **yes**
6. $\frac{17}{20}, \frac{90.1}{106}$ **yes**
7. $\frac{3.2}{4}, \frac{16}{24}$ **no**
8. $\frac{34}{50}, \frac{6.8}{10}$ **yes**

Tell whether the two rates form a proportion.

9. 28 points in 3 games; 112 points in 12 games **yes**
10. 32 notes in 4 measures; 12 notes in 2 measures **no**

11. You can type 105 words in two minutes. Your friend can type 210 words in four minutes. Are these rates proportional? Explain. **yes**

Vocabulary and Concept Check

1. Both ratios are equal.
2. Compare the ratios in simplest form and compare the cross products.
3. *Sample answer:* $\frac{6}{10}, \frac{12}{20}$
4. $\frac{3}{5}$; The others are equal to $\frac{2}{5}$.

Practice and Problem Solving

5. yes
6. yes
7. no
8. no
9. yes
10. no
11. no
12. no
13. no
14. yes
15. yes
16. no
17. no
18. no
19. yes
20. no
21. yes; Both can do 45 sit-ups per minute.
22. you: 1.1 beats per second
    friend: 1.2 beats per second
    No, the rates are not equivalent.

Practice and Problem Solving

23. yes **24.** no

25. yes

26. **a.** \$7 per hour

b. \$9 per hour

c. no; Your friend earns more money per hour.

27. yes; The ratio of height to base for both triangles is $\frac{4}{5}$.

28. **a.** x and y, x and z, y and z

b. 30

29. See *Taking Math Deeper*.

30. **a.** no

b. *Sample answer:* If the collection has 50 quarters and 30 dimes, when 10 of each coin are added, the new ratio of quarters to dimes is 3 : 2.

31. no; The ratios are not equivalent; $\frac{13}{19} \neq \frac{14}{20} \neq \frac{15}{21}$ etc.

32. See Additional Answers.

Fair Game Review

33. −13 **34.** −17

35. −18 **36.** −3

37. D

Mini-Assessment

Tell whether the ratios form a proportion.

1. $\frac{4}{12}, \frac{5}{15}$ yes **2.** $\frac{8}{4}, \frac{12}{8}$ no

3. $\frac{14}{17}, \frac{42}{51}$ yes **4.** $\frac{16}{12}, \frac{26}{22}$ no

5. You can do 40 push-ups in 2 minutes. Your friend can do 57 push-ups in 3 minutes. Are these rates proportional? no

Taking Math Deeper

Exercise 29

In this problem, students are given a mixture of red and yellow pigment and are asked to decide whether they should add more red or more yellow to get a desired shade.

1 Summarize the given information. Find each unit rate.

Ratio for desired shade: $\frac{7 \text{ parts red}}{2 \text{ parts yellow}} = \frac{3\frac{1}{2} \text{ parts red}}{1 \text{ part yellow}}$

Given mixture: $\frac{35 \text{ parts red}}{8 \text{ parts yellow}} = \frac{4\frac{3}{8} \text{ parts red}}{1 \text{ part yellow}}$

2 Interpret the given information.

The ratio for the given mixture needs more yellow to take it down to the desired ratio of $3\frac{1}{2}$: 1 (red to yellow).

3 Use a table with *Guess, Check, and Revise* to find how much yellow to add to the given mixture.

Red	Yellow	Ratio
35 qt	8 qt	$\frac{35}{8} = \frac{4\frac{3}{8} \text{ parts red}}{1 \text{ part yellow}}$
35 qt	9 qt	$\frac{35}{9} = \frac{3\frac{8}{9} \text{ parts red}}{1 \text{ part yellow}}$
35 qt	10 qt	$\frac{35}{10} = \frac{3\frac{1}{2} \text{ parts red}}{1 \text{ part yellow}}$

Add 2 quarts of yellow.

0% magenta 100% yellow | 50% magenta 50% yellow | 100% magenta 0% yellow

Here are some mixtures of yellow and magenta (red).

Reteaching and Enrichment Strategies

If students need help. . .	If students got it. . .
Resources by Chapter • Practice A and Practice B • Puzzle Time Record and Practice Journal Practice Differentiating the Lesson Lesson Tutorials Skills Review Handbook	Resources by Chapter • Enrichment and Extension • Technology Connection Start the next section

Tell whether the ratios form a proportion.

23. $\frac{2.5}{4}, \frac{7}{11.2}$

24. 2 to 4, 11 to $\frac{11}{2}$

25. $2 : \frac{4}{5}, \frac{3}{4} : \frac{3}{10}$

26. **PAY RATE** You earn \$56 walking your neighbor's dog for 8 hours. Your friend earns \$36 painting your neighbor's fence for 4 hours.

a. What is your pay rate?

b. What is your friend's pay rate?

c. Are the pay rates equivalent? Explain.

27. **GEOMETRY** Are the heights and bases of the two triangles proportional? Explain.

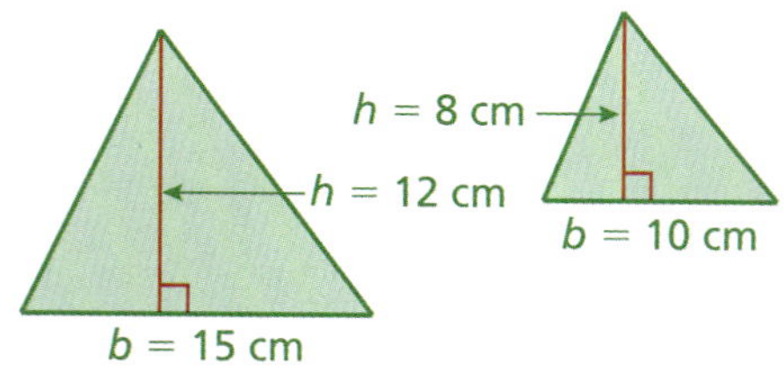

28. **BASEBALL** A pitcher coming back from an injury limits the number of pitches thrown in bull pen sessions as shown.

a. Which quantities are proportional?

b. How many pitches that are not curveballs do you think the pitcher will throw in Session 5?

Session Number, x	Pitches, y	Curveballs, z
1	10	4
2	20	8
3	30	12
4	40	16

29. **NAIL POLISH** A specific shade of red nail polish requires 7 parts red to 2 parts yellow. A mixture contains 35 quarts of red and 8 quarts of yellow. How can you fix the mixture to make the correct shade of red?

30. **COIN COLLECTION** The ratio of quarters to dimes in a coin collection is 5 : 3. You add the same number of new quarters as dimes to the collection.

a. Is the ratio of quarters to dimes still 5 : 3?

b. If so, illustrate your answer with an example. If not, show why with a "counterexample."

31. **AGE** You are 13 years old, and your cousin is 19 years old. As you grow older, is your age proportional to your cousin's age? Explain your reasoning.

32. Critical Thinking Ratio A is equivalent to Ratio B. Ratio B is equivalent to Ratio C. Is Ratio A equivalent to Ratio C? Explain.

Fair Game Review What you learned in previous grades & lessons

Add or subtract. *(Section 11.2 and Section 11.3)*

33. $-28 + 15$

34. $-6 + (-11)$

35. $-10 - 8$

36. $-17 - (-14)$

37. **MULTIPLE CHOICE** Which fraction is not equivalent to $\frac{2}{6}$? *(Skills Review Handbook)*

Ⓐ $\frac{1}{3}$ Ⓑ $\frac{12}{36}$ Ⓒ $\frac{4}{12}$ Ⓓ $\frac{6}{9}$

Extension 14.2

Graphing Proportional Relationships

Check It Out
Lesson Tutorials
BigIdeasMath.com

Recall that you can graph the values from a ratio table.

Time, *x* (seconds)	Height, *y* (meters)
3	2
6	4
9	6
12	8

+3, +3, +3 (Time); +2, +2, +2 (Height)

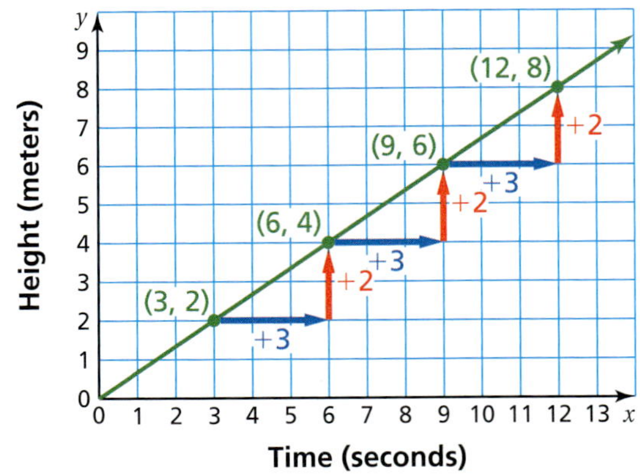

The structure in the ratio table shows why the graph has a constant *rate of change*. You can use the constant rate of change to show that the graph passes through the origin. The graph of every proportional relationship is a line through the origin.

EXAMPLE 1 Determining Whether Two Quantities Are Proportional

Use a graph to tell whether *x* and *y* are in a proportional relationship.

a.

x	2	4	6
y	6	8	10

Plot (2, 6), (4, 8), and (6, 10). Draw a line through the points.

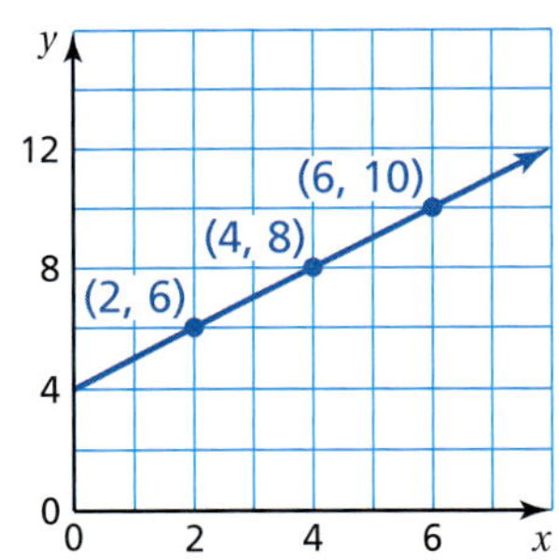

The graph is a line that does not pass through the origin.

∴ So, *x* and *y* are not in a proportional relationship.

b.

x	1	2	3
y	2	4	6

Plot (1, 2), (2, 4), and (3, 6). Draw a line through the points.

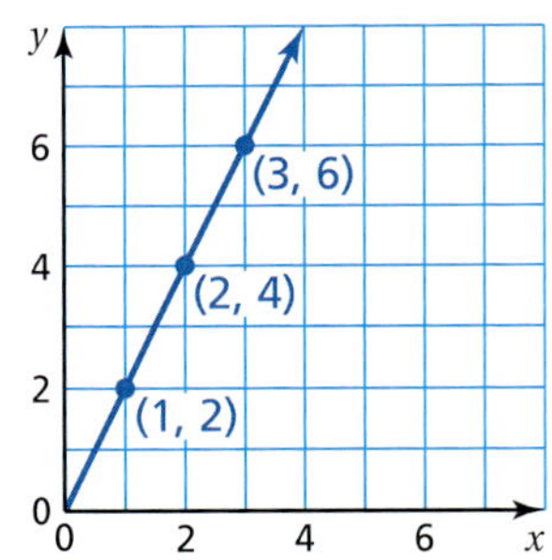

The graph is a line that passes through the origin.

∴ So, *x* and *y* are in a proportional relationship.

Proportions
In this extension, you will
- use graphs to determine whether two ratios form a proportion.
- interpret graphs of proportional relationships.

Practice

Use a graph to tell whether *x* and *y* are in a proportional relationship.

1.

x	1	2	3	4
y	3	4	5	6

2.

x	1	3	5	7
y	0.5	1.5	2.5	3.5

Laurie's Notes

Introduction

Connect

- **Yesterday:** Students determined whether two ratios are proportional.
- **Today:** Students will graph proportional relationships.

Motivate

- Share the following information with students. The Mars rover *Curiosity*, a large mobile laboratory, launched from Cape Canaveral on November 26, 2011, and landed successfully on Mars on August 6, 2012. For three minutes right before touchdown, the spacecraft used a parachute and retrorockets to slow its descent. In the final seconds before touchdown, the upper stage acted as a crane, lowering the rover to the surface on a tether.
- Explain that one of the examples in today's lesson will explore the distance *Curiosity* travels over time.

Lesson Notes

Discuss

- Introduce the lesson with the ratio table and graph shown. This is an example in which a constant amount is added to the x-values and a different constant amount is added to the y-values. The ratios are equivalent: $\frac{3}{2} = \frac{6}{4} = \frac{9}{6} = \frac{12}{8}$.
- State that the graph of every proportional relationship is a line through the origin.
- You could justify that the line goes through the origin using the arrows shown in the graph. You could also work backwards in the table to show that (0, 0) would be in the table. Students will learn more about rate of change and slope in Sections 14.5 and 14.6.

Example 1

- Write the table of values and plot the corresponding ordered pairs for each part of the example.
- **Attend to Precision:** Have students use grid paper instead of making freehand sketches.
- **Connection:** Check the tables for equivalent ratios to justify the answers. In part (a), $\frac{2}{6} \neq \frac{4}{8}$ and $\frac{4}{8} \neq \frac{6}{10}$. In part (b), $\frac{1}{2} = \frac{2}{4} = \frac{3}{6}$.

What Your Students Will Learn

- Use graphs to determine whether two quantities are proportional.
- Interpret graphs of proportional relationships.

Goal Today's lesson is graphing proportional relationships.

Technology for the Teacher

Dynamic Classroom

Lesson Tutorials
Lesson Plans
Answer Presentation Tool

Extra Example 1

Use a graph to tell whether x and y are in a proportional relationship.

a.

x	2	4	6
y	4	6	8

no

b.

x	3	6	9
y	1	2	3

yes

Record and Practice Journal Extension 14.2 Practice

1–8. See Additional Answers.

Extra Example 2

The graph shows that the distance traveled by a kayak at top speed is proportional to the time traveled. Interpret each plotted point in the graph.

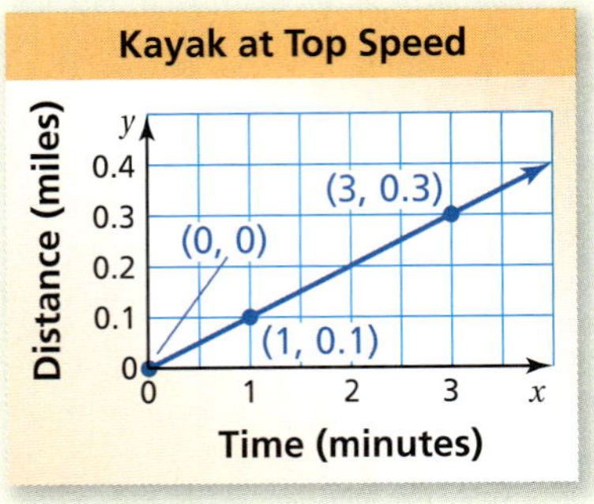

Sample Answer: The kayak travels 0 miles in 0 minutes, 0.1 mile in 1 minute, and 0.3 mile in 3 minutes. The unit rate is 0.1 mile per minute.

Practice

1.

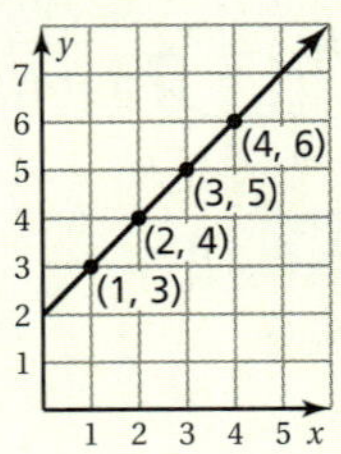

no

2.

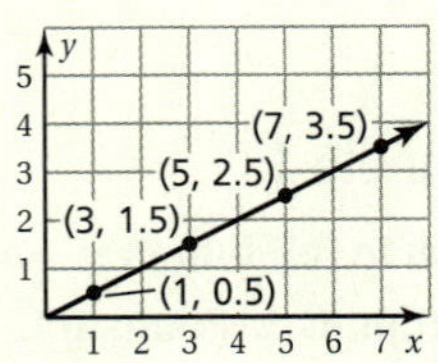

yes

3–8. See Additional Answers.

Mini-Assessment

Use a graph to tell whether *x* and *y* are in a proportional relationship.

1.

x	2	4	6
y	2	3	4

no

2.

x	2	4	6
y	3	6	9

yes

Laurie's Notes

Example 2

- Explain that this example involves the top speed of the Mars rover *Curiosity* discussed earlier.
- ? "How do you know that the graph shows a proportional relationship between distance traveled and time traveled?" The line passes through the origin.
- Ask students to read and interpret each plotted point in the graph. Note that in addition to interpreting the points, the unit rate is also found. It is a constant value of 1.5 inches per second.
- Make sure that students understand the Study Tip. If you know (1, *y*), then you know the unit rate *y*.

Practice

- In Exercises 3 and 4, students should be thinking about the labels on the axes as they interpret the plotted points.
- In Exercise 6, *x* and *y* are not in a proportional relationship, but you could still ask students whether there is a constant rate of change, and if so, what it is. This gives a preview of Section 14.5.
- **Connection:** In Exercise 7, students can draw a line through (12, 16) and (0, 0), and then think about where the ordered pair (1, *y*) would be on the line.

Closure

? Is it possible to have a constant rate of change and not have a proportional relationship? Explain.

yes; *Sample answer:* Example 1(a) has a constant rate of change, because the points lie on a line. The relationship is not proportional, because the line does not pass through the origin.

EXAMPLE 2 Interpreting the Graph of a Proportional Relationship

The graph shows that the distance traveled by the Mars rover *Curiosity* is proportional to the time traveled. Interpret each plotted point in the graph.

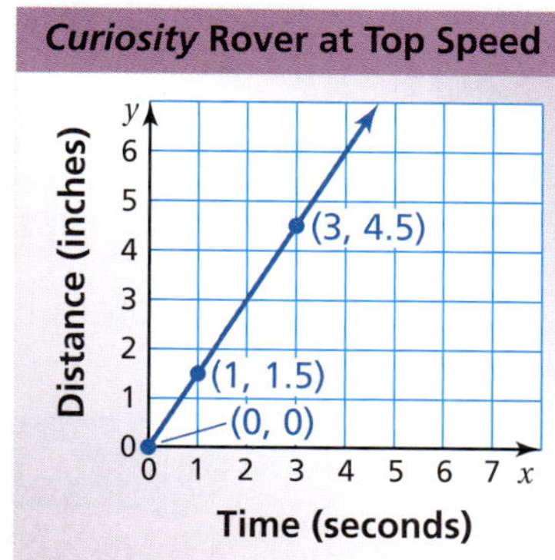

(0, 0): The rover travels 0 inches in 0 seconds.

(1, 1.5): The rover travels 1.5 inches in 1 second. So, the unit rate is 1.5 inches per second.

(3, 4.5): The rover travels 4.5 inches in 3 seconds. Because the relationship is proportional, you can also use this point to find the unit rate.

$$\frac{4.5 \text{ in.}}{3 \text{ sec}} = \frac{1.5 \text{ in.}}{1 \text{ sec}}, \text{ or } 1.5 \text{ inches per second}$$

Study Tip

In the graph of a proportional relationship, you can find the unit rate from the point $(1, y)$.

Practice

Interpret each plotted point in the graph of the proportional relationship.

3.

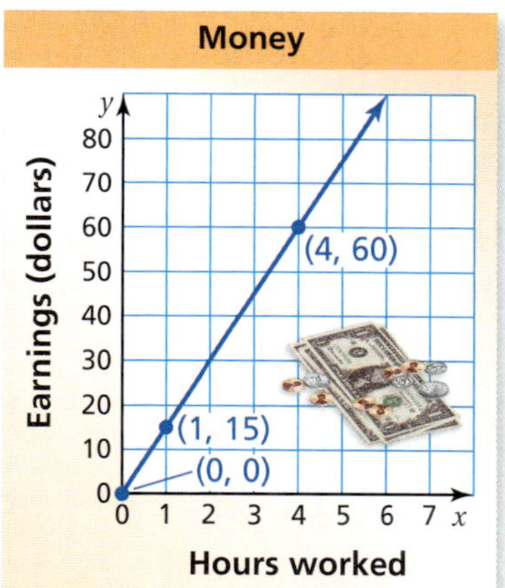

4.

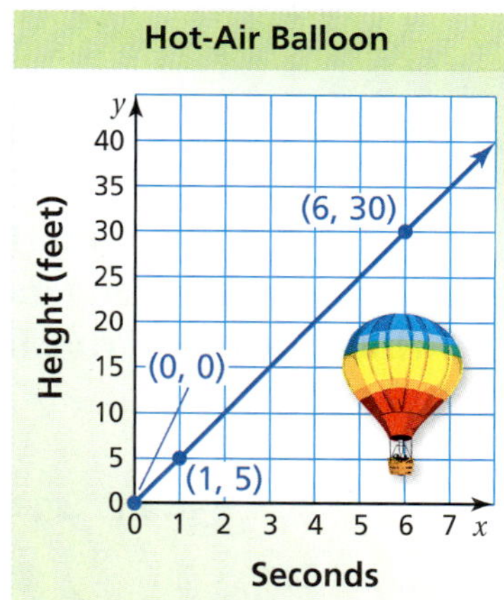

Tell whether x and y are in a proportional relationship. If so, find the unit rate.

5.

x (hours)	1	4	7	10
y (feet)	5	20	35	50

6. Let y be the temperature x hours after midnight. The temperature is 60°F at midnight and decreases 2°F every $\frac{1}{2}$ hour.

7. **REASONING** The graph of a proportional relationship passes through (12, 16) and $(1, y)$. Find y.

8. **MOVIE RENTAL** You pay \$1 to rent a movie plus an additional \$0.50 per day until you return the movie. Your friend pays \$1.25 per day to rent a movie.

   a. Make tables showing the costs to rent a movie up to 5 days.

   b. Which person pays an amount proportional to the number of days rented?

14.3 Writing Proportions

Essential Question How can you write a proportion that solves a problem in real life?

1 ACTIVITY: Writing Proportions

Work with a partner. A rough rule for finding the correct bat length is "the bat length should be half of the batter's height." So, a 62-inch-tall batter uses a bat that is 31 inches long. Write a proportion to find the bat length for each given batter height.

a. 58 inches

b. 60 inches

c. 64 inches

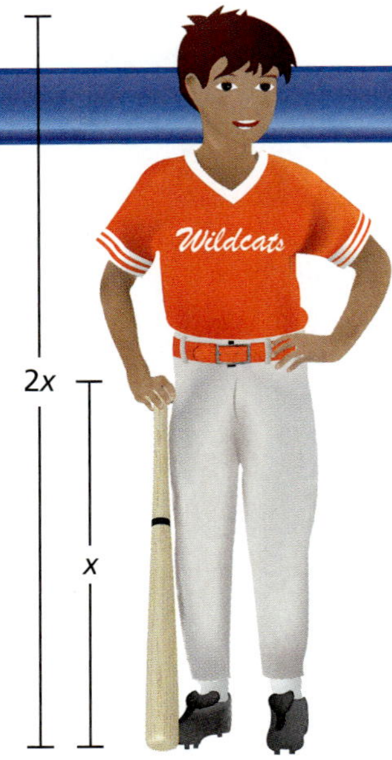

2 ACTIVITY: Bat Lengths

Work with a partner. Here is a more accurate table for determining the bat length for a batter. Find all the batter heights and corresponding weights for which the rough rule in Activity 1 is exact.

Weight of Batter (pounds)	Height of Batter (inches)							
	45–48	**49–52**	**53–56**	**57–60**	**61–64**	**65–68**	**69–72**	**Over 72**
Under 61	28	29	29					
61–70	28	29	30	30				
71–80	28	29	30	30	31			
81–90	29	29	30	30	31	32		
91–100	29	30	30	31	31	32		
101–110	29	30	30	31	31	32		
111–120	29	30	30	31	31	32		
121–130	29	30	30	31	32	33	33	
131–140	30	30	31	31	32	33	33	
141–150	30	30	31	31	32	33	33	
151–160	30	31	31	32	32	33	33	33
161–170		31	31	32	32	33	33	34
171–180				32	33	33	34	34
Over 180					33	33	34	34

Proportions

In this lesson, you will

- write proportions.
- solve proportions using mental math.

Laurie's Notes

Introduction

Applying Mathematical Practices

- **Reason Abstractly and Quantitatively:** In the first two activities, students work with three quantities: a batter's height and weight and the length of the batter's bat. Activity 1 states a rough rule for determining bat length based on the batter's height. Mathematically proficient students are able to apply the rule and also recognize that other variables are involved in bat selection.

Motivate

- **Management Tip:** You may want to pre-cut several lengths of string prior to this activity so students can join in.
- Ask for a volunteer. Say, "I can estimate the distance around your neck without actually measuring your neck!"
- Use the string to measure the distance around the student's wrist.
- Double this length, and it will be approximately the distance around his/her neck. Have the student verify this length by measuring the distance around his/her own neck.
- **Write:** $\frac{\text{distance around wrist}}{\text{distance around neck}} = \frac{1}{2}$
- Write a new proportion substituting the length around the wrist: $\frac{8.5 \text{ inches}}{x \text{ inches}} = \frac{1}{2}$
- **Solve:** The distance around the neck is two times 8.5 inches, or 17 inches.

Activity Notes

Activity 1

- Borrow a baseball bat for this activity to use as a prop.
- Help students translate the words in the activity into a proportion: $\frac{\text{length of bat}}{\text{height of batter}} = \frac{1}{2}$. Say, "The ratio of the length of the bat to the height of the batter is 1 : 2. A proportion to determine the bat length for a player 58 inches tall is $\frac{1}{2} = \frac{b}{58}$."
- You want students to understand that two measures are being compared, and the order in which you write (and say) them does matter.

Activity 2

- This activity promotes reading information from a table.
- The batter's height *and* weight factor into selecting the correct bat length.
- Practice reading information from the table.
- ? "A 50-inch tall batter weighing 95 pounds should use what length bat?" 30"
- ? "A 5-foot, 5-inch tall batter weighing 95 pounds should use what length bat?" 32"

What Your Students Will Learn

- Write proportions using tables.
- Solve proportions using mental math.

Previous Learning

Students have written and simplified ratios and determined if two ratios are equal.

Technology for the Teacher

Dynamic Classroom

Lesson Plans
Complete Materials List

14.3 Record and Practice Journal

Essential Question How can you write a proportion that solves a problem in real life?

1 ACTIVITY: Writing Proportions

Work with a partner. A rough rule for finding the correct bat length is "the bat length should be half of the batter's height." So, a 62-inch tall batter uses a bat that is 31 inches long. Write a proportion to find the bat length for each given batter height.

a. 58 inches $\frac{1}{2} = \frac{x}{58}$

b. 60 inches $\frac{1}{2} = \frac{x}{60}$

c. 64 inches $\frac{1}{2} = \frac{x}{64}$

2 ACTIVITY: Bat Lengths

Work with a partner. Here is a more accurate table for determining the bat length for a batter. Find all the batter heights and corresponding weights for which the rough rule in Activity 1 is exact.

Weight of Batter (pounds)	Height of Batter (inches)							
	45–48	49–52	53–56	57–60	61–64	65–68	69–72	Over 72
Under 61	28	29	29					
61–70	28	29	30	30				
71–80	28	29	30	30	31			
81–90	29	29	30	30	31	32		
91–100	29	30	30	31	31	32		
101–110	29	30	30	31	31	32		
111–120	29	30	30	31	31	32		
121–130	29	30	30	31	32	33	33	
131–140	30	30	31	31	32	33	33	
141–150	30	30	31	31	32	33	33	
151–160	30	31	31	32	32	33	33	33
161–170		31	31	32	32	33	33	34
171–180				32	33	33	34	34
Over 180					33	33	34	34

60 in. (61–90 lb), 62 in. (71–120 lb), 64 in. (121–170 lb), 66 in. (121+ lb)

English Language Learners

Visual Aids

English learners might find it useful to use a general template when writing a proportion problem.

$$\frac{\text{part}}{\text{whole}} = \frac{\text{part}}{\text{whole}}$$

14.3 Record and Practice Journal

3 ACTIVITY: Writing Proportions

Work with a partner. The batting average of a baseball player is the number of "hits" divided by the number of "at bats."

$$\text{batting average} = \frac{\text{hits } (H)}{\text{at Bats } (A)}$$

A player whose batting average is 0.250 is said to be "batting 250."

$$\frac{20 \text{ hits}}{80 \text{ at bats}} = 0.250 = \frac{250 \text{ hits}}{1000 \text{ at bats}}$$

Actual hits → 20 hits; Actual at bats → 80 at bats; Batting average → 0.250; Batting 250 out of 1000 → $\frac{250 \text{ hits}}{1000 \text{ at bats}}$

Write a proportion to find how many hits H a player needs to achieve the given batting average. Then solve the proportion.

a. 50 times at bat; batting average is 0.200.

$\frac{200}{1000} = \frac{H}{50}$; 10 hits

b. 84 times at bat; batting average is 0.250.

$\frac{250}{1000} = \frac{H}{84}$; 21 hits

c. 80 times at bat; batting average is 0.350.

$\frac{350}{1000} = \frac{H}{80}$; 28 hits

d. 1 time at bat; batting average is 1.000.

$\frac{1000}{1000} = \frac{H}{1}$; 1 hit

What Is Your Answer?

4. **IN YOUR OWN WORDS** How can you write a proportion that solves a problem in real life?

Check students' work.

5. Two players have the same batting average.

	At Bats	Hits	Batting Average
Player 1	132	45	
Player 2	132	45	

Player 1 gets four hits in the next five at bats. Player 2 gets three hits in the next three at bats.

a. Who has the higher batting average?

Player 1

b. Does this seem fair? Explain your reasoning.

no; Check students' work.

Laurie's Notes

Activity 3

- ? "Does anyone know what the term *batting average* means and how it is computed?" It is actually more involved than explained in the text. For instance, if a batter walks, has a sacrifice fly, or is hit by a pitch, it is not considered an "at bat."
- After the discussion of batting average, write the formula followed by the example. To have a proportion, the decimal form of the fraction is written.
- Remind students of how ratios are simplified (by dividing out common factors).
- Example: "Determine how many hits a batter has if he has 100 at bats and his batting average is 0.300.

  $\frac{H}{100} = \frac{300}{1000}$ → dividing out a common factor of 10 → $\frac{H}{100} = \frac{30}{100}$
- **Common Error:** Students have divided out common factors when simplifying a simple fraction or when multiplying two fractions. A proportion is neither of these! A common error at this point is for students to divide out a factor from the numerator on one side of the equal sign with a factor in the denominator on the other side of the equal sign. For instance, in this example a student would incorrectly write $\frac{H}{1} = \frac{3}{1000}$ and get an answer of 0.003 hit.
- **Look for and Express Regularity in Repeated Reasoning** and **Reason Abstractly and Quantitatively:** Students should be able to use the information given to understand what *batting average* means. Students should also be able to check and compare their answers to each part of the activity for reasonableness. For instance, the batter in part (c) has a better average than, and has been at bat fewer times than, the batter in part (b). So, it should make sense that the batter in part (c) has had more hits than the batter in part (b).
- Have pairs of students show their work at the board.

What Is Your Answer?

- To answer Question 5, students may need to be reminded of how to write a fraction as a decimal. A calculator will be helpful.

Closure

- Write and solve a proportion: A stadium holds approximately 45,000 people during a baseball game. If the ratio of season ticket holders to all tickets is 1 : 3, approximately how many season ticket holders are there? 15,000 season ticket holders

3 ACTIVITY: Writing Proportions

Math Practice

Evaluate Results

How do you know if your results are reasonable? Explain.

Work with a partner. The batting average of a baseball player is the number of "hits" divided by the number of "at bats."

$$\text{batting average} = \frac{\text{hits } (H)}{\text{at bats } (A)}$$

A player whose batting average is 0.250 is said to be "batting 250."

$$\frac{20 \text{ hits}}{80 \text{ at bats}} = 0.250 = \frac{250 \text{ hits}}{1000 \text{ at bats}}$$

Actual hits → 20 hits; Actual at bats → 80 at bats; Batting average → 0.250; Batting 250 out of 1000 → $\frac{250 \text{ hits}}{1000 \text{ at bats}}$

Write a proportion to find how many hits *H* a player needs to achieve the given batting average. Then solve the proportion.

a. 50 times at bat; batting average is 0.200.

b. 84 times at bat; batting average is 0.250.

c. 80 times at bat; batting average is 0.350.

d. 1 time at bat; batting average is 1.000.

What Is Your Answer?

4. IN YOUR OWN WORDS How can you write a proportion that solves a problem in real life?

5. Two players have the same batting average.

	At Bats	Hits	Batting Average
Player 1	132	45	
Player 2	132	45	

Player 1 gets four hits in the next five at bats. Player 2 gets three hits in the next three at bats.

a. Who has the higher batting average?

b. Does this seem fair? Explain your reasoning.

Practice Use what you discovered about proportions to complete Exercises 4–7 on page 618.

14.3 Lesson

Check It Out
Lesson Tutorials
BigIdeasMath.com

One way to write a proportion is to use a table.

	Last Month	This Month
Purchase	2 ringtones	3 ringtones
Total Cost	6 dollars	x dollars

Use the columns or the rows to write a proportion.

Use columns:

$$\frac{2 \text{ ringtones}}{6 \text{ dollars}} = \frac{3 \text{ ringtones}}{x \text{ dollars}}$$

Numerators have the same units.

Denominators have the same units.

Use rows:

$$\frac{2 \text{ ringtones}}{3 \text{ ringtones}} = \frac{6 \text{ dollars}}{x \text{ dollars}}$$

The units are the same on each side of the proportion.

EXAMPLE 1 Writing a Proportion

Black Bean Soup

1.5 cups black beans
0.5 cup salsa
2 cups water
1 tomato
2 teaspoons seasoning

A chef increases the amounts of ingredients in a recipe to make a proportional recipe. The new recipe has 6 cups of black beans. Write a proportion that gives the number x of tomatoes in the new recipe.

Organize the information in a table.

	Original Recipe	New Recipe
Black Beans	1.5 cups	6 cups
Tomatoes	1 tomato	x tomatoes

∴ One proportion is $\frac{1.5 \text{ cups beans}}{1 \text{ tomato}} = \frac{6 \text{ cups beans}}{x \text{ tomatoes}}$.

On Your Own

Now You're Ready
Exercises 8–11

1. Write a different proportion that gives the number x of tomatoes in the new recipe.
2. Write a proportion that gives the amount y of water in the new recipe.

Laurie's Notes

Introduction

Connect

- **Yesterday:** Students wrote and solved proportions related to baseball that used simple mental math.
- **Today:** Students will write and solve a proportion using mental math.

Motivate

? "A student is reading a 280-page book. On average he/she reads 8 pages in 3 minutes. How long will it take to read the book?"

- If students can offer strategies, work on this problem now. If not, wait until later in the class.
- There are two equivalent proportions that can be set up. Be sure to use labels.

  $\frac{8 \text{ pages}}{3 \text{ minutes}} = \frac{280 \text{ pages}}{x \text{ minutes}}$ or $\frac{8 \text{ pages}}{280 \text{ pages}} = \frac{3 \text{ minutes}}{x \text{ minutes}}$

- The Cross Products Property is not needed. Because $280 \div 8 = 35$, mental math strategies are sufficient.

Lesson Notes

Discuss

? "What is the information in the table saying?" You purchased 2 ringtones last month for \$6. This month, you purchase 3 ringtones, and the cost is unknown.

- Work through the example showing how the labels (ringtones and dollars) are used to identify the numbers.
- **Big Idea:** As identified in the text, when you use the rows or columns from the table, the proportion will be set up correctly. If the table had *not* been provided, it is possible for students to incorrectly set up the proportion:

  $\frac{2 \text{ ringtones}}{x \text{ dollars}} = \frac{3 \text{ ringtones}}{6 \text{ dollars}}$ or $\frac{2 \text{ ringtones}}{3 \text{ ringtones}} = \frac{x \text{ dollars}}{6 \text{ dollars}}$

- What students must also see besides the labels is that the information from one month (i.e., 2 ringtones, \$6) must be in one ratio *or* in the numerators, and the information from the second month (i.e., 3 ringtones, \$$x$) must be in a second ratio *or* in the denominators.
- **Model with Mathematics:** The column and the rows from a correct table ensure this happening.

Example 1

- Organizing the information in a table helps to write the proportion correctly.

? "Can the rows and columns be interchanged?" yes

On Your Own

- **Think-Pair-Share:** Students should read each question independently and then work in pairs to answer the questions. When they have answered the questions, the pair should compare their answers with another group and discuss any discrepancies.

Goal Today's lesson is writing and solving a proportion using mental math.

Technology for the Teacher

Dynamic Classroom

Lesson Tutorials
Lesson Plans
Answer Presentation Tool

Extra Example 1

The chef increases the amounts of ingredients in the recipe in Example 1 to make a proportional recipe. The new recipe has 3 cups of salsa. Write a proportion that gives the amount w of water in the new recipe.

Sample answer:

$\frac{0.5 \text{ cup salsa}}{3 \text{ cups salsa}} = \frac{2 \text{ cups water}}{w \text{ cups water}}$

On Your Own

1. $\frac{1.5 \text{ cups beans}}{6 \text{ cups beans}} = \frac{1 \text{ tomato}}{x \text{ tomatoes}}$
2. *Sample answer:* $\frac{1.5 \text{ cups beans}}{2 \text{ cups water}} = \frac{6 \text{ cups beans}}{y \text{ cups water}}$

Extra Example 2

Solve $\frac{8}{5} = \frac{n}{15}$. 24

Extra Example 3

In Extra Example 1, how much water is in the new recipe? 12 cups water

On Your Own

3. $d = 32$
4. $z = 5$
5. $x = 7$
6. $\frac{48}{95} = \frac{f}{950}$; 480 female students

Differentiated Instruction

Kinesthetic

Provide students with counters and two pieces of blank paper. Draw fraction bar lines on each of the papers. Place counters on the papers to represent the two ratios. For each ratio, rearrange the counters in the numerator and denominator in stacks of equal number. If the stacks in each individual ratio are the same size and there are the same number of stacks in each ratio, then the ratios are equal and form a proportion. For instance, in Example 2, the first ratio has 3 stacks of 1 in the numerator and 2 stacks of 1 in the denominator. The second ratio has 3 stacks of 4 in the numerator and 2 stacks of 4 in the denominator. Because there are 3 stacks over 2 stacks in each of the ratios, the ratios form a proportion.

Laurie's Notes

Example 2

- **Reason Abstractly and Quantitatively:** This example has no context. The focus is on the process and how mental math is used in solving the proportion.
- When finished, present the following problems to assess if students can distinguish when mental math is a reasonable approach.

? Tell whether you can easily use mental math to solve these problems:

a. $\frac{3}{7} = \frac{x}{27}$ **b.** $\frac{3}{7} = \frac{27}{x}$

Answers:

a. 7 is not a factor of 27, so mental math is not an easy approach.

b. 3 is a factor of 27, so mental math can be used.

Example 3

- Work through this example.
- Not all students will know that $4 \times 1.5 = 6$. If students are still struggling with multiplying decimals, you may want to review these rules prior to Example 3.
- Two ideas to help develop fluency:
  $4 \times 15 = 60$, so $4 \times 1.5 = 6$.
  $1 \times 4 = 4$, and half (0.5) of 4 is 2, so add $2 + 4$ to get $4 \times 1.5 = 6$.

On Your Own

- **Neighbor Check:** Have students work independently and then have their neighbors check their work. Have students discuss any discrepancies.

Closure

- **Exit Ticket:** The ratio of quarts to gallons is 4 : 1. If a recipe calls for 14 quarts, how many gallons would be needed? 3.5

EXAMPLE 2 Solving Proportions Using Mental Math

Solve $\frac{3}{2} = \frac{x}{8}$.

Step 1: Think: The product of 2 and what number is 8?

$$\frac{3}{2} = \frac{x}{8}$$

$2 \times ? = 8$

Step 2: Because the product of 2 and 4 is 8, multiply the numerator by 4 to find x.

$3 \times 4 = 12$

$$\frac{3}{2} = \frac{x}{8}$$

$2 \times 4 = 8$

The solution is $x = 12$.

EXAMPLE 3 Solving Proportions Using Mental Math

In Example 1, how many tomatoes are in the new recipe?

Solve the proportion $\frac{1.5}{1} = \frac{6}{x}$.

cups black beans

tomatoes

Step 1: Think: The product of 1.5 and what number is 6?

$1.5 \times ? = 6$

$$\frac{1.5}{1} = \frac{6}{x}$$

Step 2: Because the product of 1.5 and 4 is 6, multiply the denominator by 4 to find x.

$1.5 \times 4 = 6$

$$\frac{1.5}{1} = \frac{6}{x}$$

$1 \times 4 = 4$

So, there are 4 tomatoes in the new recipe.

On Your Own

Now You're Ready
Exercises 16–21

Solve the proportion.

3. $\frac{5}{8} = \frac{20}{d}$ 4. $\frac{7}{z} = \frac{14}{10}$ 5. $\frac{21}{24} = \frac{x}{8}$

6. A school has 950 students. The ratio of female students to all students is $\frac{48}{95}$. Write and solve a proportion to find the number f of students who are female.

14.3 Exercises

Check It Out
Help with Homework
BigIdeasMath.com

Vocabulary and Concept Check

1. **WRITING** Describe two ways you can use a table to write a proportion.

2. **WRITING** What is your first step when solving $\frac{x}{15} = \frac{3}{5}$? Explain.

3. **OPEN-ENDED** Write a proportion using an unknown value x and the ratio 5 : 6. Then solve it.

Practice and Problem Solving

Write a proportion to find how many points a student needs to score on the test to get the given score.

4. test worth 50 points; test score of 40%

5. test worth 50 points; test score of 78%

6. test worth 80 points; test score of 80%

7. test worth 150 points; test score of 96%

Use the table to write a proportion.

1 8.

	Game 1	Game 2
Points	12	18
Shots	14	w

9.

	May	June
Winners	n	34
Entries	85	170

10.

	Today	Yesterday
Miles	15	m
Hours	2.5	4

11.

	Race 1	Race 2
Meters	100	200
Seconds	x	22.4

12. **ERROR ANALYSIS** Describe and correct the error in writing the proportion.

	Monday	Tuesday
Dollars	2.08	d
Ounces	8	16

$$\frac{2.08}{16} = \frac{d}{8}$$

13. **T-SHIRTS** You can buy 3 T-shirts for \$24. Write a proportion that gives the cost c of buying 7 T-shirts.

14. **COMPUTERS** A school requires 2 computers for every 5 students. Write a proportion that gives the number c of computers needed for 145 students.

15. **SWIM TEAM** The school team has 80 swimmers. The ratio of seventh-grade swimmers to all swimmers is 5 : 16. Write a proportion that gives the number s of seventh-grade swimmers.

Assignment Guide and Homework Check

Level	Assignment	Homework Check
Advanced	1–7, 8–24 even, 25–30	8, 14, 18, 24

Common Errors

- **Exercises 8–11** Students may write half of the proportion using rows and the other half using columns. They will have forgotten to include one of the values. Remind students that they need to pick a method for writing proportions with tables and be consistent throughout the problem.
- **Exercises 16–21** Students may get confused with using mental math to find the value of the variable and try to multiply the numerator and denominator by different numbers. Tell students that they are finding an equivalent ratio, or an equivalent fraction. They are multiplying the original fraction by 1 $\left(\text{or } \frac{4}{4}, \text{ for example}\right)$ to find the equivalent fraction, so they must multiply the numerator and denominator by the same number.

14.3 Record and Practice Journal

Write a proportion to find how many points a student needs to score on the test to get the given score.

1. test worth 50 points; test score of 84%
   $\frac{x}{50} = \frac{84}{100}$
2. test worth 75 points; test score of 96%
   $\frac{x}{75} = \frac{96}{100}$

Use the table to write a proportion.

3.

	Trip 1	Trip 2
Miles	104	78
Gallons	4	g

$\frac{104}{4} = \frac{78}{g}$

4.

	Tree 1	Tree 2
Inches	15	x
Years	4	3

$\frac{15}{4} = \frac{x}{3}$

Solve the proportion.

5. $\frac{1}{3} = \frac{x}{12}$ $x = 4$
6. $\frac{5}{9} = \frac{25}{y}$ $y = 45$
7. $\frac{26}{z} = \frac{13}{22}$ $z = 44$
8. $\frac{b}{30} = \frac{2.6}{1.5}$ $b = 52$

9. A local Humane Society houses 300 animals. The ratio of cats to all animals is 7 : 15.
   a. Write a proportion that gives the number of cats c.
   $\frac{c}{300} = \frac{7}{15}$
   b. How many cats are in the Humane Society?
   140

Vocabulary and Concept Check

1. You can use the columns or the rows of the table to write a proportion.
2. Find the number that when multiplied by 5 is 15.
3. *Sample answer:* $\frac{x}{12} = \frac{5}{6}$; $x = 10$

Practice and Problem Solving

4. $\frac{x}{50} = \frac{40}{100}$
5. $\frac{x}{50} = \frac{78}{100}$
6. $\frac{x}{80} = \frac{80}{100}$
7. $\frac{x}{150} = \frac{96}{100}$
8. $\frac{12 \text{ points}}{14 \text{ shots}} = \frac{18 \text{ points}}{w \text{ shots}}$
9. $\frac{n \text{ winners}}{85 \text{ entries}} = \frac{34 \text{ winners}}{170 \text{ entries}}$
10. $\frac{15 \text{ miles}}{2.5 \text{ hours}} = \frac{m \text{ miles}}{4 \text{ hours}}$
11. $\frac{100 \text{ meters}}{x \text{ seconds}} = \frac{200 \text{ meters}}{22.4 \text{ seconds}}$
12. The proportion cannot be written using diagonals of the table. $\frac{2.08}{8} = \frac{d}{16}$
13. $\frac{\$24}{3 \text{ shirts}} = \frac{c}{7 \text{ shirts}}$
14. $\frac{2 \text{ computers}}{5 \text{ students}} = \frac{c \text{ computers}}{145 \text{ students}}$
15. $\frac{5 \text{ 7th grade swimmers}}{16 \text{ swimmers}} = \frac{s \text{ 7th grade swimmers}}{80 \text{ swimmers}}$

Practice and Problem Solving

16. $z = 5$ **17.** $y = 16$

18. $k = 15$ **19.** $c = 24$

20. $b = 20$ **21.** $g = 14$

22. a. $\frac{1 \text{ trombone}}{3 \text{ violas}} = \frac{t \text{ trombones}}{9 \text{ violas}}$

b. 3 trombones

23. $\frac{1}{200} = \frac{19.5}{x}$; Dimensions for the model are in the numerators and the corresponding dimensions for the actual space shuttle are in the denominators.

24. no; The solution of that equation is $x = 1.5$, but using mental math, you can see that the solution of the proportion is $x = 24$.

25. See *Taking Math Deeper*.

Fair Game Review

26. $x = 150$ **27.** $x = 9$

28. $x = 75$ **29.** $x = 140$

30. C

Mini-Assessment

Write a proportion to find how many points a student needs to score on the test to get the given score.

1. Test worth 60 points; test score of 60% $\frac{x}{60} = \frac{60}{100}$
2. Test worth 50 points; test score of 70% $\frac{x}{50} = \frac{70}{100}$
3. Test worth 100 points; test score of 85% $\frac{x}{100} = \frac{85}{100}$
4. Test worth 120 points; test score of 88% $\frac{x}{120} = \frac{88}{100}$
5. You can buy four DVDs for \$48. Write a proportion that gives the cost c of buying six DVDs. $\frac{4}{48} = \frac{6}{c}$

Taking Math Deeper

Exercise 25

Although this problem does not have difficult or messy mathematics, it is still difficult for many students to know how to start. Emphasize that it is good to just "write things down" and organize the given facts. In this problem, it might help the visual learner to sketch 3 white lockers for every 5 blue lockers.

1 Draw a diagram that shows the given values and the unknown values.

2 Write and solve a proportion.

$\frac{x \text{ blue lockers}}{180 \text{ white lockers}} = \frac{5}{3}$ Write a proportion.

$x = 300$ Use mental math.

3 Answer the question.

There are 180 white lockers and 300 blue lockers. So, there are a total of $180 + 300 = 480$ lockers in the school.

Reteaching and Enrichment Strategies

If students need help. . .	If students got it. . .
Resources by Chapter • Practice A and Practice B • Puzzle Time Record and Practice Journal Practice Differentiating the Lesson Lesson Tutorials Skills Review Handbook	Resources by Chapter • Enrichment and Extension • Technology Connection Start the next section

Solve the proportion.

16. $\frac{1}{4} = \frac{z}{20}$ **17.** $\frac{3}{4} = \frac{12}{y}$ **18.** $\frac{35}{k} = \frac{7}{3}$

19. $\frac{15}{8} = \frac{45}{c}$ **20.** $\frac{b}{36} = \frac{5}{9}$ **21.** $\frac{1.4}{2.5} = \frac{g}{25}$

22. **ORCHESTRA** In an orchestra, the ratio of trombones to violas is 1 to 3.

a. There are 9 violas. Write a proportion that gives the number t of trombones in the orchestra.

b. How many trombones are in the orchestra?

23. **ATLANTIS** Your science teacher has a 1 : 200 scale model of the space shuttle *Atlantis*. Which of the proportions can you use to find the actual length x of *Atlantis*? Explain.

$\frac{1}{200} = \frac{19.5}{x}$ $\frac{1}{200} = \frac{x}{19.5}$ $\frac{200}{19.5} = \frac{x}{1}$ $\frac{x}{200} = \frac{1}{19.5}$

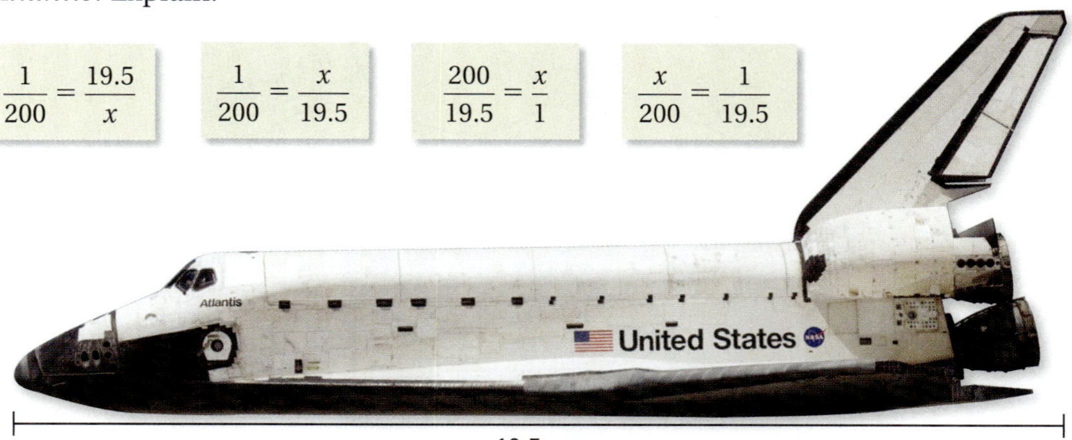

19.5 cm

24. **YOU BE THE TEACHER** Your friend says "$48x = 6 \cdot 12$." Is your friend right? Explain.

Solve $\frac{6}{x} = \frac{12}{48}$.

25. **Reasoning** There are 180 white lockers in the school. There are 3 white lockers for every 5 blue lockers. How many lockers are in the school?

Fair Game Review What you learned in previous grades & lessons

Solve the equation. *(Section 13.4)*

26. $\frac{x}{6} = 25$ **27.** $8x = 72$ **28.** $150 = 2x$ **29.** $35 = \frac{x}{4}$

30. **MULTIPLE CHOICE** What is the value of $-\frac{9}{4} + \left|-\frac{8}{5}\right| - 2\frac{1}{2}$? *(Section 12.3)*

Ⓐ $-6\frac{7}{20}$ Ⓑ $-5\frac{7}{20}$ Ⓒ $-3\frac{3}{20}$ Ⓓ $-2\frac{3}{20}$

14 Study Help

Check It Out
Graphic Organizer
BigIdeasMath.com

You can use an **information wheel** to organize information about a concept. Here is an example of an information wheel for ratio.

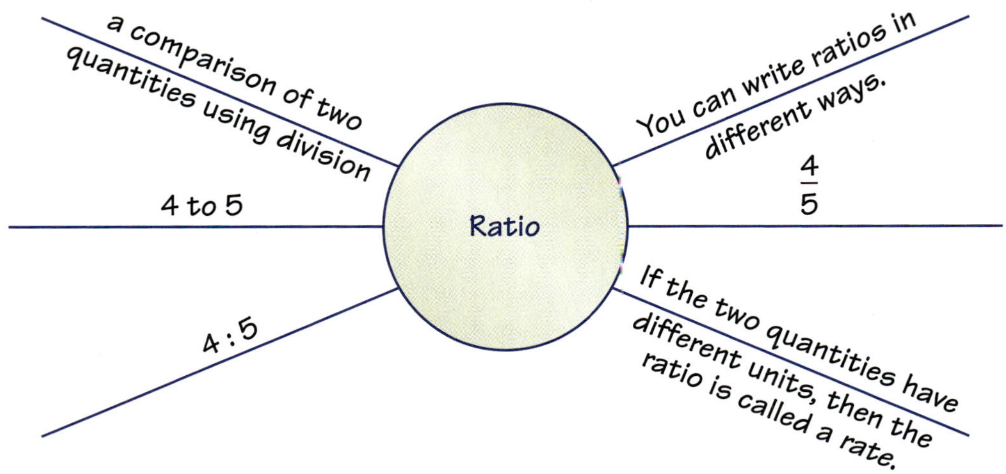

On Your Own

Make information wheels to help you study these topics.

1. rate
2. unit rate
3. proportion
4. cross products
5. graphing proportional relationships

After you complete this chapter, make information wheels for the following topics.

6. solving proportions
7. slope
8. direct variation

"My information wheel summarizes how cats act when they get baths."

Sample Answers

1.

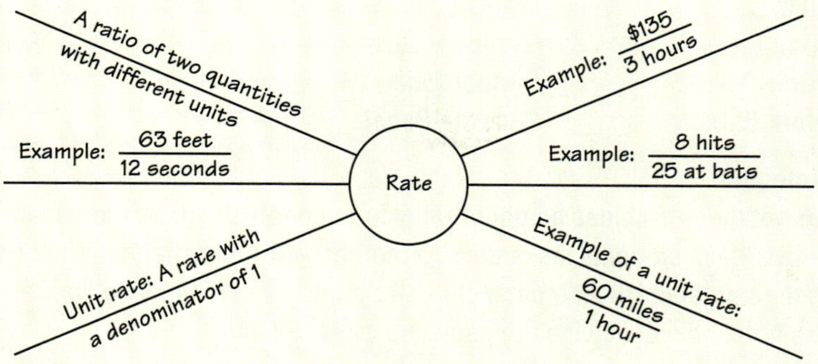

2.

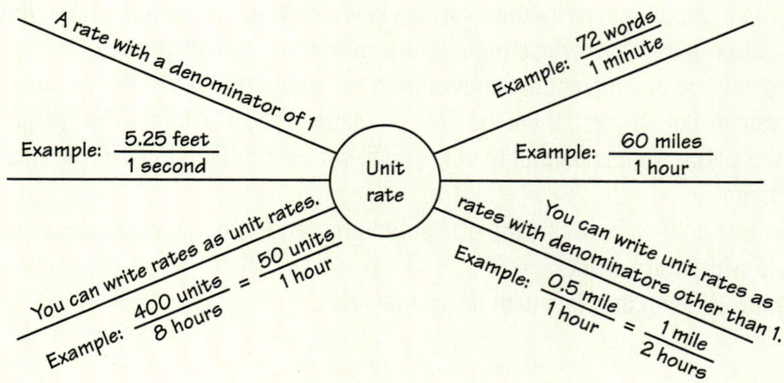

3.

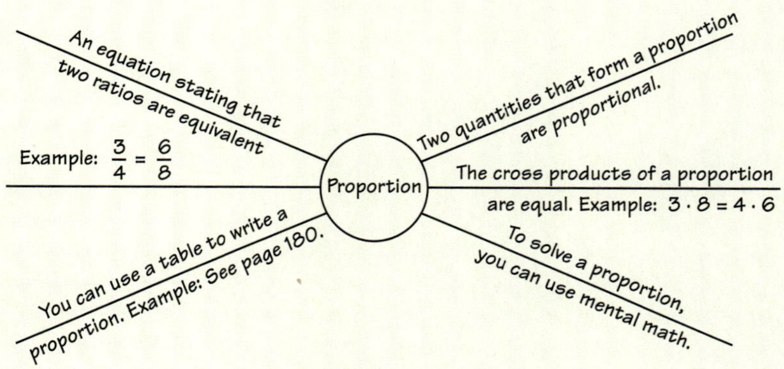

4.

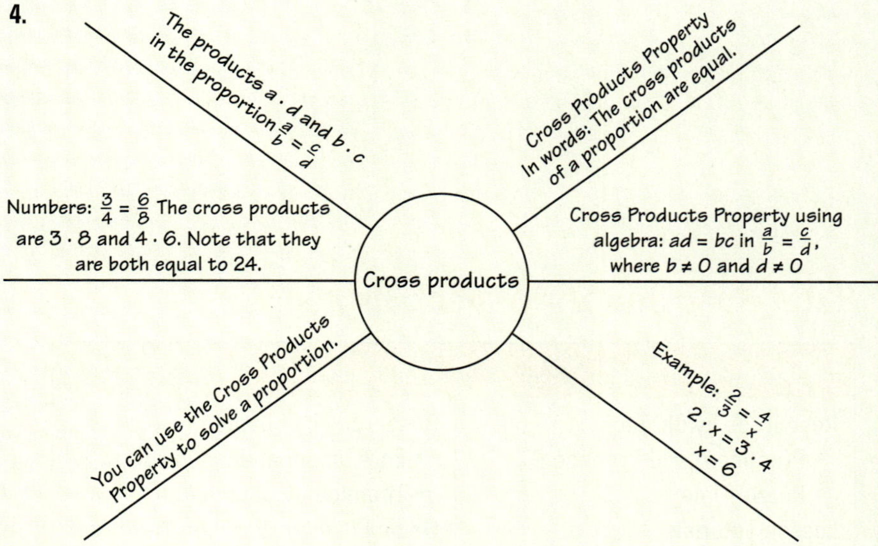

5. Available at *BigIdeasMath.com*.

List of Organizers

Available at *BigIdeasMath.com*

Comparison Chart
Concept Circle
Example and Non-Example Chart
Formula Triangle
Four Square
Idea (Definition) and Examples Chart
Information Frame
Information Wheel
Notetaking Organizer
Process Diagram
Summary Triangle
Word Magnet
Y Chart

About this Organizer

An **Information Wheel** can be used to organize information about a concept. Students write the concept in the middle of the "wheel." Then students write information related to the concept on the "spokes" of the wheel. Related information can include, but is not limited to: vocabulary words or terms, definitions, formulas, procedures, examples, and visuals. This type of organizer serves as a good summary tool because any information related to a concept can be included.

Technology for the Teacher

Editable Graphic Organizer

Answers

1. $\frac{3}{2}$
2. $\frac{15}{8}$
3. $0.99 per song
4. 3.5 gallons per hour
5. yes
6. no
7. yes
8. no
9. yes
10. yes
11. no
12. *Sample answer:* $\frac{\$56}{\$42} = \frac{h \text{ hours}}{6 \text{ hours}}$
13. *Sample answer:* $\frac{g \text{ games}}{4 \text{ wins}} = \frac{6 \text{ games}}{3 \text{ wins}}$
14. $\frac{1}{3}$ MB per second
15. 6 km per second
16. no; Your rate is 5 minutes per level and your friend's rate is 4 minutes per level.
17. $\frac{150 \text{ minutes}}{3 \text{ classes}} = \frac{x \text{ minutes}}{5 \text{ classes}}$; 250 minutes

Technology for the Teacher

Online Assessment
Assessment Book
ExamView® Assessment Suite

Alternative Quiz Ideas

100% Quiz
Error Notebook
Group Quiz
Homework Quiz
Math Log
Notebook Quiz
Partner Quiz
Pass the Paper

Error Notebook

An error notebook provides an opportunity for students to analyze and learn from their errors. Have students make an error notebook for this chapter. They should work in their notebook a little each day. Give students the following directions.

- Use a notebook and divide the page into three columns.
- Label the first column *problem*, second column *error*, and third column *correction*.
- In the first column, write the exercise in which the errors were made. Record the source of the exercise (homework, quiz, in-class assignment).
- The second column should show the exact error that was made. Include a statement of why you think the error was made. This is where the learning takes place, so it is helpful to use a different color ink for the work in this column.
- The last column contains the corrected problems and comments that will help with future work.
- Separate each problem with horizontal lines.

Reteaching and Enrichment Strategies

If students need help. . .	If students got it. . .
Resources by Chapter • Practice A and Practice B • Puzzle Time Lesson Tutorials *BigIdeasMath.com*	Resources by Chapter • Enrichment and Extension • Technology Connection Game Closet at *BigIdeasMath.com* Start the next section

14.1–14.3 Quiz

Check It Out
Progress Check
BigIdeasMath.com

Write the ratio as a fraction in simplest form. *(Section 14.1)*

1. 18 red buttons : 12 blue buttons

2. $\frac{5}{4}$ inches to $\frac{2}{3}$ inch

Use the ratio table to find the unit rate with the specified units. *(Section 14.1)*

3. cost per song

Songs	0	2	4	6
Cost	\$0	\$1.98	\$3.96	\$5.94

4. gallons per hour

Hours	3	6	9	12
Gallons	10.5	21	31.5	42

Tell whether the ratios form a proportion. *(Section 14.2)*

5. $\frac{1}{8}, \frac{4}{32}$

6. $\frac{2}{3}, \frac{10}{30}$

7. $\frac{7}{4}, \frac{28}{16}$

Tell whether the two rates form a proportion. *(Section 14.2)*

8. 75 miles in 3 hours; 140 miles in 4 hours

9. 12 gallons in 4 minutes; 21 gallons in 7 minutes

10. 150 steps in 50 feet; 72 steps in 24 feet

11. 3 rotations in 675 days; 2 rotations in 730 days

Use the table to write a proportion. *(Section 14.3)*

12.

	Monday	Tuesday
Dollars	42	56
Hours	6	h

13.

	Series 1	Series 2
Games	g	6
Wins	4	3

14. MUSIC DOWNLOAD The amount of time needed to download music is shown in the table. Find the unit rate in megabytes per second. *(Section 14.1)*

Seconds	6	12	18	24
Megabytes	2	4	6	8

15. SOUND The graph shows the distance that sound travels through steel. Interpret each plotted point in the graph of the proportional relationship. *(Section 14.2)*

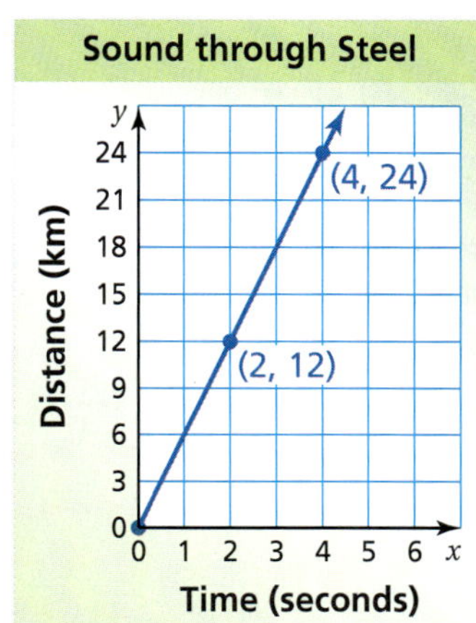

16. GAMING You advance 3 levels in 15 minutes. Your friend advances 5 levels in 20 minutes. Do these rates form a proportion? Explain. *(Section 14.2)*

17. CLASS TIME You spend 150 minutes in 3 classes. Write and solve a proportion to find how many minutes you spend in 5 classes. *(Section 14.3)*

14.4 Solving Proportions

Essential Question How can you use ratio tables and cross products to solve proportions?

1 ACTIVITY: Solving a Proportion in Science

Work with a partner. You can use ratio tables to determine the amount of a compound (like salt) that is dissolved in a solution. Determine the unknown quantity. Explain your procedure.

a. Salt Water

Salt Water	1 L	3 L
Salt	250 g	x g

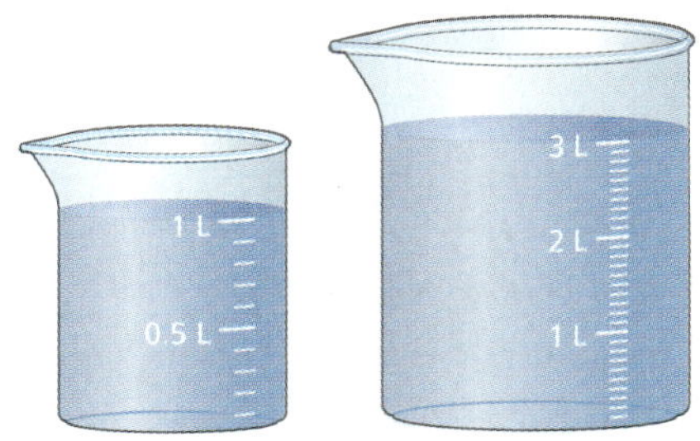

1 liter   3 liters

$\dfrac{1\text{ L}}{\square} = \dfrac{\square}{\square}$ Write proportion.

$1 \cdot \square = \square \cdot \square$ Set cross products equal.

$\square = \square$ Simplify.

There are $\square$ grams of salt in the 3-liter solution.

b. White Glue Solution

Water	$\frac{1}{2}$ cup	1 cup
White Glue	$\frac{1}{2}$ cup	x cups

c. Borax Solution

Borax	1 tsp	2 tsp
Water	1 cup	x cups

d. Slime (See recipe.)

Borax Solution	$\frac{1}{2}$ cup	1 cup
White Glue Solution	y cups	x cups

Recipe for SLIME

1. Add ½ cup of water and ½ cup white glue. Mix thoroughly. This is your white glue solution.
2. Add a couple drops of food coloring to the white glue solution. Mix thoroughly.
3. Add 1 teaspoon of borax to 1 cup of water. Mix thoroughly. This is your borax solution (about 1 cup).
4. Pour the borax solution and the glue solution into a separate bowl.
5. Place the slime that forms into a plastic bag. Squeeze the mixture repeatedly to mix it up.

Proportions

In this lesson, you will
- solve proportions using multiplication or the Cross Products Property.
- use a point on a graph to write and solve proportions.

Laurie's Notes

Introduction

Applying Mathematical Practices

- **Model with Mathematics:** Mathematically proficient students use a ratio table as a tool to show relationships between different quantities.

Motivate

- **Model:** Display two containers, each filled with water. The ratio of their volumes should be 2 : 1. Let students watch you put 4 drops of food coloring in the smaller one. Stir the water.
- ? "How many drops do I need to put in the larger container so that the water is the same darkness as the smaller vessel?" Students should note the ratio of the volumes of the two vessels. It may be helpful to label the volume of each container. Students should have no difficulty in understanding that 8 drops are needed.
- ? "Suppose that the two volumes are not in the ratio 1 : 2. The smaller container has a volume of 400 milliliters and the larger vessel has a volume of 600 milliliters. If I add 4 drops of food coloring to the smaller vessel, how much should I add to the larger vessel?" 6 drops
- This should get students thinking about a proportion.
- **Discuss:** Ask for volunteers to share their thinking.

Activity Notes

Activity 1

- Write the ratio table.
- ? "If there are 250 grams of salt in 1 liter of salt water, how many grams of salt are in 3 liters?" Students should quickly answer 750 grams, not just 750. The units are extremely important!
- Set up the proportion from the ratio table and discuss how common units divide out when comparing liters to liters and grams to grams.
- Then use the Cross Products Property to verify the answer.
- ? "Could the proportion $\frac{1\text{ L}}{250\text{ g}} = \frac{3\text{ L}}{x\text{ g}}$ be used to solve for x?" yes; Listen for an explanation that 1 liter has 250 grams of salt and you're solving for the number of grams of salt in 3 liters.
- ? "Is there another way you might solve this proportion without using the Cross Products Property?" mental math: $1 \times 250 = 250$, so $3 \times 250 = 750$
- **Model with Mathematics:** Ask students to solve for x in each of the three parts—white glue solution, borax solution, and Slime. In each part of the activity, a ratio table helps to show relationships between quantities.
- ? **Extension:** Ask the following:
  - "If you only had $\frac{1}{4}$ cup of white glue, how much water should you mix with it to make the white glue solution?" $\frac{1}{4}$ cup
  - "If you have 1 tablespoon of borax, how much water do you need to make the borax solution?" 3 cups

What Your Students Will Learn

- Solve proportions using the Multiplication Property of Equality or the Cross Products Property.
- Use a point on a graph to write and solve proportions.

Previous Learning

Students should know how to write and solve proportions using mental math and how to convert measures.

Technology for the Teacher

Dynamic Classroom

Lesson Plans
Complete Materials List

14.4 Record and Practice Journal

Essential Question How can you use ratio tables and cross products to solve proportions?

1 ACTIVITY: Solving a Proportion in Science

Work with a partner. You can use ratio tables to determine the amount of a compound (like salt) that is dissolved in a solution. Determine the unknown quantity. Explain your procedure.

a. Salt Water

Salt Water	1 L	3 L
Salt	250 g	x g

$x = 750$

There are 750 grams of salt in the 3-liter solution.

b. White Glue Solution

Water	$\frac{1}{2}$ cup	1 cup
White Glue	$\frac{1}{2}$ cup	x cups

1 cup

c. Borax Solution

Borax	1 tsp	2 tsp
Water	1 cup	x cups

2 cups

English Language Learners

Visual

When solving a proportion using the Cross Products Property, use a visual X through the equal sign.

$\frac{4}{3} \times \frac{f}{27}$	Write the proportion.
$4 \cdot 27 = 3 \cdot f$	Cross Products Property
$108 = 3f$	Multiply.
$36 = f$	Divide.

14.4 Record and Practice Journal

5.4 Solving Proportions (continued)

d. Slime (See recipe.)

Borax Solution	$\frac{1}{2}$ cup	1 cup
White Glue Solution	y cups	x cups

$y = \frac{1}{2}$ cup

$x = 1$ cup

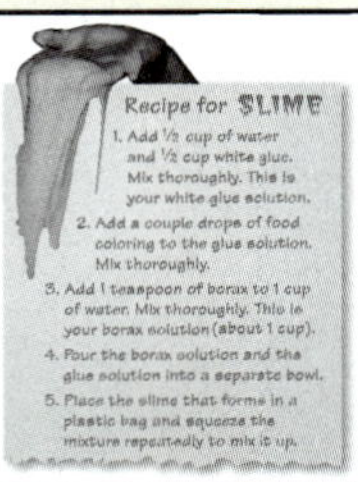

2 ACTIVITY: The Game of Criss Cross

Preparation: Check students' work.

- Cut index cards to make 48 playing cards.
- Write each number on a card.
  1, 1, 1, 2, 2, 2, 3, 3, 3, 4, 4, 4, 5, 5, 5, 6, 6, 6, 7, 7, 7, 8, 8, 8, 9, 9, 9, 10, 10, 10, 12, 12, 12, 13, 13, 13, 14, 14, 14, 15, 15, 15, 16, 16, 16, 18, 20, 25
- Make a copy of the game board.

To play:

- Play with a partner.
- Deal 8 cards to each player.
- Begin by drawing a card from the remaining cards. Use four of your cards to try and form a proportion.
- Lay the four cards on the game board. If you form a proportion, then say "Criss Cross." You earn 4 points. Place the four cards in a discard pile. Now it is your partner's turn.
- If you cannot form a proportion, then it is your partner's turn.
- When the original pile of cards is empty, shuffle the cards in the discard pile. Start again.
- The first player to reach 20 points wins.

What Is Your Answer?

3. **IN YOUR OWN WORDS** How can you use ratio tables and cross products to solve proportions? Give an example.
   Set up a proportion to mix chemical solutions.

4. **PUZZLE** Use each number once to form three proportions.

1	2	10	4	12	20
15	5	16	6	8	3

Sample answer: $\frac{1}{5} = \frac{2}{10}, \frac{4}{16} = \frac{3}{12}, \frac{8}{20} = \frac{6}{15}$

Laurie's Notes

Activity 2

- **Management Tip:** The cards can be photocopied onto heavier weight paper, laminated, cut apart, and stored in locking plastic bags for easy distribution *and* for use next year.
- **Game Notes:**
  - The number of cards in your hand will vary depending upon whether you have been able to make a proportion.
  - The game moves along fairly quickly.
  - If a pair of students finishes early, they can play again.
- After students have played for a period of time, ask what strategies they used to decide if they have a proportion or not. Make the decision on whether or not they had a proportion.
- **Construct Viable Arguments and Critique the Reasoning of Others:** Mathematically proficient students should be expected to explain their reasoning. It is also important for them to hear other students explain their reasoning.
- Some students may only think about equivalent fractions. Others may think about the Cross Products Property.

What Is Your Answer?

? Question 4 is a nice recap of the lesson. When students have finished, ask: "Do you think everyone has the same 3 proportions?" Students may likely say yes, forgetting that there are different arrangements of the 4 numbers in a proportion that will form another proportion.

Closure

- Use the information in the table to solve for x.

# of bracelets	3	x
Yellow twine	48 in.	80 in.

$\frac{3}{48} = \frac{x}{80}$

$x = 5$

five bracelets

2 ACTIVITY: The Game of Criss Cross

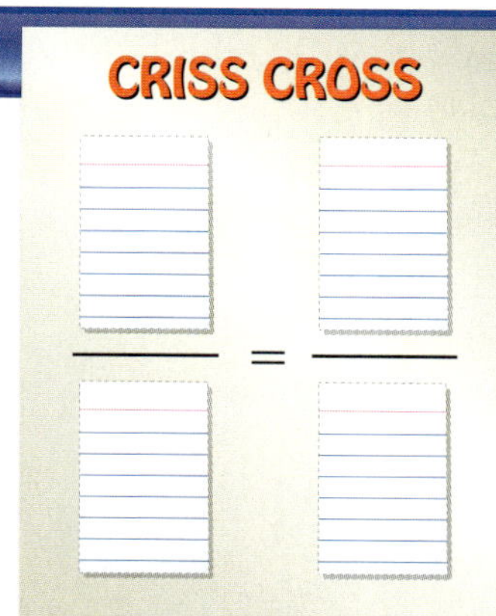

Math Practice

Use Operations

How can you use the name of the game to determine which operation to use?

Preparation:

- Cut index cards to make 48 playing cards.
- Write each number on a card.

  1, 1, 1, 2, 2, 2, 3, 3, 3, 4, 4, 4, 5, 5, 5, 6, 6, 6, 7, 7,

  7, 8, 8, 8, 9, 9, 9, 10, 10, 10, 12, 12, 12, 13, 13,

  13, 14, 14, 14, 15, 15, 15, 16, 16, 16, 18, 20, 25
- Make a copy of the game board.

To Play:

- Play with a partner.
- Deal eight cards to each player.
- Begin by drawing a card from the remaining cards. Use four of your cards to try to form a proportion.
- Lay the four cards on the game board. If you form a proportion, then say "Criss Cross." You earn 4 points. Place the four cards in a discard pile. Now it is your partner's turn.
- If you cannot form a proportion, then it is your partner's turn.
- When the original pile of cards is empty, shuffle the cards in the discard pile. Start again.
- The first player to reach 20 points wins.

What Is Your Answer?

3. **IN YOUR OWN WORDS** How can you use ratio tables and cross products to solve proportions? Give an example.

4. **PUZZLE** Use each number once to form three proportions.

1	2	10	4	12	20
15	5	16	6	8	3

Practice Use what you discovered about solving proportions to complete Exercises 10–13 on page 626.

14.4 Lesson

Check It Out Lesson Tutorials BigIdeasMath.com

Key Idea

Solving Proportions

Method 1 Use mental math. *(Section 14.3)*

Method 2 Use the Multiplication Property of Equality. *(Section 14.4)*

Method 3 Use the Cross Products Property. *(Section 14.4)*

EXAMPLE 1 Solving Proportions Using Multiplication

Solve $\frac{5}{7} = \frac{x}{21}$.

$\frac{5}{7} = \frac{x}{21}$ Write the proportion.

$21 \cdot \frac{5}{7} = 21 \cdot \frac{x}{21}$ Multiplication Property of Equality

$15 = x$ Simplify.

∴ The solution is 15.

On Your Own

Now You're Ready Exercises 4–9

Use multiplication to solve the proportion.

1. $\frac{w}{6} = \frac{6}{9}$ **2.** $\frac{12}{10} = \frac{a}{15}$ **3.** $\frac{y}{6} = \frac{2}{4}$

EXAMPLE 2 Solving Proportions Using the Cross Products Property

Solve each proportion.

a. $\frac{x}{8} = \frac{7}{10}$

$x \cdot 10 = 8 \cdot 7$ Cross Products Property

$10x = 56$ Multiply.

$x = 5.6$ Divide.

∴ The solution is 5.6.

b. $\frac{9}{y} = \frac{3}{17}$

$9 \cdot 17 = y \cdot 3$

$153 = 3y$

$51 = y$

∴ The solution is 51.

Laurie's Notes

Introduction

Connect

- **Yesterday:** Students solved proportions using the Cross Products Property.
- **Today**: Students will solve proportions using different strategies.

Motivate

- The Cross Products Property can be used to solve any proportion, but you want students to recognize when it is more efficient to use simple mental math or the Multiplication Property of Equality.
- **Make Sense of Problems and Persevere in Solving Them:** As you work through problems with students, share with them the wisdom of analyzing the problem first to decide what method makes the most sense.
- **Common Error:** Students sometimes confuse multiplication of fractions and the Cross Products Property.

Lesson Notes

Example 1

- The Multiplication Property of Equality works because the variable is in the numerator.
- If this same problem had been $\frac{7}{5} = \frac{21}{x}$, you could not solve by multiplying both sides of the equation by $\frac{1}{21}$ because that would simplify to $\frac{1}{15} = \frac{1}{x}$ and you still haven't solved for x. Be sure students understand this.
- Be sure to check for understanding with this idea.

? "Could you use another strategy such as mental math to solve this problem?" Yes. Listen for the idea of equivalent fractions.

On Your Own

- **Think-Pair-Share:** Students should read each question independently and then work in pairs to answer the questions. When they have answered the questions, the pair should compare their answers with another group and discuss any discrepancies.

? Ask students to share their strategies.

- At least one pair of students should solve Question 3 by simplifying $\frac{2}{4} = \frac{1}{2}$, and using mental math to finish.

Example 2

? "How are parts (a) and (b) different?" In part (a), the variable is in the numerator and in part (b), the variable is in the denominator. Part (b) involves one numerator that is a factor of the other numerator.

? "Can you easily use the Multiplication Property of Equality to solve both examples?" Using the Multiplication Property of Equality would be difficult in part (b) because the variable is in the denominator.

Goal Today's lesson is solving proportions using a variety of strategies.

Technology for the Teacher

Dynamic Classroom

Lesson Tutorials
Lesson Plans
Answer Presentation Tool

Extra Example 1

Solve $\frac{c}{12} = \frac{5}{3}$. 20

On Your Own

1. $w = 4$
2. $a = 18$
3. $y = 3$

Extra Example 2

Solve each proportion.

a. $\frac{3}{4} = \frac{u}{6}$ 4.5

b. $\frac{4}{13} = \frac{12}{h}$ 39

On Your Own

4. $x = 8$
5. $y = 2.5$
6. $z = 15$

Extra Example 3

In Example 3, your toll is \$9. How many kilometers did you drive? about 193 km

On Your Own

7. $\frac{7.5}{x} = \frac{1}{2.54}$; about 19.05
8. $\frac{100}{x} = \frac{1}{0.035}$; about 3.5
9. $\frac{2}{x} = \frac{1}{1.06}$; about 2.12
10. $\frac{4}{x} = \frac{1}{3.28}$; about 13.12

Differentiated Instruction

Visual

To reinforce the Cross Products Property, write each number and variable of a proportion on a card. Give the set of cards to a student. Have the student set up the proportion.

$$\frac{\boxed{3}}{\boxed{8}} = \frac{\boxed{k}}{\boxed{4}}$$

Then have the student find the cross products with the cards.

$$\boxed{3} \cdot \boxed{4} = \boxed{8} \cdot \boxed{k}$$

Finish by having the student complete the solution using paper and pencil.

Laurie's Notes

On Your Own

- **Think-Pair-Share:** Students should read each question independently and then work in pairs to answer the questions. When they have answered the questions, the pair should compare their answers with another group and discuss any discrepancies.

? Ask students to share their strategies.

- Although the directions say to solve using the Cross Products Property, one pair of students might solve Questions 4 and 5 by using mental math and recognize equivalent fractions.

Example 3

? "Which is farther, one mile or one kilometer?" one mile

- Hopefully, students will know that 1 mile ≈ 1.61 kilometers. If not, have them refer to the conversion chart in the back of the book.

? "What do the ordered pairs (100, 7.5) and (200, 15) mean on the graph?" If you drive 100 miles the toll is \$7.50, and if you drive 200 miles the toll is \$15.

- You may need to help students focus on the units for each axis.
- Explain that you want to convert miles to kilometers.
- The first method uses the conversion factor and dimensional analysis. Work through the problem as shown.
- The second method uses a proportion. Although the units for each value are not written, say aloud the units as you write the proportion: "1.61 kilometers is to 1 mile as how many kilometers is to 100 miles?" Each ratio compares kilometers to miles.
- Students may say it is not necessary to set up a proportion because the numbers are so simple. Change the problem by asking what toll would be due for driving 150 miles or 232 miles.

On Your Own

- These exercises provide a helpful review of decimal multiplication and division.
- Students may need to refer to the conversion chart in the back of the book to complete these problems.
- **Attend to Precision:** Remind students that the symbol ≈ means *approximately equal to* and that answers should be rounded to the nearest hundredth, if necessary.

Closure

- Write and solve 3 proportions. One should use mental math to solve, one should use the Multiplication Property of Equality, and one should use the Cross Products Property.

On Your Own

Now You're Ready
Exercises 10–21

Use the Cross Products Property to solve the proportion.

4. $\frac{2}{7} = \frac{x}{28}$ **5.** $\frac{12}{5} = \frac{6}{y}$ **6.** $\frac{40}{z+1} = \frac{15}{6}$

EXAMPLE 3 Real-Life Application

The graph shows the toll y due on a turnpike for driving x miles. Your toll is \$7.50. How many *kilometers* did you drive?

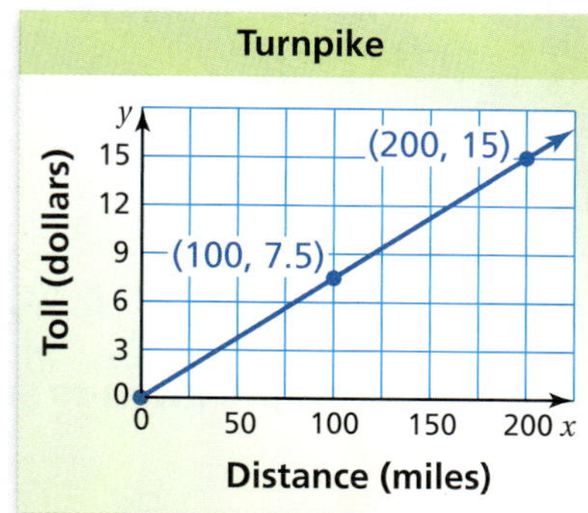

The point (100, 7.5) on the graph shows that the toll is \$7.50 for driving 100 miles. Convert 100 miles to kilometers.

Method 1: Convert using a ratio.

1 mi ≈ 1.61 km

$$100 \text{ mi} \times \frac{1.61 \text{ km}}{1 \text{ mi}} = 161 \text{ km}$$

So, you drove about 161 kilometers.

Method 2: Convert using a proportion.

Let x be the number of kilometers equivalent to 100 miles.

kilometers / miles → $\frac{1.61}{1} = \frac{x}{100}$ ← kilometers / miles   Write a proportion. Use 1.61 km ≈ 1 mi.

$1.61 \cdot 100 = 1 \cdot x$ Cross Products Property

$161 = x$ Simplify.

So, you drove about 161 kilometers.

On Your Own

Now You're Ready
Exercises 28–30

Write and solve a proportion to complete the statement. Round to the nearest hundredth, if necessary.

7. 7.5 in. ≈ ___ cm **8.** 100 g ≈ ___ oz

9. 2 L ≈ ___ qt **10.** 4 m ≈ ___ ft

14.4 Exercises

Check It Out
Help with Homework
BigIdeasMath.com

Vocabulary and Concept Check

1. **WRITING** What are three ways you can solve a proportion?

2. **OPEN-ENDED** Which way would you choose to solve $\frac{3}{x} = \frac{6}{14}$? Explain your reasoning.

3. **NUMBER SENSE** Does $\frac{x}{4} = \frac{15}{3}$ have the same solution as $\frac{x}{15} = \frac{4}{3}$? Use the Cross Products Property to explain your answer.

Practice and Problem Solving

Use multiplication to solve the proportion.

1 4. $\frac{9}{5} = \frac{z}{20}$

5. $\frac{h}{15} = \frac{16}{3}$

6. $\frac{w}{4} = \frac{42}{24}$

7. $\frac{35}{28} = \frac{n}{12}$

8. $\frac{7}{16} = \frac{x}{4}$

9. $\frac{y}{9} = \frac{44}{54}$

Use the Cross Products Property to solve the proportion.

2 10. $\frac{a}{6} = \frac{15}{2}$

11. $\frac{10}{7} = \frac{8}{k}$

12. $\frac{3}{4} = \frac{v}{14}$

13. $\frac{5}{n} = \frac{16}{32}$

14. $\frac{36}{42} = \frac{24}{r}$

15. $\frac{9}{10} = \frac{d}{6.4}$

16. $\frac{x}{8} = \frac{3}{12}$

17. $\frac{8}{m} = \frac{6}{15}$

18. $\frac{4}{24} = \frac{c}{36}$

19. $\frac{20}{16} = \frac{d}{12}$

20. $\frac{30}{20} = \frac{w}{14}$

21. $\frac{2.4}{1.8} = \frac{7.2}{k}$

22. **ERROR ANALYSIS** Describe and correct the error in solving the proportion $\frac{m}{8} = \frac{15}{24}$.

✗
$$\frac{m}{8} = \frac{15}{24}$$
$$8 \cdot m = 24 \cdot 15$$
$$m = 45$$

23. **PENS** Forty-eight pens are packaged in 4 boxes. How many pens are packaged in 9 boxes?

24. **PIZZA PARTY** How much does it cost to buy 10 medium pizzas?

Solve the proportion.

25. $\frac{2x}{5} = \frac{9}{15}$

26. $\frac{5}{2} = \frac{d-2}{4}$

27. $\frac{4}{k+3} = \frac{8}{14}$

Assignment Guide and Homework Check

Level	Assignment	Homework Check
Advanced	1–3, 4–8 even, 10–13, 14–38 even, 39–43	18, 26, 30, 32, 36

Common Errors

- **Exercises 4–9** Some students may multiply by the denominator of the fraction without the variable. Remind them that they are trying to get the variable alone, so they want to multiply both sides by the denominator of the fraction with the variable. Give students an example without a fraction on the other side of the equation to remind them of the process.
- **Exercises 10–21** Students may divide instead of multiply when finding the cross products, or they may multiply across the numerators and the denominators as if they were multiplying fractions. Remind students that the ratios have an equal sign between them, not a multiplication sign. Also tell them that when they use the Cross Products Property, it produces an "X," which means multiplication.

14.4 Record and Practice Journal

Use multiplication to solve the proportion.

1. $\frac{a}{40} = \frac{3}{10}$ **$a = 12$**
2. $\frac{6}{11} = \frac{c}{77}$ **$c = 42$**
3. $\frac{b}{65} = \frac{7}{13}$ **$b = 35$**

Use the Cross Products Property to solve the proportion.

4. $\frac{k}{6} = \frac{8}{16}$ **$k = 3$**
5. $\frac{5.4}{7} = \frac{27}{h}$ **$h = 35$**
6. $\frac{8}{11} = \frac{4}{y + 2}$ **$y = 3.5$**

Write and solve a proportion to complete the statement.

7. 42 in. = _____ cm **106.68**
8. 12.6 kg ≈ _____ lb **28**
9. 3 oz ≈ _____ g **84**

10. A cell phone company charges $5 for 250 text messages. How much does the company charge for 300 text messages? **$6**

Vocabulary and Concept Check

1. mental math; Multiplication Property of Equality; Cross Products Property
2. *Sample answer:* mental math; Because $3 \cdot 2 = 6$, the product of x and 2 is 14. So, $x = 7$.
3. yes; Both cross products give the equation $3x = 60$.

Practice and Problem Solving

4. $z = 36$
5. $h = 80$
6. $w = 7$
7. $n = 15$
8. $x = 1\frac{3}{4}$
9. $y = 7\frac{1}{3}$
10. $a = 45$
11. $k = 5.6$
12. $v = 10.5$
13. $n = 10$
14. $r = 28$
15. $d = 5.76$
16. $x = 2$
17. $m = 20$
18. $c = 6$
19. $d = 15$
20. $w = 21$
21. $k = 5.4$
22. They did not perform the cross multiplication properly.

$$\frac{m}{8} = \frac{15}{24}$$
$$m \cdot 24 = 8 \cdot 15$$
$$m = 5$$

23. 108 pens
24. \$35
25. $x = 1.5$
26. $d = 12$
27. $k = 4$

Practice and Problem Solving

28. $\frac{6}{x} = \frac{1}{0.62}$; about 3.7

29. $\frac{2.5}{x} = \frac{1}{0.26}$; about 0.65

30. $\frac{90}{x} = \frac{1}{0.45}$; about 40.5

31. true; Both cross products give the equation $3a = 2b$.

32. $769.50 **33.** 15.5 lb

34. **a.** about 7.62 cm

b. 16 mo

c. 40 mo

35. no; The relationship is not proportional. It should take more people less time to build the swing set.

36. See *Taking Math Deeper*.

37. 4 bags

38. $\frac{1}{5}$; $\frac{m}{k} = \frac{\frac{n}{2}}{\frac{5n}{2}} = \frac{n}{2} \cdot \frac{2}{5n} = \frac{1}{5}$

Fair Game Review

39–42.

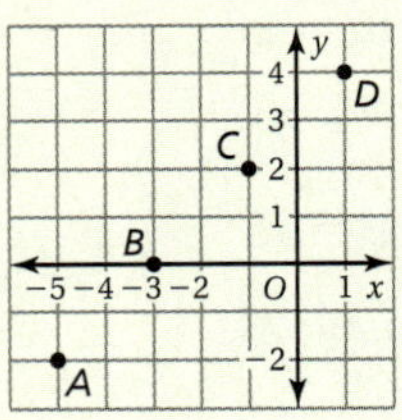

43. D

Mini-Assessment

Solve the proportion.

1. $\frac{x}{12} = \frac{3}{8}$ $x = 4.5$
2. $\frac{6}{11} = \frac{9}{m}$ $m = 16.5$
3. $\frac{6}{12} = \frac{c}{36}$ $c = 18$
4. $\frac{18}{3} = \frac{24}{b}$ $b = 4$
5. Thirty-six pencils are packed in three boxes. How many pencils are packed in five boxes? 60 pencils

Taking Math Deeper

Exercise 36

Sometimes it is a good suggestion to "forget about algebra and just answer the question." After answering the question, we might "relax" and try to look for an efficient or clever way to answer the question.

1 Start with a ratio table. Adults : children is 5 : 3.

Adults	Children	Total
5	3	8
10	6	16
15	9	24

2 I can see this will take a while. I'm going to jump ahead in the ratio table.

Adults	Children	Total
100	60	160 too many
80	48	128 too few
90	54	144 just right

3 The answer is 90 adults. I wonder if I can get the answer algebraically.

Adults = $5x$ Children = $3x$

$$5x + 3x = 144$$
$$8x = 144$$
$$x = 18$$

Adults = $5 \cdot 18 = 90$

Reteaching and Enrichment Strategies

If students need help. . .	If students got it. . .
Resources by Chapter • Practice A and Practice B • Puzzle Time Record and Practice Journal Practice Differentiating the Lesson Lesson Tutorials Skills Review Handbook	Resources by Chapter • Enrichment and Extension • Technology Connection Start the next section

Write and solve a proportion to complete the statement. Round to the nearest hundredth if necessary.

3 **28.** 6 km ≈ ___ mi **29.** 2.5 L ≈ ___ gal **30.** 90 lb ≈ ___ kg

31. TRUE OR FALSE? Tell whether the statement is *true* or *false*. Explain.

$$\text{If } \frac{a}{b} = \frac{2}{3}, \text{ then } \frac{3}{2} = \frac{b}{a}.$$

32. CLASS TRIP It costs \$95 for 20 students to visit an aquarium. How much does it cost for 162 students?

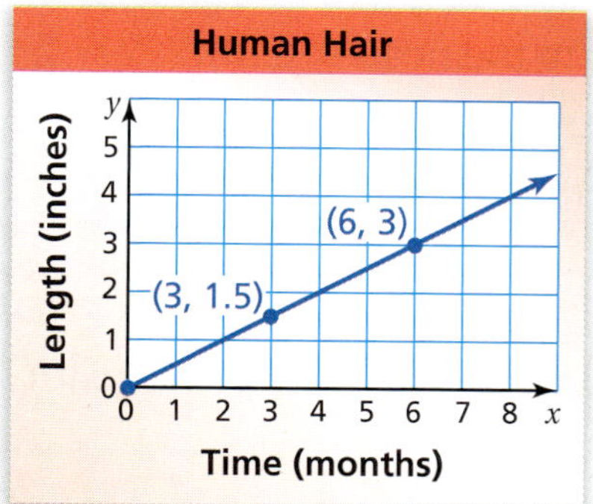

33. GRAVITY A person who weighs 120 pounds on Earth weighs 20 pounds on the Moon. How much does a 93-pound person weigh on the Moon?

34. HAIR The length of human hair is proportional to the number of months it has grown.

a. What is the hair length in *centimeters* after 6 months?

b. How long does it take hair to grow 8 inches?

c. Use a different method than the one in part (b) to find how long it takes hair to grow 20 inches.

35. SWING SET It takes 6 hours for 2 people to build a swing set. Can you use the proportion $\frac{2}{6} = \frac{5}{h}$ to determine the number of hours h it will take 5 people to build the swing set? Explain.

36. REASONING There are 144 people in an audience. The ratio of adults to children is 5 to 3. How many are adults?

37. PROBLEM SOLVING Three pounds of lawn seed covers 1800 square feet. How many bags are needed to cover 8400 square feet?

38. Critical Thinking Consider the proportions $\frac{m}{n} = \frac{1}{2}$ and $\frac{n}{k} = \frac{2}{5}$. What is the ratio $\frac{m}{k}$? Explain your reasoning.

Fair Game Review What you learned in previous grades & lessons

Plot the ordered pair in a coordinate plane. *(Section 6.5)*

39. $A(-5, -2)$ **40.** $B(-3, 0)$ **41.** $C(-1, 2)$ **42.** $D(1, 4)$

43. MULTIPLE CHOICE Which expression is equivalent to $(3w - 8) - 4(2w + 3)$? *(Section 13.2)*

Ⓐ $11w + 4$ Ⓑ $-5w - 5$ Ⓒ $-5w + 4$ Ⓓ $-5w - 20$

14.5 Slope

Essential Question How can you compare two rates graphically?

1 ACTIVITY: Comparing Unit Rates

Work with a partner. The table shows the maximum speeds of several animals.

a. Find the missing speeds. Round your answers to the nearest tenth.

b. Which animal is fastest? Which animal is slowest?

c. Explain how you convert between the two units of speed.

Animal	Speed (miles per hour)	Speed (feet per second)
Antelope	61.0	
Black mamba snake		29.3
Cheetah		102.6
Chicken		13.2
Coyote	43.0	
Domestic pig		16.0
Elephant		36.6
Elk		66.0
Giant tortoise	0.2	
Giraffe	32.0	
Gray fox		61.6
Greyhound	39.4	
Grizzly bear		44.0
Human		41.0
Hyena	40.0	
Jackal	35.0	
Lion		73.3
Peregrine falcon	200.0	
Quarter horse	47.5	
Spider		1.76
Squirrel	12.0	
Thomson's gazelle	50.0	
Three-toed sloth		0.2
Tuna	47.0	

Slope

In this lesson, you will

- find the slopes of lines.
- interpret the slopes of lines as rates.

Laurie's Notes

Introduction

Applying Mathematical Practices

- **Model with Mathematics:** Mathematically proficient students use a graph as a tool to show the relationship between two quantities.

Motivate

- If students are not comfortable with dimensional analysis from the previous lesson, you will need to help them get started with this activity.
- **Big Idea:** You begin with a speed in certain units (i.e., miles per hour) and you want a speed with different units (i.e., feet per second).
- You need to set up the factors so that the unwanted units divide out. The units you want to divide out should always appear diagonally from one another. In your final answer, one of the units should end up in the numerator and the other should end up in the denominator.
- Example: $\dfrac{61\ \cancel{\text{miles}}}{\cancel{\text{hour}}} \times \dfrac{1\ \cancel{\text{hour}}}{3600\ \text{seconds}} \times \dfrac{5280\ \text{feet}}{1\ \cancel{\text{mile}}}$
- Because 1 hour = 3600 seconds, multiplying by $\dfrac{1\ \text{hour}}{3600\ \text{seconds}}$ or $\dfrac{3600\ \text{seconds}}{1\ \text{hour}}$ is equivalent to multiplying by 1.
- Divide out by a common factor of 240, and the answer is $89\frac{7}{15}$ feet per second.
- Converting from feet per second to miles per hour will require students to multiply by the reciprocal of each of the conversion factors.

Activity Notes

Activity 1

- It would be appropriate for students to work in pairs and use a calculator to complete the activity.
- ? "Look through the list and predict the fastest animal. Mark it with the letter F. Predict the slowest animal. Mark it with the letter S."
- Did students select the Peregrine falcon or cheetah as the fastest?
- Did students select the giant tortoise or three-toed sloth as the slowest?
- **Look for and Express Regularity in Repeated Reasoning:** You may want to help students see how to convert from miles per hour to feet per second and vice versa.

$$\frac{\cancel{\text{miles}}}{\cancel{\text{hour}}} \times \underbrace{\frac{1\ \cancel{\text{hour}}}{3600\ \text{seconds}}}_{\uparrow\ \text{equals 1}} \times \underbrace{\frac{5280\ \text{feet}}{1\ \cancel{\text{mile}}}}_{\uparrow\ \text{equals 1}}$$

- When students have finished, discuss the results, the answer to part (c), and their predictions.

What Your Students Will Learn

- Find the slopes of lines using graphs.
- Interpret the slopes of lines as rates by using tables.

Previous Learning

Students should know how to write and simplify ratios, how to convert measures, and how to graph proportional relationships.

Technology for the Teacher

Dynamic Classroom

Lesson Plans
Complete Materials List

14.5 Record and Practice Journal

Essential Question How can you compare two rates graphically?

1 ACTIVITY: Comparing Unit Rates

Work with a partner. The table shows the maximum speeds of several animals.

a. Find the missing speeds. Round your answers to the nearest tenth.

b. Which animal is fastest? Which animal is slowest? **peregrine falcon; three-toed sloth**

c. Explain how you convert between the two units of speed. **See Additional Answers.**

Animal	Speed (miles per hour)	Speed (feet per second)
Antelope	61.0	**89.5**
Black mamba snake	**20.0**	29.3
Cheetah	**70.0**	102.6
Chicken	**9.0**	13.2
Coyote	43.0	**63.1**
Domestic pig	**10.9**	16.0
Elephant	**25.0**	36.6
Elk	**45.0**	66.0
Giant tortoise	0.2	**0.3**
Giraffe	32.0	**46.9**
Gray fox	**42.0**	61.6
Greyhound	39.4	**57.8**
Grizzly bear	**30.0**	44.0
Human	**28.0**	41.0
Hyena	40.0	**58.7**
Jackal	35.0	**51.3**
Lion	**50.0**	73.3
Peregrine falcon	200.0	**293.3**
Quarter horse	47.5	**69.7**
Spider	**1.2**	1.76
Squirrel	12.0	**17.6**
Thomson's gazelle	50.0	**73.3**
Three-Toed sloth	**0.1**	0.2
Tuna	47.0	**68.9**

English Language Learners

Labels

English learners may recognize the fraction bar as division, but may not be familiar with its use in the concept of rate. The following unit rates are equivalent.

$\frac{3 \text{ m}}{1 \text{ h}}$, 3 m/h, 3 meters per hour

Each of these rates can be read as "three meters *for every* hour."

14.5 Record and Practice Journal

2 ACTIVITY: Comparing Two Rates Graphically

Work with a partner. A cheetah and a Thomson's gazelle run at maximum speeds.

a. Use the table in Activity 1 to calculate the missing distances.

	Cheetah	Gazelle
Time (seconds)	Distance (feet)	Distance (feet)
0	0	0
1	102.6	73.3
2	205.2	146.6
3	307.8	219.9
4	410.4	293.2
5	513	366.5
6	615.6	439.8
7	718.2	513.1

b. Use the table to write ordered pairs. Then plot the ordered pairs and connect the points for each animal. What do you notice about the graphs?

The graphs are lines that pass through the origin.

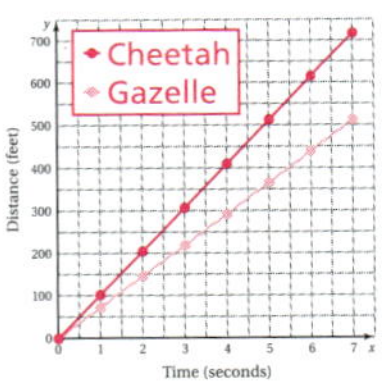

c. Which graph is steeper? The speed of which animal is greater?

cheetah; cheetah

What Is Your Answer?

3. **IN YOUR OWN WORDS** How can you compare two rates graphically? Explain your reasoning. Give some examples with your answer.

The steeper line has a higher rate.

4. **REPEATED REASONING** Choose 10 animals from Activity 1. Check students' work.

a. Make a table for each animal similar to the table in Activity 2.

b. Sketch a graph of the distances for each animal.

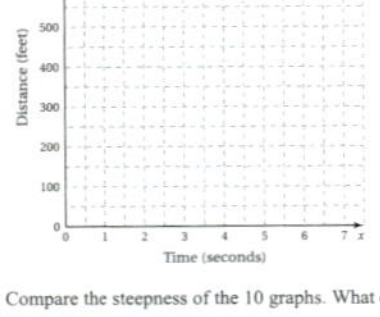

c. Compare the steepness of the 10 graphs. What can you conclude?

Laurie's Notes

Activity 2

- *Note:* The formal definition of **slope** will come in the lesson. The idea in this activity is to get students to understand that a steeper line translates to a greater speed.
- Discuss the concept of a constant speed. Talk about a car on cruise control or items traveling on an assembly line in a factory. Over short periods of time it is possible for animals to run at a constant speed.
- ? "If a cheetah is running at a constant speed and has traveled a distance of 102.6 feet in 1 second, how far will it run in two seconds?" 205.2 ft
- Have students complete the table. This is a good review of decimal multiplication or decimal addition (depending on how the student completes the work). You may find it helpful to review the rules for multiplying and adding decimals beforehand.
- ? **Discuss:** "After 3 seconds, how far has each animal run? after 7 seconds?"
- ? "Is there ever a time when the gazelle has run farther than the cheetah?"
- Students might find it helpful to use a straightedge or ruler to connect their points.
- ? "If the cheetah was not running at a constant speed, would your graph look different? Explain." Yes. The points would not be in a line. The graph would go up and down accordingly.
- **Model with Mathematics:** Discuss the relationship between the speed of animals and the steepness of the two graphs. Students should be able to tell from the graph which animal has the greater speed.
- **Model with Mathematics:** Some students may recognize which graph will be steeper after graphing the first two points for each animal. They may also remember from earlier in the chapter that in the graph of a proportional relationship, you can find the unit rate y from the point $(1, y)$.

What Is Your Answer?

- Question 4 can be done as homework.

Closure

- An airplane is traveling at a constant speed of 6 miles per minute. Make a table to show the distance traveled each minute for 8 minutes. Graph your data and connect the points. Then describe the graph.

Minutes, x	0	1	2	3	4	5	6	7	8
Miles, y	0	6	12	18	24	30	36	42	48

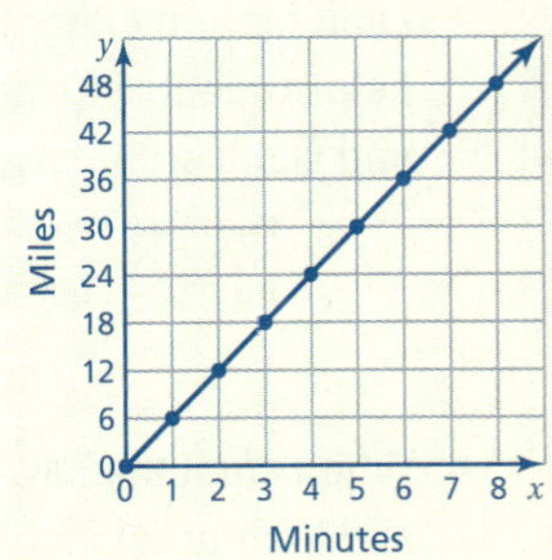

a line that passes through the origin

2 ACTIVITY: Comparing Two Rates Graphically

> **Math Practice**
>
> **Apply Mathematics**
>
> How can you use the graph to determine which animal has the greater speed?

Work with a partner. A cheetah and a Thomson's gazelle run at maximum speed.

a. Use the table in Activity 1 to calculate the missing distances.

	Cheetah	Gazelle
Time (seconds)	Distance (feet)	Distance (feet)
0		
1		
2		
3		
4		
5		
6		
7		

b. Use the table to write ordered pairs. Then plot the ordered pairs and connect the points for each animal. What do you notice about the graphs?

c. Which graph is steeper? The speed of which animal is greater?

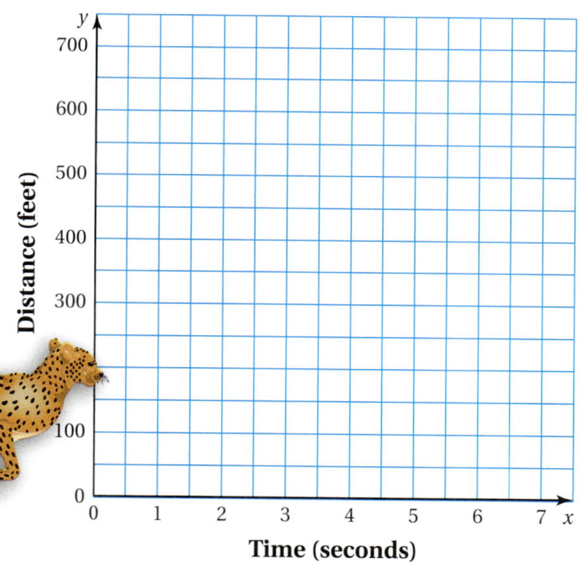

What Is Your Answer?

3. **IN YOUR OWN WORDS** How can you compare two rates graphically? Explain your reasoning. Give some examples with your answer.

4. **REPEATED REASONING** Choose 10 animals from Activity 1.

   a. Make a table for each animal similar to the table in Activity 2.

   b. Sketch a graph of the distances for each animal.

   c. Compare the steepness of the 10 graphs. What can you conclude?

14.5 Lesson

Check It Out
Lesson Tutorials
BigIdeasMath.com

Key Vocabulary
slope, *p. 630*

Study Tip
The slope of a line is the same between any two points on the line because lines have a *constant* rate of change.

Key Idea

Slope

Slope is the rate of change between any two points on a line. It is a measure of the *steepness* of a line.

To find the slope of a line, find the ratio of the change in y (vertical change) to the change in x (horizontal change).

$$\text{slope} = \frac{\text{change in } y}{\text{change in } x}$$

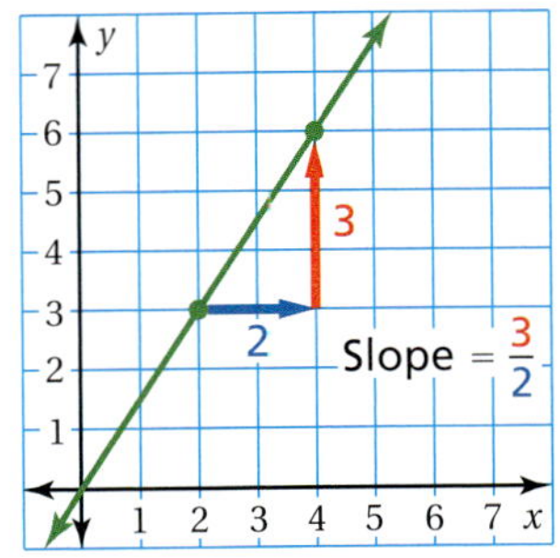

EXAMPLE 1 Finding Slopes

Find the slope of each line.

a.

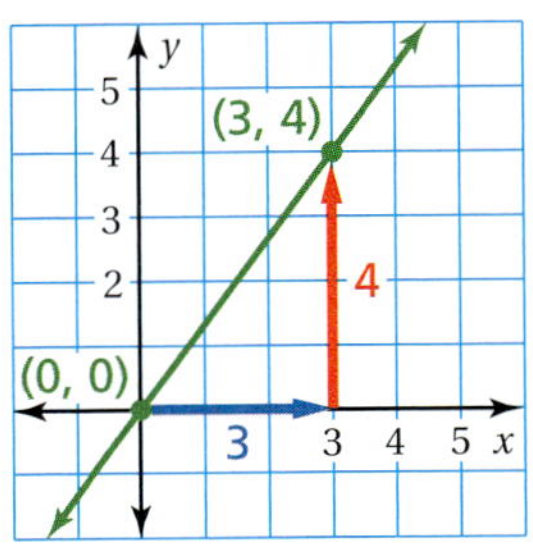

$$\text{slope} = \frac{\text{change in } y}{\text{change in } x}$$

$$= \frac{4}{3}$$

The slope of the line is $\frac{4}{3}$.

b.

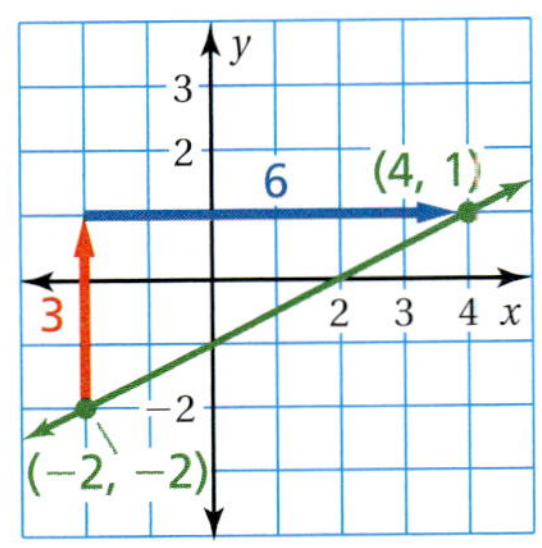

$$\text{slope} = \frac{\text{change in } y}{\text{change in } x}$$

$$= \frac{3}{6} = \frac{1}{2}$$

The slope of the line is $\frac{1}{2}$.

On Your Own

Now You're Ready
Exercises 4–9

Find the slope of the line.

1.

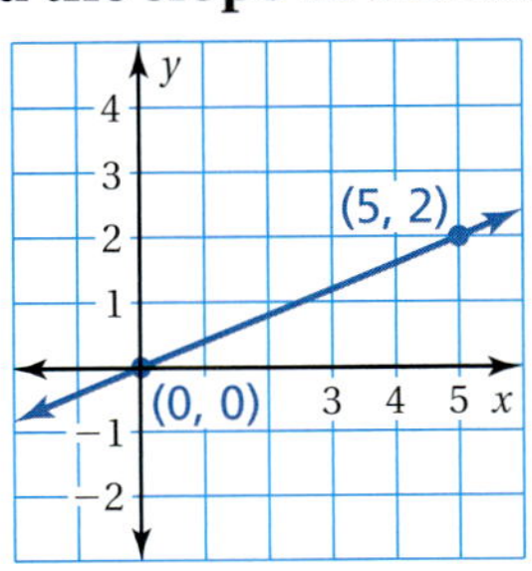

2.

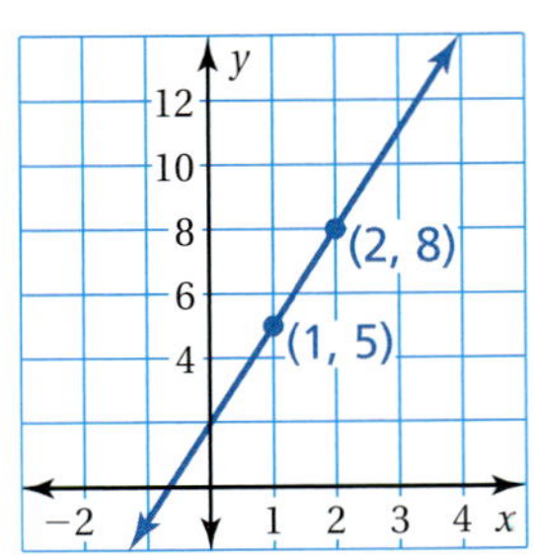

Multi-Language Glossary at BigIdeasMath.com

Laurie's Notes

Introduction

Connect

- **Yesterday:** Students explored constant rates of speed in a table and on a graph.
- **Today:** Students will define slope and determine the slope of a line from its graph.

Motivate

- Ask students about the slope of a half-pipe at a skateboard park, the slope of a local hill that they are familiar with, or the slope of a wheelchair ramp at school. (Chances are they will all have different slopes.)
- ? "What does the word slope mean when talking about a wheelchair ramp?" Listen for words such as "steepness" or "incline." Students often use their hands to demonstrate slope.

Lesson Notes

Key Idea

- Write the definition of slope on the board.
- Remind students that a rate is a ratio. Slopes are often thought of as rates.
- **FYI:** You may remember learning that the formula for the slope of a line passing through the points (x_1, y_1) and (x_2, y_2) is $\frac{y_2 - y_1}{x_2 - x_1}$. Students are not formally introduced to this formula until future courses. Try to enforce the concept rather than teach the formula.

Example 1

- **Use Appropriate Tools Strategically:** The arrows on the graphs are good visual aids to help students think about how much each variable changes.
- Writing the amount of change on the graph is good reinforcement.
- At this stage, have them think about reading the graph left-to-right.
- ? "For the second graph, what would the slope be if you used the points (0, −1) and (4, 1)?" the same, $\frac{2}{4} = \frac{1}{2}$
- **Common Error:** A very common error is for students to find slope by writing the change in x over the change in y.

On Your Own

- **Neighbor Check:** Have students work independently and then have their neighbors check their work. Have students discuss any discrepancies.

Goal Today's lesson is determining the **slope** of a line, when given a graph.

Technology for the Teacher

Dynamic Classroom

Lesson Tutorials
Lesson Plans
Answer Presentation Tool

Extra Example 1

Find the slope of each line.

a.

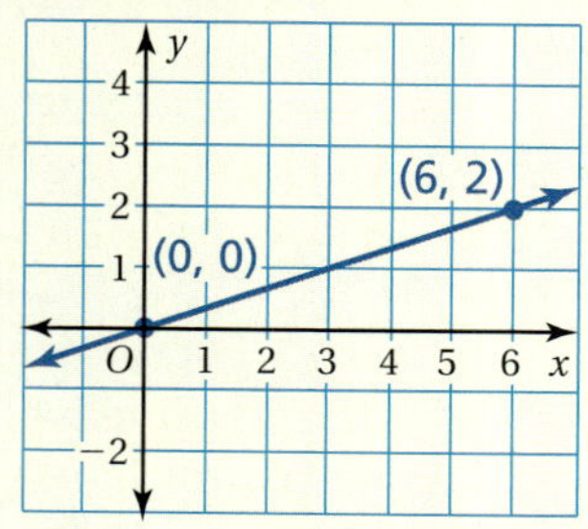

$\frac{1}{3}$

b.

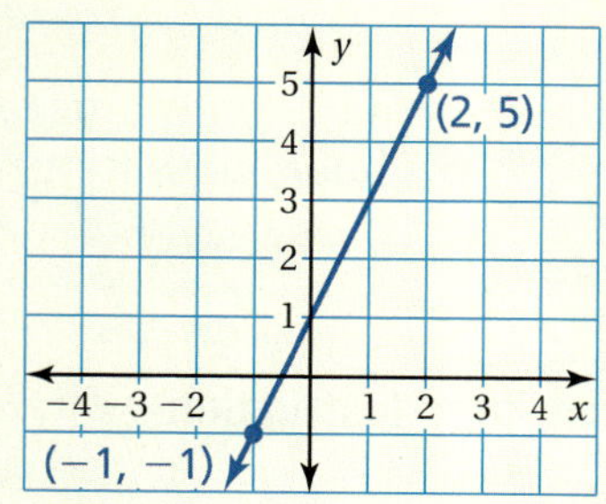

2

On Your Own

1. $\frac{2}{5}$
2. 3

Extra Example 2

The table shows your earnings for mowing lawns.

Lawns Mowed, x	3	6	9	12
Earnings, y (dollars)	45	90	135	180

a. Graph the data.

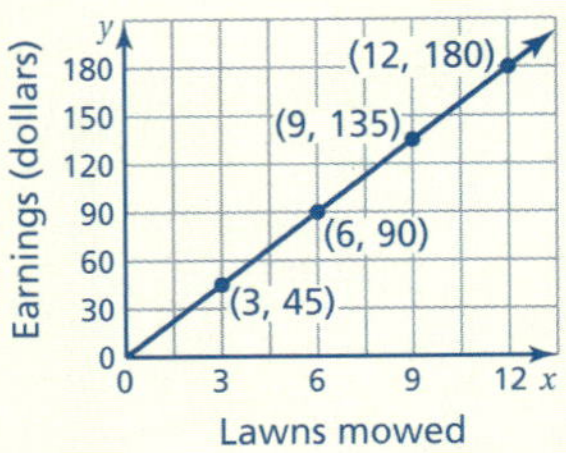

b. Find and interpret the slope of the line through the points. 15; You earn \$15 per lawn mowed.

On Your Own

3. 5; No

4. a. Your friend's line is steeper. Your friend's pay rate is greater than yours.

b. 7; Your friend earns \$7 per hour babysitting.

Differentiated Instruction

Auditory

Emphasize that slope relates to rate problems. The slope formula expresses how much the y-coordinate changes for a given change in the x-coordinate. For example, weighing out 3 pounds of apples and paying \$4.05 is equivalent to the unit rate of \$1.35 per pound. So, the change in the cost is \$1.35 for every one pound increase of apples.

Laurie's Notes

Example 2

- Students must first read and understand the data (table of ordered pairs), and then plot the data.
- Start to use language that will prepare students for future courses.
- The number of hours worked is x, and the amount of dollars earned is y. "The amount of money earned depends upon how many hours you worked."
- **Construct Viable Arguments and Critique the Reasoning of Others:** Slopes are often rates. Here, the rate is *dollars per hour.* Be sure to have students explain the meaning of this concept. Listen for an understanding of the connection between *slope* and *unit rate*.
- **Big Idea:** Students will often be asked to interpret a slope. This means to look at the context of the problem and decide what a slope means with regard to the two variables.

On Your Own

- Share answers as a class.
- **Extension:** "Read information from the graph. How many hours did you work to earn \$30?" 6

Closure

- The cost of admission to the local museum is given in the table. Graph the data and determine the slope. Interpret the slope in the context of the problem.

Number of People, x	2	3	4
Total Cost of Admission, y	\$4.50	\$6.75	\$9.00

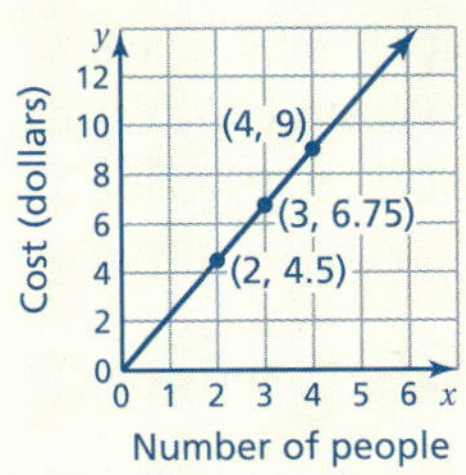

slope: 2.25; The slope represents the cost per person.

EXAMPLE 2 Interpreting a Slope

The table shows your earnings for babysitting.

a. Graph the data.

b. Find and interpret the slope of the line through the points.

Hours, x	0	2	4	6	8	10
Earnings, y (dollars)	0	10	20	30	40	50

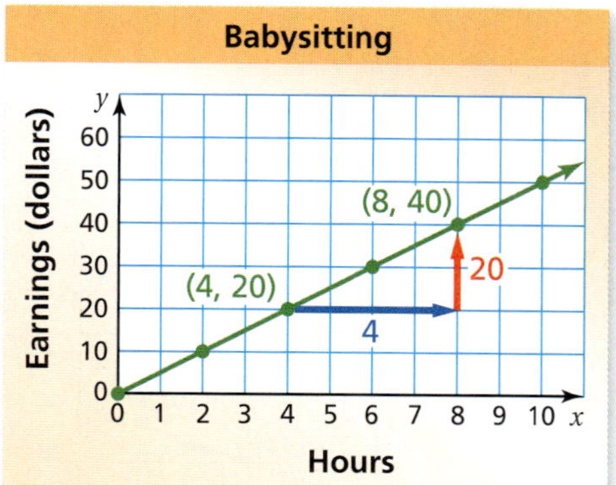

a. Graph the data. Draw a line through the points.

b. Choose any two points to find the slope of the line.

$$\text{slope} = \frac{\text{change in } y}{\text{change in } x}$$

$$= \frac{20}{4} \quad \begin{matrix}\leftarrow \text{dollars} \\ \leftarrow \text{hours}\end{matrix}$$

$$= 5$$

∴ The slope of the line represents the unit rate. The slope is 5. So, you earn $5 per hour babysitting.

On Your Own

Now You're Ready
Exercises 10 and 11

3. In Example 2, use two other points to find the slope. Does the slope change?

4. The graph shows the amounts you and your friend earn babysitting.

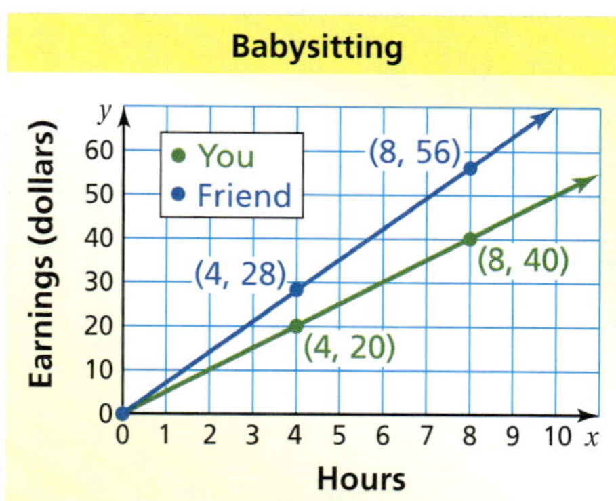

a. Compare the steepness of the lines. What does this mean in the context of the problem?

b. Find and interpret the slope of the blue line.

14.5 Exercises

Check It Out
Help with Homework
BigIdeasMath.com

Vocabulary and Concept Check

1. **VOCABULARY** Is there a connection between rate and slope? Explain.

2. **REASONING** Which line has the greatest slope?

3. **REASONING** Is it more difficult to run up a ramp with a slope of $\frac{1}{5}$ or a ramp with a slope of 5? Explain.

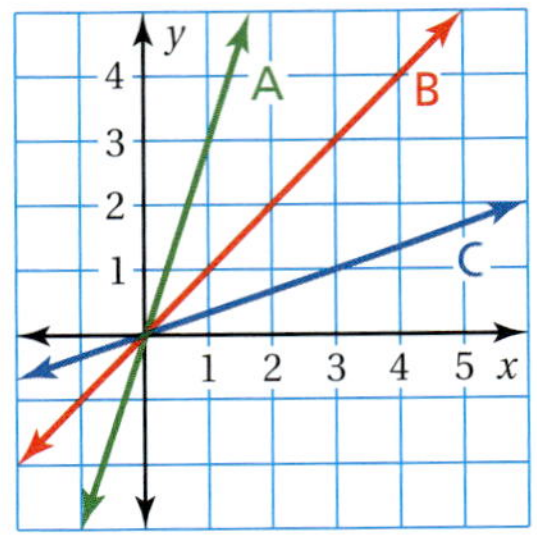

Practice and Problem Solving

Find the slope of the line.

1 4.

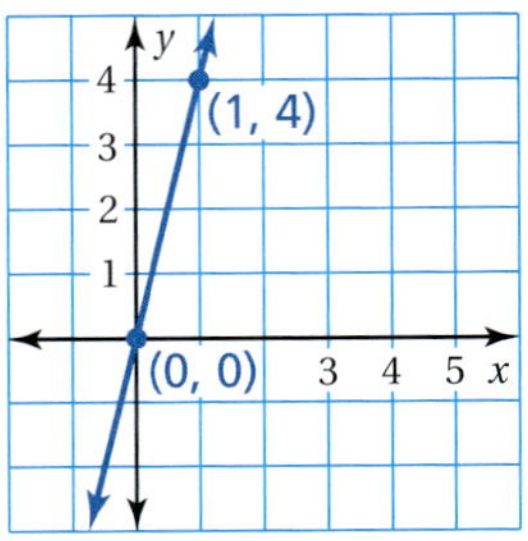

5.

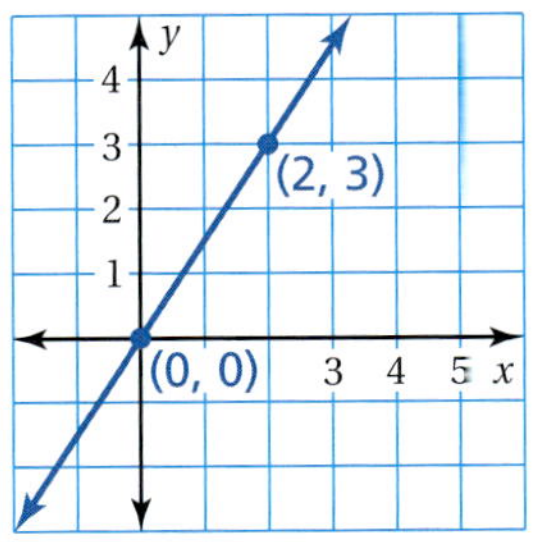

6.

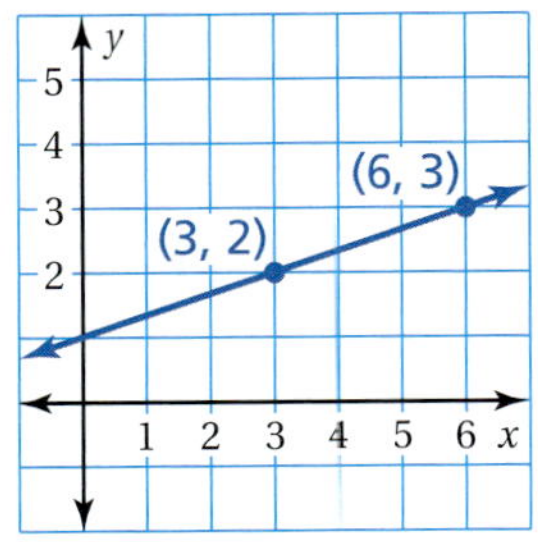

7.

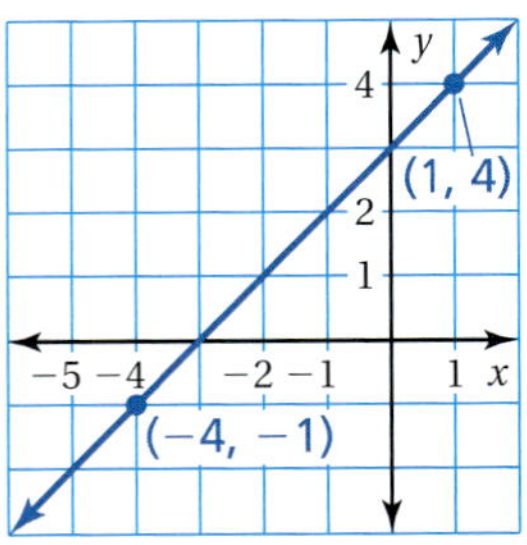

8.

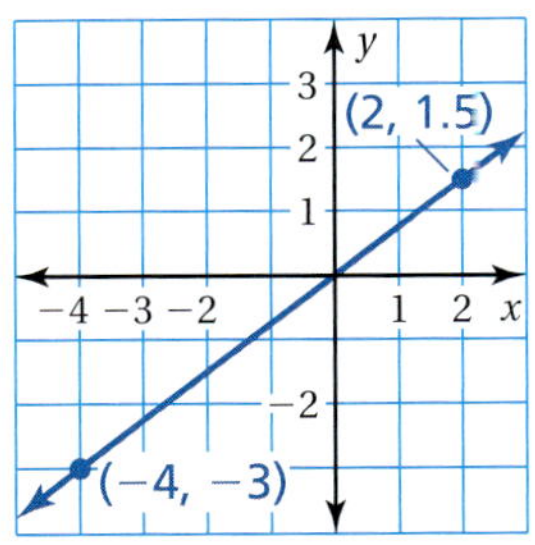

9.

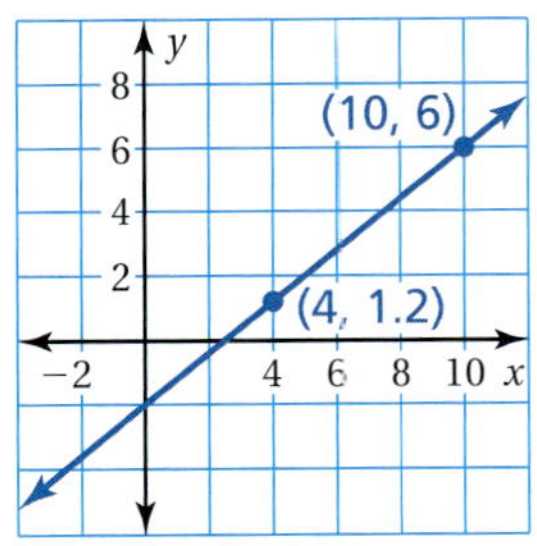

Graph the data. Then find and interpret the slope of the line through the points.

2 10.

Minutes, x	3	5	7	9
Words, y	135	225	315	405

11.

Gallons, x	5	10	15	20
Miles, y	162.5	325	487.5	650

12. **ERROR ANALYSIS** Describe and correct the error in finding the slope of the line passing through (0, 0) and (4, 5).

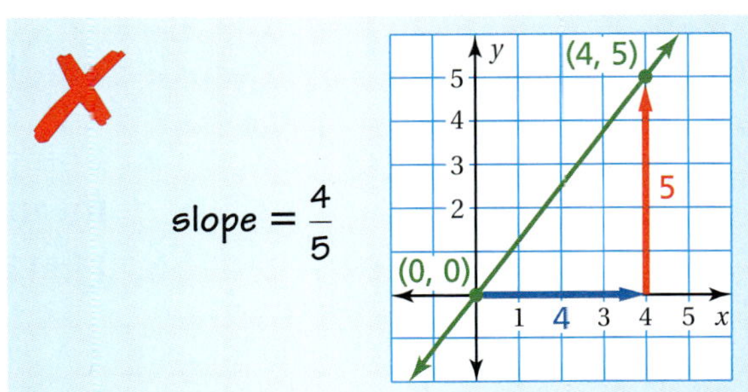

Assignment Guide and Homework Check

Level	Assignment	Homework Check
Advanced	1–3, 4–18 even, 20–23	4, 10, 16, 18

Common Errors

- **Exercises 4–11** Students may put the change in *x* over the change in *y*. Remind them that the vertical change is written over the horizontal change and that slope is a rate. Tell the students to label the axes with units that represent a common rate to help them remember which change goes on top. For example, label the *y*-axis "miles" and the *x*-axis "gallons." This should help students identify that the change in *y* goes on top because miles is first in the rate.
- **Exercises 13–15** Students may not remember how to plot ordered pairs with negative numbers. Remind students of Quadrants II, III, and IV, and the signs of the coordinates in each quadrant.

14.5 Record and Practice Journal

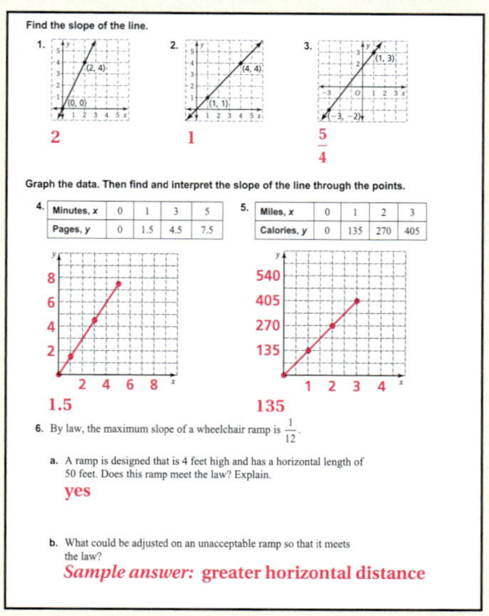
Find the slope of the line.

1. 2
2. 1
3. $\frac{5}{4}$

Graph the data. Then find and interpret the slope of the line through the points.

4.

Minutes, x	0	1	3	5
Pages, y	0	1.5	4.5	7.5

1.5

5.

Miles, x	0	1	2	3
Calories, y	0	135	270	405

135

6. By law, the maximum slope of a wheelchair ramp is $\frac{1}{12}$.

a. A ramp is designed that is 4 feet high and has a horizontal length of 50 feet. Does this ramp meet the law? Explain.
yes

b. What could be adjusted on an unacceptable ramp so that it meets the law?
Sample answer: greater horizontal distance

Vocabulary and Concept Check

1. yes; Slope is the rate of change of a line.
2. A
3. 5; A ramp with a slope of 5 increases 5 units vertically for every 1 unit horizontally. A ramp with a slope of $\frac{1}{5}$ increases 1 unit vertically for every 5 units horizontally.

Practice and Problem Solving

4. 4
5. $\frac{3}{2}$
6. $\frac{1}{3}$
7. 1
8. $\frac{3}{4}$
9. $\frac{4}{5}$
10. 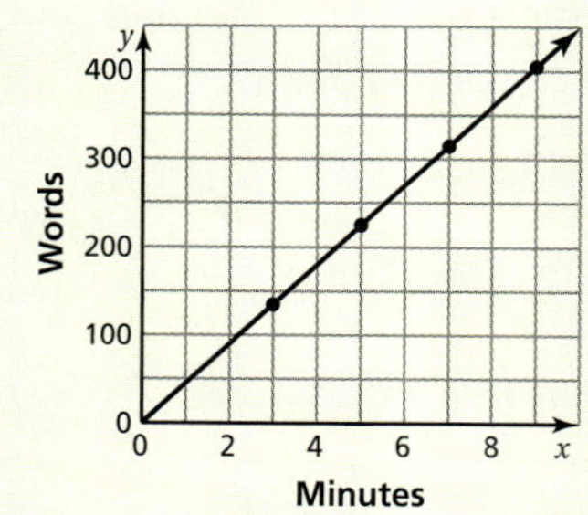

slope = 45; 45 words per minute
11. 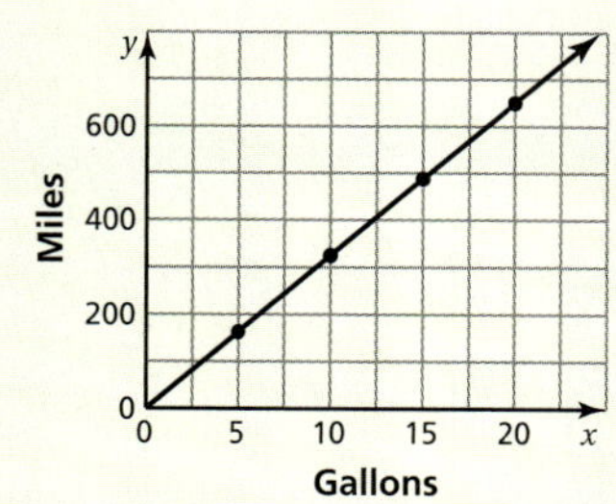

slope = 32.5; 32.5 miles per gallon
12. See Additional Answers.

Practice and Problem Solving

13–15. See Additional Answers.

16. See *Taking Math Deeper.*

17. See Additional Answers.

18. 0; The change in y is 0 because the y-values do not change. So, the slope is 0.

19. $y = 6$

Fair Game Review

20. $-\frac{4}{5}$ **21.** $-\frac{3}{5}$

22. 3 **23.** C

Mini-Assessment

Find the slope of the line that passes through the two points.

1. (0, 0), (3, 2) $\frac{2}{3}$
2. (−2, −2), (5, 5) 1
3. (−3, −4), (6, 8) $\frac{4}{3}$
4. (−4, −2), (−2, −1) $\frac{1}{2}$
5. The graph shows the amount of money you are saving for a computer.

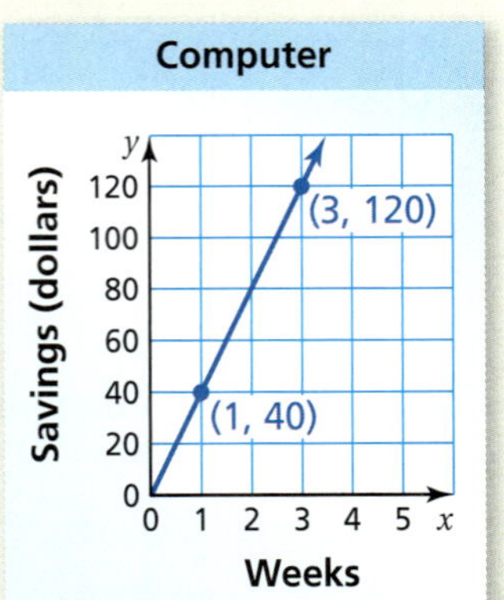

a. Find the slope of the line. 40

b. Interpret the slope of the line in the context of the problem. You are saving $40 per week.

Taking Math Deeper

Exercise 16

This is a classic type of problem that uses linear models to predict future events. Each person is saving money at a constant rate (constant slope). The fact that the rate is constant is what makes the graph a line. The prediction of when $165 will be saved assumes that the constant rate continues into the future.

1 Interpret the slope in context.

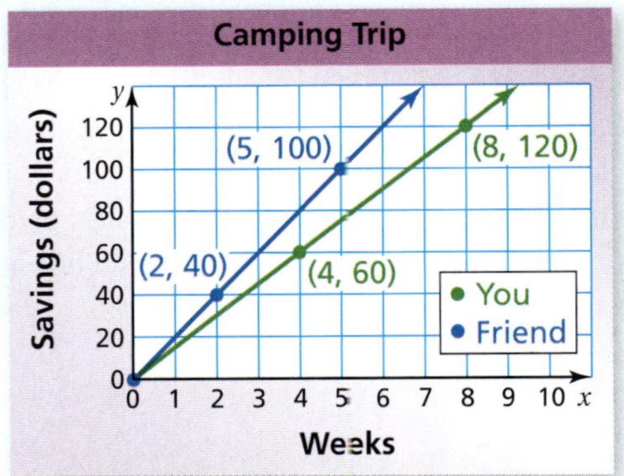

a. Your friend's graph is steeper than yours. So, your friend's saving rate (in dollars per week) is greater than yours.

2 Find the slope of each line.

b. Slopes Rates

You: $\frac{60}{4}$ = $15 per week

Friend: $\frac{60}{3}$ = $20 per week

c. Your friend saves $20 − $15 = $5 more per week.

3 How long will it take for you to save $165?

d. At $15 per week, it will take $\frac{165}{15} = 11$ weeks.

Reteaching and Enrichment Strategies

If students need help. . .	If students got it. . .
Resources by Chapter • Practice A and Practice B • Puzzle Time Record and Practice Journal Practice Differentiating the Lesson Lesson Tutorials Skills Review Handbook	Resources by Chapter • Enrichment and Extension • Technology Connection Start the next section

Graph the line that passes through the two points. Then find the slope of the line.

13. $(0, 0), \left(\frac{1}{3}, \frac{7}{3}\right)$

14. $\left(-\frac{3}{2}, -\frac{3}{2}\right), \left(\frac{3}{2}, \frac{3}{2}\right)$

15. $\left(1, \frac{5}{2}\right), \left(-\frac{1}{2}, -\frac{1}{4}\right)$

16. **CAMPING** The graph shows the amount of money you and a friend are saving for a camping trip.

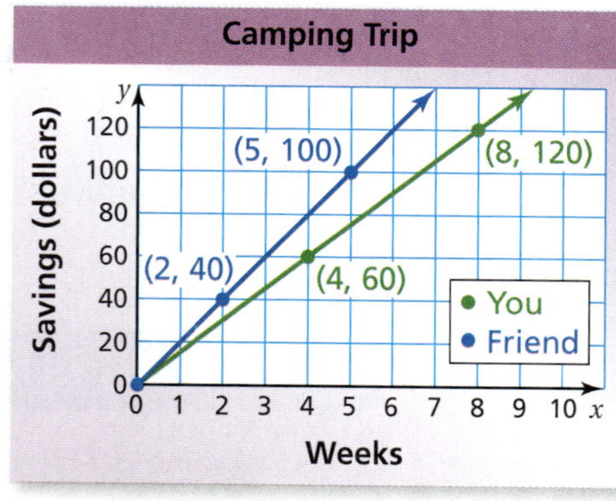

a. Compare the steepness of the lines. What does this mean in the context of the problem?

b. Find the slope of each line.

c. How much more money does your friend save each week than you?

d. The camping trip costs $165. How long will it take you to save enough money?

17. **MAPS** An atlas contains a map of Ohio. The table shows data from the key on the map.

Distance on Map (mm), x	10	20	30	40
Actual Distance (mi), y	25	50	75	100

a. Graph the data.

b. Find the slope of the line. What does this mean in the context of the problem?

c. The map distance between Toledo and Columbus is 48 millimeters. What is the actual distance?

d. Cincinnati is about 225 miles from Cleveland. What is the distance between these cities on the map?

18. **CRITICAL THINKING** What is the slope of a line that passes through the points (2, 0) and (5, 0)? Explain.

19. Number Sense A line has a slope of 2. It passes through the points (1, 2) and $(3, y)$. What is the value of y?

Fair Game Review What you learned in previous grades & lessons

Multiply. *(Section 12.4)*

20. $-\frac{3}{5} \times \frac{8}{6}$

21. $1\frac{1}{2} \times \left(-\frac{6}{15}\right)$

22. $-2\frac{1}{4} \times \left(-1\frac{1}{3}\right)$

23. **MULTIPLE CHOICE** You have 18 stamps from Mexico in your stamp collection. These stamps represent $\frac{3}{8}$ of your collection. The rest of the stamps are from the United States. How many stamps are from the United States? *(Section 13.4)*

Ⓐ 12 Ⓑ 24 Ⓒ 30 Ⓓ 48

14.6 Direct Variation

Essential Question How can you use a graph to show the relationship between two quantities that vary directly? How can you use an equation?

1 ACTIVITY: Math in Literature

Gulliver's Travels was written by Jonathan Swift and published in 1726. Gulliver was shipwrecked on the island Lilliput, where the people were only 6 inches tall. When the Lilliputians decided to make a shirt for Gulliver, a Lilliputian tailor stated that he could determine Gulliver's measurements by simply measuring the distance around Gulliver's thumb. He said "Twice around the thumb equals once around the wrist. Twice around the wrist is once around the neck. Twice around the neck is once around the waist."

Work with a partner. Use the tailor's statement to complete the table.

Thumb, t	Wrist, w	Neck, n	Waist, x
0 in.			
1 in.			
	4 in.		
		12 in.	
			32 in.
	10 in.		

Direct Variation

In this lesson, you will

- identify direct variation from graphs or equations.
- use direct variation models to solve problems.

Laurie's Notes

Introduction

Applying Mathematical Practices

- **Model with Mathematics:** In this lesson, students will use a graph as a tool to show that two quantities vary directly.

Motivate

- Write the following table.

	Paper	Pencil	$1 Bill	Stick of Gum
Length (inches)	11	8	6	?
Length (centimeters)	27.94	20.32	?	6

? "Does anyone have an idea of how to find the missing values in the table without measuring?" number of inches $\times$ 2.54 = number of centimeters

- This question is checking to see if students remember the relationship between inches and centimeters. It is also a perfect example for direct variation.

Activity Notes

Discuss

- **Big Idea:** When two quantities *vary directly*, the ratio of one quantity to another is a *constant*. The term *direct variation* will be introduced in tomorrow's lesson.
- The ordered pair (0, 0) is always a solution of an equation describing two quantities that vary directly.
- When solutions of the equation are plotted, the *constant ratio* is the slope of the line.
- When two quantities vary directly, it can also be said that they are directly proportional.

Activity 1

- Ask a student to read the introduction.

? "Have any of you read *Gulliver's Travels*?" Give students an opportunity to share information about the story if they have read it.

- Students should work with partners to complete the table.

? "Are there any patterns in the table? Describe the patterns." **Construct Viable Arguments and Critique the Reasoning of Others:** There are many patterns! Listen for patterns in a single column, between adjacent columns, and between non-adjacent columns. Students might also mention the only odd numbers in the table appear in the first column. Ask students whether they can explain each pattern observed. In many cases, they will refer to "twice around the thumb equals twice around the wrist . . .."

What Your Students Will Learn

- Identify direct variation from tables, graphs, or equations.
- Use direct variation models to solve problems.

Previous Learning

Students have used a variable to write an expression and have plotted points in the coordinate plane. Students have also written equations in two variables.

Technology for the Teacher

Dynamic Classroom

Lesson Plans
Complete Materials List

14.6 Record and Practice Journal

Essential Question How can you use a graph to show the relationship between two quantities that vary directly? How can you use an equation?

1 ACTIVITY: Math in Literature

Gulliver's Travels was written by Jonathan Swift and published in 1726. Gulliver was shipwrecked on an island in Lilliput, where the people were only 6 inches tall. When the Lilliputians decided to make a shirt for Gulliver, a Lilliputian tailor stated that he could determine Gulliver's measurements by simply measuring the distance around Gulliver's thumb. He said "Twice around the thumb equals once around the wrist. Twice around the wrist is once around the neck. Twice around the neck is once around the waist."

Work with a partner. Use the tailor's statement to complete the table.

Thumb, t	Wrist, w	Neck, n	Waist, x
0 in.	0 in.	0 in.	0 in.
1 in.	2 in.	4 in.	8 in.
2 in.	4 in.	8 in.	16 in.
3 in.	6 in.	12 in.	24 in.
4 in.	8 in.	16 in.	32 in.
5 in.	10 in.	20 in.	40 in.

Differentiated Instruction

Auditory

A common mistake when plotting points is to confuse the order of the coordinates in the ordered pair. The simple phrase "over and up" will assist the students in moving over the x-axis with the first number, and then moving up with the second number.

14.6 Record and Practice Journal

2 ACTIVITY: Drawing a Graph

Work with a partner. Use the information from Activity 1.

a. In your own words, describe the relationship between t and w.
w is two times *t*.

b. Use the table to write ordered pairs (t, w). Then plot the ordered pairs.
(0, 0), (1, 2), (2, 4), (3, 6), (4, 8), (5, 10)

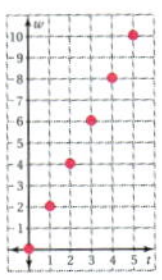

c. What do you notice about the graph of the ordered pairs?
All the points lie on a line;
The line passes through the origin.

d. Choose two points and find the slope of the line between them.
2

e. The quantities t and w are said to *vary directly*. An equation that describes the relationship is $w =$ __2__ t.

3 ACTIVITY: Drawing a Graph and Writing an Equation

Work with a partner. Use the information from Activity 1 to draw a graph of the relationship. Write an equation that describes the relationship between the two quantities.

a. Thumb t and neck n

$n = \boxed{4}\, t$

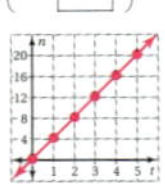

b. Wrist w and waist x

$x = \boxed{4}\, w$

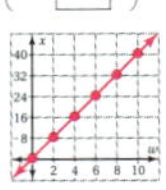

c. Wrist w and thumb t

$t = \boxed{\frac{1}{2}}\, w$

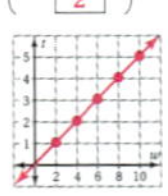

d. Waist x and wrist w

$w = \boxed{\frac{1}{4}}\, x$

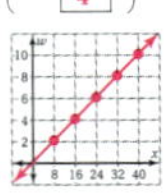

What Is Your Answer?

4. **IN YOUR OWN WORDS** How can you use a graph to show the relationship between two quantities that vary directly? How can you use an equation?
Check students' work.

5. **STRUCTURE** How are all the graphs in Activity 3 alike?
All the points in each graph lie on a line.
All the lines pass through the origin.

6. Give a real-life example of two variables that vary directly.
Sample answer: the distance and time of a car traveling at a constant rate

7. Work with a partner. Use string to find the distance around your thumb, wrist, and neck. Do your measurements agree with the tailor's statements in *Gulliver's Travels*? Explain your reasoning.
Check students' work.

Laurie's Notes

Activity 2

- Many skills and concepts from this chapter are integrated in this activity.
- When asked to describe the relationship between t and w, students may say "twice around the thumb equals once around the wrist." **Attend to Precision:** Be sure students understand that what is being referred to is the *distance around* the thumb and the wrist.

? "What does the ordered pair (0, 0) mean?" In the context of the problem, it does not make sense to talk about distances around the thumb and the wrist that are equal to 0. Each is getting closer to 0.

- Because the line passes through (0, 0) and (1, 2), some students might quickly note that the slope is 2, remembering that in the graph of a proportional relationship, you can find the unit rate y from the point $(1, y)$.

? What is the relationship between the coefficient of t in part (e) and the slope? They are the same.

Activity 3

- **Scaffolding:** If time is short, you may need to scaffold this activity. Have different groups do a different problem and record a sample of each on the board.
- **Common Error:** Students may reverse the ordered pairs (wrist length, waist length) versus (waist length, wrist length).
- Remind students that as the problems are written, the first quantity mentioned is the first coordinate.
- They should do a test point to see if it satisfies the equation they wrote.
- **Connection:** There are two pairs of equations that share a pattern: $w = 2t$ and $t = \frac{1}{2}w$; $x = 4w$ and $w = \frac{1}{4}x$. The slopes are reciprocals.

What Is Your Answer?

- Question 6: Students are sometimes stumped by this question, yet it is very common. For instance, the cost of an item varies directly with how many items are purchased (1 newspaper, \$1.25), (2 newspapers, \$2.50), and so on.
- Question 7: Leave sufficient time for this problem.

Closure

- **Exit Ticket:** Write an equation that describes the relationship between the length of objects measured in inches x and measured in centimeters y.
$y = 2.54x$

2 ACTIVITY: Drawing a Graph

Work with a partner. Use the information from Activity 1.

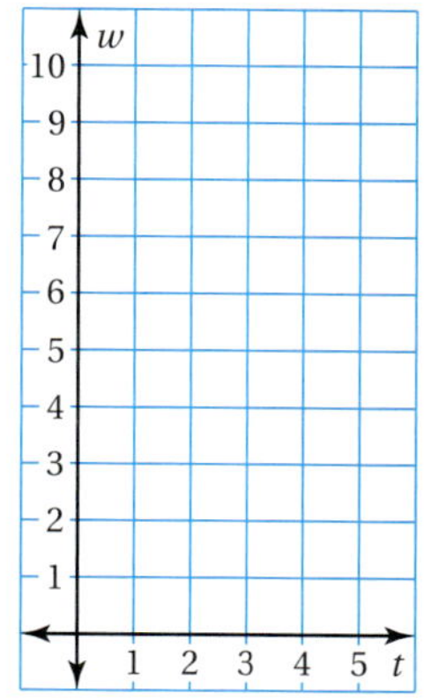

a. In your own words, describe the relationship between t and w.

b. Use the table to write the ordered pairs (t, w). Then plot the ordered pairs.

c. What do you notice about the graph of the ordered pairs?

d. Choose two points and find the slope of the line between them.

e. The quantities t and w are said to *vary directly*. An equation that describes the relationship is

$$w = \square\, t.$$

3 ACTIVITY: Drawing a Graph and Writing an Equation

Math Practice

Label Axes

How do you know which labels to use for the axes? Explain.

Work with a partner. Use the information from Activity 1 to draw a graph of the relationship. Write an equation that describes the relationship between the two quantities.

a. Thumb t and neck n $(n = \square\, t)$

b. Wrist w and waist x $(x = \square\, w)$

c. Wrist w and thumb t $(t = \square\, w)$

d. Waist x and wrist w $(w = \square\, x)$

What Is Your Answer?

4. **IN YOUR OWN WORDS** How can you use a graph to show the relationship between two quantities that vary directly? How can you use an equation?

5. **STRUCTURE** How are all the graphs in Activity 3 alike?

6. Give a real-life example of two variables that vary directly.

7. Work with a partner. Use string to find the distance around your thumb, wrist, and neck. Do your measurements agree with the tailor's statement in *Gulliver's Travels*? Explain your reasoning.

Practice Use what you learned about quantities that vary directly to complete Exercises 4 and 5 on page 638.

14.6 Lesson

Check It Out Lesson Tutorials BigIdeasMath.com

Key Vocabulary

direct variation, *p. 636*

constant of proportionality, *p. 636*

Key Idea

Direct Variation

Words Two quantities x and y show **direct variation** when $y = kx$, where k is a number and $k \neq 0$. The number k is called the **constant of proportionality**.

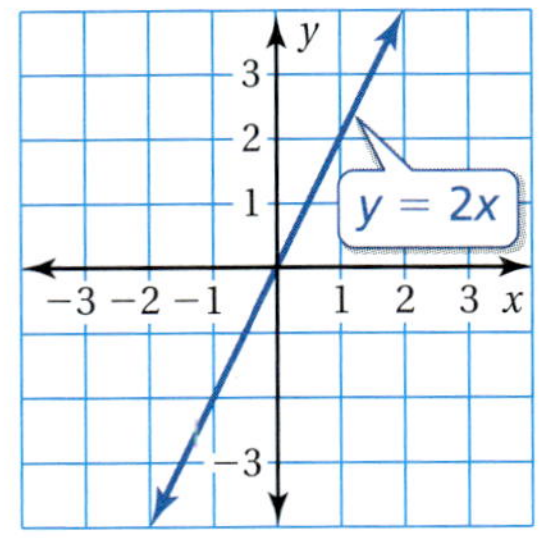

Graph The graph of $y = kx$ is a line with a slope of k that passes through the origin. So, two quantities that show direct variation are in a proportional relationship.

EXAMPLE 1 Identifying Direct Variation

Tell whether x and y show direct variation. Explain your reasoning.

a.

x	1	2	3	4
y	−2	0	2	4

Plot the points. Draw a line through the points.

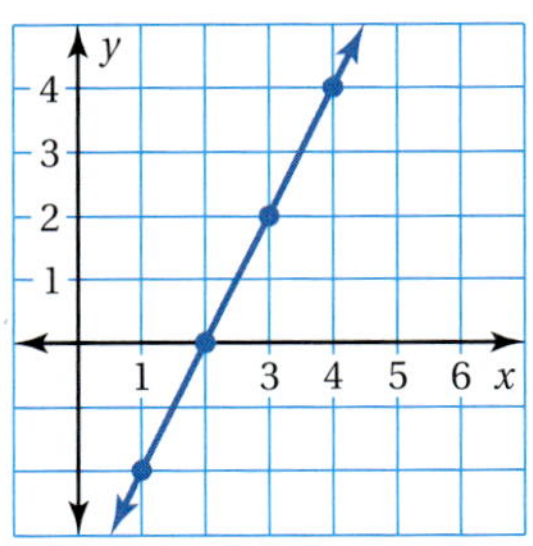

The line does *not* pass through the origin. So, x and y do *not* show direct variation.

b.

x	0	2	4	6
y	0	2	4	6

Plot the points. Draw a line through the points.

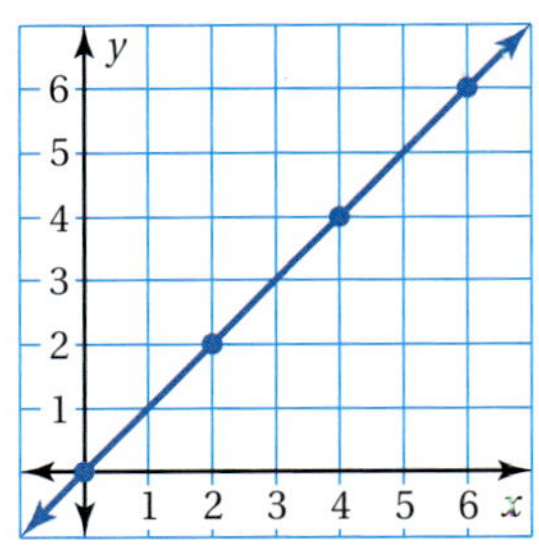

The line passes through the origin. So, x and y show direct variation.

Study Tip

Other ways to say that x and y show direct variation are "y varies directly with x" and "x and y are directly proportional."

EXAMPLE 2 Identifying Direct Variation

Tell whether x and y show direct variation. Explain your reasoning.

a. $y + 1 = 2x$

$y = 2x - 1$ Solve for y.

The equation *cannot* be written as $y = kx$. So, x and y do *not* show direct variation.

b. $\frac{1}{2}y = x$

$y = 2x$ Solve for y.

The equation can be written as $y = kx$. So, x and y show direct variation.

Multi-Language Glossary at BigIdeasMath.com

Laurie's Notes

Introduction

Connect

- **Yesterday:** Students completed a table of values, plotted ordered pairs, and wrote equations as they developed an understanding of quantities that vary directly.
- **Today:** Students will use a formal definition of direct variation.

Motivate

? Some states have returnable bottle laws. "If you receive \$0.05 for each bottle, how much money do you receive for 4 bottles? 10 bottles?" \$0.20; \$0.50

- Have students make a table to show the relationship between the number x of bottles collected and the amount y of money received.
- Then have students make a quick sketch of the ordered pairs.
- Observe that (0, 0) is on the graph and the ordered pairs lie on a line.

Lesson Notes

Key Idea

- Write the Key Idea on the board. Remind students throughout the lesson that k is the **constant of proportionality**, and it is also the slope of the line.
- The equation $y = kx$ can be confusing to students. They see three variables. Remind them that this is the *general form*. The variables y and x will remain in the final equation but k will be replaced by a number.
- Examples of equations in general form include $y = 2x$ and $y = 1.25x$. Point out to students that k has been replaced with a number.

? "Why should $k \neq 0$?" If k did equal 0, the resulting equation would be $y = 0 \cdot x$ or $y = 0$. While k has been replaced with a number and y still appears in the final equation, x no longer does. Because there is no x, the equation is not in general form and does not show direct variation.

- Mention a key feature of this graph: it passes through the origin.

Example 1

- **Common Error:** Students may look at the table of values and believe that the first example is not direct variation simply because x and y are increasing by different rates in the table. x is increasing by 1 and y is increasing by 2.
- **Connection:** When two variables vary directly, you can also say that they vary proportionally.
- **Extension:** Ask students to find the slope of the line in part (b).

Example 2

- This example requires students to recall equations in two variables.
- Students need to think about how they solved equations using the Properties of Equality.
- **Extension:** Ask students whether the point (0, 0) satisfies either equation.

Goal Today's lesson is using a formal definition of **direct variation**.

Technology for the Teacher

Dynamic Classroom

Lesson Tutorials
Lesson Plans
Answer Presentation Tool

Extra Example 1

Tell whether x and y show direct variation. Explain your reasoning.

a.

x	1	2	3	4
y	−1	0	1	2

The line does not pass through the origin. So, x and y do not show direct variation.

b.

x	0	3	6	9
y	0	1	2	3

The line passes through the origin. So, x and y show direct variation.

Extra Example 2

Tell whether x and y show direct variation. Explain your reasoning.

a. $y - 6 = 3x$

The equation cannot be written as $y = kx$. So, x and y do not show direct variation.

b. $x = 4y$

The equation can be written as $y = kx$. So, x and y show direct variation.

On Your Own

1. no; The line does not pass through the origin.
2. yes; The line passes through the origin.
3. no; The points do not lie on a line.

4–6. See Additional Answers.

Extra Example 3

The table shows the area *y* (in square feet) that a power paint sprayer can paint in *x* minutes.

x	$\frac{1}{2}$	1	$\frac{3}{2}$	2
y	10	20	30	40

a. Graph the data. Tell whether x and y are directly proportional.

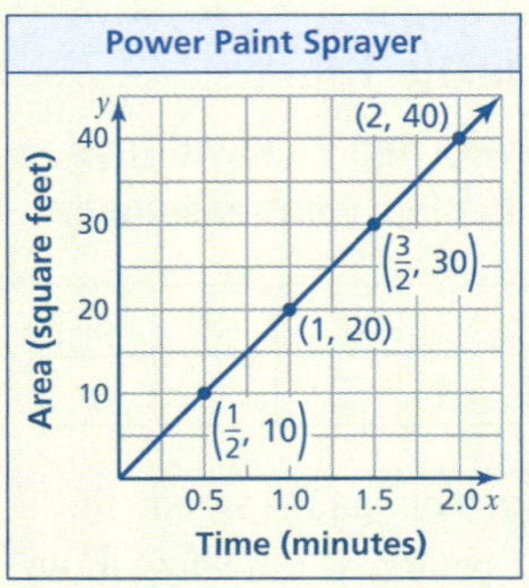

yes

b. Write an equation that represents the line. $y = 20x$

c. Use the equation to find the area painted in 10 minutes. 200 square feet

On Your Own

7. no; There is not a constant rate of change.

English Language Learners

Vocabulary

Students may confuse the words *variation* and *variable*. A variable is a number that changes and is represented by a letter. Stress that variation refers to how the variable *y varies* in relation to the variable *x*.

Laurie's Notes

On Your Own

- **Think-Pair-Share:** Students should read each question independently and then work in pairs to answer the questions. When they have answered the questions, the pair should compare their answers with another group and discuss any discrepancies.
- If students have difficulty with Questions 4–6, they could make a quick table of values and plot the ordered pairs.

Example 3

- Ask students to interpret an ordered pair in the table. For instance, $\left(\frac{1}{2}, 8\right)$ means that in $\frac{1}{2}$ minute, the robotic vacuum can clean 8 square feet.
- **Model with Mathematics:** Ask students to explain why the table of values is a ratio table, and to identify the unit rate. Plotting the ordered pairs confirms that x and y are directly proportional.

? "What is the constant of proportionality?" 16

? "What is the slope of the line?" 16

? "What is the equation of the line?" $y = 16x$

- Students can use the equation to find the area cleaned for any amount of time.

On Your Own

- **Neighbor Check:** Have students work independently and then have their neighbors check their work. Have students discuss any discrepancies.

Closure

- Tell whether x and y show direct variation. Explain your reasoning.

x	y
1	0
3	2
5	4
7	6

no; The line passes through (1, 0), so it does not pass through the origin.

On Your Own

Now You're Ready
Exercises 6–17

Tell whether x and y show direct variation. Explain your reasoning.

1.

x	y
0	−2
1	1
2	4
3	7

2.

x	y
1	4
2	8
3	12
4	16

3.

x	y
−2	4
−1	2
0	0
1	2

4. $xy = 3$

5. $x = \frac{1}{3}y$

6. $y + 1 = x$

EXAMPLE 3 Real-Life Application

x	y
$\frac{1}{2}$	8
1	16
$\frac{3}{2}$	24
2	32

The table shows the area y (in square feet) that a robotic vacuum cleans in x minutes.

a. Graph the data. Tell whether x and y are directly proportional.

Graph the data. Draw a line through the points.

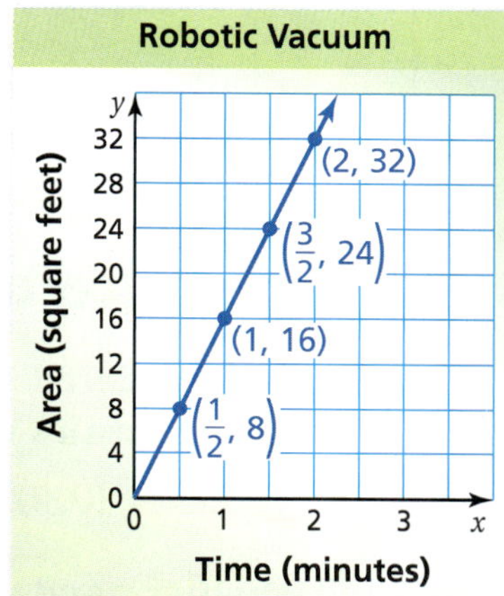

The graph is a line through the origin. So, x and y are directly proportional.

b. Write an equation that represents the line.

Choose any two points to find the slope of the line.

$$\text{slope} = \frac{\text{change in } y}{\text{change in } x} = \frac{16}{1} = 16$$

The slope of the line is the constant of proportionality, k. So, an equation of the line is $y = 16x$.

c. Use the equation to find the area cleaned in 10 minutes.

$y = 16x$	Write the equation.
$= 16(10)$	Substitute 10 for x.
$= 160$	Multiply.

So, the vacuum cleans 160 square feet in 10 minutes.

On Your Own

Now You're Ready
Exercise 19

7. WHAT IF? The battery weakens and the robot begins cleaning less and less area each minute. Do x and y show direct variation? Explain.

14.6 Exercises

Vocabulary and Concept Check

1. **VOCABULARY** What does it mean for x and y to vary directly?
2. **WRITING** What point is on the graph of every direct variation equation?
3. **DIFFERENT WORDS, SAME QUESTION** Which is different? Find "both" answers.

Do x and y show direct variation?

Are x and y in a proportional relationship?

Is the graph of the relationship a line?

Does y vary directly with x?

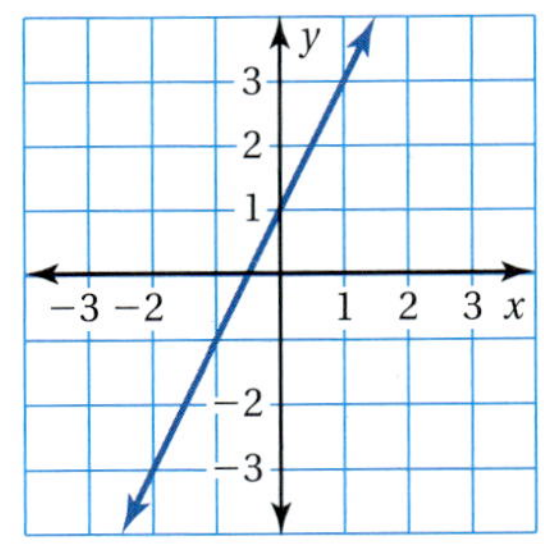

Practice and Problem Solving

Graph the ordered pairs in a coordinate plane. Do you think that graph shows that the quantities vary directly? Explain your reasoning.

4. $(-1, -1), (0, 0), (1, 1), (2, 2)$
5. $(-4, -2), (-2, 0), (0, 2), (2, 4)$

Tell whether x and y show direct variation. Explain your reasoning. If so, find k.

1 6.

x	1	2	3	4
y	2	4	6	8

7.

x	−2	−1	0	1
y	0	2	4	6

8.

x	−1	0	1	2
y	−2	−1	0	1

9.

x	3	6	9	12
y	2	4	6	8

2 10. $y - x = 4$

11. $x = \frac{2}{5}y$

12. $y + 3 = x + 6$

13. $y - 5 = 2x$

14. $x - y = 0$

15. $\frac{x}{y} = 2$

16. $8 = xy$

17. $x^2 = y$

18. **ERROR ANALYSIS** Describe and correct the error in telling whether x and y show direct variation.

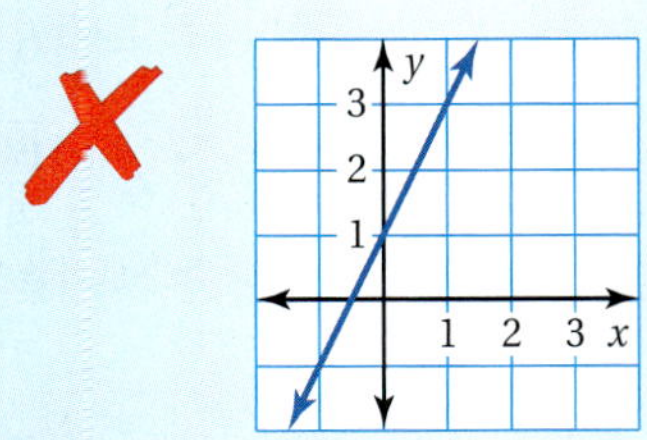

3 19. **RECYCLING** The table shows the profit y for recycling x pounds of aluminum. Graph the data. Tell whether x and y show direct variation. If so, write an equation that represents the line.

Aluminum (lb), x	10	20	30	40
Profit, y	\$4.50	\$9.00	\$13.50	\$18.00

Assignment Guide and Homework Check

Level	Assignment	Homework Check
Advanced	1–5, 8–28 even, 30–34	14, 20, 22, 24, 28

Common Errors

- **Exercises 6–9** Students may immediately state that the table does not show direct variation because (0, 0) is not listed. Encourage them to find the change in x and change in y. Then use that knowledge to go back to $x = 0$ and determine if the table satisfies both requirements for direct variation.
- **Exercises 10–17** Students may try to identify the direct variation equations without solving for y. Remind them to solve for y first.
- **Exercise 17** Students may say that it shows direct variation because the graph goes through (0, 0). Ask them if the graph will be linear.
- **Exercises 20–22** Students may not grasp how to write the equation. Remind them of slope to determine the coefficient of x.

14.6 Record and Practice Journal

Tell whether x and y show direct variation. Explain your reasoning. If so, find k.

1.

x	1	2	3	4
y	3	6	9	12

yes

2.

x	−1	0	1	2
y	1	3	7	13

no

3.

x	0	2	4	6
y	8	5	2	−1

no

4. $y + 2 = x$ — no

5. $3y = x$ — yes

6. $\frac{y}{x} = 4$ — yes

The variables x and y vary directly. Use the values to find the constant of proportionality and write an equation that relates x and y.

7. $y = 8; x = 2$ — $y = 4x$

8. $y = 14, x = 16$ — $y = \frac{7}{8}x$

9. $y = 25, x = 35$ — $y = \frac{5}{7}x$

10. The table shows the cups c of dog food needed to feed a dog that weighs p pounds. Graph the data. Tell whether p and c show direct variation. If so, write an equation that represents the line.

Pounds, p	10	20	40	70
Food, c	$\frac{3}{4}$	$1\frac{1}{4}$	2	$2\frac{3}{4}$

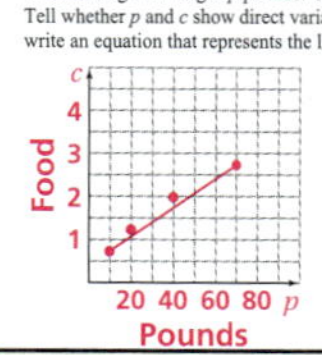

No, p and c do not show direct variation.

Vocabulary and Concept Check

1. $y = kx$, where k is a number and $k \neq 0$.
2. (0, 0)
3. Is the graph of the relationship a line?; yes; no

Practice and Problem Solving

4. 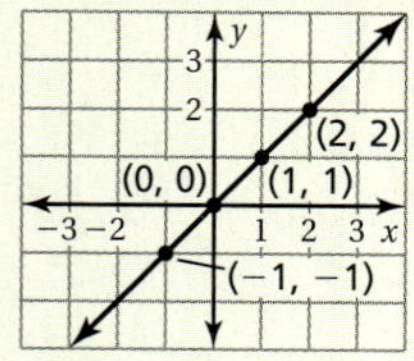

yes; All the points lie on a line and the line passes through the origin.

5. 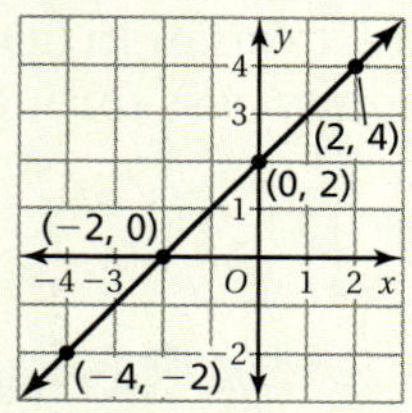

no; The line does not pass through the origin.

6. yes; The line passes through the origin; $k = 2$
7. no; The line does not pass through the origin.
8. no; The line does not pass through the origin.
9. yes; The line passes through the origin; $k = \frac{2}{3}$
10. no; The equation cannot be written as $y = kx$.
11. yes; The equation can be written as $y = kx$; $k = \frac{5}{2}$
12. no; The equation cannot be written as $y = kx$.
13. no; The equation cannot be written as $y = kx$.

14–19. See Additional Answers.

Practice and Problem Solving

20. $k = 24$; $y = 24x$
21. $k = \frac{5}{3}$; $y = \frac{5}{3}x$
22. $k = \frac{9}{8}$; $y = \frac{9}{8}x$
23. $y = 2.54x$
24. See *Taking Math Deeper*.
25. When $x = 0$, $y = 0$. So, the graph of a proportional relationship always passes through the origin.
26. yes; $k = 13$; The cost of 1 ticket is \$13; $y = 13x$; \$182
27. no
28. 76,000 mg
29. Every graph of direct variation is a line; however, not all lines show direct variation because the line must pass through the origin.

Fair Game Review

30. 0.65
31. 0.5625
32. 0.525
33. 0.96
34. D

Mini-Assessment

Tell whether *x* and *y* show direct variation.

1. $y - 3 = 4x$ no
2. $\frac{1}{4}y = x$ yes
3. $x - y = 2$ no
4. $6y + 3 = 12x + 3$ yes
5. One mile is approximately equal to 1.61 kilometers. Write a direct variation equation that relates *x* miles to *y* kilometers. $y = 1.61x$

Taking Math Deeper

Exercise 24

The problem shows students how the coordinate plane can help draw a blueprint.

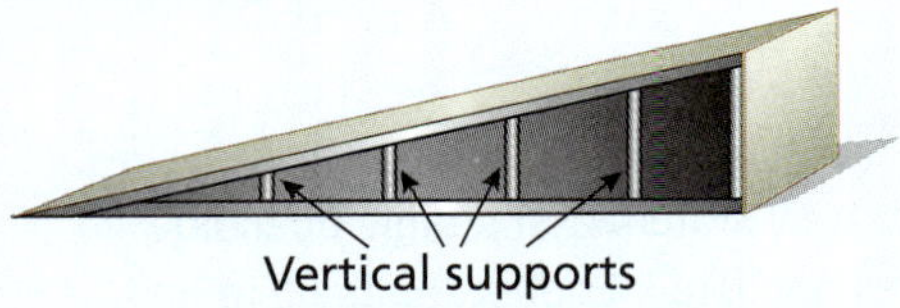

To design the waterskiing ramp, locate the beginning of the ramp at the origin. The horizontal distances are the *x*-values. The vertical distances are the *y*-values.

1 Design the ramp.

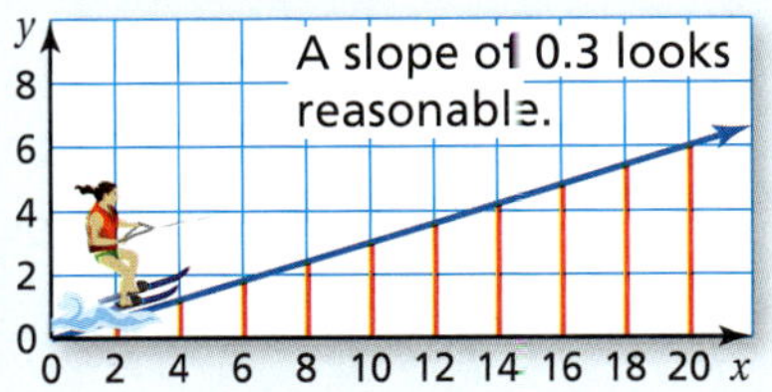

The water skier leaves this ramp at a height of 6 feet. If this seems too high, then redesign the ramp with a slope of 0.2. Then, the final height will be 4 feet.

2 Write a direct variation equation.

For the slope in the graph, $y = 0.3x$.

3 Plan 10 vertical support heights.

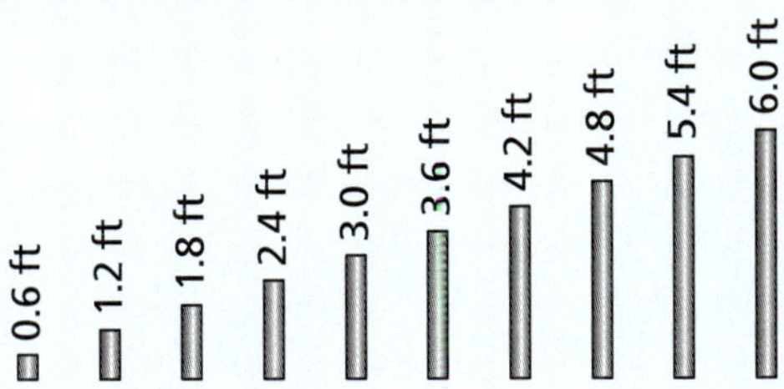

Project

Write a report comparing jumping ramps that are used in a variety of sports. Include your opinion as to why some ramps are higher and others are lower.

Reteaching and Enrichment Strategies

If students need help. . .	If students got it. . .
Resources by Chapter • Practice A and Practice B • Puzzle Time Record and Practice Journal Practice Differentiating the Lesson Lesson Tutorials Skills Review Handbook	Resources by Chapter • Enrichment and Extension • Technology Connection Start the next section

The variables x and y vary directly. Use the values to find the constant of proportionality. Then write an equation that relates x and y.

20. $y = 72; x = 3$

21. $y = 20; x = 12$

22. $y = 45; x = 40$

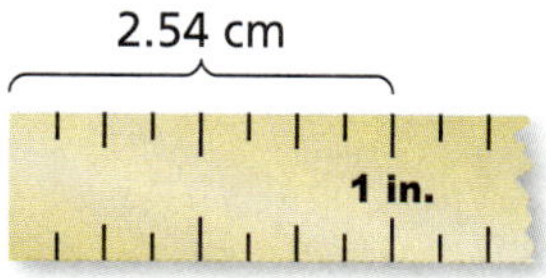

23. MEASUREMENT Write a direct variation equation that relates x inches to y centimeters.

24. MODELING Design a waterskiing ramp. Show how you can use direct variation to plan the heights of the vertical supports.

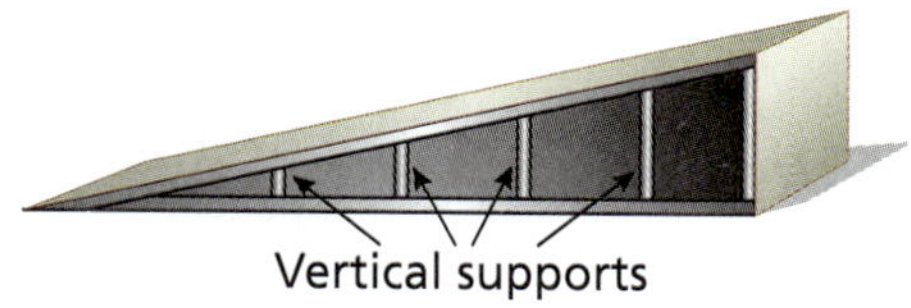

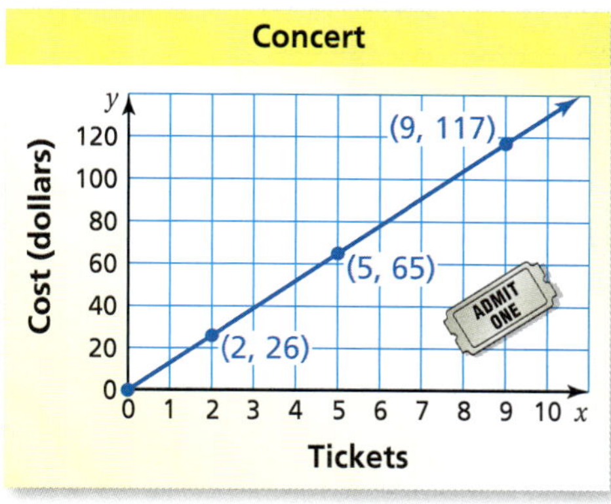

25. REASONING Use $y = kx$ to show why the graph of a proportional relationship always passes through the origin.

26. TICKETS The graph shows the cost of buying concert tickets. Tell whether x and y show direct variation. If so, find and interpret the constant of proportionality. Then write an equation and find the cost of 14 tickets.

27. CELL PHONE PLANS Tell whether x and y show direct variation. If so, write an equation of direct variation.

Minutes, *x*	500	700	900	1200
Cost, *y*	\$40	\$50	\$60	\$75

28. CHLORINE The amount of chlorine in a swimming pool varies directly with the volume of water. The pool has 2.5 milligrams of chlorine per liter of water. How much chlorine is in the pool?

29. Critical Thinking Is the graph of every direct variation equation a line? Does the graph of every line represent a direct variation equation? Explain your reasoning.

Fair Game Review What you learned in previous grades & lessons

Write the fraction as a decimal. *(Section 12.1)*

30. $\frac{13}{20}$

31. $\frac{9}{16}$

32. $\frac{21}{40}$

33. $\frac{24}{25}$

34. MULTIPLE CHOICE Which rate is *not* equivalent to 180 feet per 8 seconds? *(Section 14.1)*

Ⓐ $\frac{225 \text{ ft}}{10 \text{ sec}}$ Ⓑ $\frac{45 \text{ ft}}{2 \text{ sec}}$ Ⓒ $\frac{135 \text{ ft}}{6 \text{ sec}}$ Ⓓ $\frac{180 \text{ ft}}{1 \text{ sec}}$

14.4–14.6 Quiz

Check It Out
Progress Check
BigIdeasMath.com

Solve the proportion. *(Section 14.4)*

1. $\frac{7}{n} = \frac{42}{48}$ **2.** $\frac{x}{2} = \frac{40}{16}$ **3.** $\frac{3}{11} = \frac{27}{z}$

Find the slope of the line. *(Section 14.5)*

4.

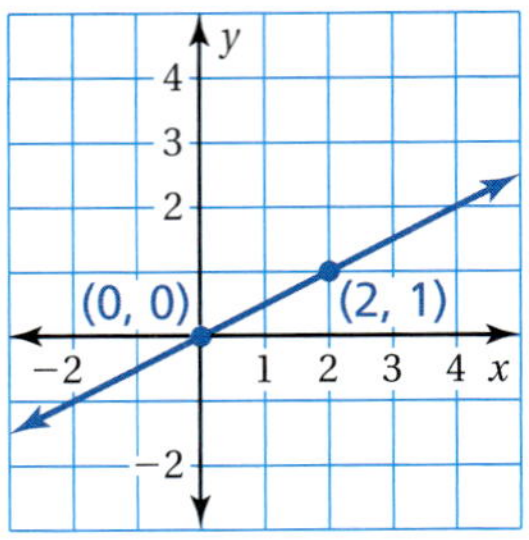

5.

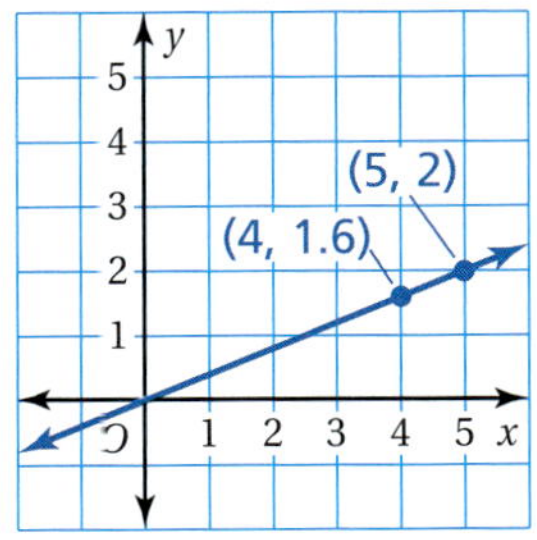

Graph the data. Then find and interpret the slope of the line through the points. *(Section 14.5)*

6.

Hours, x	2	4	6	8
Miles, y	10	20	30	40

7.

Packages, x	6	10	14	18
Servings, y	9	15	21	27

Tell whether x and y show direct variation. Explain your reasoning. *(Section 14.6)*

8. $y - 9 = 6 + x$ **9.** $x = \frac{5}{8}y$

10. CONCERT A benefit concert with three performers lasts 8 hours. At this rate, how many hours is a concert with four performers? *(Section 14.4)*

11. LAWN MOWING The graph shows how much you and your friend each earn mowing lawns. *(Section 14.5)*

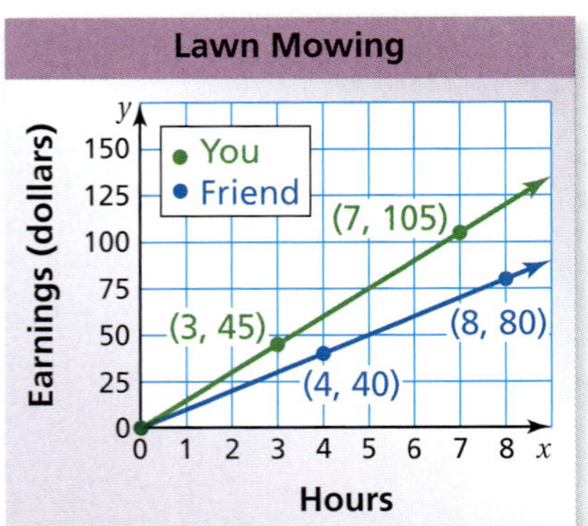

a. Compare the steepness of the lines. What does this mean in the context of the problem?

b. Find and interpret the slope of each line.

c. How much more money do you earn per hour than your friend?

12. PIE SALE The table shows the profits of a pie sale. Tell whether x and y show direct variation. If so, write the equation of direct variation. *(Section 14.6)*

Pies Sold, x	10	12	14	16
Profit, y	\$79.50	\$95.40	\$111.30	\$127.20

Alternative Assessment Options

Math Chat	Student Reflective Focus Question
Structured Interview	Writing Prompt

Structured Interview

Interviews can occur formally or informally. Ask a student to perform a task and explain it, describing his or her thought process throughout the task. Probe the student for more information. Do not ask leading questions. Keep a rubric or notes.

Teacher Prompts	Student Answers	Teacher Notes
Tell me a story about taking a vacation. Include this sentence. Three tickets cost \$420, so 5 tickets cost ? dollars.	Five friends go to Florida. All of their plane tickets are the same price. Three friends buy their tickets together. Three tickets cost \$420, so 5 tickets cost \$700.	Student can solve a proportion.
Add to your story using this phrase. 37.5 miles in ? minutes, or 450 miles in 1 hour	While at its cruising altitude, the plane flies 37.5 miles in 5 minutes, or 450 miles in 1 hour.	Student can solve a proportion.

Study Help Sample Answers

Remind students to complete Graphic Organizers for the rest of the chapter.

6.

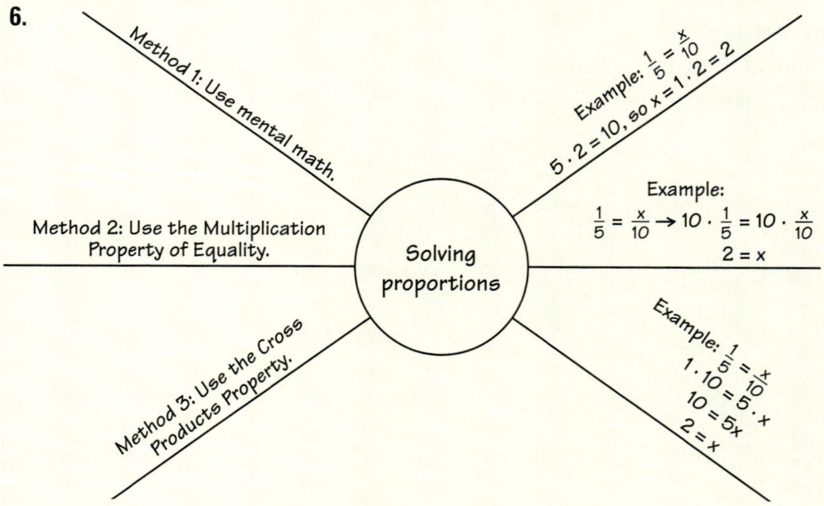

7–8. Available at *BigIdeasMath.com*.

Reteaching and Enrichment Strategies

If students need help. . .	If students got it. . .
Resources by Chapter • Practice A and Practice B • Puzzle Time Lesson Tutorials *BigIdeasMath.com*	Resources by Chapter • Enrichment and Extension • Technology Connection Game Closet at *BigIdeasMath.com* Start the Chapter Review

Answers

1. $n = 8$
2. $x = 5$
3. $z = 99$
4. $\frac{1}{2}$
5. $\frac{2}{5}$
6. 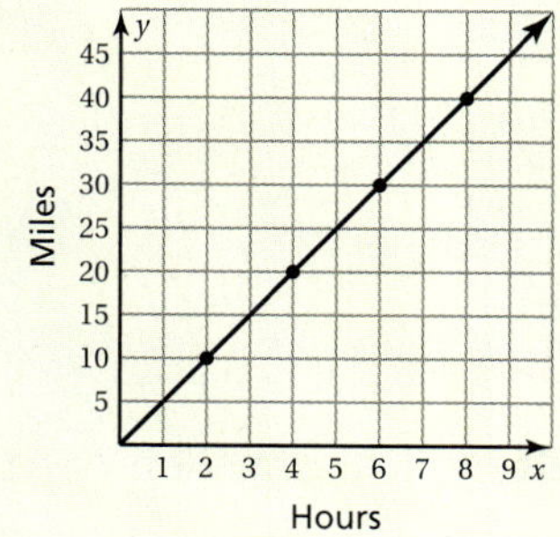

slope = 5; 5 miles per hour

7. 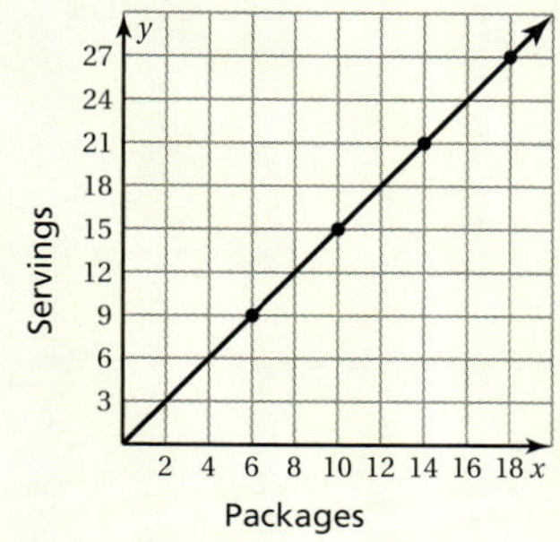

slope = $\frac{3}{2}$; 3 servings per 2 packages

8. no; The equation cannot be written as $y = kx$.
9. yes; The equation can be written as $y = kx$.
10. $10\frac{2}{3}$ hours or 10 hours 40 minutes
11. **a.** Your line is steeper; You earn more per hour than your friend.

    b. You earn \$15 per hour and your friend earns \$10 per hour.

    c. \$5
12. direct variation; $y = 7.95x$

Technology for the Teacher

Online Assessment
Assessment Book
ExamView® Assessment Suite

For the Teacher
Additional Review Options

- *BigIdeasMath.com*
- Online Assessment
- Game Closet at *BigIdeasMath.com*
- Vocabulary Help
- Resources by Chapter

Answers

1. 28.9 miles per gallon
2. 2.4 revolutions per second
3. 120 calories per serving

Review of Common Errors

Exercises 1–3

- Students may find the unit rate but forget to include the units. Remind them that the units are necessary for understanding a unit rate, or any rate.

Exercises 4–7

- Students may have difficulty understanding why you can write the ratio in simplest form. Tell students to compare the ratio to a fraction. Simplifying ratios is the same as writing equivalent fractions.

Exercise 8

- Students may immediately state that x and y are not in a proportional relationship because (0, 0) is not listed. Encourage students to use a graph to solve the problem, as stated in the directions.

Exercises 9 and 10

- Students may write half of the proportion using rows and the other half using columns. They will have forgotten to include one of the values. Remind students that they need to pick a method for writing proportions with tables and be consistent throughout the problem.

Exercises 11–14

- Students may divide instead of multiply when finding cross products, or they may multiply across the numerators and the denominators as if they were multiplying fractions. Remind students that the ratios have an equal sign between them, not a multiplication sign. Also tell them that when they use the Cross Products Property, it produces an "X," which means multiplication.

Exercises 15–17

- Students may put the change in x over the change in y. Remind them that the vertical change is written over the horizontal change and that slope is a rate.

Exercises 18–21

- Students may try to identify the direct variation equations without solving for y. Remind them to solve for y first.

14 Chapter Review

Check It Out Vocabulary Help BigIdeasMath.com

Review Key Vocabulary

ratio, *p. 600*
rate, *p. 600*
unit rate, *p. 600*
complex fraction, *p. 601*
proportion, *p. 608*
proportional, *p. 608*
cross products, *p. 609*
slope, *p. 630*
direct variation, *p. 636*
constant of proportionality, *p. 636*

Review Examples and Exercises

14.1 Ratios and Rates *(pp. 598–605)*

There are 15 orangutans and 25 gorillas in a nature preserve. One of the orangutans swings 75 feet in 15 seconds on a rope.

a. Find the ratio of orangutans to gorillas.

b. How fast is the orangutan swinging?

a. $\frac{\text{orangutans}}{\text{gorillas}} = \frac{15}{25} = \frac{3}{5}$

The ratio of orangutans to gorillas is $\frac{3}{5}$.

b. $75 \text{ feet in } 15 \text{ seconds} = \frac{75 \text{ ft}}{15 \text{ sec}}$

$= \frac{75 \text{ ft} \div 15}{15 \text{ sec} \div 15}$

$= \frac{5 \text{ ft}}{1 \text{ sec}}$

The orangutan is swinging 5 feet per second.

Exercises

Find the unit rate.

1. 289 miles on 10 gallons
2. $6\frac{2}{5}$ revolutions in $2\frac{2}{3}$ seconds
3. calories per serving

Servings	2	4	6	8
Calories	240	480	720	960

14.2 Proportions *(pp. 606–613)*

Tell whether the ratios $\frac{9}{12}$ and $\frac{6}{8}$ form a proportion.

$\frac{9}{12} = \frac{9 \div 3}{12 \div 3} = \frac{3}{4}$

$\frac{6}{8} = \frac{6 \div 2}{8 \div 2} = \frac{3}{4}$

The ratios are equivalent.

So, $\frac{9}{12}$ and $\frac{6}{8}$ form a proportion.

Exercises

Tell whether the ratios form a proportion.

4. $\frac{4}{9}, \frac{2}{3}$

5. $\frac{12}{22}, \frac{18}{33}$

6. $\frac{8}{50}, \frac{4}{10}$

7. $\frac{32}{40}, \frac{12}{15}$

8. Use a graph to determine whether x and y are in a proportional relationship.

x	1	3	6	8
y	4	12	24	32

14.3 Writing Proportions (pp. 614–619)

Write a proportion that gives the number r of returns on Saturday.

	Friday	Saturday
Sales	40	85
Returns	32	r

∴ One proportion is $\frac{40 \text{ sales}}{32 \text{ returns}} = \frac{85 \text{ sales}}{r \text{ returns}}$.

Exercises

Use the table to write a proportion.

9.

	Game 1	Game 2
Penalties	6	8
Minutes	16	m

10.

	Concert 1	Concert 2
Songs	15	18
Hours	2.5	h

14.4 Solving Proportions (pp. 622–627)

Solve $\frac{15}{2} = \frac{30}{y}$.

$15 \cdot y = 2 \cdot 30$ Cross Products Property

$15y = 60$ Multiply.

$y = 4$ Divide.

∴ The solution is 4.

Exercises

Solve the proportion.

11. $\frac{x}{4} = \frac{2}{5}$

12. $\frac{5}{12} = \frac{y}{15}$

13. $\frac{8}{20} = \frac{6}{w}$

14. $\frac{s+1}{4} = \frac{4}{8}$

Review Game

Proportions

Materials

- 1 deck of cards for each group
- paper for each student
- pencil for each student

Directions

Play in groups of 4 to 6 people. Each player is dealt four cards. Two are dealt face up. The other two are dealt face down and are held in the player's hand. The remainder of the deck is placed face down between the players.

The face up cards represent the denominators of two fractions. The object is to use the face down cards and other cards to form a proportion. Several cards can be added together in the numerator and denominator of both fractions until a proportion is obtained.

When it is a player's turn, he or she must do one of three things: lay a card down in either fraction's numerator or denominator, ask another player for a specific card, or draw from the pile. A student who draws from the pile or obtains a card from another student must wait until their next turn to lay a card down.

Card values are as follows:

2 through 10: face value
Jack: -1
Queen: -2
King: -3
Ace: -4

Points are awarded as follows:

First person to form a proportion: 10 points
Second: 9 points
Third: 8 points
Fourth: 7 points, and so on.

Who Wins?

After a set amount of time, the player with the most points wins.

For the Student Additional Practice

- Lesson Tutorials
- Multi-Language Glossary
- Self-Grading Progress Check
- *BigIdeasMath.com*
  Dynamic Student Edition
  Student Resources

Answers

4. no **5.** yes

6. no **7.** yes

8.

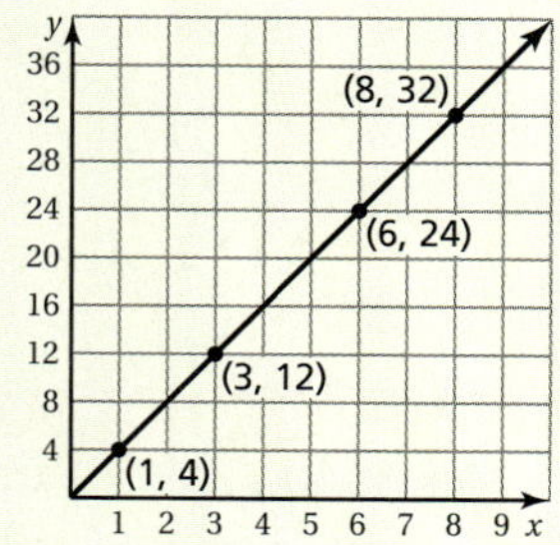

x and y are in a proportional relationship.

9. *Sample answer:* $\dfrac{8 \text{ penalties}}{6 \text{ penalties}} = \dfrac{m \text{ minutes}}{16 \text{ minutes}}$

10. *Sample answer:* $\dfrac{15 \text{ songs}}{2.5 \text{ hours}} = \dfrac{18 \text{ songs}}{h \text{ hours}}$

11. $x = 1.6$

12. $y = 6.25$

13. $w = 15$

14. $s = 1$

15. slope $= 1$

16. slope $= \dfrac{2}{3}$

17. slope $= 2$

18. no; The equation cannot be written as $y = kx$.

19. yes; The equation can be written as $y = kx$.

20. yes; The equation can be written as $y = kx$.

21. no; The equation cannot be written as $y = kx$.

My Thoughts on the Chapter

What worked. . .

Teacher Tip

Not allowed to write in your teaching edition? Use sticky notes to record your thoughts.

What did not work. . .

What I would do differently. . .

14.5 Slope (pp. 628–633)

The graph shows the number of visits your website received over the past 6 months. Find and interpret the slope.

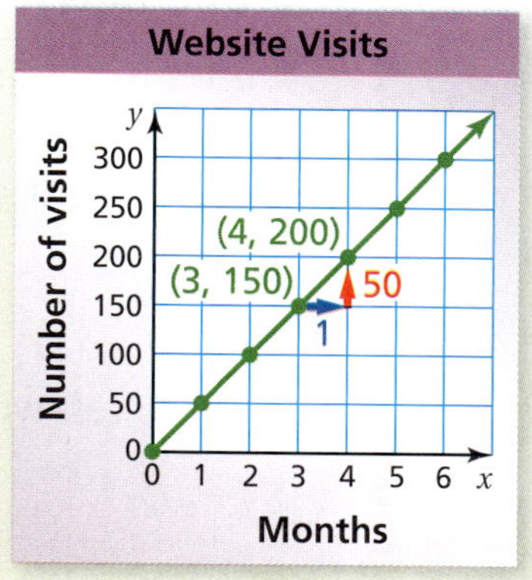

Choose any two points to find the slope of the line.

$$\text{slope} = \frac{\text{change in } y}{\text{change in } x}$$

$$= \frac{50}{1}$$ ← visits (50), months (1)

$$= 50$$

The slope of the line represents the unit rate. The slope is 50. So, the number of visits increased by 50 each month.

Exercises

Find the slope of the line.

15.

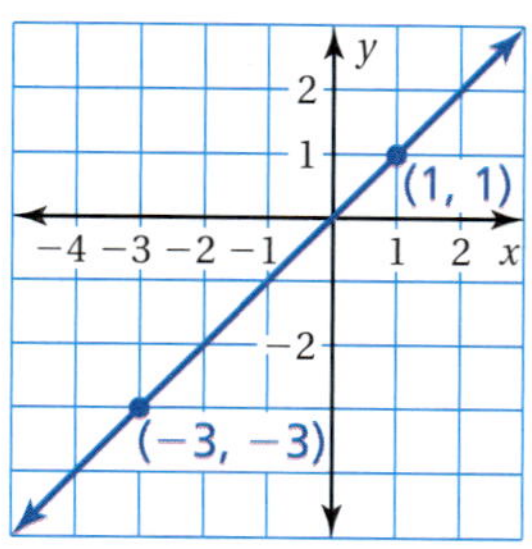

16.

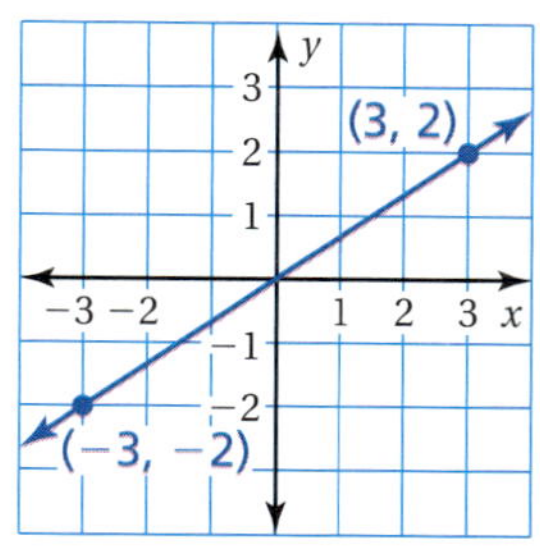

17.

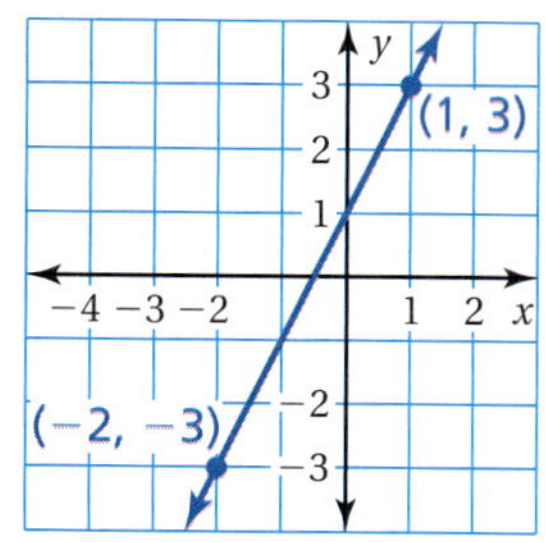

14.6 Direct Variation (pp. 634–639)

Tell whether x and y show direct variation. Explain your reasoning.

a. $x + y - 1 = 3$

$y = 4 - x$ Solve for y.

The equation *cannot* be written as $y = kx$. So, x and y do *not* show direct variation.

b. $x = 8y$

$\frac{1}{8}x = y$ Solve for y.

The equation can be written as $y = kx$. So, x and y show direct variation.

Exercises

Tell whether x and y show direct variation. Explain your reasoning.

18. $x + y = 6$

19. $y - x = 0$

20. $\frac{x}{y} = 20$

21. $x = y + 2$

14 Chapter Test

Find the unit rate.

1. 84 miles in 12 days

2. $2\frac{2}{5}$ kilometers in $3\frac{3}{4}$ minutes

Tell whether the ratios form a proportion.

3. $\frac{1}{9}, \frac{6}{54}$

4. $\frac{9}{12}, \frac{8}{72}$

Use a graph to tell whether *x* and *y* are in a proportional relationship.

5.

x	2	4	6	8
y	10	20	30	40

6.

x	1	3	5	7
y	3	7	11	15

Use the table to write a proportion.

7.

	Monday	Tuesday
Gallons	6	8
Miles	180	m

8.

	Thursday	Friday
Classes	6	c
Hours	8	4

Solve the proportion.

9. $\frac{x}{8} = \frac{9}{4}$

10. $\frac{17}{3} = \frac{y}{6}$

Graph the line that passes through the two points. Then find the slope of the line.

11. (15, 9), (−5, −3)

12. (2, 9), (4, 18)

Tell whether *x* and *y* show direct variation. Explain your reasoning.

13. $xy - 11 = 5$

14. $x = \frac{3}{y}$

15. $\frac{y}{x} = 8$

16. MOVIE TICKETS Five movie tickets cost $36.25. What is the cost of 8 movie tickets?

17. CROSSWALK The graph shows the number of cycles of a crosswalk signal during the day and during the night.

a. Compare the steepness of the lines. What does this mean in the context of the problem?

b. Find and interpret the slope of each line.

Don't Walk

Walk

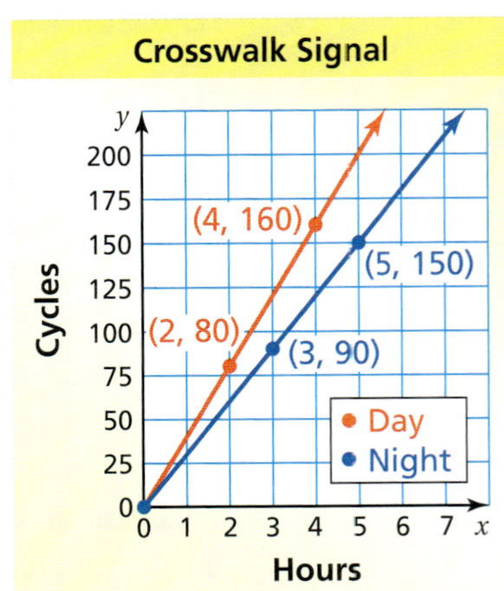

18. GLAZE A specific shade of green glaze requires 5 parts blue to 3 parts yellow. A glaze mixture contains 25 quarts of blue and 9 quarts of yellow. How can you fix the mixture to make the specific shade of green glaze?

Test Item References

Chapter Test Questions	Section to Review
1, 2	14.1
3–6	14.2
7, 8	14.3
9, 10, 16, 18	14.4
11, 12, 17	14.5
13–15	14.6

Test-Taking Strategies

Remind students to quickly look over the entire test before they start so that they can budget their time. Some students may have difficulty distinguishing between such concepts as ratios, rates, unit rates, proportions, and slopes, so encourage students to jot down definitions on the back of the test before they start. Students should use **Stop** and **Think** strategies to ensure that they understand what is being asked before they write an answer.

Common Errors

- **Exercises 1 and 2** Students may find the unit rate but forget to include the units. Remind them that the units are necessary for understanding a unit rate.
- **Exercises 3 and 4** Students may have difficulty understanding why you can write the ratio in simplest form. Tell students to compare the ratio to a fraction. Simplifying ratios is the same as writing equivalent fractions.
- **Exercises 5 and 6** Students may immediately state that *x* and *y* are not in a proportional relationship because (0, 0) is not listed. Encourage students to use a graph to solve the problem, as stated in the directions.
- **Exercises 7 and 8** Students may write half of the proportion using rows and the other half using columns. They will have forgotten to include one of the values. Remind students that they need to pick a method for writing proportions with tables and be consistent throughout the problem.
- **Exercises 13–15** Students may try to identify the direct variation equations without solving for *y*. Remind them to solve for *y* first.

Reteaching and Enrichment Strategies

If students need help. . .	If students got it. . .
Resources by Chapter • Practice A and Practice B • Puzzle Time Record and Practice Journal Practice Differentiating the Lesson Lesson Tutorials *BigIdeasMath.com* Skills Review Handbook	Resources by Chapter • Enrichment and Extension • Technology Connection Game Closet at *BigIdeasMath.com* Start Cumulative Assessment

Answers

1. 7 miles per day
2. $\frac{16}{25}$ kilometer per minute
3. yes
4. no
5. 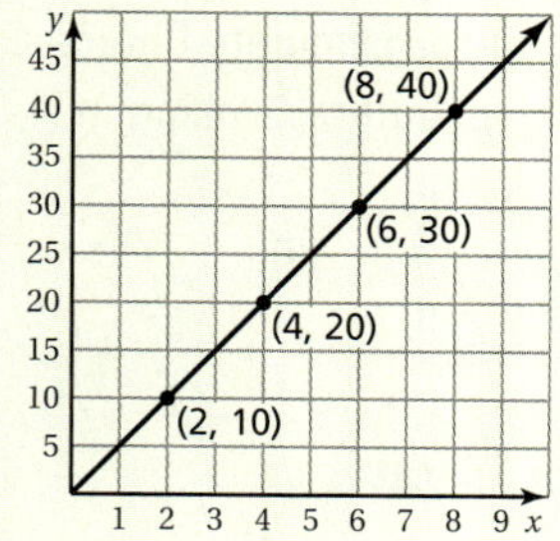

yes
6. 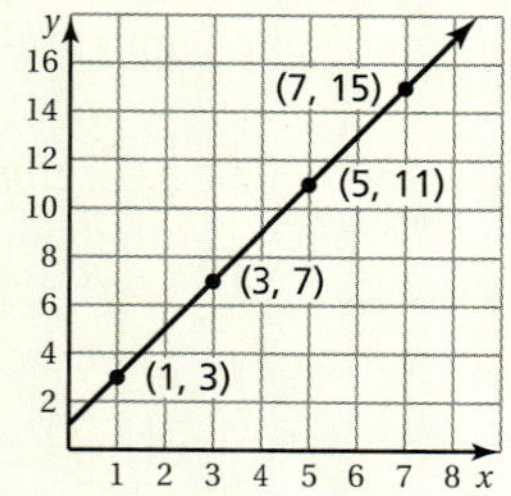

no
7. *Sample answer:* $\frac{8 \text{ gallons}}{6 \text{ gallons}} = \frac{m \text{ miles}}{180 \text{ miles}}$
8. *Sample answer:* $\frac{6 \text{ classes}}{8 \text{ hours}} = \frac{c \text{ classes}}{4 \text{ hours}}$
9. $x = 18$
10. $y = 34$
11. 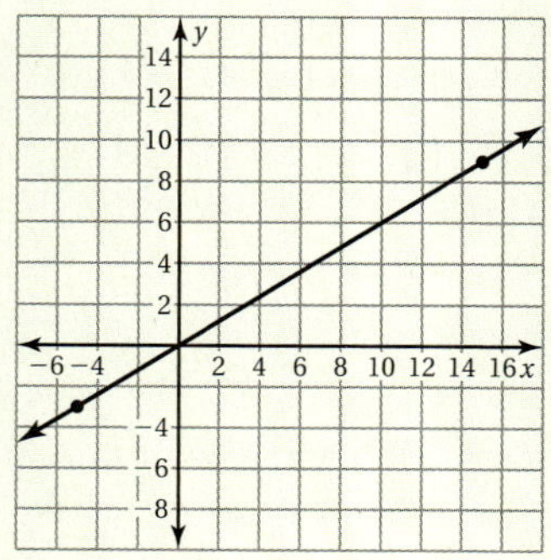
slope $= \frac{3}{5}$

12–17. See Additional Answers.

18. Add 6 quarts of yellow

Technology for the Teacher

Online Assessment
Assessment Book
ExamView® Assessment Suite

Test-Taking Strategies

Available at *BigIdeasMath.com*

After Answering Easy Questions, Relax
Answer Easy Questions First
Estimate the Answer
Read All Choices before Answering
Read Question before Answering
Solve Directly or Eliminate Choices
Solve Problem before Looking at Choices
Use Intelligent Guessing
Work Backwards

About this Strategy

When taking a multiple choice test, be sure to read each question carefully and thoroughly. It is also very important to read each answer choice carefully. Do not pick the first answer that you think is correct!

Answers

1. A
2. F
3. 29
4. B
5. H

Technology for the Teacher

Performance Tasks
Online Assessment
Assessment Book
ExamView® Assessment Suite

Item Analysis

1. **A.** Correct answer
   B. The student thinks the price of the 4 pencils is the unit price.
   C. The student multiplies 0.80 and 4.
   D. The student divides 4 by 0.80.

2. **F.** Correct answer
   G. The student applies operations incorrectly.
   H. The student applies operations incorrectly.
   I. The student applies operations incorrectly.

3. **Gridded Response:** Correct answer: 29

   Common Error: The student makes a sign error when multiplying and gets an answer of -19.

4. **A.** The student finds the change in x over the change in y.
   B. Correct answer
   C. The student finds the change in x.
   D. The student finds the change in y.

5. **F.** The student adds the two numbers.
   G. The student doesn't account for distance being positive.
   H. Correct answer
   I. The student accounts for distance being positive, but adds the two numbers.

14 Cumulative Assessment

1. The school store sells 4 pencils for $0.80. What is the unit cost of a pencil?

A. $0.20 **C.** $3.20

B. $0.80 **D.** $5.00

2. Which expressions do *not* have a value of 3?

I. $2 + (-1)$ II. $2 - (-1)$

III. $-3 \times (-1)$ IV. $-3 \div (-1)$

F. I only **H.** II only

G. III and IV **I.** I, III, and IV

Test-Taking Strategy
Read Question Before Answering

"Be sure to read the question before choosing your answer. You may find a word that changes the meaning."

3. What is the value of the expression below?

$$-4 \times (-6) - (-5)$$

4. What is the slope of the line shown?

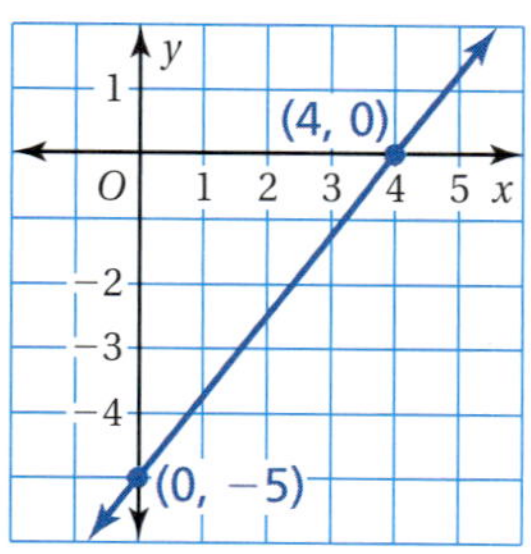

A. $\frac{4}{5}$ **C.** 4

B. $\frac{5}{4}$ **D.** 5

5. What is the distance between the two numbers on the number line?

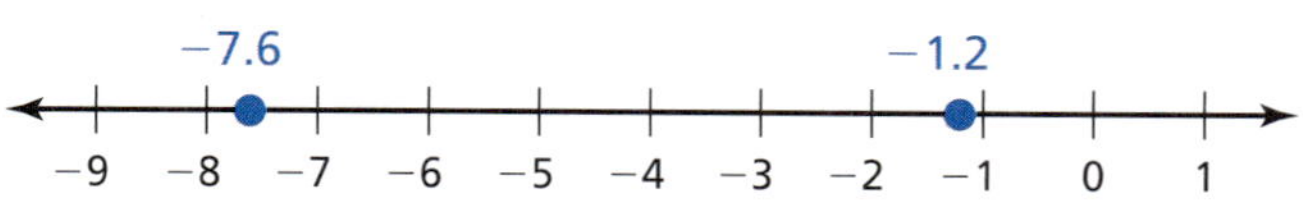

F. -8.8 **H.** 6.4

G. -6.4 **I.** 8.8

6. The quantities x and y are proportional. What is the missing value in the table?

x	y
$\frac{2}{3}$	6
$\frac{4}{3}$	12
$\frac{8}{3}$	24
5	

A. 30 **C.** 45

B. 36 **D.** 48

7. You are selling tomatoes. You have already earned \$16 today. How many additional pounds of tomatoes do you need to sell to earn a total of \$60?

F. 4 **H.** 15

G. 11 **I.** 19

8. The distance traveled by the a high-speed train is proportional to the number of hours traveled. Which of the following is *not* a valid interpretation of the graph below?

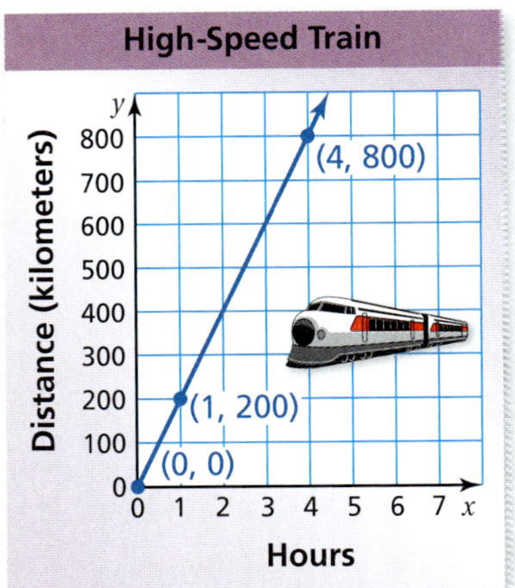

A. The train travels 0 kilometers in 0 hours.

B. The unit rate is 200 kilometers per hour.

C. After 4 hours, the train is traveling 800 kilometers per hour.

D. The train travels 800 kilometers in 4 hours.

Item Analysis (continued)

6. **A.** The student adds 6 to 24 because $12 - 6 = 6$.
   B. The student adds 12 to 24 because $24 - 12 = 12$.
   C. Correct answer
   D. The student doubles 24 because each value in the *y* column is twice the previous value.

7. **F.** The student finds the number of pounds already sold.
   G. Correct answer
   H. The student finds the total number of pounds instead of the additional number of pounds.
   I. The student adds the total number of pounds and the number of pounds already sold.

8. **A.** The student misinterprets the information in the graph.
   B. The student misinterprets the information in the graph.
   C. Correct answer
   D. The student misinterprets the information in the graph.

Answers

6. C
7. G
8. C

Answers

9. G

10. 3

11. *Part A* *Sample answer:*

$\frac{800}{15} = \frac{6000}{m}$

Part B 112.5 min

12. D

Item Analysis (continued)

9. **F.** The student takes the reciprocal of the dividend.

G. Correct answer

H. The student switches denominators instead of taking the reciprocal of the divisor.

I. The student makes a sign error.

10. **Gridded Response:** Correct answer: 3

Common Error: The student uses the wrong operations after substitution.

11. **2 points** The student demonstrates a thorough understanding of writing and solving proportions. For Part A, the student correctly writes a proportion such as $\frac{800}{15} = \frac{6000}{m}$ and provides an appropriate explanation. For Part B, the student correctly gets a value of 112.5 for m, shows appropriate work, and states that it would take 112.5 minutes.

1 point The student demonstrates a partial understanding of writing and solving proportions. The student writes a correct proportion but does not solve it successfully, or the student does not write a correct proportion but demonstrates the ability to solve a proportion.

0 points The student demonstrates insufficient understanding of writing and solving proportions. The student does not write a correct proportion and shows little or no evidence of being able to solve proportions.

12. **A.** The student makes a sign error when dividing each side of the equation by -2.

B. The student subtracts 6 from the left side of the equation but adds 6 to the right side of the equation, and makes a sign error when dividing each side of the equation by -2.

C. The student subtracts 6 from the left side of the equation but adds 6 to the right side of the equation.

D. Correct answer

9. Regina was evaluating the expression below. What should Regina do to correct the error she made?

$$-\frac{3}{2} \div \left(-\frac{8}{7}\right) = -\frac{2}{3} \times \left(-\frac{7}{8}\right)$$
$$= \frac{2 \times 7}{3 \times 8}$$
$$= \frac{14}{24}$$
$$= \frac{7}{12}$$

F. Rewrite $-\frac{3}{2} \div \left(-\frac{8}{7}\right)$ as $-\frac{2}{3} \times \left(-\frac{8}{7}\right)$.

G. Rewrite $-\frac{3}{2} \div \left(-\frac{8}{7}\right)$ as $-\frac{3}{2} \times \left(-\frac{7}{8}\right)$.

H. Rewrite $-\frac{3}{2} \div \left(-\frac{8}{7}\right)$ as $-\frac{3}{7} \times \left(-\frac{8}{2}\right)$.

I. Rewrite $-\frac{2}{3} \times \left(-\frac{7}{8}\right)$ as $-\frac{2 \times 7}{3 \times 8}$.

10. What is the value of the expression below when $x = 2$ and $y = -7$?

$$y + 5x$$

11. You can mow 800 square feet of lawn in 15 minutes. At this rate, how many minutes will you take to mow a lawn that measures 6000 square feet?

Think Solve Explain

Part A Write a proportion to represent the problem. Use m to represent the number of minutes. Explain your reasoning.

Part B Solve the proportion you wrote in Part A. Then use it to answer the problem. Show your work.

12. What value of p makes the equation below true?

$$6 - 2p = -48$$

A. -27

B. -21

C. 21

D. 27

15 Percents

75% off
$1
$0.25
"Here's my sales strategy. I buy each dog bone for $0.05."
I guess P.T. Barnum was right.
"Then I mark each one up to $1. Then, I have a 75% off sale. Cool, huh?"

"At 4 a day, I have chewed 17,536 dog biscuits. At only 99.9% pure, that means that..."
I hope the same logic doesn't apply for mice!
"I have swallowed seventeen and a half contaminated dog biscuits during the past twelve years."

What Your Students Have Learned

- Multiply and divide by powers of 10 and explain the placement of the decimal point.
- Find equivalent fractions.
- Compare decimals to the thousandths place.
- Use and interpret simple equations.
- Understand ratios and describe ratio relationships.
- Understand and find percent as a rate per 100.
- Find the part and the whole of ratio relationships.
- Identify equivalent expressions.
- Solve one-step equations.

What Your Students Will Learn

- Write percents as decimals and decimals as percents.
- Compare fractions, decimals, and percents.
- Use proportionality to solve percent problems.
- Use the percent equation.
- Solve percent problems involving percents of increase and decrease, and simple interest.

Pacing Guide for Chapter 15

Chapter Opener Advanced	1 Day
Section 1 Advanced	1 Day
Section 2 Advanced	1 Day
Section 3 Advanced	1 Day
Section 4 Advanced	1 Day
Study Help/Quiz Advanced	1 Day
Section 5 Advanced	1 Day
Section 6 Advanced	1 Day
Section 7 Advanced	1 Day
Chapter Review/ Chapter Tests Advanced	2 Days
Total Chapter 15 Advanced	11 Days
Year-to-Date Advanced	145 Days

Technology for the Teacher

BigIdeasMath.com
Chapter at a Glance
Complete Materials List
Parent Letters: English and Spanish

What Your Students Have Learned

- Write percents as fractions in simplest form.
- Write fractions and mixed numbers as percents.

Additional Topics for Review

- Compare and Order Integers
- Writing Decimals as Fractions and Percents
- Writing Fractions as Decimals
- Ratios
- Solving Proportions

Try It Yourself

1. $\frac{4}{25}$
2. $\frac{2}{5}$
3. $\frac{17}{25}$
4. $\frac{17}{20}$
5. $1\frac{12}{25}$
6. $1\frac{1}{2}$
7. $1\frac{1}{20}$
8. $2\frac{19}{25}$
9. 36%
10. 86%
11. 55%
12. 60%
13. 125%
14. 148%
15. 180%
16. 230%

Record and Practice Journal Fair Game Review

1. $\frac{1}{4}$
2. $\frac{13}{20}$
3. $1\frac{1}{10}$
4. $2\frac{1}{2}$
5. $\frac{3}{20}$
6. $\frac{3}{50}$
7. $\frac{3}{10}$
8. 20%
9. 25%
10. 84%
11. 140%
12. 265%
13. 150%
14. 60%

Math Background Notes

Vocabulary Review

- Percent
- Numerator
- Denominator
- Equivalent Fractions

Writing Percents as Fractions

- Students learned how to convert between percents and fractions.
- A 10-by-10 grid has 100 small squares making it a convenient model for percents as well as decimals.
- Be sure students understand that when modeling, for instance 45%, any 45 squares can be shaded. However, for ease of being able to read the model quickly, we generally shade four strips of 10 and one strip of 5.

? "What is 100% as a fraction?" 1

? "If the percent is greater than 100%, what do you know about the equivalent fraction?" It will be greater than 1.

Writing Fractions as Percents

- Students have learned how to write equivalent fractions.

? "What are equivalent fractions?" two fractions that represent the same amount

- Review which fractions can be written as equivalent fractions with a denominator of 100.

? "How do you write equivalent fractions with a denominator of 100?" Find a number you can multiply the denominator by so that it equals 100. Multiply the numerator by the same number.

? "If the number is greater than 1, what do you know about the percent?" It will be greater than 100%.

Reteaching and Enrichment Strategies

If students need help. . .	If students got it. . .
Record and Practice Journal • Fair Game Review Skills Review Handbook Lesson Tutorials	Game Closet at *BigIdeasMath.com* Start the next section

What You Learned Before

"The fact that these two percents do not total 100 is a sad commentary on humans."

Writing Percents as Fractions

Example 1 Write 45% as a fraction in simplest form.

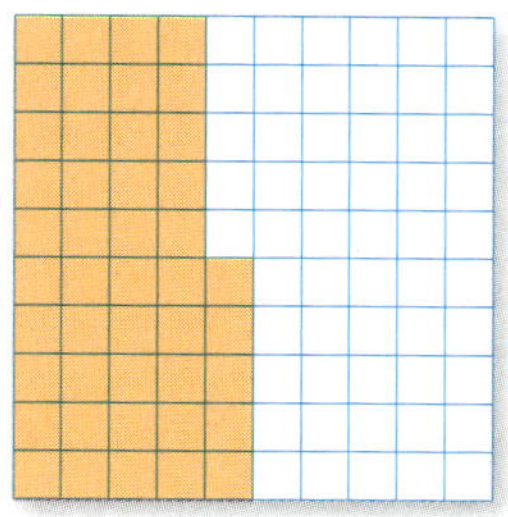

$$45\% = \frac{45}{100}$$ Write as a fraction with a denominator of 100.

$$= \frac{9}{20}$$ Simplify.

So, $45\% = \frac{9}{20}$.

Try It Yourself

Write the percent as a fraction or mixed number in simplest form.

1. 16% **2.** 40% **3.** 68% **4.** 85%

5. 148% **6.** 150% **7.** 105% **8.** 276%

Writing Fractions as Percents

Example 2 Write $\frac{3}{25}$ as a percent.

$$\frac{3}{25} \overset{\times 4}{=} \frac{12}{100} = 12\%$$

Because $25 \times 4 = 100$, multiply the numerator and denominator by 4. Write the numerator with a percent symbol.

Try It Yourself

Write the fraction or mixed number as a percent.

9. $\frac{9}{25}$ **10.** $\frac{43}{50}$ **11.** $\frac{11}{20}$ **12.** $\frac{3}{5}$

13. $1\frac{1}{4}$ **14.** $1\frac{12}{25}$ **15.** $1\frac{4}{5}$ **16.** $2\frac{3}{10}$

15.1 Percents and Decimals

Essential Question How does the decimal point move when you rewrite a percent as a decimal and when you rewrite a decimal as a percent?

1 ACTIVITY: Writing Percents as Decimals

Work with a partner. Write the percent shown by the model. Write the percent as a decimal.

a.

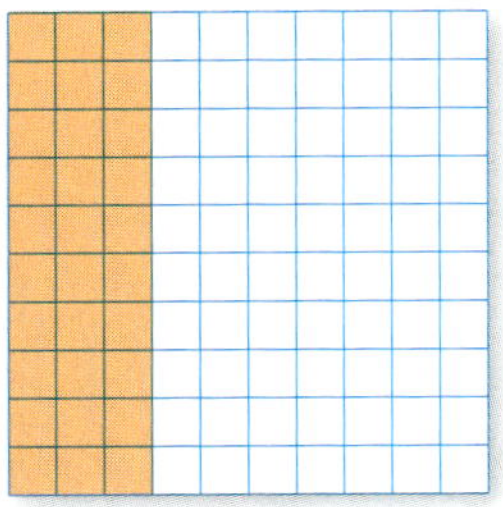

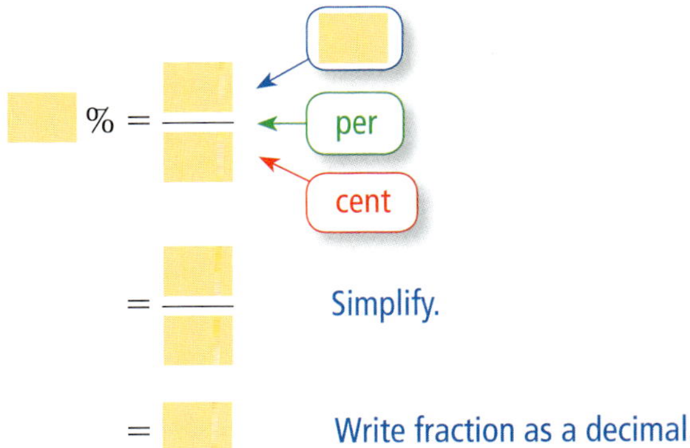

b.

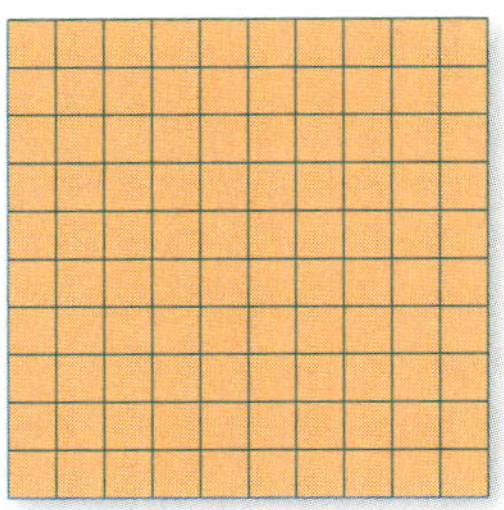

c.

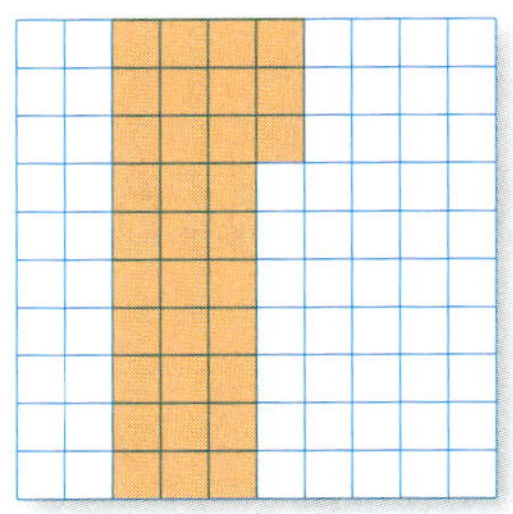

d.

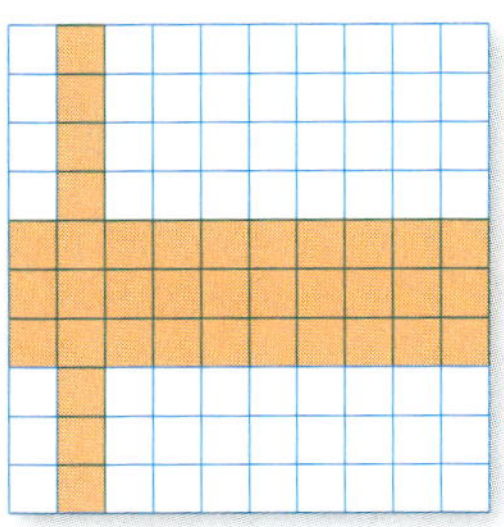

e.

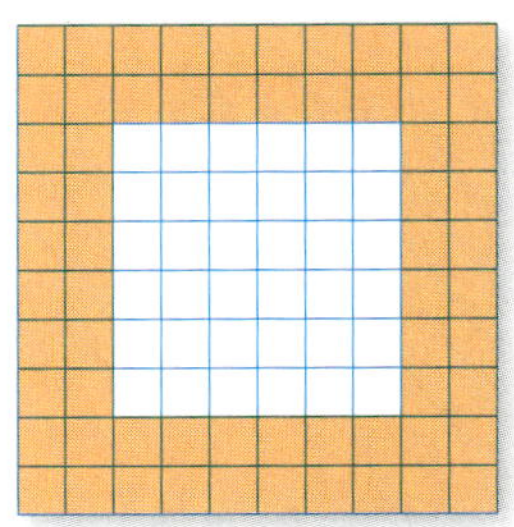

f.

g.

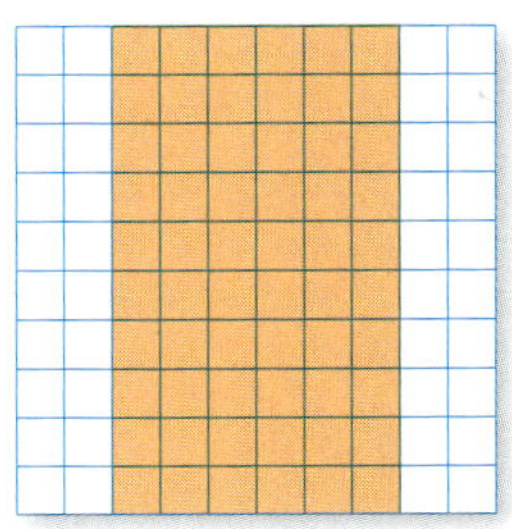

Percents and Decimals

In this lesson, you will

- write percents as decimals.
- write decimals as percents.
- solve real-life problems.

Laurie's Notes

Introduction

Applying Mathematical Practices

- **Model with Mathematics:** Mathematically proficient students use models to help make sense of different representations of numbers. In this lesson, a 100-grid is used to model percents, making the connection to the equivalent fraction and decimal forms.

Motivate

- Share a fictional story about collecting student homework on a USB drive and the need to purchase a new drive with greater capacity—for all of their homework! Work the following nonfictional facts into the story.
  - Most common USB drives hold 4 GB (gigabytes) or 8 GB.
  - A 16 GB USB drive holds 4 or 400% as much data as the 4 GB, and 2 or 200% as much data as the 8 GB.

For the Teacher

- Students have converted between fractions and decimals and between fractions and percents. This lesson completes the triangle by converting between decimals and percents.

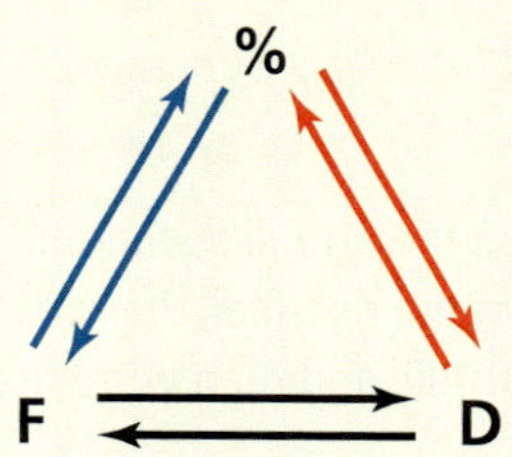

- Relating percents to money (part of a dollar) is often a helpful technique.

Activity Notes

Activity 1

- Check in with students as they work through part (a). There are 3 steps: write the percent; write the percent as a fraction; write the fraction as a decimal.
- A common question from students is whether they can leave the answer as 0.30 versus 0.3. The two decimals are equivalent. However, explain that you generally write the simplified version 0.3 for the same reason you simplify fractions.
- Students should work with partners to answer the remaining parts.
- Probe what strategies were used to count the shaded squares.
- **Common Error:** In part (b), students often write "100% = 100" instead of "100% = 1."

What Your Students Will Learn

- Write percents as decimals and decimals as percents.
- Write fractions as percents and decimals.

Previous Learning

Students should know how to write halves, fourths, tenths, and hundredths as percents. Students also need to be able to convert between fractions and decimals, and multiply and divide decimals by powers of 10.

Technology for the Teacher

Dynamic Classroom

Lesson Plans
Complete Materials List

15.1 Record and Practice Journal

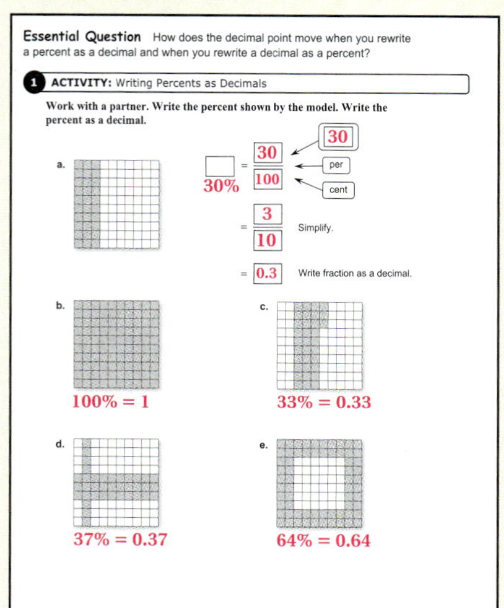
Essential Question How does the decimal point move when you rewrite a percent as a decimal and when you rewrite a decimal as a percent?

1 ACTIVITY: Writing Percents as Decimals

Work with a partner. Write the percent shown by the model. Write the percent as a decimal.

a. 30% = 30/100 (30 per cent) = 3/10 Simplify. = 0.3 Write fraction as a decimal.

b. 100% = 1

c. 33% = 0.33

d. 37% = 0.37

e. 64% = 0.64

English Language Learners

Have students make drawings to visualize percents. Have them use grid paper and draw 50%, 5%, and $\frac{5}{10}\%$ so that they can see the difference between the percents.

15.1 Record and Practice Journal

f. 
50% = 0.5

g. 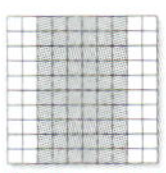
60% = 0.6

2 ACTIVITY: Writing Percents as Decimals

Work with a partner. Write the percent as a decimal.

a. $13.5\% = \frac{13.5}{100}$ (13.5, per, cent)

$= \frac{135}{1000}$ Multiply numerator and denominator by 10.

$= 0.135$ Write fraction as a decimal.

b. 12.5% 0.125

c. 3.8% 0.038

d. 0.5% 0.005

3 ACTIVITY: Writing Decimals as Percents

Work with a partner. Draw a model to represent the decimal. Write the decimal as a percent.

a. 0.1 10%

b. 0.24 24%

c. 0.58 58%

d. 0.05 5%

What Is Your Answer?

4. **IN YOUR OWN WORDS** How does the decimal point move when you rewrite a percent as a decimal and when you rewrite a decimal as a percent?
percent as decimal: move decimal point two places left
decimal to percent: move decimal point two places right

5. Explain why the decimal point moves when you rewrite a percent as a decimal and when you rewrite a decimal as a percent.
Sample answer: because you are multiplying and dividing by powers of 10

Laurie's Notes

Activity 2

- **Discuss:** Percents do not need to involve whole numbers, and in fact often do not.

? "What is different about the visual model from what you have seen previously?" One-half of a square is shaded.

- **FYI:** Students initially find it odd to have a decimal as part of a fraction. Multiplying the numerator and denominator by a power of 10 makes sense to them.
- **Model with Mathematics:** Sketching models of parts (b)–(d) helps students focus on the fractional portion of 1%. You could also "magnify" one small square and draw it off to the side to shade various portions of 1%, as shown.

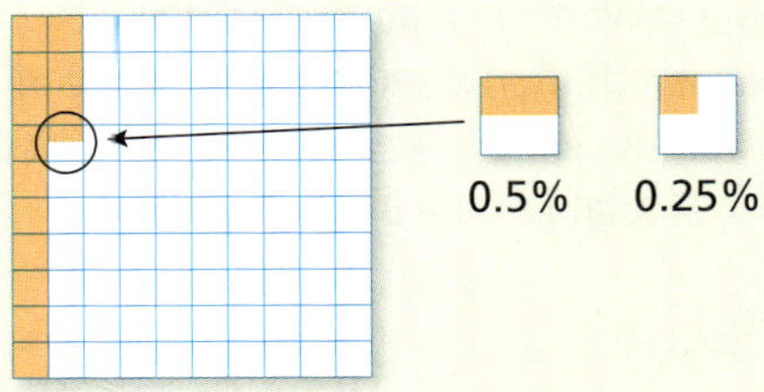

- **Common Error:** When percents are less than 1%, students often just remove the percent symbol and leave the decimal. In part (d), 0.5% does not equal 0.5.

Activity 3

- This example reverses the process students have just completed. After they sketch a model of the decimal, they write the decimal as a fraction with a denominator of 100, and then write the percent.

What Is Your Answer?

- Students should recognize that the decimal point has moved two decimal places left or right, although they may not have a good understanding of why.
- Listen for explanations about the percent symbol and its meaning.

? Probe about the location of the decimal point for percents involving whole numbers, such as 25%. "Where is the decimal point located?" after the 5

- **Construct Viable Arguments and Critique the Reasoning of Others:** Students may not be able to articulate clearly an explanation for Question 5. Listen to students' explanations and references to place value, definition of percent, and multiplication and division of powers of 10.

Closure

- **Exit Ticket:** Write each percent as a decimal.
60%, 6%, and 0.6% 0.6, 0.06, 0.006

2 ACTIVITY: Writing Percents as Decimals

Math Practice

Communicate Precisely

How can reading the fraction aloud help you write it as a decimal?

Work with a partner. Write the percent as a decimal.

a. 13.5%

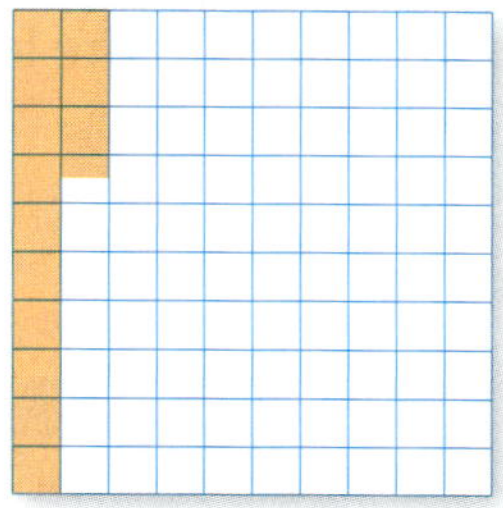

$\square\% = \dfrac{\square}{\square}$ ← per, cent

$= \dfrac{\square}{\square}$ Multiply numerator and denominator by 10.

$= \square$ Write fraction as a decimal.

b. 12.5% **c.** 3.8% **d.** 0.5%

3 ACTIVITY: Writing Decimals as Percents

Work with a partner. Draw a model to represent the decimal. Write the decimal as a percent.

a. 0.1

0.1 $=$ $0.10 = \dfrac{\square}{\square}$ $=$ $\square\%$

One $\square$

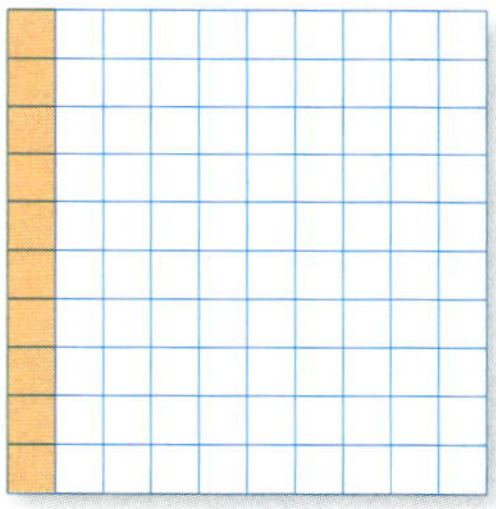

Ten $\square$

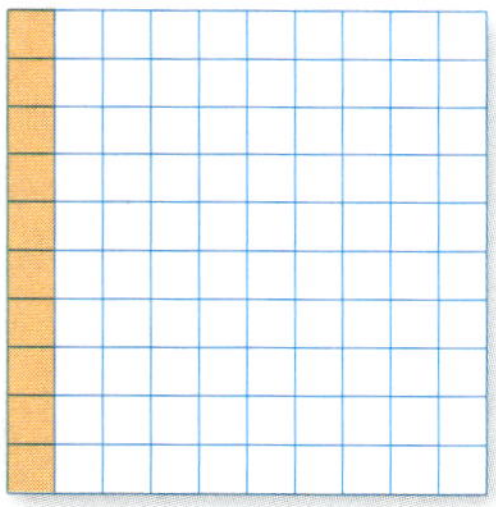

$\square$ Percent

b. 0.24 **c.** 0.58 **d.** 0.05

What Is Your Answer?

4. **IN YOUR OWN WORDS** How does the decimal point move when you rewrite a percent as a decimal and when you rewrite a decimal as a percent?

5. Explain why the decimal point moves when you rewrite a percent as a decimal and when you rewrite a decimal as a percent.

Practice Use what you learned about percents and decimals to complete Exercises 7–12 and 19–24 on page 654.

15.1 Lesson

Check It Out
Lesson Tutorials
BigIdeasMath.com

Key Idea

Writing Percents as Decimals

Words Remove the percent symbol. Then divide by 100, or just move the decimal point two places to the left.

Numbers 23% = 23.% = 0.23

EXAMPLE 1 Writing Percents as Decimals

Study Tip

When moving the decimal point, you may need to place one or more zeros in the number.

a. Write 52% as a decimal.

52% = 52.% = 0.52

Check

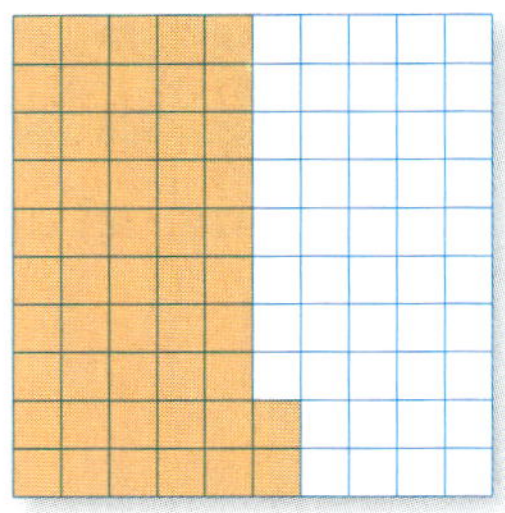

b. Write 7% as a decimal.

7% = 07.% = 0.07

Check

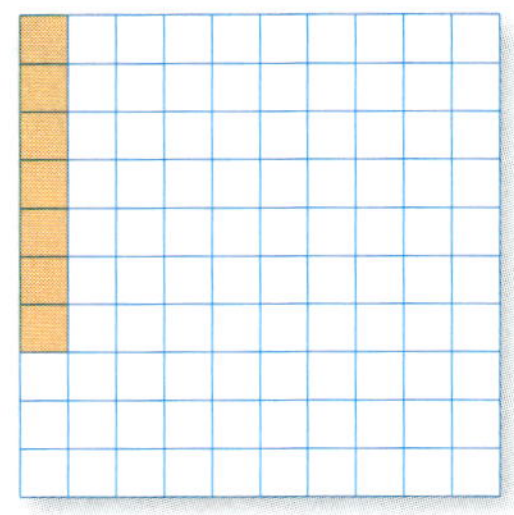

On Your Own

Now You're Ready
Exercises 7–18

Write the percent as a decimal. Use a model to check your answer.

1. 24% **2.** 3% **3.** 107% **4.** 92.7%

Key Idea

Writing Decimals as Percents

Words Multiply by 100, or just move the decimal point two places to the right. Then add a percent symbol.

Numbers 0.36 = 0.36 = 36%

EXAMPLE 2 Writing Decimals as Percents

a. Write 0.47 as a percent.

0.47 = 0.47 = 47%

b. Write 0.663 as a percent.

0.663 = 0.663 = 66.3%

c. Write 1.8 as a percent.

1.8 = 1.80 = 180%

d. Write 0.009 as a percent.

0.009 = 0.009 = 0.9%

Laurie's Notes

Introduction

Connect

- **Yesterday:** Students used a visual model to represent percents, then wrote the percents as decimals.
- **Today:** Students will convert between percents and decimals.

Motivate

- Divide the class into 4 groups by any method. Each group has a different rule to follow.
  Group A: Multiply by 100. Group B: Multiply by 0.01.
  Group C: Divide by 100. Group D: Divide by 0.01.
- Use small white boards, an electronic polling system, or scrap paper to record answers.
- Use the numbers 60, 44, and 2.5. Check results from each group after each problem.
- Students should recognize the pattern. Groups A and D have the same answers as do Groups B and C. Discuss the results.

Lesson Notes

Key Idea

- To reinforce the meaning behind moving the decimal point, say "23 percent, 23 per one hundred, or 23 hundredths."

Example 1

- Work through both parts of the example.

? "52% is close to which benchmark fraction?" $\frac{1}{2}$

- The visual model shows an amount just more than 50% or $\frac{1}{2}$.
- Point out the *Study Tip*. In part (b), 7% means 7 per one hundred or 7 hundredths. To write that, you have to place a zero to the left of the seven, 0.07.

On Your Own

- **Think-Pair-Share:** Students should read each question independently and then work in pairs to answer the questions. When they have answered the questions, the pair should compare their answers with another group and discuss any discrepancies.

Key Idea

- **Connection:** Dividing by 100 is equivalent to multiplying by 0.01. The decimal point moves 2 places to the left. Multiplying by 100 is equivalent to dividing by 0.01. The decimal point moves 2 places to the right.

Example 2

- Work through the examples.

Goal Today's lesson is converting between percents and decimals.

Technology for the Teacher

Dynamic Classroom

Lesson Tutorials
Lesson Plans
Answer Presentation Tool

Extra Example 1

a. Write 66% as a decimal. 0.66
b. Write 2% as a decimal. 0.02

On Your Own

1. 0.24 **2.** 0.03
3. 1.07 **4.** 0.927

Extra Example 2

a. Write 0.29 as a percent. 29%
b. Write 0.775 as a percent. 77.5%
c. Write 2.5 as a percent. 250%
d. Write 0.0075 as a percent. 0.75%

On Your Own

5. 94% **6.** 120%

7. 31.6% **8.** 0.5%

Extra Example 3

On a science test, you get 85 out of a possible 100 points. Write your score as a percent, a fraction, and a decimal.

85%, $\frac{17}{20}$, 0.85

Extra Example 4

In Example 4, how many times more UV rays are reflected by water than by grass? 28 times more

On Your Own

9. $18\% = \frac{9}{50} = 0.18$

10. 5.6

Differentiated Instruction

Kinesthetic

Some students may struggle with writing decimals for very small percents (i.e., 0.0075% as 0.000075) or writing percents for very small decimals (i.e., 0.0075 as 0.75%). To improve their number sense, have students draw a 10-by-10 grid with sides of 20 centimeters. Divide one of the squares to make a smaller 10-by-10 grid. Have students compare 75% and 0.75% by shading squares to represent each percent. Then have students write the decimal equivalents of the percents.

Laurie's Notes

On Your Own

- **Think-Pair-Share:** Students should read each question independently and then work in pairs to answer the questions. When they have answered the questions, the pair should compare their answers with another group and discuss any discrepancies.

Words of Wisdom

- Students quickly recognize that the decimal point moves 2 places left or right depending on the type of conversion.
- Students generally do well with 2-digit problems: 24%, or 0.24.
- Students have difficulty with percents greater than 100% or less than 1% (i.e., 250%, 0.025%). They want to quickly move the decimal point without thinking first.
- Reinforce the meaning behind moving the decimal point.
  - 25% is 25 per one hundred or 0.25.
  - 250% is 250 per one hundred or 2.5.
  - 0.25% is 25 hundredths per one hundred or 0.0025.
  - 0.025% is 25 thousandths per one hundred or 0.00025.
- When you feel that students are comfortable with percents between 1 and 100, practice converting with greater and lesser percents.

Example 3

- Note the test-taking strategy of eliminating choices.

Example 4

- **FYI:** UV rays are necessary for our bodies to produce vitamin D, a substance that helps strengthen bones. The downside: reflection of UV rays off of snow and sand are enough to cause photokeratitis, sunburn of the cornea.
- This example reviews all three forms: percents, decimals, and fractions, as well as division of decimals.

? "What is the key question being asked?" This is a language connection. "How many times more. . . ."

On Your Own

- **Construct Viable Arguments and Critique the Reasoning of Others:** There are several ways in which students may justify their answers. Take time to hear a variety of approaches.

Closure

- **Writing Prompts:**
  To write a percent as a decimal, . . .
  To write a decimal as a percent, . . .

On Your Own

Now You're Ready
Exercises 19–30

Write the decimal as a percent. Use a model to check your answer.

5. 0.94 **6.** 1.2 **7.** 0.316 **8.** 0.005

EXAMPLE 3 Writing a Fraction as a Percent and a Decimal

On a math test, you get 92 out of a possible 100 points. Which of the following is *not* another way of expressing 92 out of 100?

Ⓐ $\frac{23}{25}$ Ⓑ 92% Ⓒ $\frac{17}{20}$ Ⓓ 0.92

92 out of 100 $= \frac{92}{100}$

$= 92\%$ Eliminate Choice B.

$= \frac{23}{25}$ Eliminate Choice A.

$= 0.92$ Eliminate Choice D.

So, the correct answer is Ⓒ.

EXAMPLE 4 Real-Life Application

The figure shows the portions of ultraviolet (UV) rays reflected by four different surfaces. How many times more UV rays are reflected by water than by sea foam?

Write 25% and $\frac{21}{25}$ as decimals.

Sea foam: $25\% = 25.\% = 0.25$ **Water:** $\frac{21}{25} = \frac{84}{100} = 0.84$

Divide 0.84 by 0.25: $0.25\overline{)0.84} \rightarrow 25\overline{)84.00}$ = 3.36

So, water reflects about 3.4 times more UV rays than sea foam.

On Your Own

9. Write "18 out of 100" as a percent, a fraction, and a decimal.

10. In Example 4, how many times more UV rays are reflected by water than by sand?

15.1 Exercises

Check It Out
Help with Homework
BigIdeasMath.com

Vocabulary and Concept Check

MATCHING **Match the decimal with its equivalent percent.**

1. 0.42
2. 4.02
3. 0.042
4. 0.0402

A. 4.02%
B. 42%
C. 4.2%
D. 402%

5. **OPEN-ENDED** Write three different decimals that are between 10% and 20%.
6. **WHICH ONE DOESN'T BELONG?** Which one does *not* belong with the other three? Explain your reasoning.

70% | 0.7 | $\frac{7}{10}$ | 0.07

Practice and Problem Solving

Write the percent as a decimal.

1 7. 78%
8. 55%
9. 18.5%
10. 57.4%
11. 33%
12. 9%
13. 47.63%
14. 91.25%
15. 166%
16. 217%
17. 0.06%
18. 0.034%

Write the decimal as a percent.

2 19. 0.74
20. 0.52
21. 0.89
22. 0.768
23. 0.99
24. 0.49
25. 0.487
26. 0.128
27. 3.68
28. 5.12
29. 0.0371
30. 0.0046

31. **ERROR ANALYSIS** Describe and correct the error in writing 0.86 as a percent.

✗ 0.86 = 00.86 = 0.0086%

32. **MUSIC** Thirty-six percent of the songs on your MP3 player are pop songs. Write this percent as a decimal.
33. **CAT** About 0.34 of the length of a cat is its tail. Write this decimal as a percent.
34. **COMPUTER** Write the percent of free space on the computer as a decimal.

Volume	Capacity	Free Space	% Free Space
(C:)	149 GB	133 GB	89 %

Write the percent as a fraction in simplest form and as a decimal.

35. 36%
36. 23.5%
37. 16.24%

Assignment Guide and Homework Check

Level	Assignment	Homework Check
Advanced	1–12, 16–30 even, 31, 32–40 even, 41–50	18, 30, 34, 36

Common Errors

- **Exercises 7–30** Students may move the decimal point the wrong way, forget to place zeros as placeholders, or move the decimal point too many places (especially when the percent is greater than 100). As a class, gather some helpful information for remembering how to convert decimals and percents. For example, when the percent is greater than 100%, the decimal equivalent will be greater than 1. Because there are two zeros in 100, you need to move the decimal point two places. When converting from percents to decimals, you move the decimal point to the left because D is to the left of P in the alphabet. When converting from decimals to percents, you move the decimal point to the right because P is to the right of D in the alphabet.
- **Exercises 35–37** Students may move the decimal point to the right instead of to the left. Remind them how to get rid of the decimal point in the percent when writing a fraction.

15.1 Record and Practice Journal

Write the percent as a decimal.

1. 35% **0.35**
2. 160% **1.6**
3. 74.8% **0.748**
4. 0.3% **0.003**

Write the decimal as a percent.

5. 1.23 **123%**
6. 0.49 **49%**
7. 0.024 **2.4%**
8. 0.881 **88.1%**

Write the percent as a fraction in simplest form and as a decimal.

9. 48% $\frac{12}{25}$; 0.48
10. 15.5% $\frac{31}{200}$; 0.155
11. 84.95% $\frac{1699}{2000}$; 0.8495

12. People with severe hearing loss were given a sentence and word recognition test six months after they got implants in their ears. The patients scored an average of 82% on the test. Write this percent as a decimal.
0.82

Vocabulary and Concept Check

1. B
2. D
3. C
4. A
5. *Sample answer:* 0.11, 0.13, 0.19
6. 0.07 because it represents 7% instead of 70%.

Practice and Problem Solving

7. 0.78
8. 0.55
9. 0.185
10. 0.574
11. 0.33
12. 0.09
13. 0.4763
14. 0.9125
15. 1.66
16. 2.17
17. 0.0006
18. 0.00034
19. 74%
20. 52%
21. 89%
22. 76.8%
23. 99%
24. 49%
25. 48.7%
26. 12.8%
27. 368%
28. 512%
29. 3.71%
30. 0.46%
31. The decimal point was moved in the wrong direction. 0.86 = 0.86 = 86%
32. 0.36
33. 34%
34. 0.89
35. $\frac{9}{25} = 0.36$
36. $\frac{47}{200} = 0.235$
37. $\frac{203}{1250} = 0.1624$

Practice and Problem Solving

38. **a.** car: 0.2, school bus: 0.48, bicycle: 0.08

b. car: $\frac{1}{5}$, school bus: $\frac{12}{25}$, bicycle: $\frac{2}{25}$

c. 24%

d. *Answer should include, but is not limited to:* A bar graph showing either the *number* of students or the *portion* of students in the class that get to school in various ways.

39. 40%

40. See *Taking Math Deeper*.

41. **a.** $16.\overline{6}\%$, or $16\frac{2}{3}\%$

b. $\frac{5}{6}$

Fair Game Review

42. $\frac{23}{50}$ **43.** $\frac{31}{100}$

44. $2\frac{1}{5}$ **45.** $4\frac{8}{25}$

46. $-5x + 3$

47. $-1.6n - 1$

48. $-3y + 15$

49. $-b - \frac{3}{2}$

50. B

Mini-Assessment

Write the percent as a decimal.

1. 12% 0.12 2. 130% 1.3

Write the decimal as a percent.

3. 0.98 98% 4. 2.25 225%

5. Forty-two percent of your cell phone ringtones are pop songs. Write this percent as a decimal. 0.42

Taking Math Deeper

Exercise 40

This is a good review that all regions of a circle graph must add up to 1 or 100%. Begin by making a table to summarize the given information.

Color	Portion Who Said this Color	
	Percent	Decimal
●	26%	0.26
●	40%	0.40
●	4%	0.04
●		
●	14%	0.14

1 What % said red, blue, or yellow? "Or" means to add the percents.

a. 26% + 40% + 4% = 70%

2 "How many times more" means to divide.

b. 0.26 ÷ 0.04 = 6.5 times more.

3 One way is to use decimals. Add the decimal numbers and write the decimal sum as a percent.

0.26 + 0.40 + 0.04 + 0.14 = 0.84 or 84%

Another way is to add the percent form of the numbers.

26% + 40% + 4% + 14% = 84%

To find the percent of students who said green, subtract from 100%.

c. 100% − 84% = 16% = 0.16

Reteaching and Enrichment Strategies

If students need help...	If students got it...
Resources by Chapter • Practice A and Practice B • Puzzle Time Record and Practice Journal Practice Differentiating the Lesson Lesson Tutorials Skills Review Handbook	Resources by Chapter • Enrichment and Extension • Technology Connection Start the next section

38. SCHOOL The percents of students who travel to school by car, bus, and bicycle are shown for a school of 825 students.

a. Write the percents as decimals.

b. Write the percents as fractions.

c. What percent of students use another method to travel to school?

d. **RESEARCH** Make a bar graph that represents how the students in your class travel to school.

39. ELECTIONS In an election, the winning candidate receives 60% of the votes. What percent of the votes does the other candidate receive?

40. COLORS Students in a class were asked to tell their favorite color.

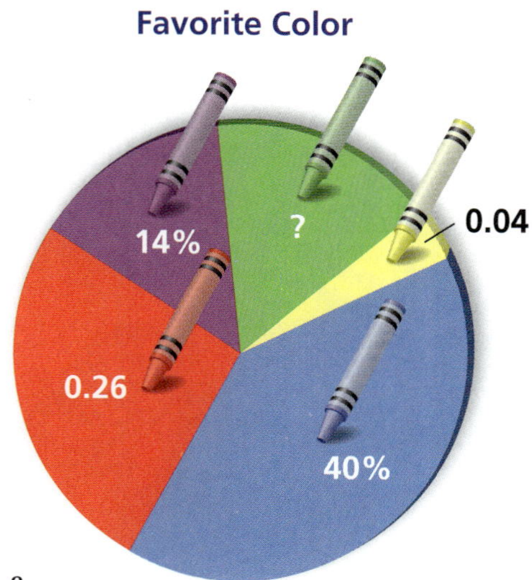

a. What percent said red, blue, or yellow?

b. How many times more students said red than yellow?

c. Use two methods to find the percent of students who said green. Which method do you prefer?

41. Problem Solving In the first 42 Super Bowls, $0.1\overline{6}$ of the MVPs (most valuable players) were running backs.

a. What percent of the MVPs were running backs?

b. What fraction of the MVPs were *not* running backs?

Fair Game Review What you learned in previous grades & lessons

Write the decimal as a fraction or mixed number in simplest form. *(Skills Review Handbook)*

42. 0.46 **43.** 0.31 **44.** 2.2 **45.** 4.32

Simplify the expression. *(Section 13.1)*

46. $4x + 3 - 9x$ **47.** $5 + 3.2n - 6 - 4.8n$

48. $2y - 5(y - 3)$ **49.** $-\frac{1}{2}(8b + 3) + 3b$

50. MULTIPLE CHOICE Ham costs $4.48 per pound. Cheese costs $6.36 per pound. You buy 1.5 pounds of ham and 0.75 pound of cheese. How much more do you pay for the ham? *(Section 2.4 and Section 2.5)*

Ⓐ $1.41 Ⓑ $1.95 Ⓒ $4.77 Ⓓ $6.18

15.2 Comparing and Ordering Fractions, Decimals, and Percents

Essential Question How can you order numbers that are written as fractions, decimals, and percents?

1 ACTIVITY: Using Fractions, Decimals, and Percents

Work with a partner. Decide which number form (fraction, decimal, or percent) is more common. Then find which is greater.

a. 7% sales tax or $\frac{1}{20}$ sales tax

b. 0.37 cup of flour or $\frac{1}{3}$ cup of flour

c. $\frac{5}{8}$-inch wrench or 0.375-inch wrench

d. $12\frac{3}{5}$ dollars or 12.56 dollars

e. 93% test score or $\frac{7}{8}$ test score

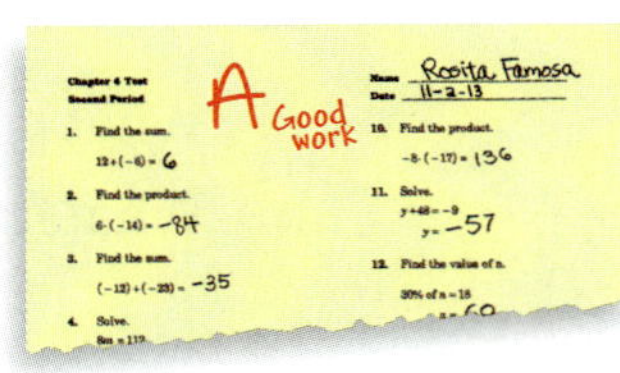

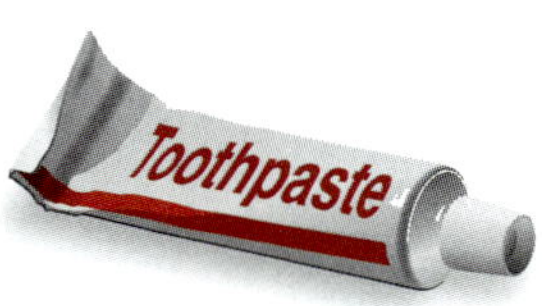

f. $5\frac{5}{6}$ fluid ounces or 5.6 fluid ounces

Fractions, Decimals, and Percents

In this lesson, you will

- compare and order fractions, decimals, and percents.
- solve real-life problems.

2 ACTIVITY: Ordering Numbers

Work with a partner to order the following numbers.

$\frac{1}{8}$ 11% $\frac{3}{20}$ 0.172 0.32 43% 7% 0.7 $\frac{5}{6}$

a. Decide on a strategy for ordering the numbers. Will you write them all as fractions, decimals, or percents?

b. Use your strategy and a number line to order the numbers from least to greatest. (Note: Label the number line appropriately.)

Laurie's Notes

Introduction

Applying Mathematical Practices

- **Reason Abstractly and Quantitatively:** In this lesson, students will be converting between different representations of a number so that numbers may be compared. Mathematically proficient students would reason that $\frac{6}{11} > 0.48$ because $\frac{6}{11} > \frac{1}{2}$ and $\frac{1}{2} > 0.48$.

Motivate

- Write each number on an index card: $\frac{1}{10}$, $\frac{13}{20}$, $\frac{4}{5}$, 0.18, 0.45, 0.5, 35%, 60%, 85%.
- Hand out the 9 cards. Ask the 3 students holding fraction cards to stand in order (least to greatest). Then ask the 3 students holding decimal cards and the 3 students holding percent cards to do the same.
- Have each group of students describe the strategies for ordering themselves.
- Now ask all 9 students to order themselves.
- Discuss strategies used when all 3 forms are used in the same problem.

Activity Notes

Activity 1

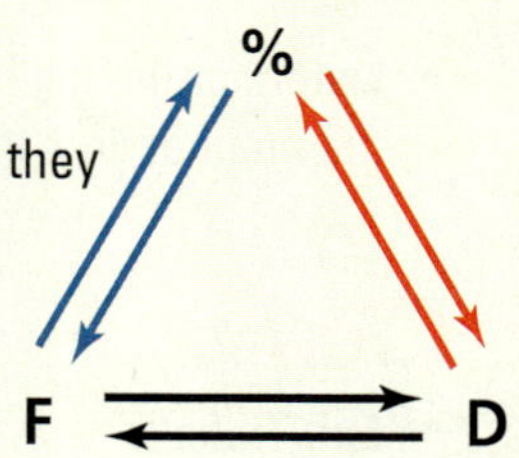

- Remind students of the conversion triangle. Today they will work with all 3 forms.
- **Reason Abstractly and Quantitatively:** Remind students that there are two parts to each question.
- Discuss the results in class.

? "Which fractions were repeating decimals?" Listen for knowledge that thirds and sixths are repeating decimals.

- **Common Error:** Some students who do not have a good understanding of how to convert a fraction to a decimal will take simple fractions such as $\frac{1}{20}$ and write 1.20 or 0.120. For $\frac{1}{3}$, they write 1.3 or 0.13.

Activity 2

- **Reason Abstractly and Quantitatively:** Have students work in groups, with each student sharing suggestions for how to order the numbers. Listen for key understanding, such as 0.7 and $\frac{5}{6}$ are the only two numbers greater than $\frac{1}{2}$.
- Note that the number line has scaling for tenths and only one number will be graphed in a tenth scale mark.
- In completing this activity, many skills are reviewed. This is an opportunity for informal assessment to guide instruction in the sections ahead.
- **Communication:** One group should present their answers to the class. They should discuss their results and the strategies they used.

What Your Students Will Learn

- Compare and order fractions, decimals, and percents by using the same number form.

Previous Learning

Students should know how to write halves, fourths, tenths, and hundredths as percents. Students also need to be able to convert between fractions and decimals, and multiply and divide decimals by powers of 10.

Technology for the Teacher

Dynamic Classroom

Lesson Plans
Complete Materials List

15.2 Record and Practice Journal

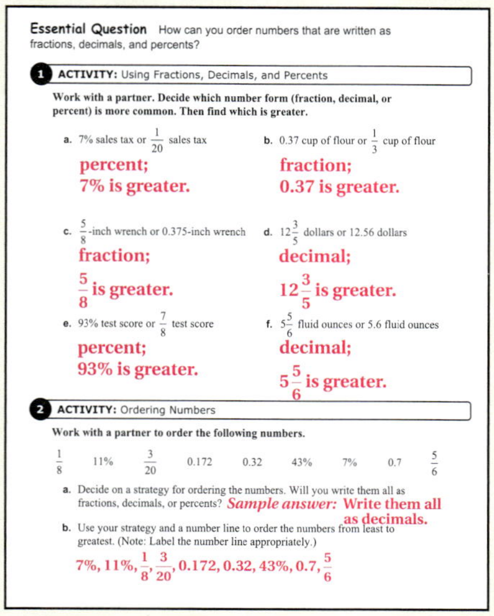

Essential Question How can you order numbers that are written as fractions, decimals, and percents?

1 ACTIVITY: Using Fractions, Decimals, and Percents

Work with a partner. Decide which number form (fraction, decimal, or percent) is more common. Then find which is greater.

a. 7% sales tax or $\frac{1}{20}$ sales tax
percent;
7% is greater.

b. 0.37 cup of flour or $\frac{1}{3}$ cup of flour
fraction;
0.37 is greater.

c. $\frac{5}{8}$-inch wrench or 0.375-inch wrench
fraction;
$\frac{5}{8}$ is greater.

d. $12\frac{3}{5}$ dollars or 12.56 dollars
decimal;
$12\frac{3}{5}$ is greater.

e. 93% test score or $\frac{7}{8}$ test score
percent;
93% is greater.

f. $5\frac{5}{6}$ fluid ounces or 5.6 fluid ounces
decimal;
$5\frac{5}{6}$ is greater.

2 ACTIVITY: Ordering Numbers

Work with a partner to order the following numbers.

$\frac{1}{8}$ 11% $\frac{3}{20}$ 0.172 0.32 43% 7% 0.7 $\frac{5}{6}$

a. Decide on a strategy for ordering the numbers. Will you write them all as fractions, decimals, or percents? *Sample answer:* Write them all as decimals.

b. Use your strategy and a number line to order the numbers from least to greatest. (Note: Label the number line appropriately.)
7%, 11%, $\frac{1}{8}$, $\frac{3}{20}$, 0.172, 0.32, 43%, 0.7, $\frac{5}{6}$

English Language Learners

Auditory

Students may struggle with the idea that three different forms of a number are useful. Have a classroom discussion where students brainstorm about when to use fractions (construction, recipes), when to use decimals (money, chemistry), and when to use percents (surveys, chance of rain).

15.2 Record and Practice Journal

3 ACTIVITY: The Game of Math Card War

Preparation:

- Cut index cards to make 40 playing cards.*
- Write each number in the table onto a card.

75%	$\frac{3}{4}$	$\frac{1}{3}$	$\frac{3}{10}$	0.3	25%	0.4	0.25	100%	0.27
0.75	$66\frac{2}{3}\%$	12.5%	40%	$\frac{1}{4}$	4%	0.5%	0.04	$\frac{1}{100}$	$\frac{2}{3}$
0	30%	5%	$\frac{27}{100}$	0.05	$33\frac{1}{3}\%$	$\frac{2}{5}$	0.333...	27%	1%
1	0.01	$\frac{1}{20}$	$\frac{1}{8}$	0.125	$\frac{1}{25}$	$\frac{1}{200}$	0.005	0.666...	0%

To Play:

- Play with a partner.
- Deal 20 cards to each player facedown.
- Each player turns one card faceup. The player with the greater number wins. The winner collects both cards and places them at the bottom of his or her cards.
- Suppose there is a tie. Each player lays three cards facedown, then a new card faceup. The player with the greater of these new cards wins. The winner collects all 10 cards and places them at the bottom of his or her cards.
- Continue playing until one player has all the cards. This player wins the game.

Check students' work.

*Cut-outs are available in the back of the Record and Practice Journal.

What Is Your Answer?

4. **IN YOUR OWN WORDS** How can you order numbers that are written as fractions, decimals, and percents? Give an example with your answer.

Convert all numbers to one form then order them.

5. All but one of the U.S. coins shown has a name that is related to its value. Which one is it? How are the names of the others related to their values?

Nickel; the name nickel comes from the metal the coin was originally made from; 1 cent = $\frac{1}{100}$; dime comes from the French word "disme" meaning tenth-part; quarter = $\frac{1}{4}$ dollar; half dollar = $\frac{1}{2}$ dollar

Laurie's Notes

Activity 3

- To preserve cards for multiple uses, make cards on colored card stock and store individual sets in plastic zipper bags.
- The card game *War* is common to most students. The question asked for each play is, "Which number is greater?" The player with the greater value collects both cards.
- If the cards have equivalent values $\left(\text{i.e., } 75\% \text{ and } \frac{3}{4}\right)$, there is a tie. As stated in the text, each player lays 3 cards face down and 1 card face up. The player with the card of greater value collects all of the cards in play.
- **Comparing Cards:** The learning component of this activity is when students actually compare two numbers. Listen to students discuss how they are comparing the numbers.
- To start the play, give students the opportunity to preview the cards. Explain the rules and let students begin.
- If one group finishes early, they should shuffle the cards and play again.
- **Extension:** You can use the same set of cards to do a matching activity with the cards either face up (easier) or down. If they make a 3-card match, such as 75%, $\frac{3}{4}$, and 0.75, they get a certain number of points. A 2-card match, such as 0 and 0%, is worth fewer points.

? What cards have no match?

What Is Your Answer?

- Listen for the big idea, namely that the farther to the right the number is on the number line, the greater the number.

Closure

- **Exit Ticket:**
  - Write a percent between 25% and 40%. Write the equivalent fraction and decimal. *Sample answer:* 30%, $\frac{3}{10}$, 0.3
  - Write a percent between 75% and 90%. Write the equivalent fraction and decimal. *Sample answer:* 85%, $\frac{17}{20}$, 0.85

3 ACTIVITY: The Game of Math Card War

Math Practice

Make Sense of Quantities

What strategies can you use to determine which number is greater?

Preparation:

- Cut index cards to make 40 playing cards.
- Write each number in the table onto a card.

To Play:

- Play with a partner.
- Deal 20 cards facedown to each player.
- Each player turns one card faceup. The player with the greater number wins. The winner collects both cards and places them at the bottom of his or her cards.
- Suppose there is a tie. Each player lays three cards facedown, then a new card faceup. The player with the greater of these new cards wins. The winner collects all 10 cards and places them at the bottom of his or her cards.
- Continue playing until one player has all the cards. This player wins the game.

75%	$\frac{3}{4}$	$\frac{1}{3}$	$\frac{3}{10}$	0.3	25%	0.4	0.25	100%	0.27
0.75	$66\frac{2}{3}\%$	12.5%	40%	$\frac{1}{4}$	4%	0.5%	0.04	$\frac{1}{100}$	$\frac{2}{3}$
0	30%	5%	$\frac{27}{100}$	0.05	$33\frac{1}{3}\%$	$\frac{2}{5}$	0.333. . .	27%	1%
1	0.01	$\frac{1}{20}$	$\frac{1}{8}$	0.125	$\frac{1}{25}$	$\frac{1}{200}$	0.005	0.666. . .	0%

What Is Your Answer?

4. **IN YOUR OWN WORDS** How can you order numbers that are written as fractions, decimals, and percents? Give an example with your answer.

5. All but one of the U.S. coins shown has a name that is related to its value. Which one is it? How are the names of the others related to their values?

Practice Use what you learned about ordering numbers to complete Exercises 4–7, 16, and 17 on page 660.

15.2 Lesson

Check It Out
Lesson Tutorials
BigIdeasMath.com

When comparing and ordering fractions, decimals, and percents, write the numbers as all fractions, all decimals, or all percents.

EXAMPLE 1 Comparing Fractions, Decimals, and Percents

a. Which is greater, $\frac{3}{20}$ or 16%?

Write $\frac{3}{20}$ as a percent: $\frac{3}{20} \overset{\times 5}{=} \frac{15}{100} = 15\%$

15% is less than 16%. So, 16% is the greater number.

b. Which is greater, 79% or 0.08?

Write 79% as a decimal: $79\% = 79.\% = 0.79$

0.79 is greater than 0.08. So, 79% is the greater number.

Study Tip

It is usually easier to order decimals or percents than to order fractions.

On Your Own

Now You're Ready
Exercises 4–15

1. Which is greater, 25% or $\frac{7}{25}$?
2. Which is greater, 0.49 or 94%?

EXAMPLE 2 Real-Life Application

You, your sister, and a friend each take the same number of shots at a soccer goal. You make 72% of your shots, your sister makes $\frac{19}{25}$ of her shots, and your friend makes 0.67 of his shots. Who made the fewest shots?

Write 72% and $\frac{19}{25}$ as decimals.

You: $72\% = 72.\% = 0.72$ **Sister:** $\frac{19}{25} \overset{\times 4}{=} \frac{76}{100} = 0.76$

Graph the decimals on a number line.

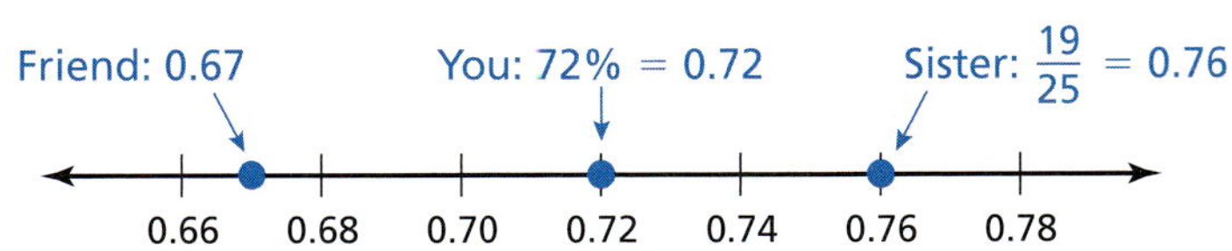

0.67 is the least number. So, your friend made the fewest shots.

Remember

To order numbers from least to greatest, write them as they appear on a number line from left to right.

Laurie's Notes

Introduction

Connect

- **Yesterday:** Students compared and ordered common percents, decimals, and fractions.
- **Today:** Students will compare and order less common percents, decimals, and fractions.

Motivate

- Read the 5 statistics about the United States and have students order the percents from least to greatest.
  1) 90% of the states joined the United States before 1900.
  2) The U.S. population represents about 4.5% of the world population.
  3) About 20% of the U.S. population is under 15 years old.
  4) 62% of the states are entirely east of the Mississippi River.
  5) Water area represents about 6.9% of the total area of the United States.
- ? "Why is it usually easier to order decimals or percents than fractions?" Listen for an understanding of unlike denominators being harder to order than place value.

Lesson Notes

Example 1

- ? "To compare $\frac{3}{20}$ and 16%, should you write $\frac{3}{20}$ as a percent or 16% as a fraction and why?" Discuss both options and perhaps work through both options, if time permits.
- ? "Could you compare $\frac{3}{20}$ and 16% by writing each as a decimal?" yes; Depending on time, show this option as well.
- ? Work through part (b). "What is 0.08 as a percent?" 8%
- **Common Error:** Students might say 80% because they are comparing it to 79%.
- **Reason Abstractly and Quantitatively:** Take time to review and analyze the efficiency of the different strategies. For instance, if the fraction in part (a) had been $\frac{3}{19}$, it would have been difficult to write an equivalent fraction with a denominator of 100. Changing the fraction and the percent to decimals would have been the preferred strategy.

On Your Own

- Discuss methods used by students to make comparisons.

Example 2

- Note the scale on the number line. A common misconception is that the scale must be in units of 1, 5, or 10. Even digits are not as common.
- ? "Why is scaling by even digits helpful in this problem? Could we use a scale of 10?" Even digits show the distance between the 3 numbers in an accurate display. You could use 10, but it would be less accurate.

Goal Today's lesson is comparing and ordering less common fractions, decimals, and percents.

Technology for the Teacher

Dynamic Classroom

Lesson Tutorials
Lesson Plans
Answer Presentation Tool

Extra Example 1

a. Which is greater, $\frac{17}{20}$ or 80%? $\frac{17}{20}$

b. Which is greater, 28% or 0.29? 0.29

On Your Own

1. $\frac{7}{25}$ **2.** 94%

Extra Example 2

You, your sister, and your friend each take the same number of shots at a soccer goal. You make 0.67 of your shots, your sister makes 68% of her shots, and your friend makes $\frac{17}{20}$ of her shots. Who made the fewest shots? You made the fewest shots.

On Your Own

3. You made the most shots.

Extra Example 3

The table shows the portions of the population of Rhode Island that live in each county. List the counties in order by population from least to greatest.

Counties	Fraction	Decimal	Percent
Bristol	$\frac{1}{20}$		
Kent		0.16	
Newport	$\frac{2}{25}$		
Providence			59%
Washington	$\frac{3}{25}$		

Bristol, Newport, Washington, Kent, Providence

On Your Own

4. Washington, Michigan, Ohio, Illinois, New York, Texas, and California

Differentiated Instruction

Kinesthetic

Some students have difficulty seeing that 60% is not only equal to "60 out of 100," but is also equal to "6 out of 10" and "3 out of 5." Give students three 100-grid squares. In the first grid, have students shade 60 out of 100 squares. With the second grid, have students draw thick lines around groups of 10 squares and then shade 6 of the groups. With the third grid, have students draw thick lines around groups of 20 squares and then shade 3 of the groups. When students compare all three grids, they should see that the same amount of 60 squares have been shaded on each of the grids. This can be seen easily if the students have grouped the squares by rows or columns.

Laurie's Notes

On Your Own

- **Think-Pair-Share:** Students should read the question independently and then work in pairs to answer the question. When they have answered the question, the pair should compare their answer with another group and discuss any discrepancies.
- Have students graph the three numbers on a scaled number line.

Example 3

- Give students time to read the information in the problem.
- ? "Interpret what it means for $\frac{1}{50}$ of the U.S. population to live in Washington." Listen for an understanding that one in every 50 U.S. residents lives in Washington.
- ? **Look for and Express Regularity in Repeated Reasoning:** "If there were 100 U.S. residents, how many would live in Washington?" 2 "If there were 1000 U.S. residents, how many would live in Washington?" 20
- Discuss the columns of the table. Note that the decimals and percents are easier to compare than the fractions.

On Your Own

- **Construct Viable Arguments and Critique the Reasoning of Others:** There are several ways in which students may justify their answers. Take time to hear a variety of approaches.

Closure

- Complete this table of common fractions, decimals, and percents.

Fraction	$\frac{1}{4}$	$\frac{3}{10}$	$\frac{1}{2}$	$\frac{2}{5}$	$\frac{1}{20}$	$\frac{3}{20}$	$\frac{7}{10}$	$\frac{3}{4}$	$\frac{1}{5}$	1
Decimal	0.25	0.3	0.5	0.4	0.05	0.15	0.7	0.75	0.2	1
Percent	25%	30%	50%	40%	5%	15%	70%	75%	20%	100%

On Your Own

Now You're Ready
Exercises 16–21

3. You make 75% of your shots, your sister makes $\frac{13}{20}$ of her shots, and your friend makes 0.7 of his shots. Who made the most shots?

EXAMPLE 3 Real-Life Application

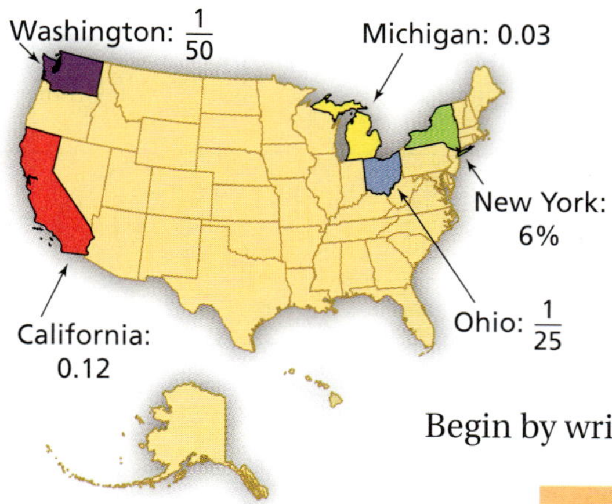

The map shows the portions of the U.S. population that live in five states.

List the five states in order by population from least to greatest.

Begin by writing each portion as a fraction, a decimal, and a percent.

State	Fraction	Decimal	Percent
Michigan	$\frac{3}{100}$	0.03	3%
New York	$\frac{6}{100}$	0.06	6%
Washington	$\frac{1}{50}$	0.02	2%
California	$\frac{12}{100}$	0.12	12%
Ohio	$\frac{1}{25}$	0.04	4%

Graph the percent for each state on a number line.

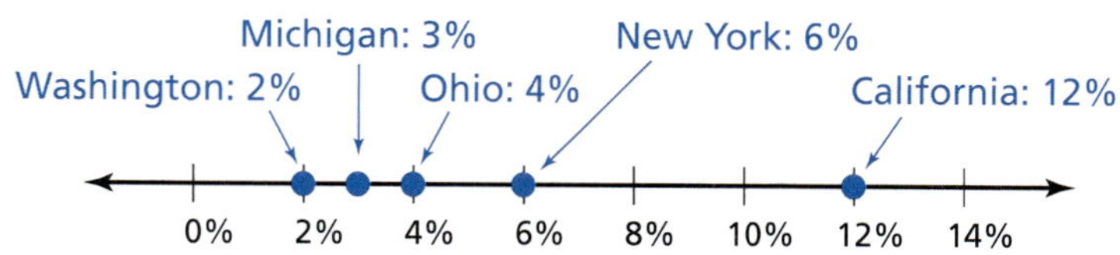

∴ The states in order by population from least to greatest are Washington, Michigan, Ohio, New York, and California.

On Your Own

4. The portion of the U.S. population that lives in Texas is $\frac{2}{25}$. The portion that lives in Illinois is 0.042. Reorder the states in Example 3 including Texas and Illinois.

15.2 Exercises

Check It Out
Help with Homework
BigIdeasMath.com

Vocabulary and Concept Check

1. **NUMBER SENSE** Copy and complete the table.

Fraction	Decimal	Percent
$\frac{18}{25}$	0.72	
$\frac{17}{20}$		85%
$\frac{13}{50}$		
	0.62	
		45%

2. **NUMBER SENSE** How would you decide whether $\frac{3}{5}$ or 59% is greater? Explain.

3. **WHICH ONE DOESN'T BELONG?** Which one does *not* belong with the other three? Explain your reasoning.

40%	$\frac{2}{5}$
0.4	0.04

Practice and Problem Solving

Tell which number is greater.

1 4. 0.9, 95%

5. 20%, 0.02

6. $\frac{37}{50}$, 37%

7. 50%, $\frac{13}{25}$

8. 0.086, 86%

9. 76%, 0.67

10. 60%, $\frac{5}{8}$

11. 0.12, 1.2%

12. 17%, $\frac{4}{25}$

13. 140%, 0.14

14. $\frac{1}{3}$, 30%

15. 80%, $\frac{7}{9}$

Use a number line to order the numbers from least to greatest.

2 16. 38%, $\frac{8}{25}$, 0.41

17. 68%, 0.63, $\frac{13}{20}$

18. $\frac{43}{50}$, 0.91, $\frac{7}{8}$, 84%

19. 0.15%, $\frac{3}{20}$, 0.015

20. 2.62, $2\frac{2}{5}$, 26.8%, 2.26, 271%

21. $\frac{87}{200}$, 0.44, 43.7%, $\frac{21}{50}$

22. **TEST** You answered 21 out of 25 questions correctly on a test. Did you reach your goal of getting at least 80%?

23. **POPULATION** The table shows the portions of the world population that live in four countries. Order the countries by population from least to greatest.

Country	Brazil	India	Russia	United States
Portion of World Population	2.8%	$\frac{7}{40}$	$\frac{1}{50}$	0.044

Assignment Guide and Homework Check

Level	Assignment	Homework Check
Advanced	1–7, 8–30 even, 31–36	18, 22, 24, 26

Common Errors

- **Exercises 4–21** Students may try to order the numbers without converting them or will only convert them mentally and do so incorrectly. Tell them that it is necessary to convert all the numbers to one form and that they should write out the steps to make sure that they are converting correctly.
- **Exercises 24 and 25** Students may try to round the numbers that have repeating decimals and incorrectly order the numbers. Remind them that even though you often round repeating decimals, the decimal is actually less than or greater than the rounded decimal. For example, $0.\overline{6}$ is less than 0.667, but greater than 0.666.

15.2 Record and Practice Journal

Circle the number that is greater.

1. 0.06, (60%) 2. (78%), $\frac{19}{25}$ 3. ($\frac{23}{20}$), 110% 4. (0.23), 2.3%

Use a number line to order the numbers from least to greatest.

5. 44.5%, 0.4445, $\frac{4}{9}$, 0.44 **0.44, $\frac{4}{9}$, 0.4445, 44.5%**

6. $\frac{5}{12}$, 0.4, 42%, 0.416 **0.4, 0.416, $\frac{5}{12}$, 42%**

7. The table shows the portion of each age group that recycles plastic. Order the groups by the portion that recycle from least to greatest.

Age Group	Echo Boomers	Gen X	Baby Boomers	Matures
Portion that Recycle	51%	0.57	0.61	$\frac{6}{10}$

Echo Boomers, Gen X, Matures, Baby Boomers

Vocabulary and Concept Check

1.

$\frac{18}{25}$	0.72	72%
$\frac{17}{20}$	0.85	85%
$\frac{13}{50}$	0.26	26%
$\frac{31}{50}$	0.62	62%
$\frac{9}{20}$	0.45	45%

2. *Sample answer*: Write $\frac{3}{5}$ as a percent, then compare 59% to determine which is greater.

3. 0.04; 0.04 = 4%, but 40%, $\frac{2}{5}$, and 0.4 are all equal to 40%.

Practice and Problem Solving

4. 95%
5. 20%
6. $\frac{37}{50}$
7. $\frac{13}{25}$
8. 86%
9. 76%
10. $\frac{5}{8}$
11. 0.12
12. 17%
13. 140%
14. $\frac{1}{3}$
15. 80%

16.

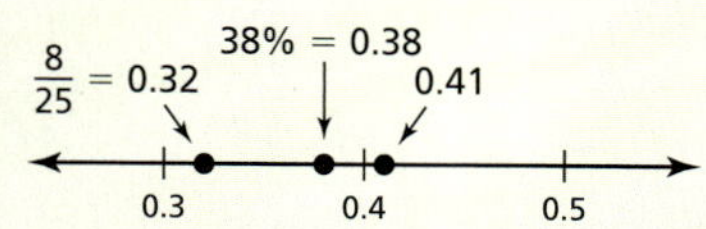

17.

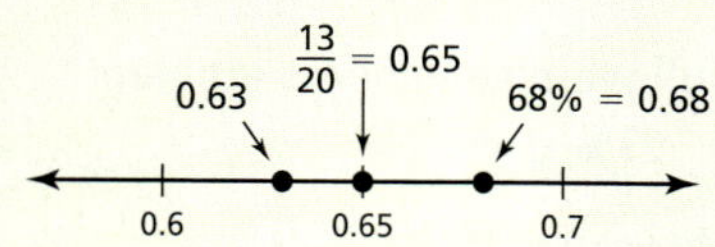

18.

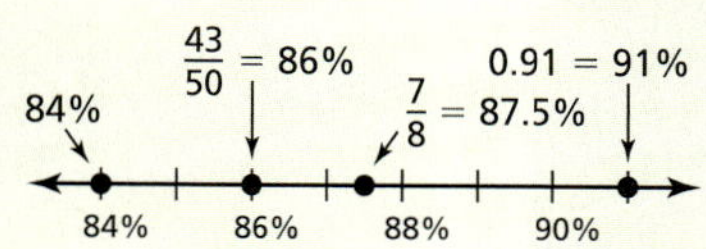

19.

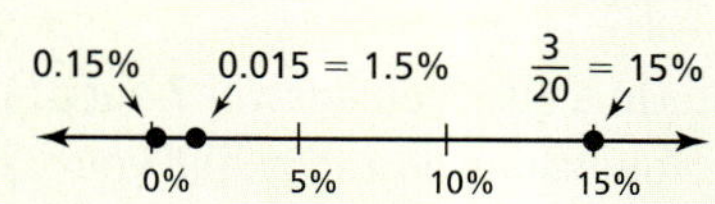

20. See Additional Answers.

Practice and Problem Solving

21.

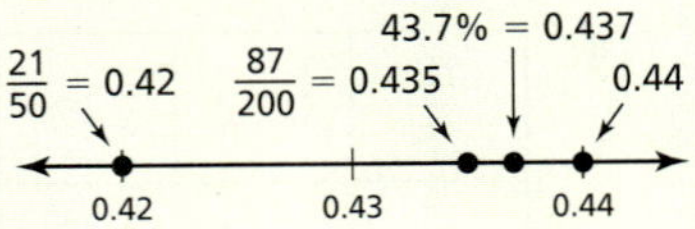

22. yes

23. Russia, Brazil, United States, India

24. 0.66, 66.1%, $\frac{2}{3}$, 0.667

25. 21%, $0.2\overline{1}$, $\frac{11}{50}$, $\frac{2}{9}$

26. *A** **27.** *D*

28. *B* **29.** *C*

30. Stage 21, Stage 7, Stage 8, Stage 1, Stage 17

31. See *Taking Math Deeper*.

32. **a.** 7

b. There is none. $\frac{1}{a}$ is less than 33% when a is greater than 3, but when a is greater than 3, $\frac{a}{8}$ is greater than 33%.

Fair Game Review

33. yes **34.** no

35. yes **36.** D

Mini-Assessment

Tell which number is greater.

1. 27%, 0.48 0.48
2. $\frac{2}{5}$, 0.125 $\frac{2}{5}$
3. $\frac{3}{4}$, 76% 76%
4. $\frac{5}{8}$, 60% $\frac{5}{8}$
5. On a quiz, you answer 7 out of 10 questions correctly. Did you reach your goal of getting 80% or better? no

Taking Math Deeper

Exercise 31

The problem doesn't tell students which form to use to order the numbers. It is up to the students to decide when percents or decimals are easier to use than fractions.

1 Use a table to organize the information.

Animal	Portion of Day Sleeping		
	Percent	Decimal	Fraction
Dolphin	43.3%	0.433	
Lion	56.3%	0.563	
Rabbit	47.5%	0.475	$\frac{19}{40}$
Squirrel	62.0%	0.620	$\frac{31}{50}$
Tiger	65.8%	0.658	

2 Use a number line to order the decimals.

a.

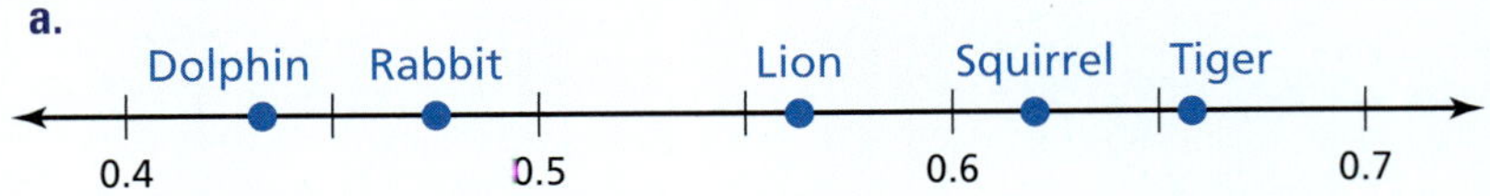

3 Using an estimate of 8 hours a day, the portion of the day a teen sleeps is

b. $\frac{8}{24} = \frac{1}{3} = 33.\overline{3}\%$.

c. On the ordered list, the teen would be first!

Project

Use a chart to keep track of the number of hours you sleep each day for a week. Compare your chart with three of your classmates. Find the average number of hours each of you sleep each day. Which day seems to be the sleepiest day of the week?

Reteaching and Enrichment Strategies

If students need help. . .	If students got it. . .
Resources by Chapter • Practice A and Practice B • Puzzle Time Record and Practice Journal Practice Differentiating the Lesson Lesson Tutorials Skills Review Handbook	Resources by Chapter • Enrichment and Extension • Technology Connection Start the next section

PRECISION **Order the numbers from least to greatest.**

24. 66.1%, 0.66, $\frac{2}{3}$, 0.667

25. $\frac{2}{9}$, 21%, $0.2\overline{1}$, $\frac{11}{50}$

Tell which letter shows the graph of the number.

26. $\frac{2}{5}$

27. 45.2%

28. 0.435

29. $\frac{4}{9}$

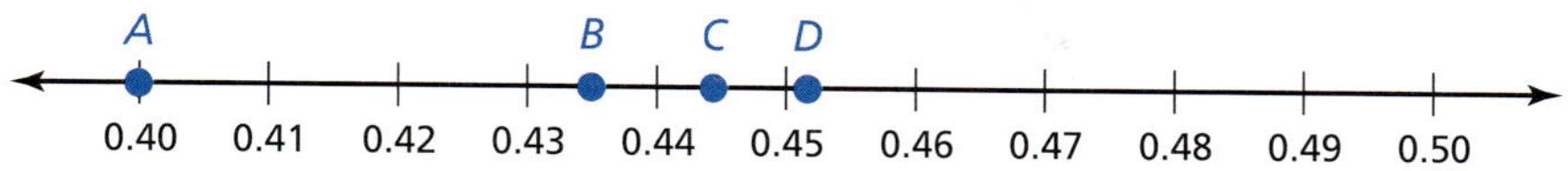

30. **TOUR DE FRANCE** The Tour de France is a bicycle road race. The whole race is made up of 21 small races called *stages*. The table shows how several stages compare to the whole Tour de France in a recent year. Order the stages from shortest to longest.

Stage	1	7	8	17	21
Portion of Total Distance	$\frac{11}{200}$	0.044	$\frac{6}{125}$	0.06	4%

31. **SLEEP** The table shows the portions of the day that several animals sleep.

a. Order the animals by sleep time from least to greatest.

b. Estimate the portion of the day that you sleep.

c. Where do you fit on the ordered list?

Animal	Portion of Day Sleeping
Dolphin	0.433
Lion	56.3%
Rabbit	$\frac{19}{40}$
Squirrel	$\frac{31}{50}$
Tiger	65.8%

32. Number Sense Tell what whole number you can substitute for a in each list so the numbers are ordered from least to greatest. If there is none, explain why.

a. $\frac{2}{a}$, $\frac{a}{22}$, 33%

b. $\frac{1}{a}$, $\frac{a}{8}$, 33%

Fair Game Review What you learned in previous grades & lessons

Tell whether the ratios form a proportion. *(Section 14.2)*

33. $\frac{6}{10}$, $\frac{9}{15}$

34. $\frac{7}{16}$, $\frac{28}{80}$

35. $\frac{20}{12}$, $\frac{35}{21}$

36. **MULTIPLE CHOICE** What is the solution of $7 < n - 3$? *(Section 7.6)*

Ⓐ $n < 4$ Ⓑ $n < 10$ Ⓒ $n > 4$ Ⓓ $n > 10$

15.3 The Percent Proportion

Essential Question How can you use models to estimate percent questions?

The statement "25% of 12 is 3" has three numbers. In real-life problems, any one of these numbers can be unknown.

Question	Which number is missing?	Type of Question
What is 25% of 12?	3	Find a part of a number.
3 is what percent of 12?	25%	Find a percent.
3 is 25% of what?	12	Find the whole.

1 ACTIVITY: Estimating a Part

Work with a partner. Use a model to estimate the answer to each question.

a. What number is 50% of 30?

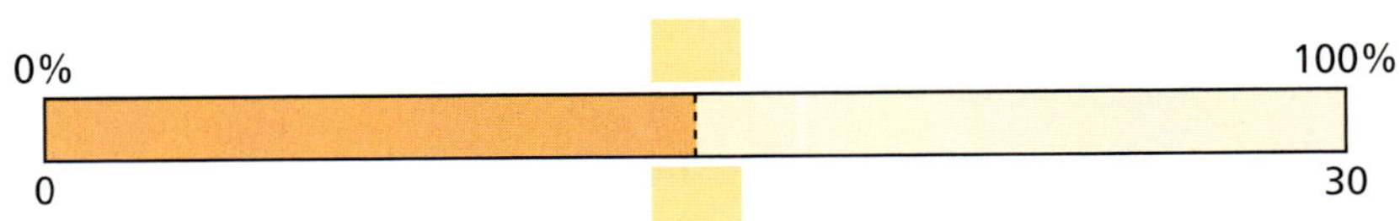

So, from the model, ______ is 50% of 30.

b. What number is 75% of 30?

c. What number is 40% of 30?

d. What number is 6% of 30?

e. What number is 65% of 30?

2 ACTIVITY: Estimating a Percent

Percent Proportion
In this lesson, you will
- use the percent proportion to find parts, wholes, and percents.

Work with a partner. Use a model to estimate the answer to each question.

a. 15 is what percent of 75?

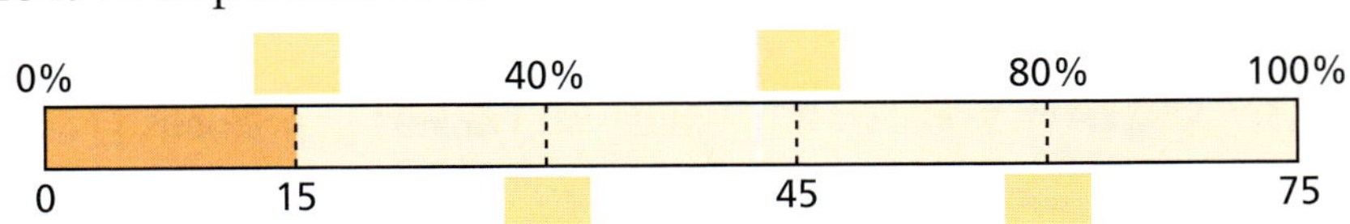

So, from the model, 15 is ______ of 75.

b. 5 is what percent of 20?

c. 18 is what percent of 40?

d. 50 is what percent of 80?

e. 75 is what percent of 50?

Laurie's Notes

Introduction

Applying Mathematical Practices

- **Reason Abstractly and Quantitatively** and **Model with Mathematics:** In this lesson, students use the concept of a proportion to solve different types of percent problems. The percent bar model and ratio tables help student reasoning from a visual and numeric display.

Motivate

- Share with students that sometimes their thinking can get *scrambled up* while solving percent problems, so an egg model would be a good way to introduce the chapter!

? Use an egg carton to help visualize a few simple percent problems.
  - "What is 75% of 12?" 9
  - "3 is what percent of 12?" 25%
  - "12 is 50% of what number?" 24

Activity Notes

Activity 1

- **FYI:** You may want to begin with a quick review of fractional equivalents of the following common percents:
  10%, 20%, 30%, 40%, 60%, 70%, 80%, 90%, 25%, 50%, 75%, $33\frac{1}{3}\%$, $66\frac{2}{3}\%$
- **Use Appropriate Tools Strategically:** The percent bar model is an effective tool for estimating an answer, or judging the reasonableness of an answer if students have an understanding of fractional parts of a whole.
- The length of the bar is 100%, the whole. Percents near 50% are about $\frac{1}{2}$ of the whole.
- Students should be able to judge percents near 25% $\left(\frac{1}{4}\right)$ and 75% $\left(\frac{3}{4}\right)$.
- Students should locate the percents on the same model.
- When students have finished, draw a percent bar model on the board. Have volunteers share their answers.
- Remind students that these are approximations. Check for reasonableness in their approximations. For example, 40% is closer to 50% than 25%.

Activity 2

- **FYI:** Some students may find it helpful to use a long strip of paper that they can fold or write on when answering questions.
- **Reason Abstractly and Quantitatively:** Students may wonder why the percent bar model was divided into 5 parts versus 2 parts or 4 parts. Students should reason that if you divide into quarters, half of 75 is about 37, and half of 37 is about 18. Because you want to know what percent 15 is of 75, a percent bar model divided into 4 parts would allow you to estimate that it is less than 25%. To get a closer estimate you would try smaller parts, such as fifths, as shown.
- Students should be able to use mental math to find 10% of any number. Knowing 10%, it is easy to find 20% (double 10%).

What Your Students Will Learn

- Use the percent proportion to find parts, wholes, and percents.

Previous Learning

Students should know how to solve simple percent problems and how to use ratio tables.

Technology for the Teacher

Dynamic Classroom

Lesson Plans
Complete Materials List

15.3 Record and Practice Journal

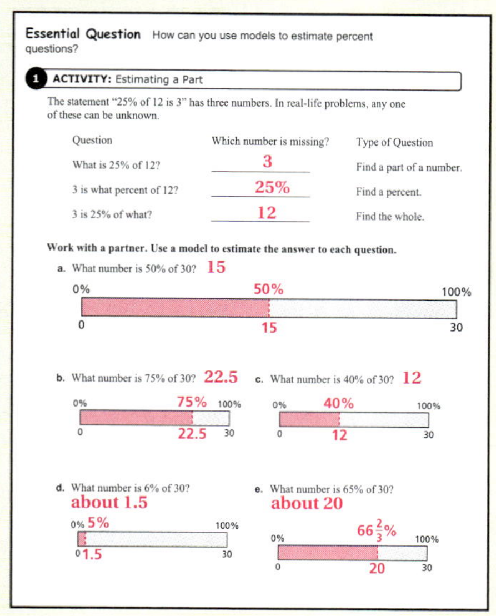

Essential Question How can you use models to estimate percent questions?

1 ACTIVITY: Estimating a Part

The statement "25% of 12 is 3" has three numbers. In real-life problems, any one of these can be unknown.

Question	Which number is missing?	Type of Question
What is 25% of 12?	3	Find a part of a number.
3 is what percent of 12?	25%	Find a percent.
3 is 25% of what?	12	Find the whole.

Work with a partner. Use a model to estimate the answer to each question.

a. What number is 50% of 30? 15

b. What number is 75% of 30? 22.5

c. What number is 40% of 30? 12

d. What number is 6% of 30? about 1.5

e. What number is 65% of 30? about 20

Differentiated Instruction

Visual

Have students use colored pencils to write the percent proportion in words in their notebooks. Then use the colored pencils to underline or circle the corresponding numbers in the problem statement.

$$\frac{\text{part}}{\text{whole}} = \frac{\text{percent}}{100}$$

18 is what percent of 40?

$$\frac{18}{40} = \frac{p}{100}$$

15.3 Record and Practice Journal

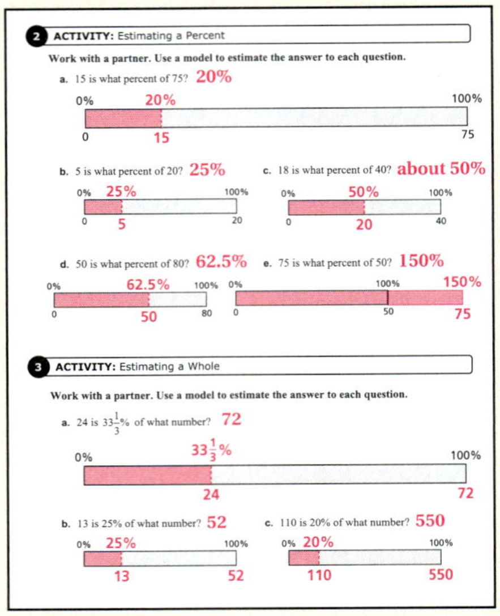

2 ACTIVITY: Estimating a Percent

Work with a partner. Use a model to estimate the answer to each question.

a. 15 is what percent of 75? 20%

b. 5 is what percent of 20? 25%

c. 18 is what percent of 40? about 50%

d. 50 is what percent of 80? 62.5%

e. 75 is what percent of 50? 150%

3 ACTIVITY: Estimating a Whole

Work with a partner. Use a model to estimate the answer to each question.

a. 24 is $33\frac{1}{3}\%$ of what number? 72

b. 13 is 25% of what number? 52

c. 110 is 20% of what number? 550

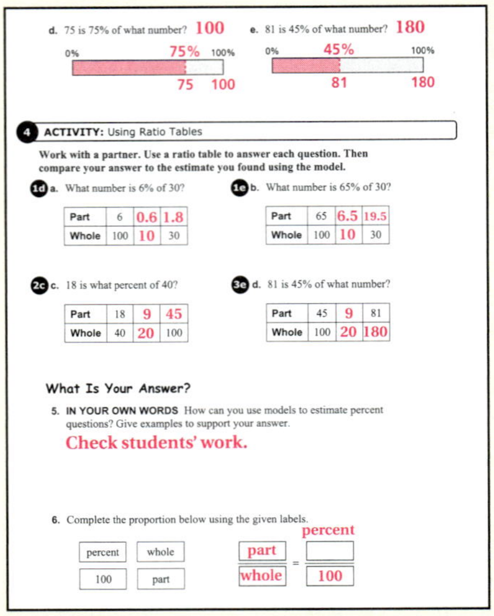

d. 75 is 75% of what number? 100

e. 81 is 45% of what number? 180

4 ACTIVITY: Using Ratio Tables

Work with a partner. Use a ratio table to answer each question. Then compare your answer to the estimate you found using the model.

1d a. What number is 6% of 30?

Part	6	0.6	1.8
Whole	100	10	30

1e b. What number is 65% of 30?

Part	65	6.5	19.5
Whole	100	10	30

2c c. 18 is what percent of 40?

Part	18	9	45
Whole	40	20	100

3e d. 81 is 45% of what number?

Part	45	9	81
Whole	100	20	180

What Is Your Answer?

5. IN YOUR OWN WORDS How can you use models to estimate percent questions? Give examples to support your answer.

Check students' work.

6. Complete the proportion below using the given labels.

percent | whole | 100 | part

$$\frac{\text{part}}{\text{whole}} = \frac{\text{percent}}{100}$$

Laurie's Notes

Activity 3

- Encourage students to draw a percent bar model for each problem. You could also provide strips of paper that students can fold or write on.
- It is important that students be able to approximate the part in the whole model. Ask questions such as, "Is it greater than or less than one-half? Is it greater than or less than one-quarter?"
- **Model with Mathematics:** Estimating the whole is generally easier for students. You can draw a model of what you have, as a part and as a percent. Because you know the percent, you know how the percent bar model should be scaled to find the whole.

Activity 4

- Percent bar models help you estimate answers. Ratio tables can be used to find the exact answers. Remind students how they found the percent of a number in Section 5.6 (multiplication) and how they found the whole (division). There are numerous ways to use ratio tables to find these values.
- Break down the ratios shown in each of the tables and compare. Discuss the "common proportion" to help lead into Question 6 and the Key Idea in tomorrow's lesson.

What Is Your Answer?

- **Think-Pair-Share:** Students should read each question independently and then work in pairs to answer the questions. When they have answered the questions, the pair should compare their answers with another group and discuss any discrepancies.

Closure

- Use the model shown. What 3 questions could be asked? "24 is what percent of 60?"; "What is 40% of 60?"; "24 is 40% of what number?"

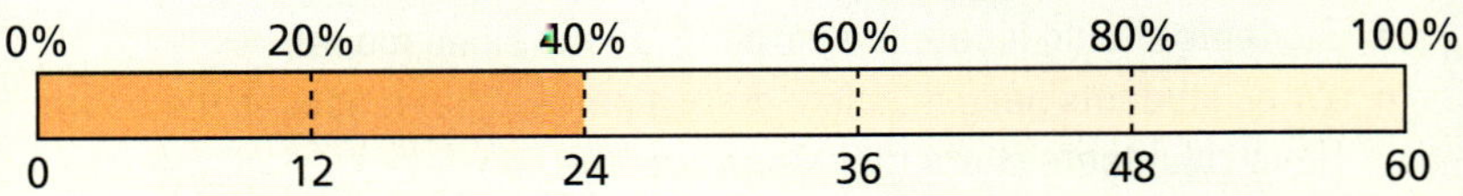

3 ACTIVITY: Estimating a Whole

Math Practice

Use a Model

What quantities are given? How can you use the model to find the unknown quantity?

Work with a partner. Use a model to estimate the answer to each question.

a. 24 is $33\frac{1}{3}\%$ of what number?

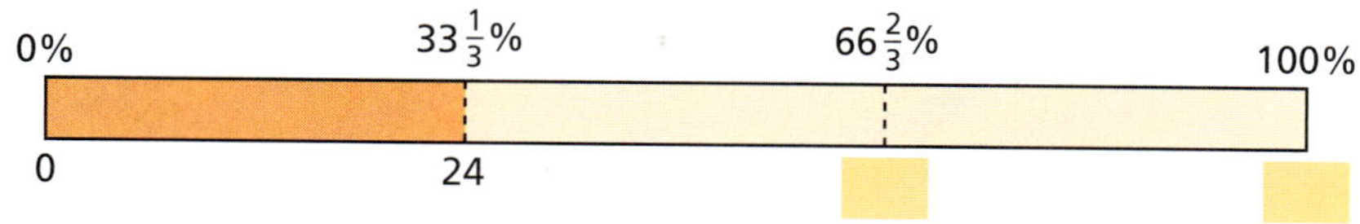

So, from the model, 24 is $33\frac{1}{3}\%$ of ___.

b. 13 is 25% of what number?

c. 110 is 20% of what number?

d. 75 is 75% of what number?

e. 81 is 45% of what number?

4 ACTIVITY: Using Ratio Tables

Work with a partner. Use a ratio table to answer each question. Then compare your answer to the estimate you found using the model.

1d **a.** What number is 6% of 30?

Part	6		
Whole	100		30

1e **b.** What number is 65% of 30?

Part	65		
Whole	100		30

2c **c.** 18 is what percent of 40?

Part	18		
Whole	40		100

3e **d.** 81 is 45% of what number?

Part	45		81
Whole	100		

What Is Your Answer?

5. IN YOUR OWN WORDS How can you use models to estimate percent questions? Give examples to support your answer.

6. Complete the proportion below using the given labels.

percent

whole

100

part

Practice Use what you learned about estimating percent questions to complete Exercises 5–10 on page 666.

15.3 Lesson

Check It Out
Lesson Tutorials
BigIdeasMath.com

Key Idea

The Percent Proportion

Words You can represent "a is p percent of w" with the proportion

$$\frac{a}{w} = \frac{p}{100}$$

where a is part of the whole w, and $p\%$, or $\frac{p}{100}$, is the percent.

Numbers 3 out of 4 is 75%.

$$\frac{3}{4} = \frac{75}{100}$$

part → 3; whole → 4; percent → 75/100

Study Tip

In percent problems, the word *of* is usually followed by the whole.

EXAMPLE 1 Finding a Percent

What percent of 15 is 12?

$$\frac{a}{w} = \frac{p}{100}$$ Write the percent proportion.

$$\frac{12}{15} = \frac{p}{100}$$ Substitute 12 for a and 15 for w.

$$100 \cdot \frac{12}{15} = 100 \cdot \frac{p}{100}$$ Multiplication Property of Equality

$$80 = p$$ Simplify.

So, 80% of 15 is 12.

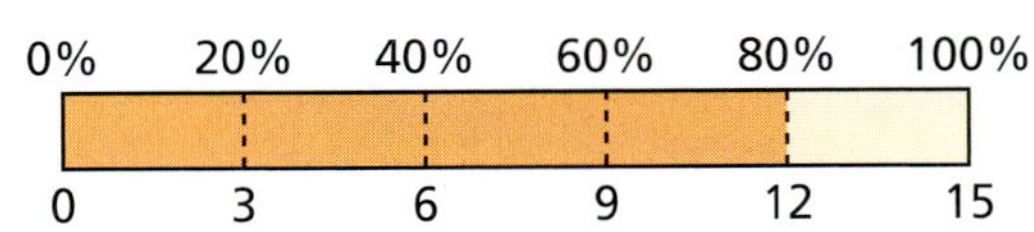

EXAMPLE 2 Finding a Part

What number is 36% of 50?

$$\frac{a}{w} = \frac{p}{100}$$ Write the percent proportion.

$$\frac{a}{50} = \frac{36}{100}$$ Substitute 50 for w and 36 for p.

$$50 \cdot \frac{a}{50} = 50 \cdot \frac{36}{100}$$ Multiplication Property of Equality

$$a = 18$$ Simplify.

So, 18 is 36% of 50.

Laurie's Notes

Introduction

Connect

- **Yesterday:** Students used the percent bar model to explore three types of percent problems.
- **Today:** Students will use the percent proportion to solve three types of percent problems.

Motivate

- Share information about the Enhanced Fujita Scale (EF Scale) used to rate the strength of a tornado based on estimated wind speeds and related damage.
- April through July are the four months with the highest frequency of tornadoes.
- Briefly discuss any experiences students have had with tornadoes and explain that they will come back to tornadoes at the end of class.

Enhanced Fujita Scale	
EF Number	**3 Second Gust (mph)**
0	65–85
1	86–110
2	111–135
3	**136–165**
4	**166–200**
5	**Over 200**

Lesson Notes

Key Idea

- Students should be familiar with the vocabulary *part*, *whole*, and *percent* from yesterday's activity.

Words of Wisdom

- **Reason Abstractly and Quantitatively** and **Model with Mathematics:** In all of the examples today, draw a percent bar model off to the side to help student reasoning. The model helps students estimate and consider the reasonableness of the answer.

Example 1

? "What is the whole? 15 What is the part?" 12.

? "Is 12 more or less than 50% of 15?" more

- Set up the percent proportion, substitute the known quantities, and solve.

? "What other strategies could be used to solve the proportion?" Simplify $\frac{12}{15} = \frac{4}{5}$, then write $\frac{4}{5}$ as a percent.

Example 2

? "Are you looking for a whole or a part?" looking for the part

- Draw a percent bar model with 50 as the whole. Divide into 5 parts and label the corresponding percents and amounts. Because 36% is closest to 40%, ask students to estimate an answer, which would be close to 20.
- Set up the percent proportion, substitute the known quantities, and solve.
- Refer back to the model to confirm that the answer makes sense.

Goal Today's lesson is finding percents using the percent proportion.

Technology for the Teacher

Dynamic Classroom

Lesson Tutorials
Lesson Plans
Answer Presentation Tool

English Language Learners

Vocabulary

English language learners need to be able to distinguish between the *whole* and the *part of the whole* when setting up a percent proportion. Have students write the percent proportion $\frac{a}{w} = \frac{p}{100}$ and work with the following two statements.

35 is 25% of 140

175% of 36 is 63

Students should easily be able to substitute the number for the percent *p*. The whole *w* is the number after the word *of*. The remaining number is the part of the whole *a*. So, the percent proportions are

$\frac{35}{140} = \frac{25}{100}$ and $\frac{63}{36} = \frac{175}{100}$.

Extra Example 1

What percent of 20 is 8?

40%

Extra Example 2

What number is 45% of 80?

36

Extra Example 3

210% of what number is 84?
40

On Your Own

1. $\frac{3}{5} = \frac{p}{100}$; $p = 60$
2. $\frac{25}{20} = \frac{p}{100}$; $p = 125$
3. $\frac{a}{60} = \frac{80}{100}$; $a = 48$
4. $\frac{a}{40.5} = \frac{10}{100}$; $a = 4.05$
5. $\frac{4}{w} = \frac{0.1}{100}$; $w = 4000$
6. $\frac{\frac{1}{2}}{w} = \frac{25}{100}$; $w = 2$

Extra Example 4

Using the bar graph from Example 4, what percent of the tornadoes were EF2s?
20%

On Your Own

7. 29 tornadoes

Laurie's Notes

Example 3

? "Are you looking for a whole or a part?" 24 is the part, so you are looking for the whole.

- Draw a percent bar model The whole is the number associated with the 100%, which is not obvious to all students. Working with percents greater than 100% is more challenging.
- Set up the percent proportion, substitute the known quantities, and solve.
- Refer back to the percent bar model to confirm that the answer makes sense.

On Your Own

- The decimals and fractions included in these exercises can present problems for some students. Encourage students to set up the percent proportion and take time to estimate an answer. For instance, Question 5 says you take a very small percent of a number, and you get 4. You have to start with a fairly large number because 1% of 100 is 4.
- **Neighbor Check:** Have students work independently and then have their neighbors check their work. Have students discuss any discrepancies.

Example 4

- Have students read the bar graph and discuss the information displayed.

? "How can you find the percent of tornadoes that were EF1s?" The whole would be the total number of tornadoes (145) and the part would be the number that were EF1s (58).

? "Will the answer be more or less than 50%? Explain." Less, because half of 145 is about 72 and 58 < 72.

- Set up the percent proportion, substitute the known quantities, and solve.

On Your Own

- **Neighbor Check:** Have students work independently and then have their neighbors check their work. Have students discuss any discrepancies.

Closure

- An average of 1253 tornadoes occur in the U.S. each year, and about 12.5% of them are in Texas. About how many tornadoes does Texas have each year? about 157

EXAMPLE 3 Finding a Whole

150% of what number is 24?

$\frac{a}{w} = \frac{p}{100}$	Write the percent proportion.
$\frac{24}{w} = \frac{150}{100}$	Substitute 24 for *a* and 150 for *p*.
$24 \cdot 100 = w \cdot 150$	Cross Products Property
$2400 = 150w$	Multiply.
$16 = w$	Divide each side by 150.

So, 150% of 16 is 24.

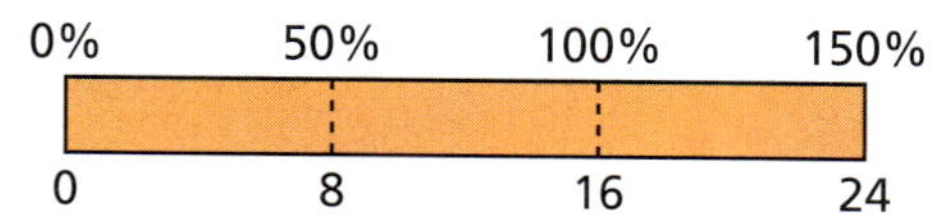

On Your Own

Now You're Ready
Exercises 11–18

Write and solve a proportion to answer the question.

1. What percent of 5 is 3?
2. 25 is what percent of 20?
3. What number is 80% of 60?
4. 10% of 40.5 is what number?
5. 0.1% of what number is 4?
6. $\frac{1}{2}$ is 25% of what number?

EXAMPLE 4 Real-Life Application

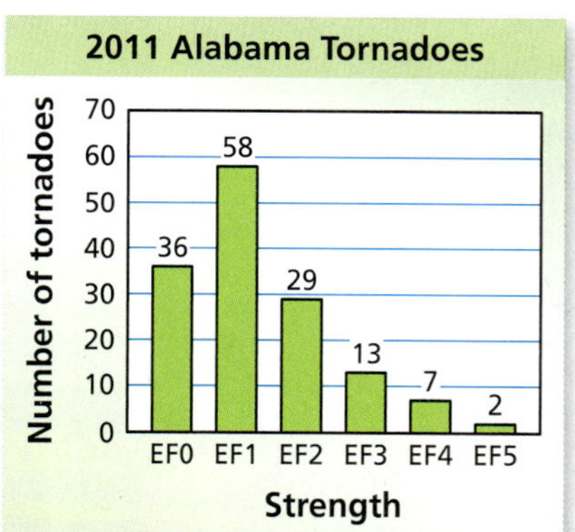

The bar graph shows the strengths of tornadoes that occurred in Alabama in 2011. What percent of the tornadoes were EF1s?

The total number of tornadoes, 145, is the *whole*, and the number of EF1 tornadoes, 58, is the *part*.

$\frac{a}{w} = \frac{p}{100}$	Write the percent proportion.
$\frac{58}{145} = \frac{p}{100}$	Substitute 58 for *a* and 145 for *w*.
$100 \cdot \frac{58}{145} = 100 \cdot \frac{p}{100}$	Multiplication Property of Equality
$40 = p$	Simplify.

So, 40% of the tornadoes were EF1s.

On Your Own

7. Twenty percent of the tornadoes occurred in central Alabama on April 27. How many tornadoes does this represent?

15.3 Exercises

Check It Out
Help with Homework
BigIdeasMath.com

Vocabulary and Concept Check

1. **VOCABULARY** Write the percent proportion in words.
2. **WRITING** Explain how to use a proportion to find 30% of a number.
3. **NUMBER SENSE** Write and solve the percent proportion represented by the model.

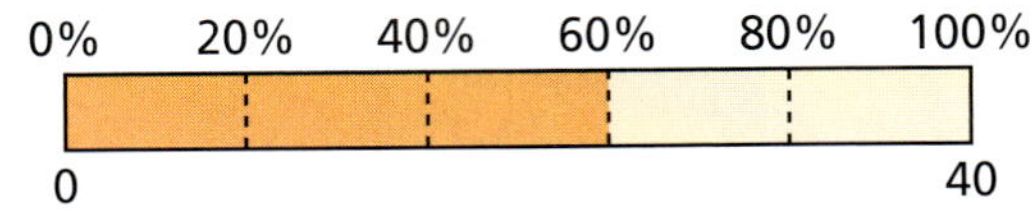

4. **WHICH ONE DOESN'T BELONG?** Which proportion does *not* belong with the other three? Explain your reasoning.

$\frac{15}{w} = \frac{50}{100}$ $\frac{12}{15} = \frac{40}{n}$ $\frac{15}{25} = \frac{p}{100}$ $\frac{a}{20} = \frac{35}{100}$

Practice and Problem Solving

Use a model to estimate the answer to the question. Use a ratio table to check your answer.

5. What number is 24% of 80?
6. 15 is what percent of 40?
7. 15 is 30% of what number?
8. What number is 120% of 70?
9. 20 is what percent of 52?
10. 48 is 75% of what number?

Write and solve a proportion to answer the question.

1 11. What percent of 25 is 12?
12. 14 is what percent of 56?
2 13. 25% of what number is 9?
14. 36 is 0.9% of what number?
3 15. 75% of 124 is what number?
16. 110% of 90 is what number?
17. What number is 0.4% of 40?
18. 72 is what percent of 45?

✗

$$\frac{a}{w} = \frac{p}{100}$$
$$\frac{a}{34} = \frac{40}{100}$$
$$a = 13.6$$

19. **ERROR ANALYSIS** Describe and correct the error in using the percent proportion to answer the question below.

"40% of what number is 34?"

20. **FITNESS** Of 140 seventh-grade students, 15% earn the Presidential Physical Fitness Award. How many students earn the award?

21. **COMMISSION** A salesperson receives a 3% commission on sales. The salesperson receives $180 in commission. What is the amount of sales?

Assignment Guide and Homework Check

Level	Assignment	Homework Check
Advanced	1–10, 12–18 even, 19, 20–30 even, 31–35	12, 14, 16, 24, 28

For Your Information

- **Exercise 21** Students may get confused by the word *commission*. Tell students that commission is a fee or percentage allowed to a sales representative or an agent for services rendered.

Common Errors

- **Exercises 5–18** Students may not know what number to substitute for each variable. Walk through each type of question with the students. Emphasize that the word *is* means *equals*, and *of* means *to multiply*. Tell students to write the question and then write the meaning of each word or group of words underneath.
- **Exercises 20 and 21** Students will mix up the whole and the part when trying to write the percent proportion for the word problems. Ask them to identify each part of the proportion before writing it in the proportion format.
- **Exercise 29** Students may struggle with this exercise because there is no vertical scale. Tell students to think of the bars as a model. This will get them heading in the right direction.

15.3 Record and Practice Journal

Write and solve a proportion to answer the question.

1. 40% of 60 is what number?
   $a = 0.4 \cdot 60; 24$
2. 17 is what percent of 50?
   $17 = p \cdot 50; 34\%$
3. 38% of what number is 57?
   $57 = 0.38 \cdot w; 150$
4. 44% of 25 is what number?
   $a \cdot 0.44 \cdot 25; 11$
5. 52 is what percent of 50?
   $52 = p \cdot 50; 104\%$
6. 150% of what number is 18?
   $18 = 1.5 \cdot w; 12$
7. You put 60% of your paycheck into your savings account. Your paycheck is $235. How much money do you put in your savings account?
   $141

Vocabulary and Concept Check

1. The percent proportion is $\frac{a}{w} = \frac{p}{100}$ where a is part of the whole w, and $p\%$, or $\frac{p}{100}$, is the percent.
2. 30% is $\frac{30}{100}$ and the number w is the whole. Set up the percent proportion as $\frac{a}{w} = \frac{30}{100}$ and solve for a.
3. $\frac{a}{40} = \frac{60}{100}$; $a = 24$
4. $\frac{12}{15} = \frac{40}{n}$; This proportion is not a percent proportion.

Practice and Problem Solving

5. 20
6. 37.5%
7. 50
8. 84
9. about 37.5%
10. 64
11. $\frac{12}{25} = \frac{p}{100}$; $p = 48$
12. $\frac{14}{56} = \frac{p}{100}$; $p = 25$
13. $\frac{9}{w} = \frac{25}{100}$; $w = 36$
14. $\frac{36}{w} = \frac{0.9}{100}$; $w = 4000$
15. $\frac{a}{124} = \frac{75}{100}$; $a = 93$
16. $\frac{a}{90} = \frac{110}{100}$; $a = 99$
17. $\frac{a}{40} = \frac{0.4}{100}$; $a = 0.16$
18. $\frac{72}{45} = \frac{p}{100}$; $p = 160$
19. See Additional Answers.
20. 21 students
21. $6000

Practice and Problem Solving

22. $\frac{0.5}{20} = \frac{p}{100}$; $p = 2.5$

23. $\frac{14.2}{w} = \frac{35.5}{100}$; $w = 40$

24. $\frac{\frac{3}{4}}{w} = \frac{60}{100}$; $w = 1\frac{1}{4}$

25. $\frac{a}{\frac{7}{8}} = \frac{25}{100}$; $a = \frac{7}{32}$

26. 4 left

27. $8.40

28. $66\frac{2}{3}\%$

29. See Additional Answers.

30. See *Taking Math Deeper.*

31. **a.** 62.5%

b. $52x$

Fair Game Review

32. 3

33. -0.6

34. -2.5

35. B

Mini-Assessment

Write and solve a proportion to answer the question.

1. What percent of 35 is 28?

   $\frac{29}{35} = \frac{p}{100}$; $p = 80$

2. What number is 28% of 50?

   $\frac{a}{50} = \frac{28}{100}$; $a = 14$

3. 160% of what number is 144?

   $\frac{144}{w} = \frac{160}{100}$; $w = 90$

4. 0.15% of what number is 10.5?

   $\frac{10.5}{w} = \frac{0.15}{100}$; $w = 7000$

5. You score an 80% on your test. You answer 44 questions correctly. How many questions were on the test?

   $\frac{44}{w} = \frac{80}{100}$; $w = 55$

Taking Math Deeper

Exercise 30

You can write and solve an equation to find the answer to this problem. You can also use other ways, such as models, to solve this problem.

1 Use a circular model.

Let the circle represent 100% of the number.

You know that 20% of the number is x. Draw a section that represents 20% of the circle and label it x.

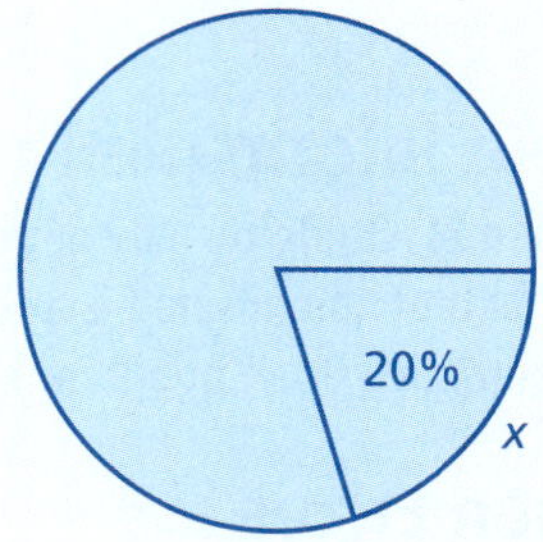

2 Notice that you can draw four more 20% sections in the circle, each representing x. The circle contains five of these sections.

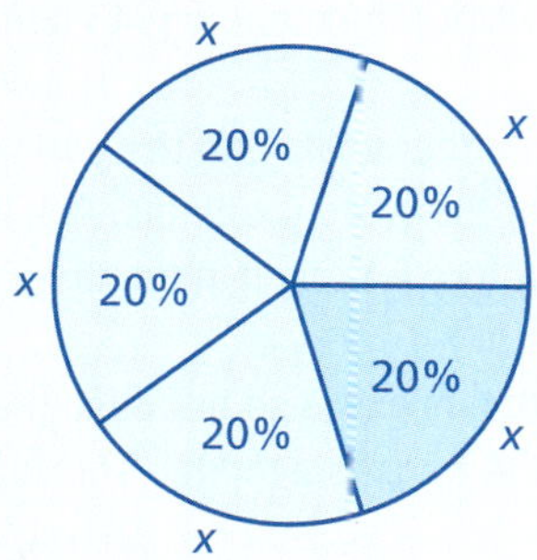

So, 100% of the number is $5x$.

3 You can also use a rectangular model.

You know that 20% of the number is x. So, for each 20% you add on the model, add another x. Stop when you reach 100%.

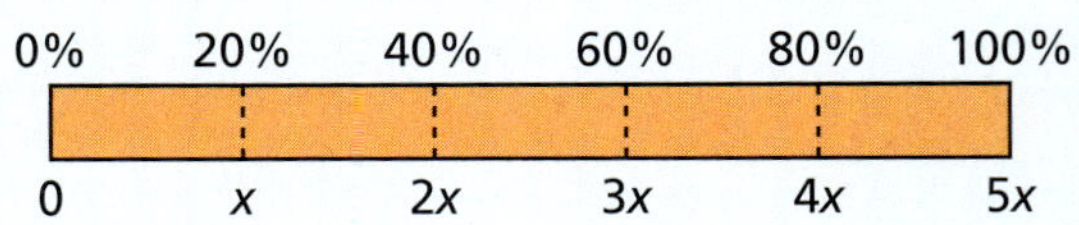

So, 100% of the number is $5x$.

Reteaching and Enrichment Strategies

If students need help. . .	If students got it. . .
Resources by Chapter • Practice A and Practice B • Puzzle Time Record and Practice Journal Practice Differentiating the Lesson Lesson Tutorials Skills Review Handbook	Resources by Chapter • Enrichment and Extension • Technology Connection Start the next section

Write and solve a proportion to answer the question.

22. 0.5 is what percent of 20?

23. 14.2 is 35.5% of what number?

24. $\frac{3}{4}$ is 60% of what number?

25. What number is 25% of $\frac{7}{8}$?

26. HOMEWORK You are assigned 32 math exercises for homework. You complete 87.5% of these before dinner. How many do you have left to do after dinner?

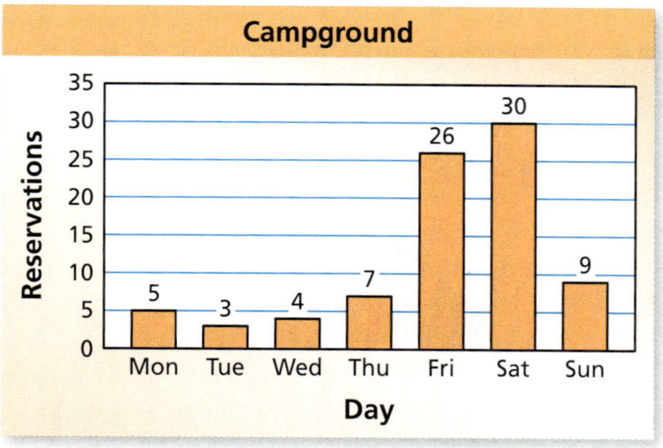

27. HOURLY WAGE Your friend earns \$10.50 per hour. This is 125% of her hourly wage last year. How much did your friend earn per hour last year?

28. CAMPSITE The bar graph shows the numbers of reserved campsites at a campground for one week. What percent of the reservations were for Friday or Saturday?

29. PROBLEM SOLVING A classmate displays the results of a class president election in the bar graph shown.

a. What is missing from the bar graph?

b. What percent of the votes does the last-place candidate receive? Explain your reasoning.

c. There are 124 votes total. How many votes does Chloe receive?

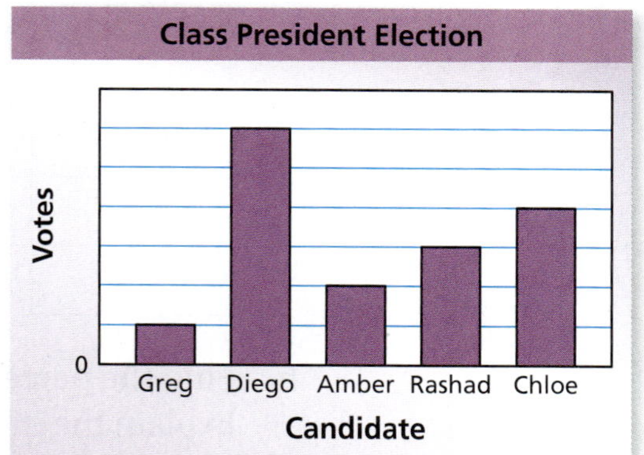

30. REASONING 20% of a number is x. What is 100% of the number? Assume $x > 0$.

31. Structure Answer each question. Assume $x > 0$.

a. What percent of $8x$ is $5x$?

b. What is 65% of $80x$?

Fair Game Review What you learned in previous grades & lessons

Evaluate the expression when $a = -15$ and $b = -5$. *(Section 11.5)*

32. $a \div b$

33. $\frac{b + 14}{a}$

34. $\frac{b^2}{a + 5}$

35. MULTIPLE CHOICE What is the solution of $9x = -1.8$? *(Section 13.4)*

Ⓐ $x = -5$ Ⓑ $x = -0.2$ Ⓒ $x = 0.2$ Ⓓ $x = 5$

15.4 The Percent Equation

Essential Question How can you use an equivalent form of the percent proportion to solve a percent problem?

1 ACTIVITY: Solving Percent Problems Using Different Methods

Work with a partner. The circle graph shows the number of votes received by each candidate during a school election. So far, only half the students have voted.

Votes Received by Each Candidate

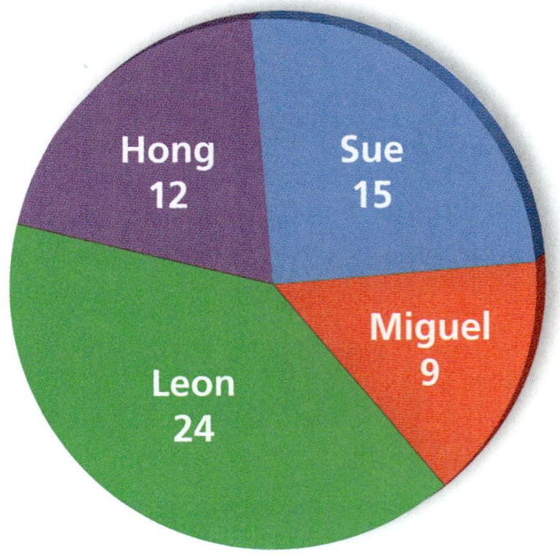

a. Complete the table.

Candidate	$\frac{\text{Number of votes received}}{\text{Total number of votes}}$
Sue	
Miguel	
Leon	
Hong	

b. Find the percent of students who voted for each candidate. Explain the method you used to find your answers.

c. Compare the method you used in part (b) with the methods used by other students in your class. Which method do you prefer? Explain.

2 ACTIVITY: Finding Parts Using Different Methods

Work with a partner. The circle graph shows the final results of the election.

Final Results

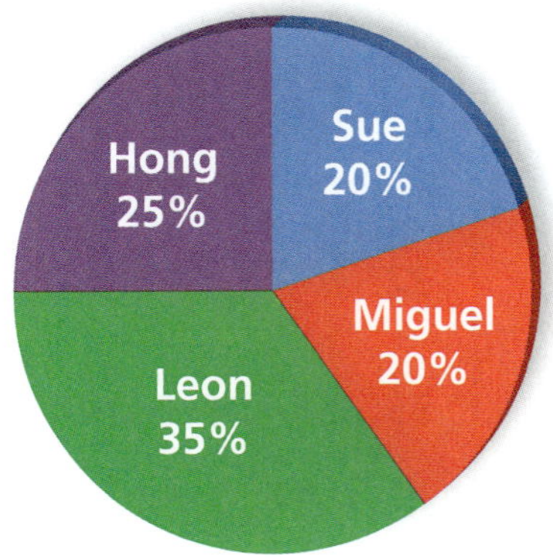

a. Find the number of students who voted for each candidate. Explain the method you used to find your answers.

b. Compare the method you used in part (a) with the methods used by other students in your class. Which method do you prefer? Explain.

Percent Equation

In this lesson, you will

- use the percent equation to find parts, wholes, and percents.
- solve real-life problems.

Laurie's Notes

Introduction

Applying Mathematical Practices

- **Construct Viable Arguments and Critique the Reasoning of Others:** Mathematically proficient students are able to explain their reasoning in a way in which others can understand. They are also able to compare different solution methods and analyze benefits of each, or why different methods might be used for certain types of problems.

Motivate

- Do a quick review of benchmark percents.
- Write the different forms of each benchmark on index cards.

  Example: | 50% | | 0.5 | | ½ |

- Distribute the cards so that each student has one card. If the number of students is not a multiple of 3, make one or two duplicate cards.
- Without speaking, students should walk around and find the other forms equivalent to their numbers.
- Debrief by having each group display their three representations.

Activity Notes

FYI

- In the first two activities, students are solving percent problems using different methods. It is important to make time for students to share different strategies versus having only one method presented.

Activity 1

- There are several ways in which students could find the percent of votes received by each of the four candidates. Methods include:
  - Simplify the fraction and then write the percent.
  - Double a previous answer. 24 votes is twice as many as 12 votes.
  - Write a proportion and solve.
  - Change the fraction to a decimal and then to a percent.
- **Construct Viable Arguments and Critique the Reasoning of Others:** When students have finished part (b), listen to several students explain their method(s), then give students time to answer part (c).

Activity 2

? "How many students voted? Explain." 120; In Activity 1, half of the students had voted, and there were 60. Therefore 100% would be 120 students.

- There are several ways in which students could find the number of votes received by each of the four candidates. Methods include:
  - Mental math
  - Write a proportion and solve.
  - Multiply.
- **Extension:** Have students explore whether voting patterns changed from the first 60 voters to the last 60 voters.

What Your Students Will Learn

- Use the percent equation to find parts, percents, and wholes.

Previous Learning

Students should know how to solve simple percent problems.

Technology for the Teacher

Dynamic Classroom

Lesson Plans

Complete Materials List

15.4 Record and Practice Journal

Essential Question How can you use an equivalent form of the percent proportion to solve a percent problem?

1 ACTIVITY: Solving Percent Problems Using Different Methods

Work with a partner. The circle graph shows the number of votes received by each candidate during a school election. So far, only half the students have voted.

Votes Received by Each Candidate

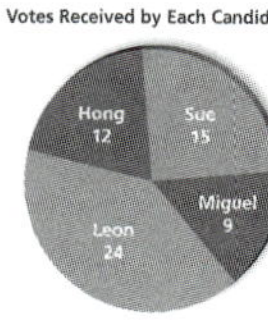

a. Complete the table.

b. Find the percent of students who voted for each candidate. Explain the method you used to find your answers.

Sue: 25%
Miguel: 15%
Leon: 40%
Hong: 20%

Candidate	Number of votes received / Total number of votes	
Sue		$\frac{15}{60}$, or $\frac{1}{4}$
Miguel		$\frac{9}{60}$, or $\frac{3}{20}$
Leon		$\frac{24}{60}$, or $\frac{2}{5}$
Hong		$\frac{12}{60}$, or $\frac{1}{5}$

c. Compare the method you used in part (b) with the methods used by other students in your class. Which method do you prefer? Explain.

Check students' work.

Differentiated Instruction

Visual

Some students will benefit from seeing how fractions and percents relate. Draw a circle on the board. Write 100% on top of the circle and 1 underneath. Explain that both of these values describe the area of the circle. Draw one-half of a circle and one-fourth of a circle and ask students to give you two representations.

$\frac{1}{2}$ and 50%, $\frac{1}{4}$ and 25%

15.4 Record and Practice Journal

2 ACTIVITY: Finding Parts Using Different Methods

Work with a partner. The circle graph shows the final results of the election.

a. Find the number of students who voted for each candidate. Explain the method you used to find your answers.

Sue: 24 students; Miguel: 24 students; Leon: 42 students; Hong: 30 students

Final Results

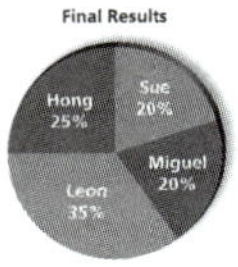

b. Compare the method you used in part (a) with the methods used by other students in your class. Which method do you prefer? Explain.

Check students' work.

3 ACTIVITY: Deriving the Percent Equation

Work with a partner. In Section 15.3, you used the percent proportion to find the missing percent, part, or whole. You can also use the *percent equation* to find these missing values.

a. Complete the steps below to find the percent equation.

$\frac{\text{part}}{\text{whole}} = \text{percent}$ — Definition of percent

$\frac{\text{part}}{\text{whole}} \bullet \boxed{\text{whole}} = \boxed{\text{percent}} \bullet \boxed{\text{whole}}$ — Multiply each side by the $\boxed{\text{whole}}$.

$\text{part} = \boxed{\text{percent}} \bullet \boxed{\text{whole}}$ — Divide out common factors. This is the percent equation.

b. You used two methods in Activity 2 to find the number of students who voted for each candidate. Do you prefer the percent proportion or the percent equation method?

Check students' work.

4 ACTIVITY: Identifying Different Equations

Work with a partner. Without doing any calculations, choose the equation that you cannot use to answer each question.

a. What number is 55% of 80?

$a = 0.55 \bullet 80$ $a = \frac{11}{20} \bullet 80$ $80a = 0.55$ $\frac{a}{80} = \frac{55}{100}$

$80a = 0.55$

b. 24 is 60% of what number?

$\frac{24}{w} = \frac{60}{100}$ $24 = 0.6 \bullet w$ $\frac{24}{60} = w$ $24 = \frac{3}{5} \bullet w$

$\frac{24}{60} = w$

What Is Your Answer?

5. **IN YOUR OWN WORDS** How can you use an equivalent form of the percent proportion to solve a percent problem?

Use the percent equation which can be derived from the percent proportion.

6. Write a percent proportion and a percent equation that you can use to answer the question below.

16 is what percent of 250?

$\frac{16}{250} = \frac{p}{100}$; $16 = p \cdot 250$

Laurie's Notes

Activity 3

- ❓ "How did we solve percent equations in the last lesson?" Students should describe the percent proportion. In particular, they should mention that percents were written as a fraction out of 100.
- In this activity, students derive the percent equation.
- **Note:** In Section 15.3, *p* is the actual percent. In Section 15.4, *p* is the percent written as a decimal or fraction.
- When the percent proportion is multiplied by the whole on both sides of the equation, the percent equation is found.
- In Section 5.6, students multiplied to find the percent of a number so the percent equation should not feel very new.
- **Connection:** This activity shows the connection between the percent proportion and percent equation.

Activity 4

- **Construct Viable Arguments and Critique the Reasoning of Others:** Listen for student justification for their answers.
- **Extension:** Have students answer each question.

What Is Your Answer?

- **Neighbor Check:** Have students work independently and then have their neighbors check their work. Have students discuss any discrepancies.

Closure

- What percent problem is suggested by the following? $12 = \frac{1}{4}w$

  12 is 25% of what number?; 48

3 ACTIVITY: Deriving the Percent Equation

Work with a partner. In Section 15.3, you used the percent proportion to find the missing percent, part, or whole. You can also use the *percent equation* to find these missing values.

a. Complete the steps below to find the percent equation.

$$\frac{\text{part}}{\text{whole}} = \text{percent}$$ Definition of percent

$$\frac{\text{part}}{\text{whole}} \cdot \square = \square \cdot \square$$ Multiply each side by the $\square$.

$$\text{part} = \square \cdot \square$$ Divide out common factors. This is the percent equation.

b. Use the percent equation to find the number of students who voted for each candidate in Activity 2. How does this method compare to the percent proportion?

4 ACTIVITY: Identifying Different Equations

Work with a partner. Without doing any calculations, choose the equation that you cannot use to answer each question.

Math Practice

Justify Conclusions

How can you justify the equations that you chose?

a. What number is 55% of 80?

$a = 0.55 \cdot 80$	$a = \frac{11}{20} \cdot 80$	$80a = 0.55$	$\frac{a}{80} = \frac{55}{100}$

b. 24 is 60% of what number?

$\frac{24}{w} = \frac{60}{100}$	$24 = 0.6 \cdot w$	$\frac{24}{60} = w$	$24 = \frac{3}{5} \cdot w$

What Is Your Answer?

5. **IN YOUR OWN WORDS** How can you use an equivalent form of the percent proportion to solve a percent problem?

6. Write a percent proportion and a percent equation that you can use to answer the question below.

   16 is what percent of 250?

Practice Use what you learned about solving percent problems to complete Exercises 4–9 on page 672.

15.4 Lesson

Check It Out
Lesson Tutorials
BigIdeasMath.com

Key Idea

The Percent Equation

Words To represent "*a* is *p* percent of *w*," use an equation.

$$a = p \cdot w$$

percent in fraction or decimal form (p)

part of the whole (a)

whole (w)

Numbers $15 = 0.5 \cdot 30$

EXAMPLE 1 Finding a Part of a Number

What number is 24% of 50?

Estimate

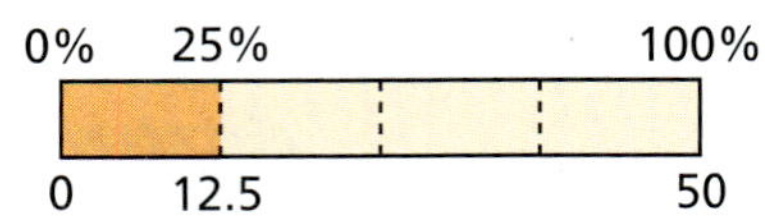

$a = p \cdot w$ — Write percent equation.

$= \dfrac{24}{100} \cdot 50$ — Substitute $\dfrac{24}{100}$ for p and 50 for w.

$= 12$ — Simplify.

So, 12 is 24% of 50. **Reasonable?** $12 \approx 12.5$ ✓

Common Error

Remember to convert a percent to a fraction or a decimal before using the percent equation. For Example 1, write 24% as $\dfrac{24}{100}$.

EXAMPLE 2 Finding a Percent

9.5 is what percent of 25?

Estimate

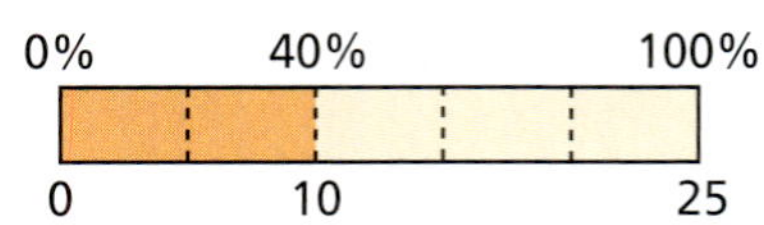

$a = p \cdot w$ — Write percent equation.

$9.5 = p \cdot 25$ — Substitute 9.5 for a and 25 for w.

$\dfrac{9.5}{25} = \dfrac{p \cdot 25}{25}$ — Division Property of Equality

$0.38 = p$ — Simplify.

Because 0.38 equals 38%, 9.5 is 38% of 25. **Reasonable?** $38\% \approx 40\%$ ✓

Laurie's Notes

Introduction

Connect

- **Yesterday:** Students explored the connection between the percent proportion and the percent equation.
- **Today:** Students will use the percent equation to solve three types of percent problems.

Motivate

- The 2010 population of the United States was approximately 309 million (*Source:* U.S. Census Bureau) with about 24% being under 18 years old. About how many people in the U.S. are under the age of 18?
  about 74,160,000

Lesson Notes

Key Idea

- **Connection:** Students should know how to find a percent of a number by multiplying. The percent equation builds upon this idea to find the missing percent or the unknown whole. When you know two of the three quantities in this equation, you can solve for the third.
- **FYI:** Students often get lost in the language of these problems. It is important to help students translate the problems and make sense of the information that is given.

Words of Wisdom

- **Reason Abstractly and Quantitatively** and **Model with Mathematics:** In all of the examples today, draw a percent bar model off to the side to help student reasoning. The model helps students estimate and consider the reasonableness of the answer.

Example 1

- Another way to phrase this question is "24% of 50 is what number?"
- **Estimate:** 24% is close to 25%, and 25% is $\frac{1}{4}$.
- ? "What is $\frac{1}{4}$ of 50?" 12.5
- If time permits, write 24% as a decimal and work the problem again.

Example 2

- Read the example as "9.5 is a part of 25."
- ? "Is 9.5 more or less than half of 25?" less
- ? Draw the percent bar model explaining that it represents 25. Draw the half mark (50%) and ask how much that would represent. 12.5
- ? Now draw the quarter mark (25%) and ask how much that would represent. 6.25 Through this process, students should recognize that 9.5 is between 25% and 50% of 25.
- **Common Error:** Students may forget that the decimal answer to the division problem needs to be rewritten as a percent.

Goal Today's lesson is finding percents using the percent equation.

Technology for the Teacher

Dynamic Classroom

Lesson Tutorials
Lesson Plans
Answer Presentation Tool

Extra Example 1

What number is 73% of 200? 146

Extra Example 2

36.4 is what percent of 40? 91%

Extra Example 3

18 is 15% of what number? 120

On Your Own

1. $a = 0.1 \cdot 20$; 2
2. $a = 1.5 \cdot 40$; 60
3. $3 = p \cdot 600$; 0.5%
4. $18 = p \cdot 20$; 90%
5. $8 = 0.8 \cdot w$; 10
6. $90 = 0.18 \cdot w$; 500

Extra Example 4

Your total cost for lunch is $18.50 for food and $1.48 for tax.

a. Find the percent of sales tax on the food total. 8%

b. Find the amount of an 18% tip on the food total. $3.33

On Your Own

7. $5.50

English Language Learners

Vocabulary

English learners may have trouble identifying which is the *whole* and which is the *part of the whole* in a percent equation. Have students write percent equations for the statements "20% of 300 is 60" and "125% of 50 is 62.5." Suggest that they start by substituting the percent *p* into the equation. Next, substitute the whole *w*. In most cases, this is the number after the word *of*. The remaining number is the part of the whole *a*.

Laurie's Notes

Example 3

- This type of problem, finding a whole, is a bit harder. Knowing fractional equivalents is extremely helpful in developing a sense about the size of the answer.
- ? "What is the part?" 39 "So, 39 is a part of something."
- ? "How big of a part is it, approximately?" 52%, about half
- **Construct Viable Arguments and Critique the Reasoning of Others:** Help students reason that if 39 is half of something, the whole must be about 80. Only at this point does it make sense to translate what is known into an equation. 39 is 52% of some number.
- **Common Error:** Students may divide 39 by 52 and ignore the decimals completely.

On Your Own

- Have students work with partners on these problems. Encourage students to sketch the percent bar model and record the information they know. Then write the percent equation.
- Have students put their work on the board.

Example 4

- Have students read the bar graph and discuss the information displayed.
- ? "In addition to paying for what you ordered (food and drink), what other costs are there when you eat at a restaurant?" sales tax and tip
- Review decimal operations as you work through each part.

On Your Own

- Model finding 10%, and then double for 20%.

Closure

- **Exit Ticket:** Use the percent equation to answer the question, 12 is what percent of 48? 25%

EXAMPLE 3 Finding a Whole

39 is 52% of what number?

Estimate

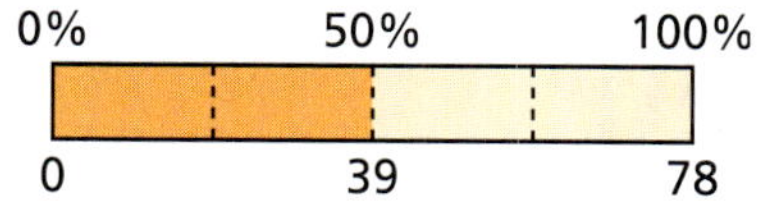

$a = p \cdot w$	Write percent equation.
$39 = 0.52 \cdot w$	Substitute 39 for *a* and 0.52 for *p*.
$75 = w$	Divide each side by 0.52.

So, 39 is 52% of 75.

Reasonable? $75 \approx 78$ ✓

On Your Own

Now You're Ready
Exercises 10–17

Write and solve an equation to answer the question.

1. What number is 10% of 20?
2. What number is 150% of 40?
3. 3 is what percent of 600?
4. 18 is what percent of 20?
5. 8 is 80% of what number?
6. 90 is 18% of what number?

EXAMPLE 4 Real-Life Application

8th Street Cafe

DATE: MAY04'13 05:45PM
TABLE: 29
SERVER: JANE

Food Total	27.50
Tax	1.65
Subtotal	29.15

TIP: ____________

TOTAL: ____________

Thank You

a. Find the percent of sales tax on the food total.

Answer the question: $1.65 is what percent of $27.50?

$a = p \cdot w$	Write percent equation.
$1.65 = p \cdot 27.50$	Substitute 1.65 for *a* and 27.50 for *w*.
$0.06 = p$	Divide each side by 27.50.

Because 0.06 equals 6%, the percent of sales tax is 6%.

b. Find the amount of a 16% tip on the food total.

Answer the question: What tip amount is 16% of $27.50?

$a = p \cdot w$	Write percent equation.
$= 0.16 \cdot 27.50$	Substitute 0.16 for *p* and 27.50 for *w*.
$= 4.40$	Multiply.

So, the amount of the tip is $4.40.

On Your Own

7. **WHAT IF?** Find the amount of a 20% tip on the food total.

15.4 Exercises

Check It Out
Help with Homework
BigIdeasMath.com

Vocabulary and Concept Check

1. **VOCABULARY** Write the percent equation in words.
2. **REASONING** A number n is 150% of number m. Is n *greater than, less than,* or *equal to* m? Explain your reasoning.
3. **DIFFERENT WORDS, SAME QUESTION** Which is different? Find "both" answers.

What number is 20% of 55?	55 is 20% of what number?
20% of 55 is what number?	0.2 • 55 is what number?

Practice and Problem Solving

Answer the question. Explain the method you chose.

4. What number is 24% of 80?
5. 15 is what percent of 40?
6. 15 is 30% of what number?
7. What number is 120% of 70?
8. 20 is what percent of 52?
9. 43 is 75% of what number?

Write and solve an equation to answer the question.

1 10. 20% of 150 is what number?
11. 45 is what percent of 60?
2 12. 35% of what number is 35?
13. 0.8% of 150 is what number?
3 14. 29 is what percent of 20?
15. 0.5% of what number is 12?
16. What percent of 300 is 51?
17. 120% of what number is 102?

ERROR ANALYSIS **Describe and correct the error in using the percent equation.**

18. What number is 35% of 20?

✗
$$\begin{aligned} a &= p \cdot w \\ &= 35 \cdot 20 \\ &= 700 \end{aligned}$$

19. 30 is 60% of what number?

✗
$$\begin{aligned} a &= p \cdot w \\ &= 0.6 \cdot 30 \\ &= 18 \end{aligned}$$

20. **COMMISSION** A salesperson receives a 2.5% commission on sales. What commission does the salesperson receive for $8000 in sales?
21. **FUNDRAISING** Your school raised 125% of its fundraising goal. The school raised $6750. What was the goal?
22. **SURFBOARD** The sales tax on a surfboard is $12. What is the percent of sales tax?

Assignment Guide and Homework Check

Level	Assignment	Homework Check
Advanced	1–9, 10–30 even, 31–36	24, 26, 28, 30

Common Errors

- **Exercises 4–17** Students may not know what number to substitute for each variable. Walk through each type of question with the students. Emphasize that the word *is* means *equals*, and *of* means *to multiply*. Tell students to write the question and then write the meaning of each word or group of words underneath.
- **Exercises 20–22** Students may mix up the whole and the part when trying to write the percent equation for the word problems. Ask them to identify each part of the equation before writing it in the equation format. For example, in Exercise 20, ask, "What is the salesperson's total sales, in dollars? 8000 "Which variable in the percent equation does this number represent?" The whole Continue to ask questions for each of the variables.
- **Exercise 28** Students may not realize that the sum of the parts of a circle graph equals 100%.

15.4 Record and Practice Journal

Write and solve an equation to answer the question.

1. What number is 35% of 80?
   $a = 0.35 \cdot 80$; 28
2. 8 is what percent of 5?
   $8 = p \cdot 5$; 160%
3. What percent of 125 is 50?
   $50 = p \cdot 125$; 40%
4. 12% of what number is 48?
   $48 = 0.12 \cdot w$; 400
5. 12 is what percent of 50?
   $12 = p \cdot 50$; 24%
6. What percent of 12 is 3?
   $3 = p \cdot 12$; 25%
7. You receive 15% of the profit from a car wash. How much money do you receive from a profit of $300?
   $45

Vocabulary and Concept Check

1. A part of the whole is equal to a percent times the whole.
2. greater than; Because 150% = 1.5, $n = 1.5 \cdot m$.
3. 55 is 20% of what number?; 275; 11

Practice and Problem Solving

4. 19.2
5. 37.5%
6. 50
7. 84
8. about 38.5%
9. 64
10. $a = 0.2 \cdot 150$; 30
11. $45 = p \cdot 60$; 75%
12. $35 = 0.35 \cdot w$; 100
13. $a = 0.008 \cdot 150$; 1.2
14. $29 = p \cdot 20$; 145%
15. $12 = 0.005 \cdot w$; 2400
16. $51 = p \cdot 300$; 17%
17. $102 = 1.2 \cdot w$; 85
18. The percent was not converted to a decimal or fraction.
    $a = p \cdot w$
    $= 0.35 \cdot 20$
    $= 7$
19. 30 represents the part of the whole.
    $30 = 0.6 \cdot w$
    $50 = w$
20. $200
21. $5400
22. 5%

Practice and Problem Solving

23. 26 years old

24. 70 years old

25. 56 signers

26. 70%

27. If the percent is less than 100%, the percent of a number is less than the number; 50% of 80 is 40; If the percent is equal to 100%, the percent of a number is equal to the number; 100% of 80 is 80; If the percent is greater than 100%, the percent of a number is greater than the number; 150% of 80 is 120.

28. **a.** 80 students

b. 30 students

29. See *Taking Math Deeper*.

30. false; If W is 25% of Z, then $Z : W$ is $100 : 25$, because Z represents the whole.

31. 92%

Fair Game Review

32. 0.6 **33.** 0.88

34. 0.25 **35.** 0.36

36. A

Mini-Assessment

Write and solve an equation to answer the question.

1. 52 is what percent of 80? 65%
2. 28 is 35% of what number? 80
3. What number is 25% of 92? 23
4. What percent of 250 is 60? 24%
5. A new laptop computer costs $800. The sales tax on the computer is $48. What is the percent of sales tax? 6%

Taking Math Deeper

Exercise 29

Any problem that has this much given information is difficult for students. Encourage students to begin by organizing the information with a table or a diagram. When organizing the information, it is a good idea to add as much other information as you can find .. *before looking at the questions.*

1. Organize the given information.
2. Add other information.

Capacity	550 gal	750 gal
Water in tank	363 gal	397.5 gal

$\frac{15}{11} \cdot 550$

$0.66 \cdot 550$ $0.53 \cdot 750$

3. Now the questions are easy.
   - **a.** Tank A has 363 gallons of water.
   - **b.** The capacity of tank B is 750 gallons.
   - **c.** Tank B has 397.5 gallons of water.

Project

Use your school library or the Internet to research how a water tower works. How does the water get into the tower? How long does it take for the water to drain out? How often is the water completely exchanged; in other words, if a gallon goes in today when will that gallon be draining out? What other interesting things did you discover?

Reteaching and Enrichment Strategies

If students need help. . .	If students got it. . .
Resources by Chapter • Practice A and Practice B • Puzzle Time Record and Practice Journal Practice Differentiating the Lesson Lesson Tutorials Skills Review Handbook	Resources by Chapter • Enrichment and Extension • Technology Connection Start the next section

PUZZLE **There were n signers of the Declaration of Independence. The youngest was Edward Rutledge, who was x years old. The oldest was Benjamin Franklin, who was y years old.**

23. x is 25% of 104. What was Rutledge's age?

24. 7 is 10% of y. What was Franklin's age?

25. n is 80% of y. How many signers were there?

26. y is what percent of $(n + y - x)$?

Favorite Sport

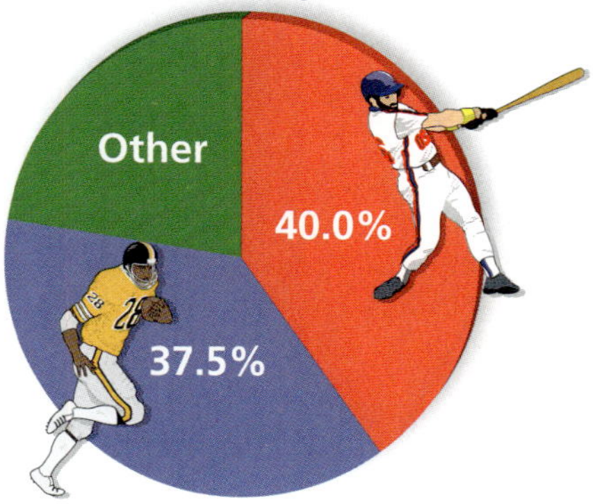

27. **LOGIC** How can you tell whether the percent of a number will be *greater than*, *less than*, or *equal to* the number? Give examples to support your answer.

28. **SURVEY** In a survey, a group of students were asked their favorite sport. Eighteen students chose "other" sports.

a. How many students participated?

b. How many chose football?

29. **WATER TANK** Water tank A has a capacity of 550 gallons and is 66% full. Water tank B is 53% full. The ratio of the capacity of Tank A to Tank B is 11 : 15.

a. How much water is in Tank A?

b. What is the capacity of Tank B?

c. How much water is in Tank B?

30. **TRUE OR FALSE?** Tell whether the statement is *true* or *false*. Explain your reasoning.

If W is 25% of Z, then $Z : W$ is 75 : 25.

31. **Reasoning** The table shows your test results for math class. What test score do you need on the last exam to earn 90% of the total points?

Test Score	Point Value
83%	100
91.6%	250
88%	150
?	300

Fair Game Review What you learned in previous grades & lessons

Simplify. Write the answer as a decimal. *(Skills Review Handbook)*

32. $\dfrac{10 - 4}{10}$ **33.** $\dfrac{25 - 3}{25}$ **34.** $\dfrac{105 - 84}{84}$ **35.** $\dfrac{170 - 125}{125}$

36. **MULTIPLE CHOICE** There are 160 people in a grade. The ratio of boys to girls is 3 to 5. Which proportion can you use to find the number x of boys? *(Section 14.3)*

Ⓐ $\dfrac{3}{8} = \dfrac{x}{160}$ Ⓑ $\dfrac{3}{5} = \dfrac{x}{160}$ Ⓒ $\dfrac{5}{8} = \dfrac{x}{160}$ Ⓓ $\dfrac{3}{5} = \dfrac{160}{x}$

15 Study Help

Check It Out
Graphic Organizer
BigIdeasMath.com

You can use a **summary triangle** to explain a concept. Here is an example of a summary triangle for writing a percent as a decimal.

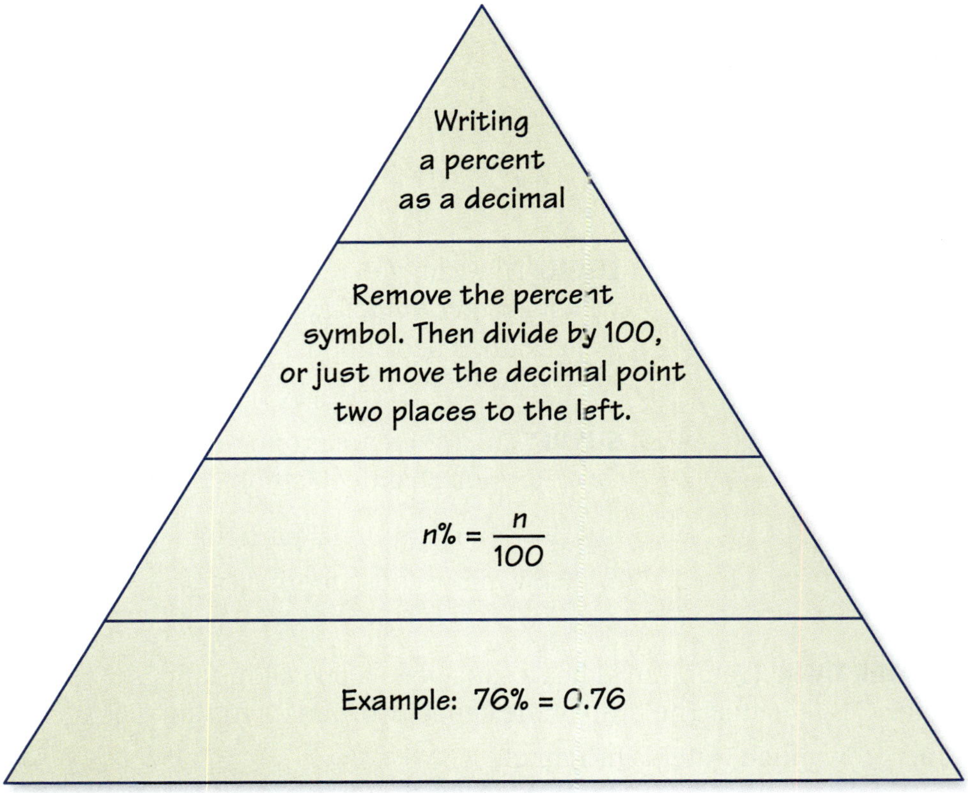

On Your Own

Make summary triangles to help you study these topics.

1. writing a decimal as a percent
2. comparing and ordering fractions, decimals, and percents
3. the percent proportion
4. the percent equation

After you complete this chapter, make summary triangles for the following topics.

5. percent of change
6. discount
7. markup
8. simple interest

"I found this great summary triangle in my *Beautiful Beagle Magazine*."

Sample Answers

1.

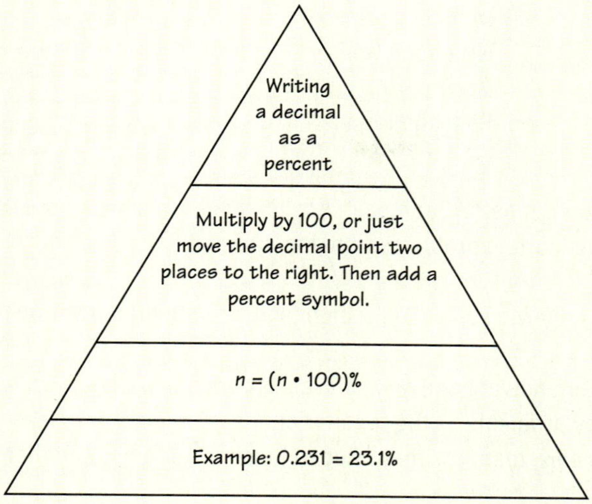

2.

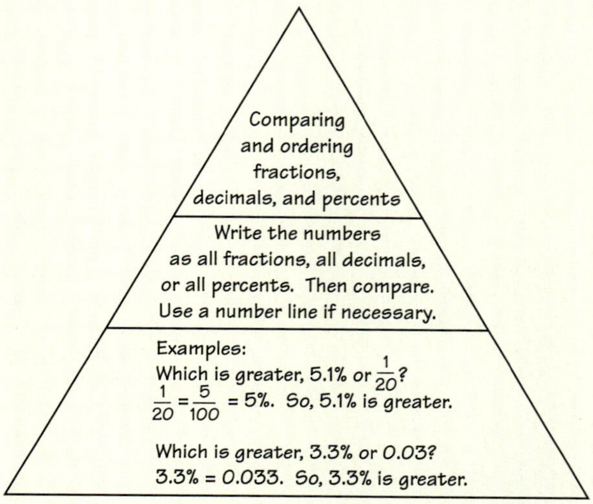

3.

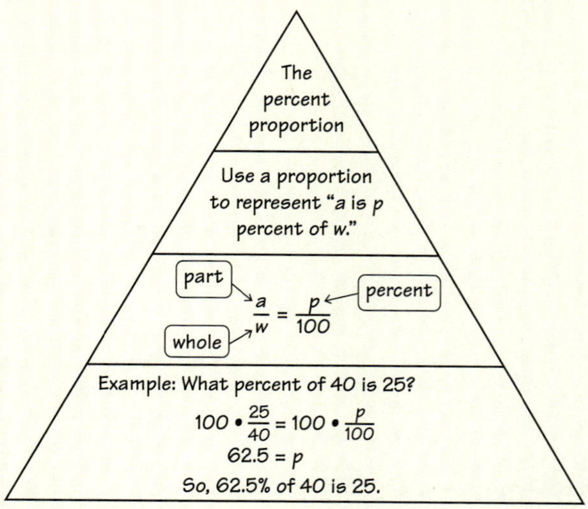

4. Available at *BigIdeasMath.com*.

List of Organizers

Available at *BigIdeasMath.com*

Comparison Chart
Concept Circle
Definition (Idea) and Example Chart
Example and Non-Example Chart
Formula Triangle
Four Square
Information Frame
Information Wheel
Notetaking Organizer
Process Diagram
Summary Triangle
Word Magnet
Y Chart

About this Organizer

A **Summary Triangle** can be used to explain a concept. Typically, the summary triangle is divided into 3 or 4 parts. In the top part, students write the concept being explained. In the middle part(s), students write any procedure, explanation, description, definition, theorem, and/or formula(s). In the bottom part, students write an example to illustrate the concept. A summary triangle can be used as an assessment tool, in which blanks are left for students to complete. Also, students can place their summary triangles on note cards to use as a quick study reference.

Technology for the Teacher

Editable Graphic Organizer

Answers

1. 0.34
2. 0.0012
3. 0.625
4. 67%
5. 535%
6. 68.5%
7. 74%
8. 0.3
9. 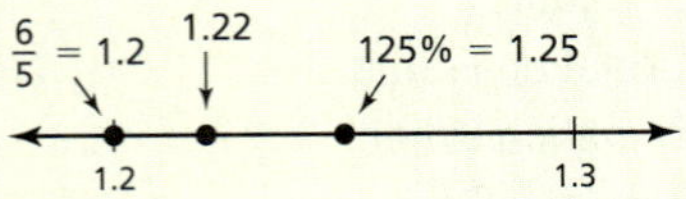

10. 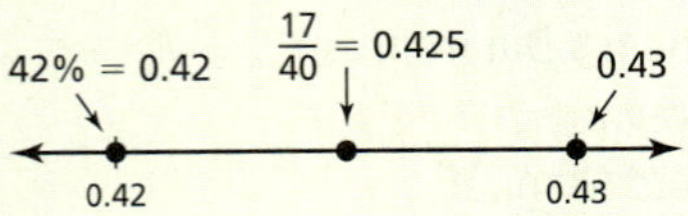

11. $\frac{6}{15} = \frac{p}{100}$; $p = 40$
12. $\frac{35}{25} = \frac{p}{100}$; $p = 140$
13. $\frac{a}{50} = \frac{40}{100}$; $a = 20$
14. $\frac{5}{w} = \frac{0.5}{100}$; $w = 1000$
15. $a = 0.28 \cdot 75$; 21
16. $42 = 0.21 \cdot w$; 200
17. 0.38
18. Team 4; Team 5
19. 17 passes
20. 50 messages

Technology for the Teacher

Online Assessment
Assessment Book
ExamView® Assessment Suite

Alternative Quiz Ideas

100% Quiz
Error Notebook
Group Quiz
Homework Quiz
Math Log
Notebook Quiz
Partner Quiz
Pass the Paper

Partner Quiz

- Students should work in pairs. Each pair should have a small white board.
- The teacher selects certain problems from the quiz and writes one on the board.
- The pairs work together to solve the problem and write their answer on the white board.
- Students show their answers and, as a class, discuss any differences.
- Repeat for as many problems as the teacher chooses.
- For the word problems, teachers may choose to have students read them out of the book.

Reteaching and Enrichment Strategies

If students need help...	If students got it...
Resources by Chapter • Practice A and Practice B • Puzzle Time Lesson Tutorials *BigIdeasMath.com*	Resources by Chapter • Enrichment and Extension • Technology Connection Game Closet at *BigIdeasMath.com* Start the next section

15.1–15.4 Quiz

Check It Out
Progress Check
BigIdeasMath.com

Write the percent as a decimal. *(Section 15.1)*

1. 34%
2. 0.12%
3. 62.5%

Write the decimal as a percent. *(Section 15.1)*

4. 0.67
5. 5.35
6. 0.685

Tell which number is greater. *(Section 15.2)*

7. $\frac{11}{15}$, 74%
8. 3%, 0.3

Use a number line to order the numbers from least to greatest. *(Section 15.2)*

9. 125%, $\frac{6}{5}$, 1.22
10. 42%, 0.43, $\frac{17}{40}$

Write and solve a proportion to answer the question. *(Section 15.3)*

11. What percent of 15 is 6?
12. 35 is what percent of 25?
13. What number is 40% of 50?
14. 0.5% of what number is 5?

Write and solve an equation to answer the question. *(Section 15.4)*

15. What number is 28% of 75?
16. 42 is 21% of what number?

17. **FISHING** On a fishing trip, 38% of the fish that you catch are perch. Write this percent as a decimal. *(Section 15.1)*

18. **SCAVENGER HUNT** The table shows the results of 8 teams competing in a scavenger hunt. Which team collected the most items? Which team collected the fewest items? *(Section 15.2)*

Team	1	2	3	4	5	6	7	8
Portion Collected	$\frac{3}{4}$	0.8	77.5%	0.825	$\frac{29}{40}$	76.25%	$\frac{63}{80}$	81.25%

19. **COMPLETIONS** A quarterback completed 68% of his passes in a game. He threw 25 passes. How many passes did the quarterback complete? *(Section 15.3)*

20. **TEXT MESSAGES** You have 44 text messages in your inbox. How many messages can your cell phone hold? *(Section 15.4)*

15.5 Percents of Increase and Decrease

Essential Question What is a percent of decrease? What is a percent of increase?

1 ACTIVITY: Percent of Decrease

Work with a partner.

Each year in the Columbia River Basin, adult salmon swim upriver to streams to lay eggs and hatch their young.

To go up the river, the adult salmon use fish ladders. But to go down the river, the young salmon must pass through several dams.

At one time, there were electric turbines at each of the eight dams on the main stem of the Columbia and Snake Rivers. About 88% of the young salmon passed through these turbines unharmed.

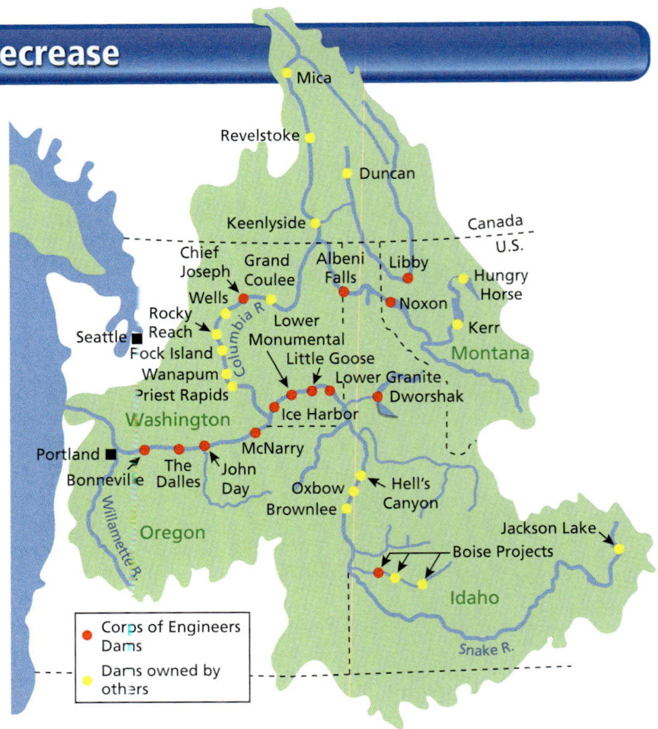

a. Copy and complete the table to show the number of young salmon that made it through the dams.

Dam	0	1	2	3	4	5	6	7	8
Salmon	1000	880	774						

88% of 1000 = 0.88 • 1000
= 880

88% of 880 = 0.88 • 880
= 774.4
≈ 774

b. Display the data in a bar graph.

c. By what percent did the number of young salmon decrease when passing through each dam?

Percents

In this lesson, you will

- find percents of increase.
- find percents of decrease.

Laurie's Notes

Introduction

Applying Mathematical Practices

- **Use Appropriate Tools Strategically** and **Look for and Express Regularity in Repeated Reasoning:** Use of calculators allows students to draw important conclusions. When there is a repeated percent change (increase or decrease), the percent remains constant while the amount changes.

Motivate

- Talk about compact fluorescent light bulbs. Fluorescent light bulbs use 75% less energy than incandescent light bulbs (percent decrease) and last up to 900% longer (percent increase).
- If your school has replaced incandescent light bulbs with fluorescent light bulbs, discuss the potential savings.

Activity Notes

Activity 1

- **Representation:** Although the difference between the decimal point and the multiplication dot are clear in the textbook, it may not be as clear when you write it on the board. You may consider using parentheses to show the multiplication: (0.88)(1000).
- Have a student read the story information.
- ? "What percent of salmon makes it through each dam?" 88%
- ? "What percent of salmon does not make it through each dam?" 12%
- Discuss the general concept of fewer salmon at dam 2 than dam 1.
- **FYI:** Electric turbines in the dams generate electricity. These turbines are what affect the survival rate of the young salmon.
- Students should follow the two calculations shown. Remind students to round their answers to a whole number of salmon at each dam.
- **Use Appropriate Tools Strategically:** Use of a calculator will help facilitate the computation so that students can focus on how the numbers are changing.
- Check students' results before completing the graph.
- **Look for and Express Regularity in Repeated Reasoning** and **Big Idea:** Each entry is 12% less than the previous entry. The *amount* of decrease is changing, but the *percent* is not.
- Have students describe patterns they observe in the numbers and the bar graph.

What Your Students Will Learn

- Find percents of increase, percents of decrease, or percent error.

Previous Learning

Students should be able to find a percent of a number, round decimal values, and convert between fractions, decimals, and percents.

Technology for the Teacher

Dynamic Classroom

Lesson Plans
Complete Materials List

15.5 Record and Practice Journal

Essential Question What is a percent of decrease? What is a percent of increase?

1 ACTIVITY: Percent of Decrease

Work with a partner.

Each year in the Columbia River Basin, adult salmon swim upriver to streams to lay eggs and hatch their young.

To go up river, the adult salmon use fish ladders. But to go down the river, the young salmon must pass through several dams.

At one time, there were electric turbines at each of the eight dams on the main stem of the Columbia and Snake Rivers. About 88% of the young salmon passed through these turbines unharmed.

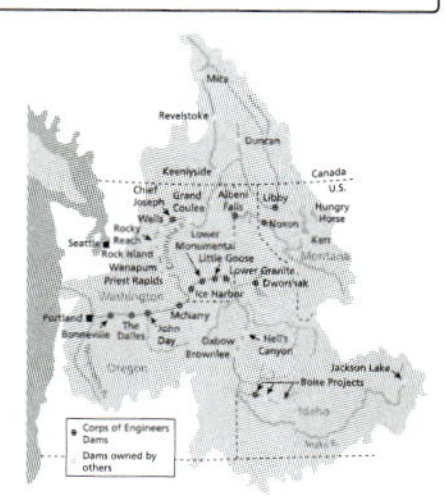

a. Complete the table to show the number of young salmon that made it through the dams.

Dam	0	1	2	3	4	5	6	7	8
Salmon	1000	880	774	681	599	527	464	408	359

English Language Learners

Visual Aid

Demonstrate writing a percent as a decimal. Locate the decimal point in the 7%.

7.%

Draw two arrows to show that the decimal point moves two places *left*.

7.%

Write zeros to the left of the number if needed.

007.%

Rewrite as a decimal with the decimal point two places to the left and without the percent sign.

0.07

15.5 Record and Practice Journal

b. Display the data in a bar graph.

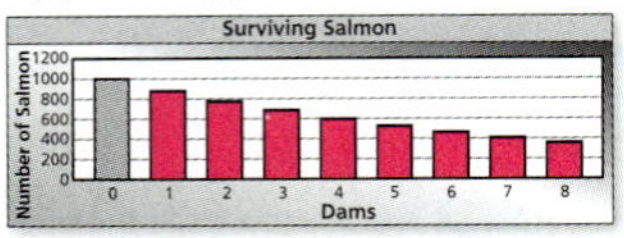

c. By what percent did the number of young salmon decrease when passing through each dam?

12%

2 ACTIVITY: Percent of Increase

Work with a partner. In 2013, the population of a city was 18,000 people.

a. An organization projects that the population will increase by 2% each year for the next 7 years. Complete the table to find the populations of the city for 2014 through 2020. Then display the data in a bar graph.

For 2014: 2% of 18,000 = 0.02 • 18,000 = 360

18,000 + 360 = 18,360 ← 2014 Population

(2013 Population) (Increase)

Year	Population
2013	18,000
2014	18,360
2015	18,727
2016	19,102
2017	19,484
2018	19,874
2019	20,271
2020	20,676

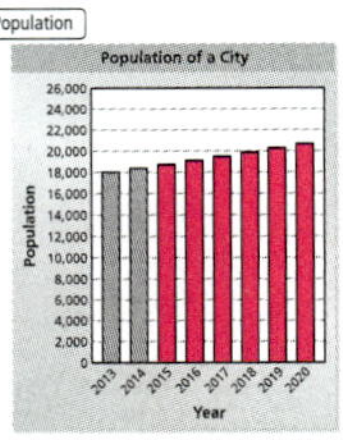

b. Another organization projects that the population will increase by 3% each year for the next 7 years. Repeat part (a) using this percent.

Year	Population
2013	18,000
2014	18,540
2015	19,096
2016	19,669
2017	20,259
2018	20,867
2019	21,493
2020	22,138

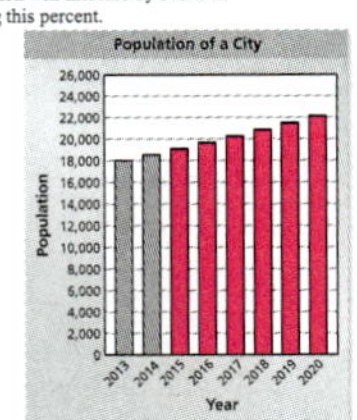

c. Which organization projects the larger populations? How many more people do they project for 2020?

the organization in part (b); 1462 people

What Is Your Answer?

3. **IN YOUR OWN WORDS** What is a percent of decrease? What is a percent of increase?

percent of change when the original amount decreases or increases

4. Describe real-life examples of a percent of decrease and a percent of increase.

Check students' work.

Laurie's Notes

Activity 2

- Ask a student to read the problem.
- Work through the first year with the students. There are two steps involved: 1) find the amount the population has increased, and 2) add this amount to the current population.
- ? "How much did the population increase in 2014?" 360
- ? "What percent did the population increase in 2014?" 2%
- Remind students to round their answers to a whole number and add this number to the current population.
- **Use Appropriate Tools Strategically:** Use of a calculator will help facilitate the computation so that students can focus on how the numbers are changing.
- Check students' results before completing the graph.
- **Look for and Express Regularity in Repeated Reasoning** and **Big Idea:** Each entry is 2% more than the previous entry. The *amount* of increase is changing, but the *percent* is not.
- **Extension:** Discuss how projections are made based upon current trends.

What Is Your Answer?

- For Question 4, students could discuss this at home and bring ideas to class.

Closure

- "You scored 80 points on your first test. If your score increased 10% on the next test, what is your score?" 88 points

2 ACTIVITY: Percent of Increase

Math Practice

Consider Similar Problems

How is this activity similar to the previous activity?

Work with a partner. In 2013, the population of a city was 18,000 people.

a. An organization projects that the population will increase by 2% each year for the next 7 years. Copy and complete the table to find the populations of the city for 2014 through 2020. Then display the data in a bar graph.

For 2014:

$$2\% \text{ of } 18{,}000 = 0.02 \cdot 18{,}000$$
$$= 360$$

$$18{,}000 + 360 = 18{,}360$$

2013 Population; Increase; 2014 Population

Year	Population
2013	18,000
2014	18,360
2015	
2016	
2017	
2018	
2019	
2020	

b. Another organization projects that the population will increase by 3% each year for the next 7 years. Repeat part (a) using this percent.

c. Which organization projects the larger populations? How many more people do they project for 2020?

What Is Your Answer?

3. IN YOUR OWN WORDS What is a percent of decrease? What is a percent of increase?

4. Describe real-life examples of a percent of decrease and a percent of increase.

Practice Use what you learned about percent of increase and percent of decrease to complete Exercises 4–7 on page 680.

15.5 Lesson

Check It Out
Lesson Tutorials
BigIdeasMath.com

Key Vocabulary
percent of change, *p. 678*
percent of increase, *p. 678*
percent of decrease, *p. 678*
percent error, *p. 679*

A **percent of change** is the percent that a quantity changes from the original amount.

$$\text{percent of change} = \frac{\text{amount of change}}{\text{original amount}}$$

Key Idea

Percents of Increase and Decrease

When the original amount increases, the percent of change is called a **percent of increase**.

$$\text{percent of increase} = \frac{\text{new amount} - \text{original amount}}{\text{original amount}}$$

When the original amount decreases, the percent of change is called a **percent of decrease**.

$$\text{percent of decrease} = \frac{\text{original amount} - \text{new amount}}{\text{original amount}}$$

EXAMPLE 1 Finding a Percent of Increase

The table shows the numbers of hours you spent online last weekend. What is the percent of change in your online time from Saturday to Sunday?

Day	Hours Online
Saturday	2
Sunday	4.5

The number of hours on Sunday is greater than the number of hours on Saturday. So, the percent of change is a percent of increase.

$$\text{percent of increase} = \frac{\text{new amount} - \text{original amount}}{\text{original amount}}$$

$$= \frac{4.5 - 2}{2} \quad \text{Substitute.}$$

$$= \frac{2.5}{2} \quad \text{Subtract.}$$

$$= 1.25, \text{ or } 125\% \quad \text{Write as a percent.}$$

So, your online time increased 125% from Saturday to Sunday.

On Your Own

Find the percent of change. Round to the nearest tenth of a percent if necessary.

1. 10 inches to 25 inches
2. 57 people to 65 people

Multi-Language Glossary at BigIdeasMath.com

Laurie's Notes

Introduction

Connect

- **Yesterday:** Students explored two real-life problems with quantities that decreased or increased by a percent.
- **Today:** Students will use a percent of change formula to solve problems.

Motivate

- Pose a question such as: "If 400 people in your neighborhood had a cell phone last year and one year later 500 people had a cell phone, what percent has cell phone ownership increased?"

Lesson Notes

Key Idea

- Explain the difference between *amount* of change and *percent* of change. Refer to the salmon and population activities.
- Use the cell phone example to help identify vocabulary:

original amount $= 400$,

amount of change $= 100$,

percent of change $= \frac{100}{400} = 25\%$.

Example 1

? "Did the online use increase or decrease from Saturday to Sunday?" increase

? "How much did the online use increase from Saturday to Sunday?" 2.5 h

- Have students write the equation, substitute the values, and then simplify. The original amount is 2. The new amount is 4.5. Because the number of hours increased, you are finding a **percent of increase**. Percent of increase $= \frac{4.5 - 2}{2} = 1.25 = 125\%$.
- **Common Error:** Students think the answer is 1.25. This decimal must still be converted to a percent. This often happens when the percent answer is greater than 100%.
- **Connection:** Draw a percent bar model of this problem.

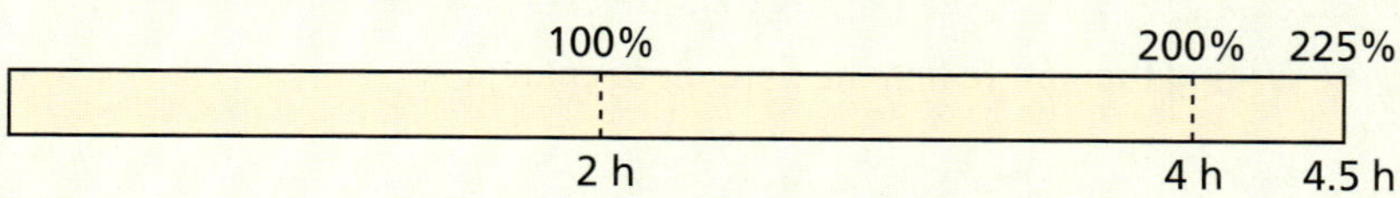

The percent of increase is 125% beyond the 100%.

- **Model with Mathematics** and **Big Idea:** From the percent bar model you can see that a 100% increase doubles the number. Another way of saying this is that when a number doubles, it has increased 100%.

On Your Own

- In Question 1, the length has more than doubled, so the percent of increase is greater than 100%.
- In Question 2, the number of people has not doubled, so the percent of increase is less than 100%.

Goal Today's lesson is using a **percent of change** formula to solve problems.

Technology for the Teacher

Dynamic Classroom

Lesson Tutorials
Lesson Plans
Answer Presentation Tool

Extra Example 1

Find the percent of change from 40 hours to 50 hours. increase of 25%

On Your Own

1. 150% increase
2. about 14.0% increase

Extra Example 2

Find the percent of change from 20 days to 12 days. decrease of 40%

Extra Example 3

You estimate that the number of text messages you sent last month is 135. When the bill comes, you find out you only sent 108 text messages. Find the percent error. 25%

On Your Own

3. decrease of about 44.4%
4. you; Your percent error is about 23.8% and your friend's percent error is about 9.5%.

Differentiated Instruction

Vocabulary

Make sure students understand the difference between *increased by 150%* and *increased to 150%*. The percent of increase in the first case is 150%. The percent of increase in the second case is 50%, because the increase does not include the original amount. By the same token, there is a difference between *decreased by 25%* and *decreased to 25%*. In the first case, it means taking 25% and leaving 75%. In the second case, it means to leave 25% and take away 75%.

Laurie's Notes

Example 2

- "How much did the number of home runs change each year?" decrease of 8, increase of 18, decrease of 8
- ? "What is the original amount?" 28
- ? "What is the new amount?" 20
- Students should now use the **percent of decrease** formula.
- This problem involves a number of skills: reading a bar graph, using the percent of decrease formula, converting a fraction to a decimal, and converting a decimal to a percent.
- The answer is rounded to the nearest tenth of a percent.
- ? "Is the percent change from 2011–2012 more or less than 100%? How do you know?" The number of home runs more than doubled, so the increase is greater than 100%.

Key Idea

- *Percent error* may be a new topic in your curriculum, but it is really an application of percent change.
- Write the Key Idea. The percent error compares the amount of error to the actual amount.
- An estimate can be too high or too low when compared to the actual, but generally **percent error** is referred to as a positive amount.

Example 3

- Before doing this example you could ask students to write down their estimates for the length of the classroom. These estimates could be used after doing this example, with students computing their percent error.
- **Use Appropriate Tools Strategically:** Calculators would be appropriate for this problem.

On Your Own

- If you have small white boards available, have each student solve Question 3 on a white board. Have students hold up their white boards and then have students determine their mistakes.

Closure

- **Writing Prompt:** To find the percent of change…

EXAMPLE 2 Finding a Percent of Decrease

The bar graph shows a softball player's home run totals. What was the percent of change from 2012 to 2013?

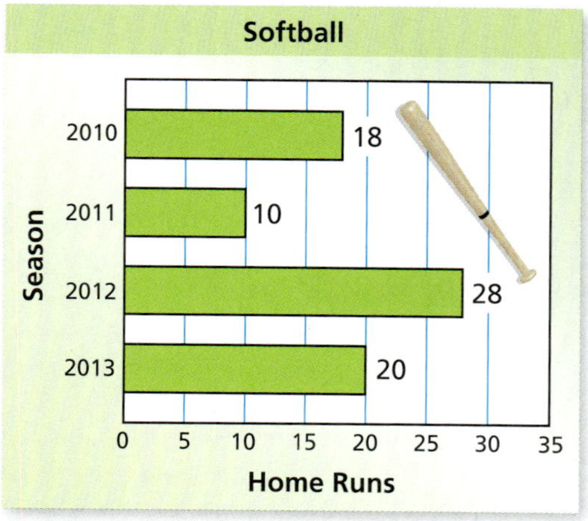

The number of home runs decreased from 2012 to 2013. So, the percent of change is a percent of decrease.

$$\text{percent of decrease} = \frac{\text{original amount} - \text{new amount}}{\text{original amount}}$$

$$= \frac{28 - 20}{28} \qquad \text{Substitute.}$$

$$= \frac{8}{28} \qquad \text{Subtract.}$$

$$\approx 0.286, \text{ or } 28.6\% \qquad \text{Write as a percent.}$$

So, the number of home runs decreased about 28.6%.

Key Idea

Study Tip

The amount of error is always positive.

Percent Error

A **percent error** is the percent that an estimated quantity differs from the actual amount.

$$\text{percent error} = \frac{\text{amount of error}}{\text{actual amount}}$$

EXAMPLE 3 Finding a Percent Error

You estimate that the length of your classroom is 16 feet. The actual length is 21 feet. Find the percent error.

The amount of error is $21 - 16 = 5$ feet.

$$\text{percent error} = \frac{\text{amount of error}}{\text{actual amount}} \qquad \text{Write percent error equation.}$$

$$= \frac{5}{21} \qquad \text{Substitute.}$$

$$\approx 0.238, \text{ or } 23.8\% \qquad \text{Write as a percent.}$$

The percent error is about 23.8%.

On Your Own

Now You're Ready
Exercises 8–15 and 18

3. In Example 2, what was the percent of change from 2010 to 2011?
4. **WHAT IF?** In Example 3, your friend estimates that the length of the classroom is 23 feet. Who has the greater percent error? Explain.

15.5 Exercises

Check It Out
Help with Homework
BigIdeasMath.com

Vocabulary and Concept Check

1. **VOCABULARY** How do you know whether a percent of change is a *percent of increase* or a *percent of decrease*?
2. **NUMBER SENSE** Without calculating, which has a greater percent of increase?
   - 5 bonus points on a 50-point exam
   - 5 bonus points on a 100-point exam
3. **WRITING** What does it mean to have a 100% decrease?

Practice and Problem Solving

Find the new amount.

4. 8 meters increased by 25%
5. 15 liters increased by 60%
6. 50 points decreased by 26%
7. 25 penalties decreased by 32%

Identify the percent of change as an *increase* or a *decrease*. Then find the percent of change. Round to the nearest tenth of a percent if necessary.

1 2 8. 12 inches to 36 inches
9. 75 people to 25 people
10. 50 pounds to 35 pounds
11. 24 songs to 78 songs
12. 10 gallons to 24 gallons
13. 72 paper clips to 63 paper clips
14. 16 centimeters to 44.2 centimeters
15. 68 miles to 42.5 miles

16. **ERROR ANALYSIS** Describe and correct the error in finding the percent increase from 18 to 26.

✗ $$\frac{26 - 18}{26} \approx 0.31 = 31\%$$

17. **VIDEO GAME** Last week, you finished Level 2 of a video game in 32 minutes. Today, you finish Level 2 in 28 minutes. What is your percent of change?

3 18. **PIG** You estimate that a baby pig weighs 20 pounds. The actual weight of the baby pig is 16 pounds. Find the percent error.

19. **CONCERT** You estimate that 200 people attended a school concert. The actual attendance was 240 people.
    a. Find the percent error.
    b. What other estimate gives the same percent error? Explain your reasoning.

Assignment Guide and Homework Check

Level	Assignment	Homework Check
Advanced	1–7, 8–30 even, 31–36	12, 18, 24, 28, 30

Common Errors

- **Exercises 4–7** Students may find the percent of the number and forget to add or subtract from the original amount. Remind them that these are two-step problems. Before evaluating, tell students to write down what needs to be done for each step.
- **Exercises 8–15** Students may mix up where to place the numbers in the equation to find percent of change. When they do not put the numbers in the right place, they might find a negative number in the numerator. First, emphasize that students must know if it is increasing or decreasing before they start the problem. Next, tell students that the number in the denominator is going to be the original or starting number given for both increasing and decreasing percents of change. Finally, the numerator should never have a negative answer. If students get a negative number, it is because they found the wrong difference. The numerator is always the greater number minus the lesser number.

15.5 Record and Practice Journal

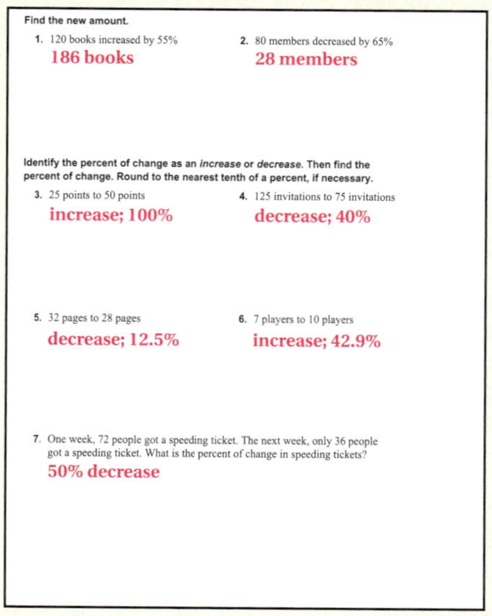

Find the new amount.

1. 120 books increased by 55%
   186 books
2. 80 members decreased by 65%
   28 members

Identify the percent of change as an *increase* or *decrease*. Then find the percent of change. Round to the nearest tenth of a percent, if necessary.

3. 25 points to 50 points
   increase; 100%
4. 125 invitations to 75 invitations
   decrease; 40%
5. 32 pages to 28 pages
   decrease; 12.5%
6. 7 players to 10 players
   increase; 42.9%
7. One week, 72 people got a speeding ticket. The next week, only 36 people got a speeding ticket. What is the percent of change in speeding tickets?
   50% decrease

Vocabulary and Concept Check

1. If the original amount decreases, the percent of change is a percent of decrease. If the original amount increases, the percent of change is a percent of increase.

2. 5 bonus points on a 50-point exam

3. The new amount is now 0.

Practice and Problem Solving

4. 10 m

5. 24 L

6. 37 points

7. 17 penalties

8. increase; 200%

9. decrease; 66.7%

10. decrease; 30%

11. increase; 225%

12. increase; 140%

13. decrease; 12.5%

14. increase; 176.3%

15. decrease; 37.5%

16. The denominator should be 18, which is the original amount.

$$\frac{26-18}{18} \approx 0.44 = 44\%$$

17. 12.5% decrease

18. 25%

19. **a.** about 16.7%

b. 280 people; To get the same percent error, the amount of error needs to be the same. Because your estimate was 40 people below the actual attendance, an estimate of 40 people above the actual attendance will give the same percent error.

Practice and Problem Solving

20. increase; 100%

21. decrease; 25%

22. increase; 133.3%

23. decrease; 70%

24–25. See Additional Answers.

26. **a.** 100% increase

b. 300% increase

27. 15.6 ounces; 16.4 ounces

28. about 24.52% decrease

29. See Additional Answers.

30. See *Taking Math Deeper*.

31. 10 girls

Fair Game Review

32. $a = 0.25 \cdot 64$; 16

33. $39.2 = p \cdot 112$; 35%

34. $5 = 0.05 \cdot w$; 100

35. $18 = 0.32 \cdot w$; 56.25

36. B

Mini-Assessment

Identify the percent of change as an *increase* or *decrease*. Then find the percent of change.

1. 15 meters to 36 meters increase; 140%
2. 20 songs to 70 songs increase; 250%
3. 90 people to 45 people decrease; 50%
4. Yesterday, it took 40 minutes to drive to school. Today, it took 32 minutes to drive to school. What is your percent of change? The number of minutes it took to get to school decreased by 20%.
5. You estimate that a box contains 141 envelopes. The actual number of envelopes is 150. Find the percent error. 6%

Taking Math Deeper

Exercise 30

This exercise is difficult because the percent of increase is given with the *new amount*, rather than the *original amount*. A good way to start is to use a table to organize the given information.

1 Organize given information.

	Donation	Increase over previous year
This year	$10,120	15%
1 year ago	x	10%
2 years ago	y	

2 Find last year's donation.

$x + 0.15x = 10{,}120$ Write an equation.

$1.15x = 10{,}120$ Combine like terms.

$x = \$8800$ Divide each side by 1.15.

3 Find donation from 2 years ago.

$y + 0.1y = 8800$ Write an equation.

$1.1y = 8800$ Combine like terms.

$y = \$8000$ Divide each side by 1.1.

Project

Plan a fundraiser for your school. Write a proposal that includes the purpose of the fundraiser, the type of activity, the length of time, and the amount of money you would like to raise. Be prepared to present your proposal to the class.

Reteaching and Enrichment Strategies

If students need help. . .	If students got it. . .
Resources by Chapter • Practice A and Practice B • Puzzle Time Record and Practice Journal Practice Differentiating the Lesson Lesson Tutorials Skills Review Handbook	Resources by Chapter • Enrichment and Extension • Technology Connection Start the next section

Identify the percent of change as an *increase* or a *decrease*. Then find the percent of change. Round to the nearest tenth of a percent if necessary.

20. $\frac{1}{4}$ to $\frac{1}{2}$ **21.** $\frac{4}{5}$ to $\frac{3}{5}$ **22.** $\frac{3}{8}$ to $\frac{7}{8}$ **23.** $\frac{5}{4}$ to $\frac{3}{8}$

24. CRITICAL THINKING Explain why a change from 20 to 40 is a 100% increase, but a change from 40 to 20 is a 50% decrease.

25. POPULATION The table shows population data for a community.

Year	Population
2007	118,000
2013	138,000

a. What is the percent of change from 2007 to 2013?

b. Use this percent of change to predict the population in 2019.

26. GEOMETRY Suppose the length and the width of the sandbox are doubled.

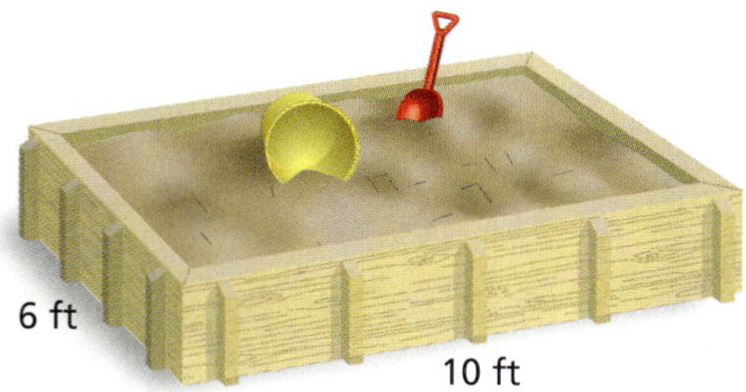

a. Find the percent of change in the perimeter.

b. Find the percent of change in the area.

27. CEREAL A cereal company fills boxes with 16 ounces of cereal. The acceptable percent error in filling a box is 2.5%. Find the least and the greatest acceptable weights.

28. PRECISION Find the percent of change from June to September in the time to run a mile.

29. CRITICAL THINKING A number increases by 10%, and then decreases by 10%. Will the result be *greater than*, *less than*, or *equal to* the original number? Explain.

30. DONATIONS Donations to an annual fundraiser are 15% greater this year than last year. Last year, donations were 10% greater than the year before. The amount raised this year is $10,120. How much was raised 2 years ago?

31. Reasoning Forty students are in the science club. Of those, 45% are girls. This percent increases to 56% after new girls join the club. How many new girls join?

Fair Game Review What you learned in previous grades & lessons

Write and solve an equation to answer the question. *(Section 15.4)*

32. What number is 25% of 64? **33.** 39.2 is what percent of 112?

34. 5 is 5% of what number? **35.** 18 is 32% of what number?

36. MULTIPLE CHOICE Which set of ratios does *not* form a proportion? *(Section 14.2)*

Ⓐ $\frac{1}{4}, \frac{6}{24}$ Ⓑ $\frac{4}{7}, \frac{7}{10}$ Ⓒ $\frac{16}{24}, \frac{2}{3}$ Ⓓ $\frac{36}{10}, \frac{18}{5}$

15.6 Discounts and Markups

Essential Question How can you find discounts and selling prices?

1 ACTIVITY: Comparing Discounts

Work with a partner. The same pair of sneakers is on sale at three stores. Which one is the best buy? Explain.

a. Regular Price: $45

b. Regular Price: $49

c. Regular Price: $39

a.

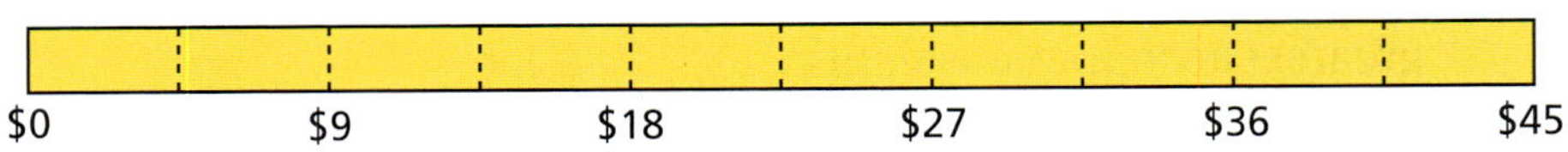

b.

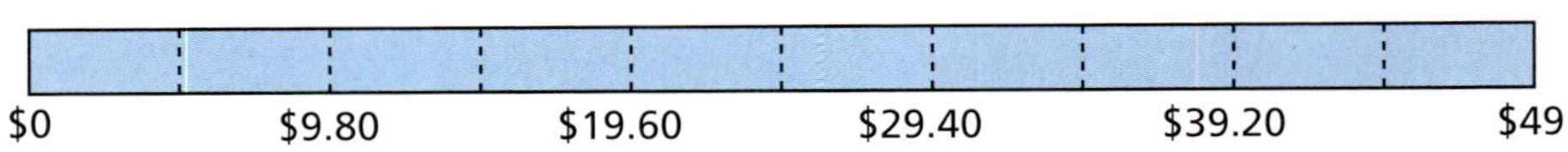

c.

2 ACTIVITY: Finding the Original Price

Work with a partner.

a. You buy a shirt that is on sale for 30% off. You pay $22.40. Your friend wants to know the original price of the shirt. Show how you can use the model below to find the original price.

b. Explain how you can use the percent proportion to find the original price.

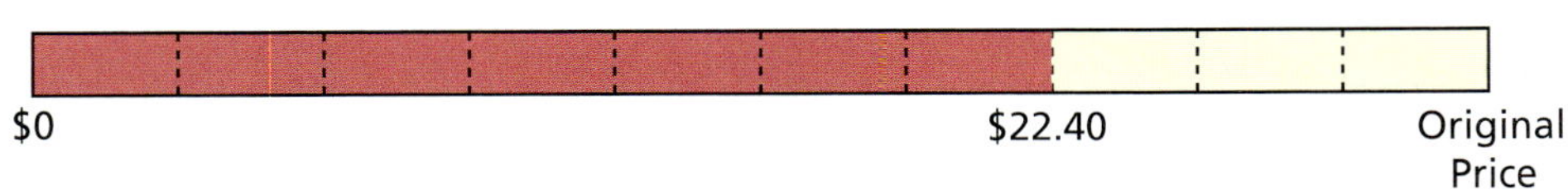

Percents

In this lesson, you will

- use percent of discounts to find prices of items.
- use percent of markups to find selling prices of items.

Laurie's Notes

Introduction

Applying Mathematical Practices

- **Model with Mathematics:** Percent applications are abundant. The percent bar model helps students visualize the problem and check on the reasonableness of the answer.

Motivate

- Show a newspaper circular that advertises a discount (sale).

Activity Notes

Activity 1

- Explain that sale items involve a *percent* discount and the *amount* of discount. If possible, use the newspaper circular to make this distinction.
- The percent bar models are divided into 10 equal parts. Dollar amounts for items are shown on the bars.
- Discuss how the dollar amounts can be computed. Students can use mental math to find 10% and multiply by the correct amount.
- **Big Idea:** When you *save* 40% ($18), you *pay* 60% ($27). Starting at $45, move to the left 40%, that is the savings.
- **Extension:** Determine the amount you save *and* the price paid. This will not be possible for the last example because the amount of discount varies.

? "How do you decide the best buy?" Listen for the lowest final price instead of the greatest savings because the original prices may vary.

? "What does the phrase, "up to 70% off" mean?" Percent off will vary from 0% up to 70%.

Activity 2

- **Connection:** Finding the original price is the same as finding the whole. $22.40 is the part.

? "What percent does $22.40 represent of the original price?" 70%

? "How does the percent bar model help you think about the original price?" Students might describe the $22.40 as 70% or about $\frac{2}{3}$ of the original price. So, another $\frac{1}{3}$ has to be added on to find the original price.

- Use the percent equation: $22.40 is 70% of what number?

? "Why is 70% used instead of 30%?" Because $22.40 is the part and it is 70% of the whole, or original price.

- **Struggling Students:** Students sometimes struggle with this concept. Reinforce by constantly telling students "30% off the original price is the same as paying 70% of the original price."
- **Construct Viable Arguments and Critique the Reasoning of Others:** Ask volunteers to explain each method—the percent equation and the percent proportion. Discuss with students the use of each method. Some students may prefer one method over the other.

What Your Students Will Learn

- Use percents of discounts to find sale prices or original prices of items.
- Use percents of markups to find selling prices of items.

Previous Learning

Students should be able to find a percent of a number, round decimal values, and convert between fractions, decimals, and percents.

Technology for the Teacher

Dynamic Classroom

Lesson Plans
Complete Materials List

15.6 Record and Practice Journal

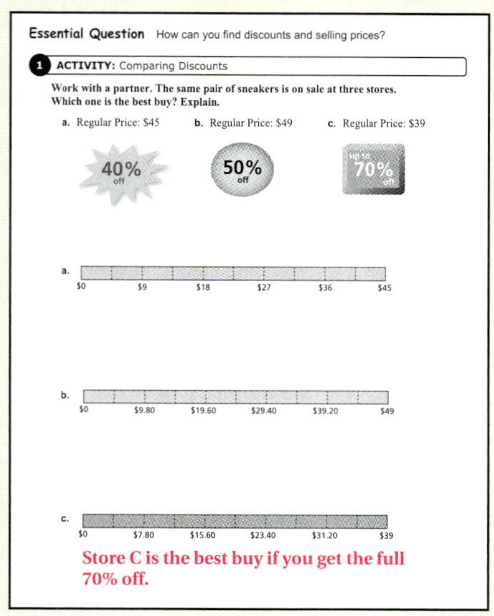

Essential Question How can you find discounts and selling prices?

1 ACTIVITY: Comparing Discounts

Work with a partner. The same pair of sneakers is on sale at three stores. Which one is the best buy? Explain.

a. Regular Price: $45 b. Regular Price: $49 c. Regular Price: $39

Store C is the best buy if you get the full 70% off.

English Language Learners

Vocabulary

English learners may not be familiar with terms used in business, such as *discount, markup, purchase price,* and *selling price*. Take time to explain these terms.

15.6 Record and Practice Journal

2 ACTIVITY: Finding the Original Price

Work with a partner.

a. You buy a shirt that is on sale for 30% off. You pay \$22.40. Your friend wants to know the original price of the shirt. Show how you can use the model to find the original price.

\$32

b. Explain how you can use the percent proportion to find the original price.

$$\frac{22.4}{w} = \frac{70}{100}$$

$$w = 32$$

3 ACTIVITY: Finding Selling Prices

You own a small jewelry store. You increase the price of jewelry by 125%.

Work with a partner. Use a model to estimate the selling price of the jewelry. Then use a calculator to find the selling price.

a. Your cost is \$250.

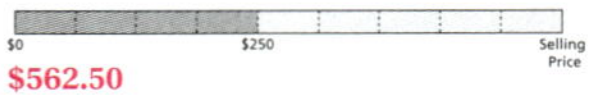

\$562.50

b. Your cost is \$50.

\$112.50

c. Your cost is \$170.

\$382.50

What Is Your Answer?

4. **IN YOUR OWN WORDS** How can you find discounts and selling prices? Give examples of each.

Check students' work.

Laurie's Notes

Activity 3

- **Discuss:** A store purchases an item for *x* dollars. The store needs to sell this item for more than *x* dollars to cover operating costs and to make a profit.
- **Explain:** A store purchases an item for \$2 and sells it for \$4. This represents a 100% increase (\$2 + 100% of \$2).
- **Tip:** Have students put 100% above the \$250 store cost in part (a).
- ? "How does the selling price compare to the price the store paid for the item?" 125% greater, or 225% of the store's cost
- **Reason Abstractly and Quantitatively:** If a \$10 item sells for \$25, the \$10 item was *increased by 150%* and the selling price is *250% of \$10*. One way to show this to students is to write: (100% of \$10) + (150% of \$10) = (250% of \$10) = \$25.

Words of Wisdom

- Be careful with language. This is not an obvious concept for students. Try to use consistent language with every example; store purchase price, store selling price, original price, increased amount, discount amount, and sale price.

What Is Your Answer?

- You want students to discover that they can find the selling price after a 25% discount by multiplying by 0.75 (one step) *or* by multiplying by 0.25 and then subtracting the result from the original price (two steps).
- Similarly, for markups, you can multiply the cost by 1.75 (one step) for a 75% markup *or* multiply by 0.75 and then add to the cost (two steps).

Closure

- You purchased an item marked 25% off. What percent of the original price did you pay? 75%

3 ACTIVITY: Finding Selling Prices

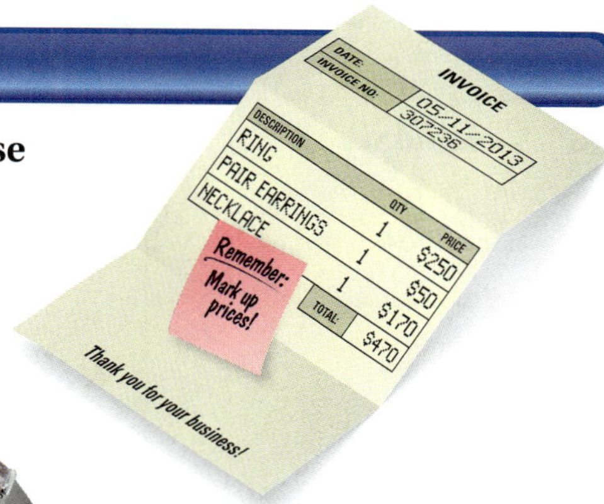

Math Practice

Make Sense of Quantities

What do the quantities represent? What is the relationship between the quantities?

You own a small jewelry store. You increase the price of the jewelry by 125%.

Work with a partner. Use a model to estimate the selling price of the jewelry. Then use a calculator to find the selling price.

a. Your cost is $250.

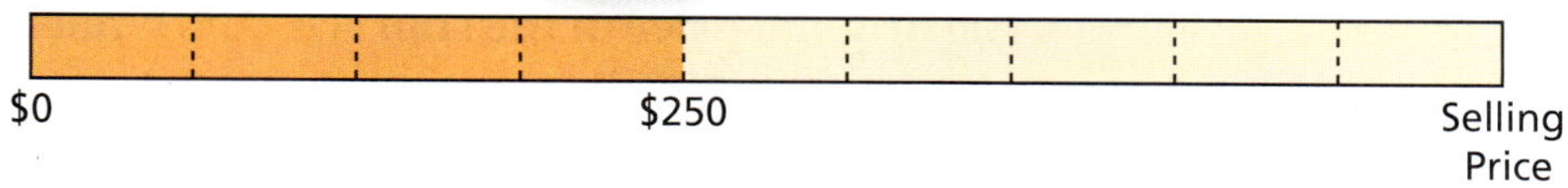

b. Your cost is $50.

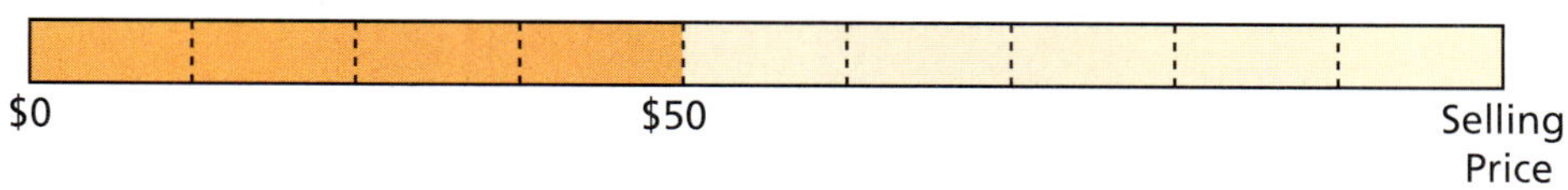

c. Your cost is $170.

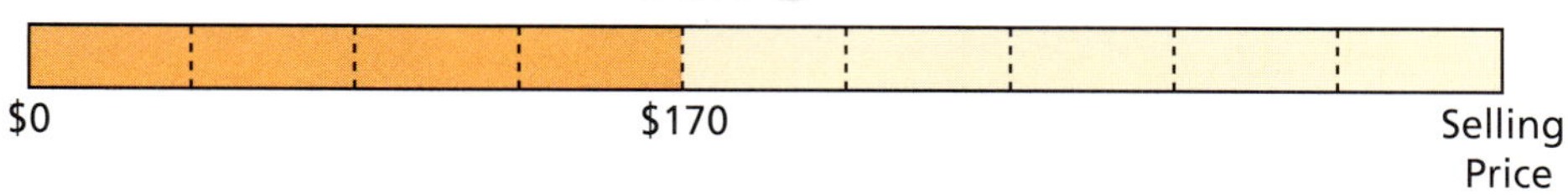

What Is Your Answer?

4. **IN YOUR OWN WORDS** How can you find discounts and selling prices? Give examples of each.

Practice Use what you learned about discounts to complete Exercises 4, 9, and 14 on page 686.

15.6 Lesson

Check It Out
Lesson Tutorials
BigIdeasMath.com

Key Vocabulary
discount, *p. 684*
markup, *p. 684*

Key Ideas

Discounts

A **discount** is a decrease in the original price of an item.

Markups

To make a profit, stores charge more than what they pay. The increase from what the store pays to the selling price is called a **markup**.

EXAMPLE 1 Finding a Sale Price

The original price of the shorts is \$35. What is the sale price?

Method 1: First, find the discount. The discount is 25% of \$35.

$a = p \cdot w$ Write percent equation.

$= 0.25 \cdot 35$ Substitute 0.25 for *p* and 35 for *w*.

$= 8.75$ Multiply.

Next, find the sale price.

sale price = original price − discount

$= 35 - 8.75$

$= 26.25$

So, the sale price is \$26.25.

Method 2: First, find the percent of the original price.

$100\% - 25\% = 75\%$

Next, find the sale price.

sale price = 75% of \$35

$= 0.75 \cdot 35$

$= 26.25$

So, the sale price is \$26.25.

Check

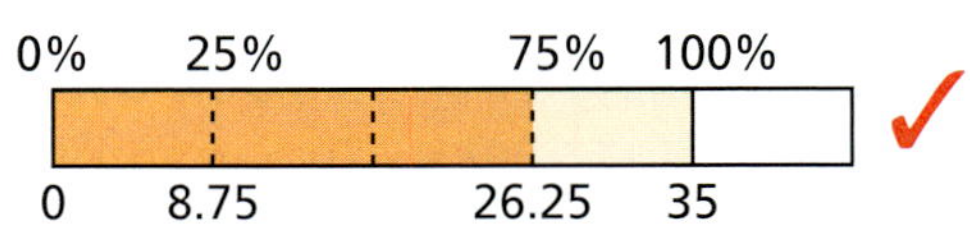

Study Tip

A 25% discount is the same as paying 75% of the original price.

On Your Own

Now You're Ready
Exercises 4–8

1. The original price of a skateboard is \$50. The sale price includes a 20% discount. What is the sale price?

Multi-Language Glossary at BigIdeasMath.com

Laurie's Notes

Introduction

Connect

- **Yesterday:** Students explored discounts and selling prices using a percent bar model.
- **Today:** Students will use the percent equation to find discounts and markups of items.

Motivate

? **Story Time:** "A store buys an MP3 player for $100 and marks it up 50%. The store has a 50% off sale. You purchase the MP3 player. What do you pay?" $75 "Did the store lose money?" yes

Lesson Notes

Key Ideas

- Discuss each concept using examples from the previous day's activity.
- Use the following to help students understand the vocabulary.

wholesale price + markup = retail price (or selling price)

↑ what a store pays — ↑ increase in price — ↑ price you pay

- **FYI:** Some students may choose to use the percent proportion instead of the percent equation. Remind them of these *equivalent* methods if desired. Examples in this lesson are solved using the percent equation.

Example 1

- Two methods are shown, work through each method. Both methods require two steps. In the first method, you multiply to find the amount of discount, then you subtract to find the sale price. In the second method, you subtract first to find the percent of the original price you will pay, then you use the percent equation to find the sale price.
- **Connection:** The amount of discount is a *part* of the *whole* original price. The percent equation is used to find the amount of the discount.
- **Common Error:** Students find the discount or the amount saved ($8.75) instead of the sale price ($26.25).
- Discuss the *Study Tip.* Try other discounts (i.e., 30%) and ask what percent you are paying (70%).

? "Why is the percent bar model divided into 4 parts?"
Because the discount is 25% or $\frac{1}{4}$.

On Your Own

? "How should the percent bar model be divided and why?" 5 parts because $20\% = \frac{1}{5}$

Goal Today's lesson is using the percent equation to find **discounts** and **markups**.

Technology for the Teacher

Dynamic Classroom

Lesson Tutorials
Lesson Plans
Answer Presentation Tool

Extra Example 1

The original price of a T-shirt is $15. The sale price includes a 35% discount. What is the sale price? $9.75

On Your Own

1. $40

Extra Example 2

The discount on a package of athletic socks is 15%. It is on sale for \$17. What is the original price of the package of athletic socks? \$20

Extra Example 3

A store pays \$15 for a baseball cap. The percent markup is 60%. What is the selling price? \$24

On Your Own

2. \$20

3. \$90

Differentiated Instruction

Visual

Some students may have a hard time remembering the relationships between *sale price*, *selling price*, *discount*, and *markup*. Have them copy the verbal models into their notebooks.

Discount

sale price = original price − discount amount

= original price − some % • original price

Markup

selling price = cost to store + markup amount

= cost to store + some % • cost to store

Laurie's Notes

Example 2

- "What is the percent equation?" $a = p \cdot w$
- ? "What do you know in this problem?" 33 is the part and 60% is the percent.
- **Common Error:** Students multiply 33 by 60% (or 40%). Students need to remember that 33 is a *part* of the original price, it's not the *whole*.

Example 3

- Work through the problem as shown. Encourage students to use mental math to find 20% of \$70. 10% of 70 is 7. So, 20% of 70 is 2(7), or 14.
- Two steps were used to answer the question: 1) Find 20% of \$70 and 2) add this amount to the original amount of \$70.
- ? "Could this problem be done in one step? Explain." yes; 120% of \$70 = \$84
- ? "Explain why the 120% makes sense." You pay 100% of the store's cost plus an additional 20% markup for a total of 120%.
- Method 2 uses the fact that the selling price is 120% of what the store paid. Make sure students understand this. Show students how to use a proportion, $\frac{a}{70} = \frac{120}{100}$ and a one-step percent equation, $a = 1.2(70)$. The ratio table shows the division and multiplication to get to 120% and that this is also the sum of the two rows, which is comparable to the procedure in Method 1.
- **Reason Abstractly and Quantitatively** and **Construct Viable Arguments and Critique the Reasoning of Others:** Now go back and ask students what other ways they can solve the previous examples. Deepening student understanding results from considering different ways in which to solve problems.
- **Common Error:** Students find the markup (\$14) instead of the selling price (\$84).
- **Extension:** Have students draw a percent bar model for this problem. The model will be divided into fifths.

On Your Own

- **Neighbor Check:** Have students work independently and then have their neighbors check their work. Have students discuss any discrepancies.

Closure

- **Writing Prompt:** Explain two ways to find the sale price for an item marked 30% off. 1) Find the amount of discount and subtract from the original price. 2) Find the percent of the original price and multiply the percent by the original price.
- **Extension:** "If an item is marked up and then discounted the same percent, will the store make a profit? Explain." no; the *amount of markup* will be less than the *amount of discount*, so the store will sell the item for less than what it paid.
- **Extension:** "Is a 25% discount followed by a 10% discount the same as a 35% discount? Explain." no; The sale price for an item discounted 25% followed by a 10% discount would be 0.75(0.9) = 0.675, or 67.5% of the original price. The sale price for an item discounted 35% would be 65% of the original price.

EXAMPLE 2 Finding an Original Price

What is the original price of the shoes?

The sale price is $100\% - 40\% = 60\%$ of the original price.

Answer the question: 33 is 60% of what number?

$a = p \cdot w$	Write percent equation.
$33 = 0.6 \cdot w$	Substitute 33 for *a* and 0.6 for *p*.
$55 = w$	Divide each side by 0.6.

∴ So, the original price of the shoes is $55.

Check

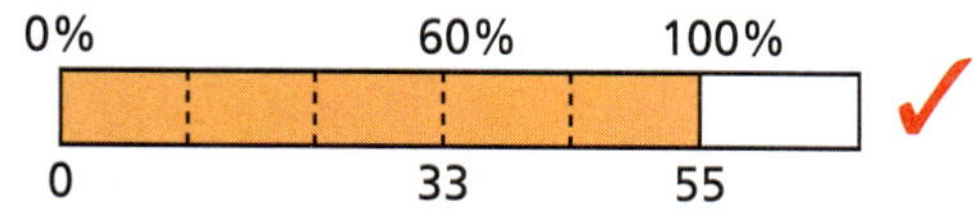

EXAMPLE 3 Finding a Selling Price

A store pays $70 for a bicycle. The percent of markup is 20%. What is the selling price?

Method 1: First, find the markup. The markup is 20% of $70.

$a = p \cdot w$

$= 0.20 \cdot 70$

$= 14$

Next, find the selling price.

selling price = cost to store + markup

$= 70 + 14$

$= 84$

∴ So, the selling price is $84.

Check

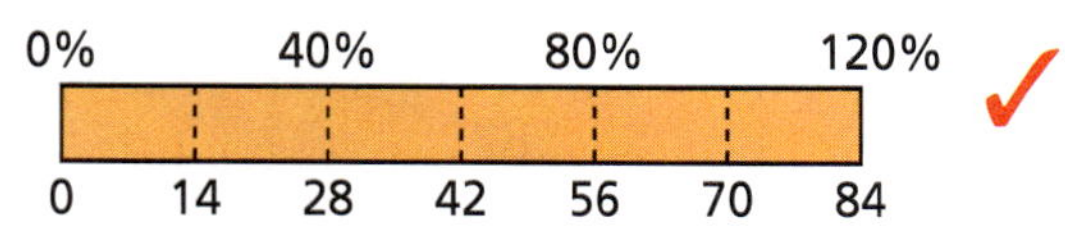

Method 2: Use a ratio table. The selling price is 120% of the cost to the store.

Percent	Dollars
100%	$70
20%	$14
120%	$84

(÷ 5 from 100% to 20% and from $70 to $14; × 6 from 20% to 120% and from $14 to $84)

∴ So, the selling price is $84.

On Your Own

Now You're Ready
Exercises 9–13 and 17–19

2. The discount on a DVD is 50%. It is on sale for $10. What is the original price of the DVD?

3. A store pays $75 for an aquarium. The markup is 20%. What is the selling price?

15.6 Exercises

Check It Out
Help with Homework
BigIdeasMath.com

Vocabulary and Concept Check

1. **WRITING** Describe how to find the sale price of an item that has been discounted 25%.
2. **WRITING** Describe how to find the selling price of an item that has been marked up 110%.
3. **REASONING** Which would you rather pay? Explain your reasoning.

   a. 6% tax on a discounted price or 6% tax on the original price

   b. 30% markup on a $30 shirt or $30 markup on a $30 shirt

Practice and Problem Solving

Copy and complete the table.

	Original Price	Percent of Discount	Sale Price
(1) **4.**	$80	20%	
5.	$42	15%	
6.	$120	80%	
7.	$112	32%	
8.	$69.80	60%	
(2) **9.**		25%	$40
10.		5%	$57
11.		80%	$90
12.		64%	$72
13.		15%	$146.54
14.	$60		$45
15.	$82		$65.60
16.	$95		$61.75

Find the selling price.

(3) **17.** Cost to store: $50
Markup: 10%

18. Cost to store: $80
Markup: 60%

19. Cost to store: $140
Markup: 25%

Assignment Guide and Homework Check

Level	Assignment	Homework Check
Advanced	1–4, 8, 9, 12–24 even, 25–29	8, 12, 16, 20, 24

Common Errors

- **Exercises 4–8** Students may write the discount amount as the sale price instead of subtracting it from the original amount. When students copy the table, ask them to add another column titled "Discount Amount." Remind them to subtract the discount amount from the original price.
- **Exercises 9–16** Remind students that there is an extra step in the problem. They should subtract the percent of discount from 100% to find the percent of the original price of the item.
- **Exercises 17–19** Students may find the markup and not the selling price. Remind them that they must add the markup to the cost to obtain the selling price.

15.6 Record and Practice Journal

Complete the table.

	Original Price	Percent of Discount	Sale Price
1.	\$20	20%	**\$16**
2.	\$95	35%	**\$61.75**
3.	**\$222**	75%	\$55.50
4.	**\$130**	40%	\$78

Find the selling price.

5. Cost to store: \$20
   Markup: 15%
   \$23
6. Cost to store: \$56
   Markup: 80%
   \$100.80
7. Cost to store: \$110
   Markup: 140%
   \$264
8. A store buys an item for \$10. To earn a profit of \$25, what percent does the store need to markup the item?
   250%

Vocabulary and Concept Check

1. *Sample answer:* Multiply the original price by 100% − 25% = 75% to find the sale price.
2. Find the markup by taking 110% of the amount. Then add the amount and the markup to find the selling price.
3. a. 6% tax on a discounted price; The discounted price is less, so the tax is less.
   b. 30% markup on a \$30 shirt; 30% of \$30 is less than \$30.

Practice and Problem Solving

4. \$64
5. \$35.70
6. \$24
7. \$76.16
8. \$27.92
9. \$53.33
10. \$60
11. \$450
12. \$200
13. \$172.40
14. 25%
15. 20%
16. 35%
17. \$55
18. \$128
19. \$175

Practice and Problem Solving

20. no; Only the amount of markup should be in the numerator, $\frac{105 - 60}{60} = 0.75$. So, the percent of markup is 75%.

21. "Multiply \$45.85 by 0.1" and "Multiply \$45.85 by 0.9, then subtract from \$45.85." Both will give the sale price of \$4.59. The first method is easier because it is only one step.

22. **a.** Store C

b. at least 11.82%

23. no; \$31.08

24. See *Taking Math Deeper*.

25. \$30

Fair Game Review

26. 170 **27.** 180

28. 1152 **29.** C

Mini-Assessment

Find the price, discount, markup, or cost to store.

1. Original price: \$50
   Discount: 15%
   Sale price: ? \$42.50
2. Original price: \$35
   Discount: ?
   Sale price: \$31.50 10%
3. Cost to store: \$75
   Markup: ?
   Selling price: \$112.50 50%
4. Cost to store: ?
   Markup: 15%
   Selling price: \$85.10 \$74
5. The sale price for a bicycle is \$89.90. The sale price includes a discount of 20%. What is the original price of the bicycle? \$112.38

Taking Math Deeper

Exercise 24

A good way to approach this problem is to take things one step at a time. Also, in problems like this, it is much easier to round up to \$40 and \$30 for easier calculations.

Jeans	\$40	39.99
Discount		-10.00
Subtotal		29.99
Sales Tax		1.95
Total		31.94

① Find the percent of discount.

a. 10 is 25% of 40.

It is easier to round \$39.99 to \$40 before doing the calculations.

② Find the percent of sales tax.

1.95 is what % of 30?

$$1.95 = p \cdot 30$$
$$0.065 = p$$

b. Sales tax = 6.5%.

Jeans		39.99
Discount		-10.00
Subtotal	\$30	29.99
Sales Tax		1.95
Total		31.94

③ Find the actual markup.

$$x + 0.6x = 40$$
$$1.6x = 40$$
$$x = \$25 \quad \text{Wholesale}$$

The \$40 jeans cost the store \$25. After the discount of \$10, the markup is \$5. Find the percent of markup by answering "5 is what % of 25?"

c. \$5 is a 20% markup on \$25.

Project

Check the newspaper or local advertisements for a store near you. Select five items that are on sale. Prepare a chart that shows the original price, the percent of discount, and the sale price. How much would you save if you purchased all five items at the sale price?

Reteaching and Enrichment Strategies

If students need help...	If students got it...
Resources by Chapter • Practice A and Practice B • Puzzle Time Record and Practice Journal Practice Differentiating the Lesson Lesson Tutorials Skills Review Handbook	Resources by Chapter • Enrichment and Extension • Technology Connection Start the next section

20. **YOU BE THE TEACHER** The cost to a store for an MP3 player is \$60. The selling price is \$105. A classmate says that the markup is 175% because $\frac{\$105}{\$60} = 1.75$. Is your classmate correct? If not, explain how to find the correct percent of markup.

21. **SCOOTER** The scooter is on sale for 90% off the original price. Which of the methods can you use to find the sale price? Which method do you prefer? Explain.

Multiply \$45.85 by 0.9.

Multiply \$45.85 by 0.1.

Multiply \$45.85 by 0.9, then add to \$45.85.

Multiply \$45.85 by 0.9, then subtract from \$45.85.

22. **GAMING** You are shopping for a video game system.

Store	Cost to Store	Markup
A	\$162	40%
B	\$155	30%
C	\$160	25%

   a. At which store should you buy the system?

   b. Store A has a weekend sale. What discount must Store A offer for you to buy the system there?

23. **STEREO** A \$129.50 stereo is discounted 40%. The next month, the sale price is discounted 60%. Is the stereo now "free"? If not, what is the sale price?

24. **CLOTHING** You buy a pair of jeans at a department store.

   a. What is the percent of discount to the nearest percent?

   b. What is the percent of sales tax to the nearest tenth of a percent?

   c. The price of the jeans includes a 60% markup. After the discount, what is the percent of markup to the nearest percent?

25. **Critical Thinking** You buy a bicycle helmet for \$22.26, which includes 6% sales tax. The helmet is discounted 30% off the selling price. What is the original price?

Fair Game Review What you learned in previous grades & lessons

Evaluate. *(Section 2.5)*

26. 2000(0.085)

27. 1500(0.04)(3)

28. 3200(0.045)(8)

29. **MULTIPLE CHOICE** Which measurement is greater than 1 meter? *(Section 5.7)*

Ⓐ 38 inches Ⓑ 1 yard Ⓒ 3.4 feet Ⓓ 98 centimeters

15.7 Simple Interest

Essential Question How can you find the amount of simple interest earned on a savings account? How can you find the amount of interest owed on a loan?

Simple interest is money earned on a savings account or an investment. It can also be money you pay for borrowing money.

Write the annual interest rate in decimal form.

Simple interest	=	Principal	×	Annual interest rate	×	Time
($)		($)		(% per yr)		(Years)

$I = Prt$

1 ACTIVITY: Finding Simple Interest

Work with a partner. You put \$100 in a savings account. The account earns 6% simple interest per year. (a) Find the interest earned and the balance at the end of 6 months. (b) Copy and complete the table. Then make a bar graph that shows how the balance grows in 6 months.

a. $I = Prt$ Write simple interest formula.

$= \square$ Substitute values.

$= \square$ Multiply.

At the end of 6 months, you earn \$ ___ in interest. So, your balance is \$ ___.

b.

Time	Interest	Balance
0 month	\$0	\$100
1 month		
2 months		
3 months		
4 months		
5 months		
6 months		

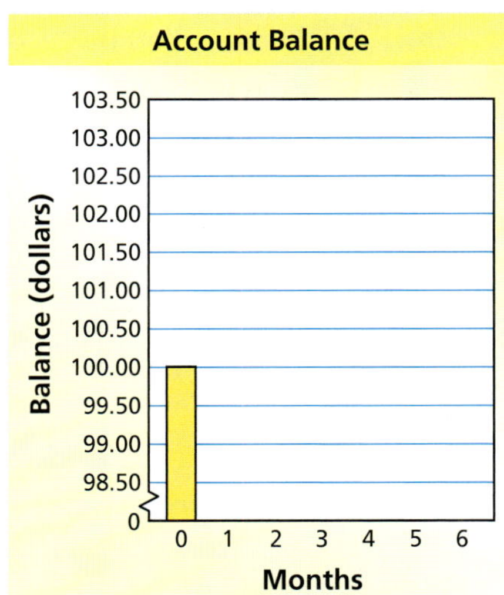

Percents

In this lesson, you will

- use the simple interest formula to find interest earned or paid, annual interest rates, and amounts paid on loans.

Laurie's Notes

Introduction

Applying Mathematical Practices

- **Model with Mathematics:** Percent applications are abundant in the financial world. Interest rates are stated as percents and students need to understand how interest is paid or charged to consumers.

Motivate

- **Story Time:** Tell students that you just saw an ad for the latest smart phone and you really want to buy it. "It's only $400, but unfortunately I have a few other bills this month and can't afford $400 all at once."

? "What can I do?" Students may suggest that you go to a bank and borrow the money.

? "Will a bank just give me $400?" Hopefully students will know that there is a fee you have to pay to borrow the money.

Activity Notes

Discuss

- Today's investigation involves three activities. Given time constraints and your own students, you may not complete all three.
- **Financial Literacy:** You want students to have some understanding of the cost of borrowing money or the ability to earn money when it is deposited in a bank, not to become trained loan officers.
- **Discuss:** When you *deposit* money, you should *earn* money. When you *borrow* money, you should *pay* money.
- Define *simple interest formula.*
- **Discuss:** Interest earned/owed is influenced by how much money is involved (principal), the rate you pay/earn, and the amount of time.
- Make clear that it is an *annual* interest rate and the time is in *years.*
- Students should assume that deposits are made at the beginning of the interest period in all banking problems, unless otherwise stated.

Activity 1

- This activity uses the simple interest formula. The principal stays the same for each month's calculation. Interest paid is *not* being compounded.
- **Demonstrate:** After one month, you earn $\$100(0.06)\left(\frac{1}{12}\right) = \$0.50$. This \$0.50 is added to the principal.
- Get students started on month 2. Students should use \$100 for the principal and $\frac{2}{12}$ for the time. Interest earned $= \$100(0.06)\left(\frac{2}{12}\right) = \$1.00$.
- Students should work with partners to complete the table and the graph.
- **Attend to Precision:** Note the use of the broken axis on the bar graph. In order to show the constant growth of \$0.50 per month, the vertical scale has to be quite small.

What Your Students Will Learn

- Use the simple interest formula to find interest earned or paid, annual interest rates, length of loan, or amounts paid on loans.

Previous Learning

Students should be familiar with finding a percent of a number, rounding decimal values, and converting between fractions, decimals, and percents.

Technology for the Teacher

Dynamic Classroom

Lesson Plans
Complete Materials List

15.7 Record and Practice Journal

Essential Question How can you find the amount of simple interest earned on a savings account? How can you find the amount of interest owed on a loan?

Simple interest is money earned on a savings account or an investment. It can also be money you pay for borrowing money.

Write the annual interest rate in decimal form.

Simple interest ($) = Principal ($) × Annual interest rate (% per yr) × Time (Years)

$I = Prt$

1 ACTIVITY: Finding Simple Interest

Work with a partner. You put \$100 in a savings account. The account earns 6% simple interest per year. (a) Find the interest earned and the balance at the end of 6 months. (b) Complete the table. Then make a bar graph that shows how the balance grows in 6 months.

a. $I = Prt$

\$3; \$103

b.

Time	Interest	Balance
0 month	\$0	\$100
1 month	\$0.50	\$100.50
2 months	\$1	\$101
3 months	\$1.50	\$101.50
4 months	\$2	\$102
5 months	\$2.50	\$102.50
6 months	\$3	\$103

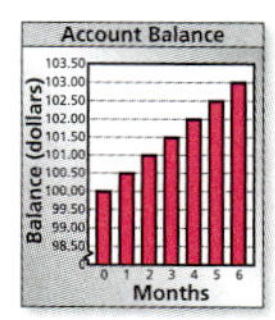

Differentiated Instruction

Auditory

Discuss the meaning of the word *interest*. An interest rate is often expressed as an annual percentage of the principal.

15.7 Record and Practice Journal

2 ACTIVITY: Financial Literacy

Work with a partner. Use the following information to write a report about credit cards. In the report, describe how a credit card works. Include examples that show the amount of interest paid each month on a credit card.

U.S. Credit Card Data

- A typical household with credit card debt in the United States owes about $16,000 to credit card companies.
- A typical credit card interest rate is 14% to 16% per year. This is called the annual percentage rate.

Check students' work.

3 ACTIVITY: The National Debt

Work with a partner. In 2012, the United States owed about $16 trillion in debt. The interest rate on the national debt is about 1% per year.

a. Write $16 trillion in decimal form. How many zeros does this number have?
$16,000,000,000,000; 12 zeros

b. How much interest does the United States pay each year on its national debt?
$160 billion

c. How much interest does the United States pay each day on its national debt?
$438,356,164.40

d. The United States has a population of about 314 million people. Estimate the amount of interest that each person pays per year toward interest on the national debt.
about $509.55

What Is Your Answer?

4. IN YOUR OWN WORDS How can you find the amount of simple interest earned on a savings account? How can you find the amount of interest owed on a loan? Give examples with your answer.
Use $I = Prt$.

Laurie's Notes

Activity 2

- You may wish to use the information to demonstrate the impact of carrying a large credit card debt.
- **Discuss:** How a credit card operates, how you can get one, and how it works (consumer, store, bank).
- **Community:** If your local bank has an education or outreach coordinator, consider having them come in as a guest speaker.
- Read through the information given. Calculate the interest owed for one month, $\$16{,}000(0.14)\left(\frac{1}{12}\right) \approx \$186.67$, or at the higher interest rate, $\$16{,}000(0.16)\left(\frac{1}{12}\right) \approx \$213.33$.
- **Reason Abstractly and Quantitatively:** Why should you shop around for the lowest interest rates? Why should you keep your principal as small as possible?
- Remind the students that the consumer needs to pay the interest ($186.67 or $213.33) *plus* they need to be paying off the principal.

Activity 3

- The national debt is a complicated concept. The intent of this activity is to raise awareness and to use the simple interest formula with a really large number.
- **Caution:** If you use a calculator for this problem, the debt and simple interest will appear in scientific notation.
- **Representation:** It will be helpful to write out the simple interest formula using the decimal numbers so that students can see all of the zeros: $I = (16{,}000{,}000{,}000{,}000)(0.01)(1)$. This has a greater impact than scientific notation.
- **Extension:** Many local newspapers print the national debt and the approximate per person debt each day. Record this information once a week for about 2 months to get a sense for how the numbers are changing. You can even do this for the entire school year.

Closure

- **Exit Ticket:** What do you need to know in order to compute simple interest? Principal, annual interest rate, and time

2 ACTIVITY: Financial Literacy

Work with a partner. Use the following information to write a report about credit cards. In the report, describe how a credit card works. Include examples that show the amount of interest paid each month on a credit card.

Math Practice

Use Other Resources

What resources can you use to find more information about credit cards?

U.S. Credit Card Data

- A typical household with credit card debt in the United States owes about \$16,000 to credit card companies.
- A typical credit card interest rate is 14% to 16% per year. This is called the annual percentage rate.

3 ACTIVITY: The National Debt

Work with a partner. In 2012, the United States owed about \$16 trillion in debt. The interest rate on the national debt is about 1% per year.

a. Write \$16 trillion in decimal form. How many zeros does this number have?

b. How much interest does the United States pay each year on its national debt?

c. How much interest does the United States pay each day on its national debt?

d. The United States has a population of about 314 million people. Estimate the amount of interest that each person pays per year toward interest on the national debt.

What Is Your Answer?

4. **IN YOUR OWN WORDS** How can you find the amount of simple interest earned on a savings account? How can you find the amount of interest owed on a loan? Give examples with your answer.

Practice Use what you learned about simple interest to complete Exercises 4–7 on page 692.

15.7 Lesson

Check It Out
Lesson Tutorials
BigIdeasMath.com

Key Vocabulary
interest, *p. 690*
principal, *p. 690*
simple interest, *p. 690*

Interest is money paid or earned for the use of money. The **principal** is the amount of money borrowed or deposited.

Key Idea

Simple Interest

Words **Simple interest** is money paid or earned only on the principal.

Algebra $I = Prt$

- Simple interest: I
- Principal: P
- Annual interest rate (in decimal form): r
- Time (in years): t

EXAMPLE 1 Finding Interest Earned

You put \$500 in a savings account. The account earns 3% simple interest per year. (a) What is the interest earned after 3 years? (b) What is the balance after 3 years?

a. $I = Prt$ Write simple interest formula.

$= 500(0.03)(3)$ Substitute 500 for P, 0.03 for r, and 3 for t.

$= 45$ Multiply.

So, the interest earned is \$45 after 3 years.

b. To find the balance, add the interest to the principal.

So, the balance is \$500 + \$45 = \$545 after 3 years.

EXAMPLE 2 Finding an Annual Interest Rate

You put \$1000 in an account. The account earns \$100 simple interest in 4 years. What is the annual interest rate?

$I = Prt$ Write simple interest formula.

$100 = 1000(r)(4)$ Substitute 100 for I, 1000 for P, and 4 for t.

$100 = 4000r$ Simplify.

$0.025 = r$ Divide each side by 4000.

So, the annual interest rate of the account is 0.025, or 2.5%.

Multi-Language Glossary at BigIdeasMath.com

Laurie's Notes

Introduction

Connect

- **Yesterday:** Students explored the simple interest formula, applying it to several consumer applications.
- **Today:** Students will use the simple interest formula and knowledge of equation solving to solve for different variables in the formula.

Motivate

- Just imagine that when you are older, you win a $5 million lottery. If you deposit that money for 10 years at 6% simple interest, how much will you have at the end of 10 years? $8,000,000

Lesson Notes

Key Idea

- **Vocabulary:** interest, money paid or earned, principal, amount of money borrowed or deposited, balance
- **Representation:** Write the formula in words first.
  Simple Interest = (Principal)(Annual interest rate)(Time)
- **Explain:** Simple interest is only one type of interest. There are also compound and exponential interest calculations. The interest rate is written as a decimal. Time is written in terms of years. When time is given in months, remember to express it as a fraction of a year or as a decimal. For example, 9 months $= \frac{9}{12}$ or 0.75 year.
- **Reason Abstractly and Quantitatively:** This formula is similar to the volume formula for a rectangular prism; three variables are multiplied together. Knowing 3 of the 4 variables, you can solve for the fourth.

Example 1

- There are two parts to the problem: Calculate the interest earned and then determine the amount (balance) in the account.
- ? "What operation is performed in writing *Prt*?" multiplication
- ? "In calculating 500(0.03)(3), what order is the multiplication performed?" Order doesn't matter, multiplication is commutative.
- **Explain:** Your balance is the original principal *plus* the interest earned.
- **Reason Abstractly and Quantitatively:** If time permits, "What would your balance be if the interest rate had been 6% instead of 3%?" $590 Doubling the interest rate doubles the amount earned. This can be shown in the equation $I = 500(0.06)(3) = 500(0.03)(2)(3)$.

Example 2

- This example uses the Division Property of Equality to solve for the interest rate.
- ? "Why does $1000(r)(4) = 4000r$?" Commutative Property of Multiplication
- **Common Error:** Students divide 4000 by 100 instead of 100 by 4000.
- ? "How do you write a decimal as a percent?" Move the decimal point two places to the right. (Multiply by 100.) Then add a percent symbol.

Goal Today's lesson is using the **simple interest** formula.

Technology for the Teacher

Dynamic Classroom

Lesson Tutorials
Lesson Plans
Answer Presentation Tool

Extra Example 1

You put $200 in a savings account. The account earns 2% simple interest per year.

a. What is the interest earned after 5 years? $20

b. What is the balance after 5 years? $220

Extra Example 2

You put $700 in an account. The account earns $224 simple interest in 8 years. What is the annual interest rate? 4%

On Your Own

1. $511.25
2. 2%

Extra Example 3

Using the pictograph in Example 3, how long does it take an account with a principal of $400 to earn $36 interest? 6 years

Extra Example 4

You borrow $300 to buy a guitar. The simple interest rate is 12%. You pay off the loan after 4 years. How much do you pay for the loan? $444

On Your Own

3. 2.5 yr
4. $270

English Language Learners

Vocabulary

Review with English learners the mathematical meanings of principal, interest, and balance because these words have multiple meanings in the English language. They should understand that interest is paid to customers when they deposit money into an account. When a person borrows money, the person pays interest to the bank.

Laurie's Notes

On Your Own

- **Neighbor Check**: Have students work independently and then have their neighbors check their work. Have students discuss any discrepancies.
- Check accuracy of decimals in these problems.

Example 3

- Discuss the diagram.

? "Why would a bank offer different interest rates for different principals?" Students may not understand that banks are using deposited money to loan to other people.

- Work through the problem.

? "What is 6.25 as a mixed number?" $6\frac{1}{4}$

- **Connection:** Students may wonder why anyone would want to know how long it takes to earn $100 in interest. Use an example of depositing money for a future purchase (car, house, college education).

Example 4

- Remind students that the simple interest formula is used to calculate interest *earned* when you *deposit* money and to calculate interest *owed* when you *borrow* money.
- **Discuss:** There are two parts to the problem: 1) Calculate the interest owed and 2) determine the total cost you must pay back for the loan.
- **Extension:** Have students find the monthly payment. $1050 ÷ 60 = $17.50

On Your Own

- **Neighbor Check:** Have students work independently and then have their neighbors check their work. Have students discuss any discrepancies.

Closure

- **Exit Ticket:** Assume $1000 was deposited at 5% simple interest when you were born. Approximately how much is the account worth today? age 11: $1550, age 12: $1600, age 13: $1650, age 14: $1700

On Your Own

Now You're Ready
Exercises 4–16

1. In Example 1, what is the balance of the account after 9 months?
2. You put \$350 in an account. The account earns \$17.50 simple interest in 2.5 years. What is the annual interest rate?

EXAMPLE 3 Finding an Amount of Time

A bank offers three savings accounts. The simple interest rate is determined by the principal. How long does it take an account with a principal of \$800 to earn \$100 in interest?

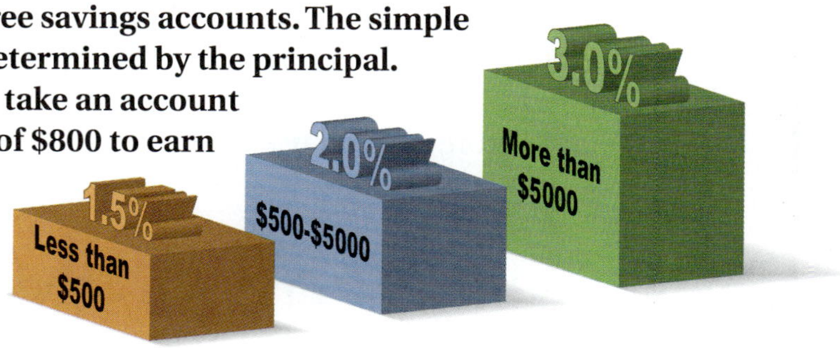

The pictogram shows that the interest rate for a principal of \$800 is 2%.

$I = Prt$ — Write simple interest formula.

$100 = 800(0.02)(t)$ — Substitute 100 for I, 800 for P, and 0.02 for r.

$100 = 16t$ — Simplify.

$6.25 = t$ — Divide each side by 16.

So, the account earns \$100 in interest in 6.25 years.

EXAMPLE 4 Finding an Amount Paid on a Loan

You borrow \$600 to buy a violin. The simple interest rate is 15%. You pay off the loan after 5 years. How much do you pay for the loan?

$I = Prt$ — Write simple interest formula.

$= 600(0.15)(5)$ — Substitute 600 for P, 0.15 for r, and 5 for t.

$= 450$ — Multiply.

To find the amount you pay, add the interest to the loan amount.

So, you pay \$600 + \$450 = \$1050 for the loan.

On Your Own

Now You're Ready
Exercises 17–20 and 24–27

3. In Example 3, how long does it take an account with a principal of \$10,000 to earn \$750 in interest?
4. **WHAT IF?** In Example 4, you pay off the loan after 2 years. How much money do you save?

15.7 Exercises

Check It Out
Help with Homework
BigIdeasMath.com

Vocabulary and Concept Check

1. **VOCABULARY** Define each variable in $I = Prt$.

2. **WRITING** In each situation, tell whether you would want a *higher* or *lower* interest rate. Explain your reasoning.

   a. you borrow money

   b. you open a savings account

3. **REASONING** An account earns 6% simple interest. You want to find the interest earned on \$200 after 8 months. What conversions do you need to make before you can use the formula $I = Prt$?

Practice and Problem Solving

An account earns simple interest. (a) Find the interest earned. (b) Find the balance of the account.

1

4. \$600 at 5% for 2 years

5. \$1500 at 4% for 5 years

6. \$350 at 3% for 10 years

7. \$1800 at 6.5% for 30 months

8. \$700 at 8% for 6 years

9. \$1675 at 4.6% for 4 years

10. \$925 at 2% for 2.4 years

11. \$5200 at 7.36% for 54 months

12. **ERROR ANALYSIS** Describe and correct the error in finding the simple interest earned on \$500 at 6% for 18 months.

$$I = (500)(0.06)(18)$$
$$= \$540$$

Find the annual interest rate.

2

13. $I = \$24$, $P = \$400$, $t = 2$ years

14. $I = \$562.50$, $P = \$1500$, $t = 5$ years

15. $I = \$54$, $P = \$900$, $t = 18$ months

16. $I = \$160.67$, $P = \$2000$, $t = 8$ months

Find the amount of time.

3

17. $I = \$30$, $P = \$500$, $r = 3\%$

18. $I = \$720$, $P = \$1000$, $r = 9\%$

19. $I = \$54$, $P = \$800$, $r = 4.5\%$

20. $I = \$450$, $P = \$2400$, $r = 7.5\%$

21. **BANKING** A savings account earns 5% simple interest per year. The principal is \$1200. What is the balance after 4 years?

22. **SAVINGS** You put \$400 in an account. The account earns \$18 simple interest in 9 months. What is the annual interest rate?

23. **CD** You put \$3000 in a CD (certificate of deposit) at the promotional rate. How long will it take to earn \$336 in interest?

Assignment Guide and Homework Check

Level	Assignment	Homework Check
Advanced	1–7, 12–36 even, 37–41	24, 30, 34, 36

Common Errors

- **Exercises 4–11** Students may forget to change the percent to a decimal. Remind them that before they can put the percent into the equation, they must change the percent to a fraction or a decimal.
- **Exercises 7 and 11** Students may not change months into years and calculate a much greater interest amount. Remind them that the simple interest formula is for *years* and that the time must be changed to years.
- **Exercises 15 and 16** Students may not change the time from months to years. Remind them that the time is in years.
- **Exercises 24–27** Students may only find the amount of interest paid for the loan. Remind them that the total amount paid on a loan is the original principal plus the interest.

15.7 Record and Practice Journal

An account earns simple interest. (a) Find the interest earned. (b) Find the balance of the account.

1. \$400 at 7% for 3 years
   a. \$84
   b. \$484
2. \$1200 at 5.6% for 4 years
   a. \$268.80
   b. \$1468.80

Find the annual interest rate.

3. I = \$18, P = \$200, t = 18 months
   6%
4. I = \$310, P = \$1000, t = 5 years
   6.2%

Find the amount of time.

5. I = \$60, P = \$750, r = 4%
   2 years
6. I = \$825, P = \$2500, r = 5.5%
   6 years

7. You put \$500 in a savings account. The account earns \$15.75 simple interest in 6 months. What is the annual interest rate?
   6.3%

Vocabulary and Concept Check

1. I = simple interest,
   P = principal,
   r = annual interest rate (in decimal form),
   t = time (in years)
2. **a.** lower interest rate because you would pay less
   b. higher interest rate because you would receive more
3. You have to change 6% to a decimal and 8 months to a fraction of a year.

Practice and Problem Solving

4. **a.** \$60 **b.** \$660
5. **a.** \$300 **b.** \$1800
6. **a.** \$105 **b.** \$455
7. **a.** \$292.50 **b.** \$2092.50
8. **a.** \$336 **b.** \$1036
9. **a.** \$308.20 **b.** \$1983.20
10. **a.** \$44.40 **b.** \$969.40
11. **a.** \$1722.24 **b.** \$6922.24
12. They did not convert 18 months to years.
    $I = 500(0.06)\left(\frac{18}{12}\right)$
    $= \$45$
13. 3%
14. 7.5%
15. 4%
16. 12.05%
17. 2 yr
18. 8 yr
19. 1.5 yr
20. 2.5 yr
21. \$1440
22. 6%
23. 2 yr

Practice and Problem Solving

24. \$1770 **25.** \$2720

26. \$3660 **27.** \$6700.80

28. \$2550 **29.** \$8500

30. 4 yr **31.** 5.25%

32. See *Taking Math Deeper.*

33. 4 yr

34. \$77.25

35. 12.5 yr; Substitute \$2000 for P and I, 0.08 for r, and solve for t.

36. \$300

37. Year 1 = \$520
Year 2 = \$540.80
Year 3 = \$562.43

Fair Game Review

38. $x < 7$;

39. $b \geq 1$;

40. $w \leq 5$;

41. A

Mini-Assessment

Find the annual simple interest rate.

1. $I = \$60$, $P = \$500$, $t = 3$ years 4%
2. $I = \$45$, $P = \$600$, $t = 2$ years 3.75%

Find the amount of time.

3. $I = \$117$, $P = \$1300$, $r = 3\%$ 3 yr
4. $I = \$71.50$, $P = \$1100$, $r = 3.25\%$ 2 yr
5. A savings account earns 4.5% annual simple interest. The principal is \$1300. What is the balance after 3 years? \$1475.50

Taking Math Deeper

Exercise 32

This problem isn't particularly difficult. However, it is a good opportunity for students to pick up some **financial literacy**. That is, when you pay for items with a credit card, you often have to pay interest. In other words, you are taking out a loan.

Zoo Trip

Tickets	67.70
Food	62.34
Gas	45.50
Total Cost	175.54

1 Find the amount spent.

Total = \$175.54

2 Find the interest paid.

$I = Prt$ Write the formula.

$= 175.54 \cdot 0.12 \cdot \frac{3}{12}$ Substitute amounts.

$\approx \$5.27$ Simplify.

3 Find the total cost of the trip.

Total = 175.54 + 5.27
= \$180.81

How much interest would I pay if I didn't pay the charge for 1 year? for 2 years?

Project

Many credit cards charge different rates of interest. Use the school library or the Internet to research the amount of interest charged by three different credit card companies. Compare the cost of the trip to the zoo based on the different interest rates. Why should you be careful when selecting a credit card and charging items to the card?

Reteaching and Enrichment Strategies

If students need help. . .	If students got it. . .
Resources by Chapter • Practice A and Practice B • Puzzle Time Record and Practice Journal Practice Differentiating the Lesson Lesson Tutorials Skills Review Handbook	Resources by Chapter • Enrichment and Extension • Technology Connection Start the next section

Find the amount paid for the loan.

4 **24.** \$1500 at 9% for 2 years

25. \$2000 at 12% for 3 years

26. \$2400 at 10.5% for 5 years

27. \$4800 at 9.9% for 4 years

Copy and complete the table.

	Principal	Interest Rate	Time	Simple Interest
28.	\$12,000	4.25%	5 years	
29.		6.5%	18 months	\$828.75
30.	\$15,500	8.75%		\$5425.00
31.	\$18,000		54 months	\$4252.50

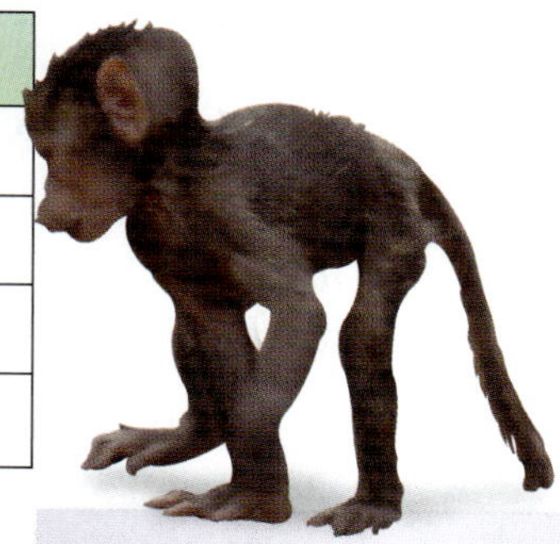

32. ZOO A family charges a trip to the zoo on a credit card. The simple interest rate is 12%. The charges are paid after 3 months. What is the total amount paid for the trip?

Zoo Trip

Tickets	67.70
Food	62.34
Gas	45.50
Total Cost	?

33. MONEY MARKET You deposit \$5000 in an account earning 7.5% simple interest. How long will it take for the balance of the account to be \$6500?

34. LOANS A music company offers a loan to buy a drum set for \$1500. What is the monthly payment?

35. REASONING How many years will it take for \$2000 to double at a simple interest rate of 8%? Explain how you found your answer.

36. PROBLEM SOLVING You have two loans, for 2 years each. The total interest for the two loans is \$138. On the first loan, you pay 7.5% simple interest on a principal of \$800. On the second loan, you pay 3% simple interest. What is the principal for the second loan?

37. Critical Thinking You put \$500 in an account that earns 4% annual interest. The interest earned each year is added to the principal to create a new principal. Find the total amount in your account after each year for 3 years.

Fair Game Review What you learned in previous grades & lessons

Solve the inequality. Graph the solution. *(Section 7.6)*

38. $x - 2 < 5$

39. $b + 3 \geq 4$

40. $w + 6 \leq 11$

41. MULTIPLE CHOICE What is the solution of $4x + 5 = -11$? *(Section 13.5)*

Ⓐ $x = -4$ Ⓑ $x = -1.5$ Ⓒ $x = 1.5$ Ⓓ $x = 4$

15.5–15.7 Quiz

Check It Out
Progress Check
BigIdeasMath.com

Identify the percent of change as an *increase* or a *decrease*. Then find the percent of change. Round to the nearest tenth of a percent if necessary. *(Section 15.5)*

1. 8 inches to 24 inches
2. 300 miles to 210 miles

Find the original price, discount, sale price, or selling price. *(Section 15.6)*

3. Original price: \$30
   Discount: 10%
   Sale price: ?
4. Original price: \$55
   Discount: ?
   Sale price: \$46.75
5. Original price: ?
   Discount: 75%
   Sale price: \$74.75
6. Cost to store: \$152
   Markup: 50%
   Selling price: ?

An account earns simple interest. Find the interest earned, principal, interest rate, or time. *(Section 15.7)*

7. Interest earned: ?
   Principal: \$1200
   Interest rate: 2%
   Time: 5 years
8. Interest earned: \$25
   Principal: \$500
   Interest rate: 5%
   Time: ?
9. Interest earned: \$76
   Principal: \$800
   Interest rate: ?
   Time: 2 years
10. Interest earned: \$119.88
    Principal: ?
    Interest rate: 3.6%
    Time: 3 years

11. **HEIGHT** You estimate that your friend is 50 inches tall. The actual height of your friend is 54 inches. Find the percent error. *(Section 15.5)*

12. **DIGITAL CAMERA** A digital camera costs \$230. The camera is on sale for 30% off, and you have a coupon for an additional 15% off the sale price. What is the final price? *(Section 15.6)*

13. **WATER SKIS** The original price of the water skis was \$200. What is the percent of discount? *(Section 15.6)*

14. **SAXOPHONE** A saxophone costs \$1200. A store offers two loan options. Which option saves more money if you pay the loan in 2 years? *(Section 15.7)*

15. **LOAN** You borrow \$200. The simple interest rate is 12%. You pay off the loan after 2 years. How much do you pay for the loan? *(Section 15.7)*

Alternative Assessment Options

Math Chat
Structured Interview
Student Reflective Focus Question
Writing Prompt

Writing Prompt

Ask students to write a story about making purchases and saving money. The students should include discounts and markups in the story. If they have money left over from their purchases, they should place it in a savings account. The students should include simple interest in the story. Then have students share their stories with the class.

Study Help Sample Answers

Remind students to complete Graphic Organizers for the rest of the chapter.

5.

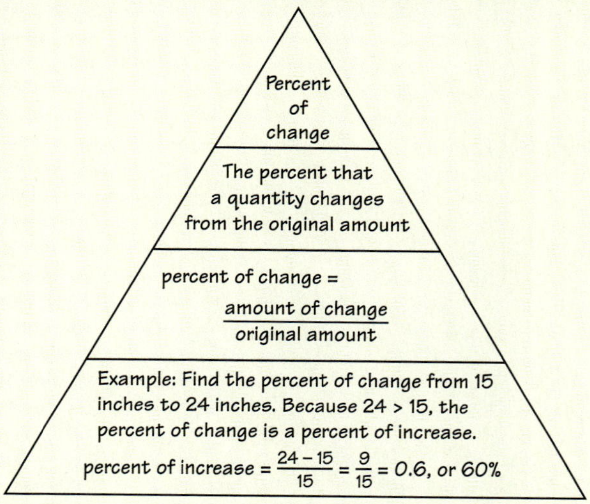

6–8. Available at *BigIdeasMath.com.*

Reteaching and Enrichment Strategies

If students need help. . .	If students got it. . .
Resources by Chapter • Practice A and Practice B • Puzzle Time Lesson Tutorials *BigIdeasMath.com*	Resources by Chapter • Enrichment and Extension • Technology Connection Game Closet at *BigIdeasMath.com* Start the Chapter Review

Answers

1. increase; 200%
2. decrease; 30%
3. $27
4. 15%
5. $299
6. $228
7. $120
8. 1 year
9. 4.75%
10. $1110
11. about 7.4%
12. $136.85
13. 25%
14. Option 2
15. $248

Technology for the Teacher

Online Assessment
Assessment Book
ExamView® Assessment Suite

Review of Common Errors

Exercises 1–6

- Students may move the decimal point the wrong way, forget to insert zeros as placeholders, or move the decimal too many places (especially when the percent is greater than 100).

Exercises 7–14

- Students may try to order the numbers without converting them, or will only convert them mentally and do so incorrectly.

For the Teacher
Additional Review Options

- *BigIdeasMath.com*
- Online Assessment
- Game Closet at *BigIdeasMath.com*
- Vocabulary Help
- Resources by Chapter

Answers

1. 0.76
2. 0.06
3. 3.34
4. 15%
5. 124%
6. 9.7%
7. 52%
8. 245%
9. 0.46
10. 22%
11. 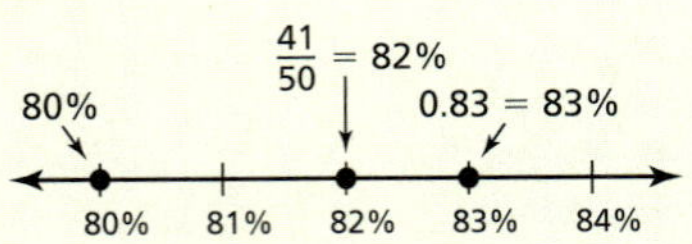

12. 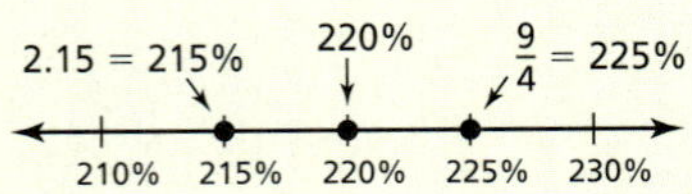

13. 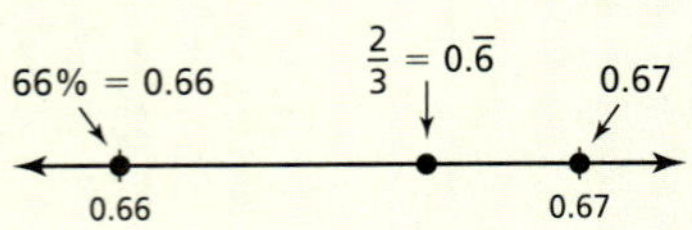

14. 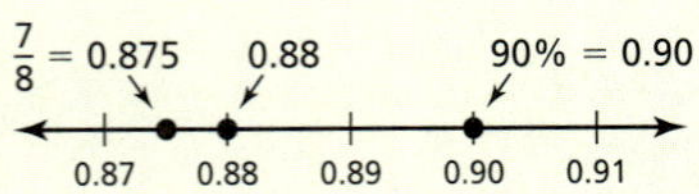

15 Chapter Review

Check It Out
Vocabulary Help
BigIdeasMath.com

Review Key Vocabulary

percent of change, *p. 678*
percent of increase, *p. 678*
percent of decrease, *p. 678*
percent error, *p. 679*
discount, *p. 684*
markup, *p. 684*
interest, *p. 690*
principal, *p. 690*
simple interest, *p. 690*

Review Examples and Exercises

15.1 Percents and Decimals *(pp. 650–655)*

a. Write 64% as a decimal.

$64\% = 64.\% = 0.64$

b. Write 0.023 as a percent.

$0.023 = 0.023 = 2.3\%$

Exercises

Write the percent as a decimal. Use a model to check your answer.

1. 76%
2. 6%
3. 334%

Write the decimal as a percent. Use a model to check your answer.

4. 0.15
5. 1.24
6. 0.097

15.2 Comparing and Ordering Fractions, Decimals, and Percents *(pp. 656–661)*

Which is greater, $\frac{9}{10}$ or 88%?

Write $\frac{9}{10}$ as a percent: $\frac{9}{10} = \frac{90}{100} = 90\%$

∴ 88% is less than 90%. So, $\frac{9}{10}$ is the greater number.

Exercises

Tell which number is greater.

7. $\frac{1}{2}$, 52%
8. $\frac{12}{5}$, 245%
9. 0.46, 43%
10. 0.023, 22%

Use a number line to order the numbers from least to greatest.

11. $\frac{41}{50}$, 0.83, 80%
12. $\frac{9}{4}$, 220%, 2.15
13. 0.67, 66%, $\frac{2}{3}$
14. 0.88, $\frac{7}{8}$, 90%

15.3 The Percent Proportion (pp. 662–667)

a. What percent of 24 is 9?

$\frac{a}{w} = \frac{p}{100}$ Write the percent proportion.

$\frac{9}{24} = \frac{p}{100}$ Substitute 9 for *a* and 24 for *w*.

$100 \cdot \frac{9}{24} = 100 \cdot \frac{p}{100}$ Multiplication Property of Equality

$37.5 = p$ Simplify.

So, 37.5% of 24 is 9.

b. What number is 15% of 80?

$\frac{a}{w} = \frac{p}{100}$ Write the percent proportion.

$\frac{a}{80} = \frac{15}{100}$ Substitute 80 for *w* and 15 for *p*.

$80 \cdot \frac{a}{80} = 80 \cdot \frac{15}{100}$ Multiplication Property of Equality

$a = 12$ Simplify.

So, 12 is 15% of 80.

c. 120% of what number is 54?

$\frac{a}{w} = \frac{p}{100}$ Write the percent proportion.

$\frac{54}{w} = \frac{120}{100}$ Substitute 54 for *a* and 120 for *p*.

$54 \cdot 100 = w \cdot 120$ Cross Products Property

$5400 = 120w$ Multiply.

$45 = w$ Divide each side by 120.

So, 120% of 45 is 54.

Exercises

Write and solve a proportion to answer the question.

15. What percent of 60 is 18?

16. 40 is what percent of 32?

17. What number is 70% of 70?

18. $\frac{3}{4}$ is 75% of what number?

Review of Common Errors (continued)

Exercises 15–18

- Students may not know what number to substitute for each variable. Walk through each type of question with the students. Emphasize that the word "is" means "equals," and "of" means "multiplied by."

Answers

15. $\frac{18}{60} = \frac{p}{100}$; $p = 30$

16. $\frac{40}{32} = \frac{p}{100}$; $p = 125$

17. $\frac{a}{70} = \frac{70}{100}$; $a = 49$

18. $\frac{\frac{3}{4}}{w} = \frac{75}{100}$; $w = 1$

Answers

19. $a = 0.24 \cdot 25$; 6

20. $9 = p \cdot 20$; 45%

21. $60.8 = p \cdot 32$; 190%

22. $91 = 1.3 \cdot w$; 70

23. $10.2 = 0.85 \cdot w$; 12

24. $a = 0.83 \cdot 20$; 16.6

25. 120 parking spaces

26. 64%

Review of Common Errors (continued)

Exercises 19–26

- Students may not know what number to substitute for each variable. Walk through each type of question with the students. Emphasize that the word "is" means "equals," and "of" means "multiplied by."
- Students may mix up the whole and the part when trying to write the percent equation for the word problems. Ask students to identify each part of the equation before writing it in the equation format.

Exercises 27–29

- Students may mix up where to place the numbers in the equation to find percent of change. When students do not put the numbers in the right place, they might find a negative number in the numerator. Emphasize that students must know if it is increasing or decreasing before they can do anything else. The numerator should never have a negative answer. If students get a negative number, then they need to switch the order of the numbers in the problem and then subtract.

Exercises 30 and 31

- Students may just find the markup and not the selling price. Remind them that they must add the markup to the cost to store.
- Remind students that the sale price is not the percent of discount multiplied by the original price.

Exercises 32–38

- Students may forget to change the percent to a decimal. Remind them that before they can put the percent into the equation, they must change the percent to a fraction or a decimal.

15.4 The Percent Equation (pp. 668–673)

a. What number is 72% of 25?

$a = p \cdot w$	Write percent equation.
$= 0.72 \cdot 25$	Substitute 0.72 for p and 25 for w.
$= 18$	Multiply.

So, 72% of 25 is 18.

b. 28 is what percent of 70?

$a = p \cdot w$	Write percent equation.
$28 = p \cdot 70$	Substitute 28 for a and 70 for w.
$\frac{28}{70} = \frac{p \cdot 70}{70}$	Division Property of Equality
$0.4 = p$	Simplify.

Because 0.4 equals 40%, 28 is 40% of 70.

c. 22.1 is 26% of what number?

$a = p \cdot w$	Write percent equation.
$22.1 = 0.26 \cdot w$	Substitute 22.1 for a and 0.26 for p.
$85 = w$	Divide each side by 0.26.

So, 22.1 is 26% of 85.

Exercises

Write and solve an equation to answer the question.

19. What number is 24% of 25?

20. 9 is what percent of 20?

21. 60.8 is what percent of 32?

22. 91 is 130% of what number?

23. 85% of what number is 10.2?

24. 83% of 20 is what number?

25. PARKING 15% of the school parking spaces are handicap spaces. The school has 18 handicap spaces. How many parking spaces are there?

26. FIELD TRIP Of the 25 students on a field trip, 16 students bring cameras. What percent of the students bring cameras?

15.5 Percents of Increase and Decrease (pp. 676–681)

The table shows the numbers of skim boarders at a beach on Saturday and Sunday. What was the percent of change in boarders from Saturday to Sunday?

The number of skim boarders on Sunday is less than the number of skim boarders on Saturday. So, the percent of change is a percent of decrease.

$$\text{percent of decrease} = \frac{\text{original amount} - \text{new amount}}{\text{original amount}}$$

Day	Number of Skim Boarders
Saturday	12
Sunday	9

$= \frac{12 - 9}{12}$ Substitute.

$= \frac{3}{12}$ Subtract.

$= 0.25 = 25\%$ Write as a percent.

So, the number of skim boarders decreased by 25% from Saturday to Sunday.

Exercises

Identify the percent of change as an *increase* or a *decrease*. Then find the percent of change. Round to the nearest tenth of a percent if necessary.

27. 6 yards to 36 yards

28. 120 meals to 52 meals

29. **MARBLES** You estimate that a jar contains 68 marbles. The actual number of marbles is 60. Find the percent error.

15.6 Discounts and Markups (pp. 682–687)

What is the original price of the tennis racquet?

The sale price is 100% − 30% = 70% of the original price.

Answer the question: 21 is 70% of what number?

$a = p \cdot w$ Write percent equation.

$21 = 0.7 \cdot w$ Substitute 21 for a and 0.7 for p.

$30 = w$ Divide each side by 0.7.

So, the original price of the tennis racquet is $30.

Exercises

Find the sale price or original price.

30. Original price: $50
Discount: 15%
Sale price: ?

31. Original price: ?
Discount: 20%
Sale price: $75

Review Game

Percents of Increase and Decrease

Materials per Group

- 1 deck of cards with the jacks, queens, kings, and aces removed
- paper
- pencil
- calculator

Directions

Each group starts with 108 points. The cards are placed face down in the middle of the group. One member of the group turns a card over. If the card is red, the face value of the card is subtracted from the number of points. If the card is black, the face value of the card is added to the number of points. Group members take turns calculating the percent increase or decrease and turning cards over. The starting number of points at each player's turn is the same as the ending number of points at the previous player's turn. The group should be back to 108 points after going through all of the cards.

Who Wins?

The group with the highest mean percent increase wins. To find the mean percent increase, add the percent increases and divide the sum by 18.

For the Student Additional Practice

- Lesson Tutorials
- Multi-Language Glossary
- Self-Grading Progress Check
- *BigIdeasMath.com*
  Dynamic Student Edition
  Student Resources

Answers

27. increase; 500%

28. decrease; 56.7%

29. about 13.3%

30. \$42.50

31. \$93.75

32. **a.** \$36

b. \$336

33. **a.** \$280

b. \$2280

34. 1.7%

35. 7.1%

36. 3 years

37. 6 years

38. 4%

My Thoughts on the Chapter

Teacher Tip

Not allowed to write in your teaching edition? Use sticky notes to record your thoughts.

What worked. . .

What did not work. . .

What I would do differently. . .

15.7 Simple Interest (pp. 688–693)

You put \$200 in a savings account. The account earns 2% simple interest per year.

a. What is the interest earned after 4 years?

b. What is the balance after 4 years?

a. $I = Prt$ — Write simple interest formula.

$= 200(0.02)(4)$ — Substitute 200 for P, 0.02 for r, and 4 for t.

$= 16$ — Multiply.

So, the interest earned is \$16 after 4 years.

b. To find the balance, add the interest to the principal.

So, the balance is \$200 + \$16 = \$216 after 4 years.

You put \$500 in an account. The account earns \$55 simple interest in 5 years. What is the annual interest rate?

$I = Prt$ — Write simple interest formula.

$55 = 500(r)(5)$ — Substitute 55 for I, 500 for P, and 5 for t.

$55 = 2500r$ — Simplify.

$0.022 = r$ — Divide each side by 2500.

So, the annual interest rate of the account is 0.022, or 2.2%.

Exercises

An account earns simple interest.

a. Find the interest earned.

b. Find the balance of the account.

32. \$300 at 4% for 3 years

33. \$2000 at 3.5% for 4 years

Find the annual simple interest rate.

34. $I = \$17$, $P = \$500$, $t = 2$ years

35. $I = \$426$, $P = \$1200$, $t = 5$ years

Find the amount of time.

36. $I = \$60$, $P = \$400$, $r = 5\%$

37. $I = \$237.90$, $P = \$1525$, $r = 2.6\%$

38. SAVINGS You put \$100 in an account. The account earns \$2 simple interest in 6 months. What is the annual interest rate?

15 Chapter Test

Check It Out
Test Practice
BigIdeasMath.com

Write the percent as a decimal.

1. 0.96%
2. 65%
3. 25.7%

Write the decimal as a percent.

4. 0.42
5. 7.88
6. 0.5854

Tell which number is greater.

7. $\frac{16}{25}$, 65%
8. 56%, 5.6

Use a number line to order the numbers from least to greatest.

9. 85%, $\frac{15}{18}$, 0.84
10. 58.3%, 0.58, $\frac{7}{12}$

Answer the question.

11. What percent of 28 is 21?
12. 64 is what percent of 40?
13. What number is 80% of 45?
14. 0.8% of what number is 6?

Identify the percent of change as an *increase* or a *decrease*. Then find the percent of change. Round to the nearest tenth of a percent if necessary.

15. 4 strikeouts to 10 strikeouts
16. \$24 to \$18

Find the sale price or selling price.

17. Original price: \$15
    Discount: 5%
    Sale price: ?
18. Cost to store: \$5.50
    Markup: 75%
    Selling price: ?

An account earns simple interest. Find the interest earned or the principal.

19. Interest earned: ?
    Principal: \$450
    Interest rate: 6%
    Time: 8 years
20. Interest earned: \$27
    Principal: ?
    Interest rate: 1.5%
    Time: 2 years

21. **BASKETBALL** You, your cousin, and a friend each take the same number of free throws at a basketball hoop. Who made the most free throws?

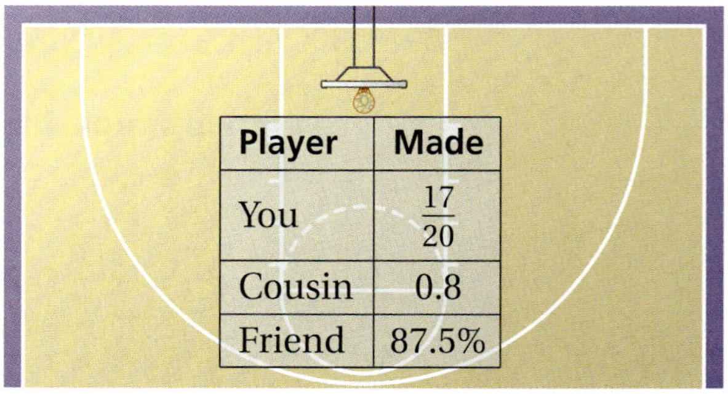

Player	Made
You	$\frac{17}{20}$
Cousin	0.8
Friend	87.5%

22. **PARKING LOT** You estimate that there are 66 cars in a parking lot. The actual number of cars is 75.
    a. Find the percent error.
    b. What other estimate gives the same percent error? Explain your reasoning.

23. **INVESTMENT** You put \$800 in an account that earns 4% simple interest. Find the total amount in your account after each year for 3 years.

Test Item References

Chapter Test Questions	Section to Review
1–6	15.1
7–10, 21	15.2
11–14	15.3
11–14	15.4
15, 16, 22	15.5
17, 18	15.6
19, 20, 23	15.7

Test-Taking Strategies

Remind students to quickly look over the entire test before they start so they can budget their time. Remind them that the test is on fractions, decimals, *and* percents and that they need to read the problems carefully. Students need to use the **Stop** and **Think** strategy before they answer a question. Students need to remember to think of the different representations of each number as they work through the test, such as 0.5, $\frac{1}{2}$, and 50%.

Common Errors

- **Exercises 7–10** Students may try to order the numbers without converting them or will only convert them mentally and do so incorrectly. Tell students that it is necessary to convert all the numbers to one form.
- **Exercises 11–14** Students may not know what numbers to substitute for the variables. Review each type of question with students. Emphasize that the word "is" means "equals" and "of" means "multiplied by." Ask students to identify the whole, the part of the whole, and the percent.
- **Exercises 15 and 16** Students might place the numbers in the percent of change formulas incorrectly. Remind them that they should have the difference between the greater amount and the lesser amount in the numerator, so the numerator should never be negative. Also point out that the original amount should always be in the denominator.

Reteaching and Enrichment Strategies

If students need help. . .	If students got it. . .
Resources by Chapter • Practice A and Practice B • Puzzle Time Record and Practice Journal Practice Differentiating the Lesson Lesson Tutorials *BigIdeasMath.com* Skills Review Handbook	Resources by Chapter • Enrichment and Extension • Technology Connection Game Closet at *BigIdeasMath.com* Start Cumulative Assessment

Answers

1. 0.0096 **2.** 0.65

3. 0.257 **4.** 42%

5. 788% **6.** 58.54%

7. 65% **8.** 5.6

9.

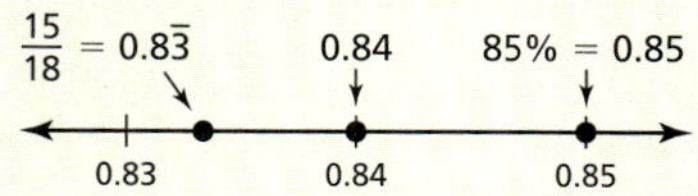

10.

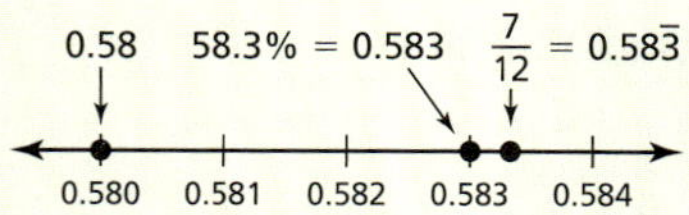

11. 75%

12. 160%

13. 36

14. 750

15. increase; 150%

16. decrease; 25%

17. \$14.25

18. \$9.63

19. \$216

20. \$900

21. Your friend

22. **a.** 12%

b. 84 cars; To get the same percent error, the amount of error needs to be the same. Because your estimate was 9 cars below the actual number, an estimate of 9 cars above the actual number will give the same percent error.

23. Year 1: \$832
Year 2: \$864
Year 3: \$896

Technology for the Teacher

Online Assessment
Assessment Book
ExamView® Assessment Suite

Test-Taking Strategies

Available at *BigIdeasMath.com*

After Answering Easy Questions, Relax
Answer Easy Questions First
Estimate the Answer
Read All Choices before Answering
Read Question before Answering
Solve Directly or Eliminate Choices
Solve Problem before Looking at Choices
Use Intelligent Guessing
Work Backwards

About this Strategy

When taking a multiple choice test, be sure to read each question carefully and thoroughly. It is also very important to read each answer choice carefully. Do not pick the first answer you think is correct! If two answer choices are the same, eliminate them both. There can only be one correct answer.

Answers

1. C
2. G
3. 6
4. D

Technology for the Teacher

Performance Tasks
Online Assessment
Assessment Book
ExamView® Assessment Suite

Item Analysis

1. **A.** The student finds 30% of \$8.50 but does not subtract this amount from \$8.50.

   B. The student thinks that 30% is equivalent to \$3.00 and subtracts this amount from \$8.50.

   C. Correct answer

   D. The student thinks that 30% is equivalent to \$0.30 and subtracts this amount from \$8.50.

2. **F.** The student divides incorrectly or converts measures incorrectly to choose an incorrect box.

   G. Correct answer

   H. The student divides incorrectly or converts measures incorrectly to choose an incorrect box.

   I. The student divides incorrectly or converts measures incorrectly to choose an incorrect box.

3. **Gridded Response:** Correct answer: 6

   Common Error: The student makes a sign error when dividing and gets an answer of $x = -6$.

4. **A.** The student chooses a proportion that will find what percent 17 is of 43.

   B. The student chooses a proportion that will find 43% of 17.

   C. The student chooses a proportion that will find 17% of 43.

   D. Correct answer

15 Cumulative Assessment

Test-Taking Strategy

Read All Choices Before Answering

"Reading all choices before answering can really pay off!"

1. A movie theater offers 30% off the price of a movie ticket to students from your school. The regular price of a movie ticket is \$8.50. What is the discounted price that you would pay for a ticket?

A. \$2.55 **C.** \$5.95

B. \$5.50 **D.** \$8.20

2. You are comparing the prices of four boxes of cereal. Two of the boxes contain free extra cereal.

- Box F costs \$3.59 and contains 16 ounces.
- Box G costs \$3.79 and contains 16 ounces, plus an additional 10% for free.
- Box H costs \$4.00 and contains 500 grams.
- Box I costs \$4.69 and contains 500 grams, plus an additional 20% for free.

Which box has the least unit cost? (1 ounce = 28.35 grams)

F. Box F **H.** Box H

G. Box G **I.** Box I

3. What value makes the equation $11 - 3x = -7$ true?

4. Which proportion represents the problem below?

"17% of a number is 43. What is the number?"

A. $\frac{17}{43} = \frac{n}{100}$ **C.** $\frac{n}{43} = \frac{17}{100}$

B. $\frac{n}{17} = \frac{43}{100}$ **D.** $\frac{43}{n} = \frac{17}{100}$

5. Which list of numbers is in order from least to greatest?

F. $0.8, \frac{5}{8}, 70\%, 0.09$

G. $0.09, \frac{5}{8}, 0.8, 70\%$

H. $\frac{5}{8}, 70\%, 0.8, 0.09$

I. $0.09, \frac{5}{8}, 70\%, 0.8$

6. What is the value of $\frac{9}{8} \div \left(-\frac{11}{4}\right)$?

7. A pair of running shoes is on sale for 25% off the original price.

Which price is closest to the sale price of the running shoes?

A. $93

B. $99

C. $124

D. $149

8. What is the slope of the line?

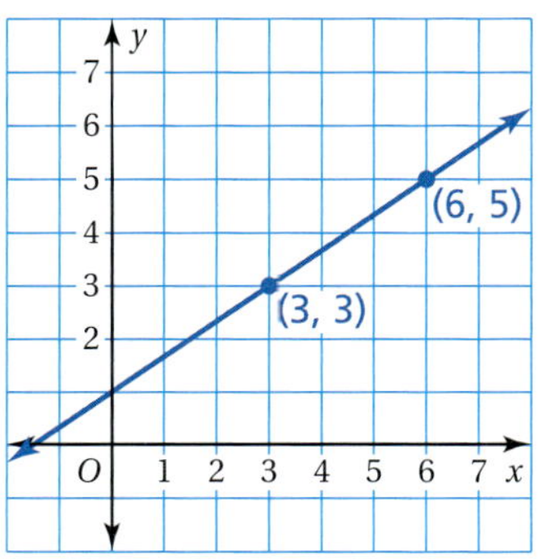

F. $\frac{2}{3}$

G. $\frac{3}{2}$

H. 2

I. 3

Item Analysis (continued)

5. **F.** The student thinks that 0.8 is less than 0.09 and does not know how to compare decimal numbers with fractions or percents.

   G. The student orders the decimal numbers and fractions correctly but thinks that 70% equals 70.

   H. The student orders the numbers using either the numerator or the leading digit.

   I. Correct answer

6. **Gridded Response:** Correct answer: $-\frac{9}{22}$

   Common Error: The student takes the reciprocal of the wrong fraction and gets an answer of $-\frac{22}{9}$.

7. **A.** Correct answer

   B. The student thinks that 25% = \$25 and subtracts \$25 from \$123.75.

   C. The student thinks that 25% = \$0.25, and then either adds \$0.25 to or subtracts \$0.25 from \$123.75.

   D. The student thinks that 25% = \$25 and adds \$25 to \$123.75.

8. **F.** Correct answer

   G. The student finds the change in x over the change in y.

   H. The student finds the change in y.

   I. The student finds the change in x.

Answers

5. I
6. $-\frac{9}{22}$
7. A
8. F

Answers

9. D

10. *Part A* \$371 at the hardware store; \$355.20 at the online store

Part B The hardware store offers the better final cost by \$5.30.

11. F

Item Analysis (continued)

9. **A.** The student does not perform the correct operation.

B. The student does not perform the correct operation.

C. The student does not perform the correct operation.

D. Correct answer

10. **4 points** The student demonstrates a thorough understanding of solving problems involving finding percents of numbers. In Part A, the student correctly calculates the cost of the ladder at each store, getting \$371 at the hardware store and \$355.20 at the online store. In Part B, the student correctly subtracts 10% of the cost of the ladder at the hardware store before adding the tax. The student also correctly recalculates the cost of the ladder at the online store without the shipping and handling charge. The student then compares the final costs, showing that the hardware store offers the better final cost by \$5.30. The student shows accurate, complete work for both parts and provides clear and complete explanations.

3 points The student demonstrates an understanding of solving problems involving finding percents of numbers, but the student's work and explanations demonstrate an essential but less than thorough understanding.

2 points The student demonstrates a partial understanding of solving problems involving finding percents of numbers. The student's work and explanations demonstrate a lack of essential understanding.

1 point The student demonstrates very limited understanding of solving problems involving finding percents of numbers. The student's response is incomplete and exhibits many flaws.

0 points The student provided no response, a completely incorrect or incomprehensible response, or a response that demonstrates insufficient understanding of solving problems involving finding percents of numbers.

11. **F.** Correct answer

G. The student realizes there is a constant rate of change, 2, but doesn't notice that the data does not go through the origin.

H. The student realizes there is a constant rate of change, 0, but doesn't notice that the data does not go through the origin.

I. The student realizes there is a constant rate of change, 1, but doesn't notice that the data does not go through the origin.

9. Brad solved the equation in the box shown.

What should Brad do to correct the error that he made?

$$-3(2 + w) = -45$$
$$2 + w = -15$$
$$w = -17$$

A. Multiply -45 by -3 to get $2 + w = 135$.

B. Add 3 to -45 to get $2 + w = -42$.

C. Add 2 to -15 to get $w = -13$.

D. Divide -45 by -3 to get 15.

10. You are comparing the costs of a certain model of ladder at a hardware store and at an online store.

Think Solve Explain

Part A What is the cost of the ladder at each of the stores? Show your work and explain your reasoning.

Part B Suppose that the hardware store is offering 10% off the price of the ladder and that the online store is offering free shipping and handling. Which store offers the better final cost? by how much? Show your work and explain your reasoning.

11. Which table shows a proportional relationship between x and y?

F.

x	0	1	2	3
y	0	3	6	9

G.

x	0	1	2	3
y	2	4	6	8

H.

x	1	2	3	4
y	5	5	5	5

I.

x	1	2	3	4
y	0	1	2	3

Key Vocabulary Index

Mathematical terms are best understood when you see them used and defined *in context*. This index lists where you will find key vocabulary. A full glossary is available in your Record and Practice Journal and at *BigIdeasMath.com*.

Student Index

This student-friendly index will help you find vocabulary, key ideas, and concepts. It is easily accessible and designed to be a reference for you whether you are looking for a definition, real-life application, or help with avoiding common errors.

D

E

N

O

P

Q

R

S

W

Additional Answers

Chapter 11

Try It Yourself

1. $11 + b$;

$3 + (b + 8) = 3 + (8 + b)$ Comm. Prop. of Add.
$= (3 + 8) + b$ Assoc. Prop. of Add.
$= 11 + b$ Add 3 and 8.

2. $d + 10$;

$(d + 4) + 6 = d + (4 + 6)$ Assoc. Prop. of Add.
$= d + 10$ Add 4 and 6.

3. $30p$;

$6(5p) = (6 \cdot 5)p$ Assoc. Prop. of Mult.
$= 30p$ Multiply 6 and 5.

4. 0;

$13 \cdot m \cdot 0 = 13 \cdot 0 \cdot m$ Comm. Prop. of Mult.
$= (13 \cdot 0) \cdot m$ Assoc. Prop. of Mult.
$= 0 \cdot m$ Mult. Prop. of Zero
$= 0$ Mult. Prop. of Zero

5. $29x$;

$1 \cdot x \cdot 29 = 1 \cdot 29 \cdot x$ Comm. Prop. of Mult.
$= (1 \cdot 29) \cdot x$ Assoc. Prop. of Mult.
$= 29x$ Mult. Prop. of One

6. $n + 14$;

$(n + 14) + 0 = n + (14 + 0)$ Assoc. Prop. of Add.
$= n + 14$ Add. Prop. of Zero

Record and Practice Journal Fair Game Review

6. $32k$;

$8(4k) = (8 \cdot 4)k$ Assoc. Prop. of Mult.
$= 32k$ Multiply 8 and 4.

7. 0;

$13 \cdot 0 \cdot p = (13 \cdot 0)p$ Assoc. Prop. of Mult.
$= 0 \cdot p = 0$ Mult. Prop. of Zero

8. 0;

$7 \cdot z \cdot 0 = 7 \cdot 0 \cdot z$ Comm. Prop. of Mult.
$= (7 \cdot 0) \cdot z$ Assoc. Prop. of Mult.
$= 0 \cdot z = 0$ Mult. Prop. of Zero

9. $2.5w$;

$2.5 \cdot w \cdot 1 = 2.5 \cdot (w \cdot 1)$ Assoc. Prop. of Mult.
$= 2.5 \cdot w$ Mult. Prop. of One
$= 2.5w$

10. $19x$;

$1 \cdot x \cdot 19 = 1 \cdot 19 \cdot x$ Comm. Prop. of Mult.
$= (1 \cdot 19) \cdot x$ Assoc. Prop. of Mult.
$= 19x$ Mult. Prop. of One

11. $t + 3$;

$(t + 3) + 0 = t + (3 + 0)$ Assoc. Prop. of Add.
$= t + 3$ Add. Prop. of Zero

12. $4 + g$;

$0 + (g + 4) = 0 + (4 + g)$ Comm. Prop. of Add.
$= (0 + 4) + g$ Assoc. Prop. of Add.
$= 4 + g$ Add. Prop. of Zero

Section 11.2

Practice and Problem Solving

48. **a.** point C; E is $15 + (-13) = 2$ higher than C, so C is deeper.

b. point B; D is $-18 + 15 = -3$ from B, so D is 3 units lower than B.

Section 11.5

Fair Game Review

42. 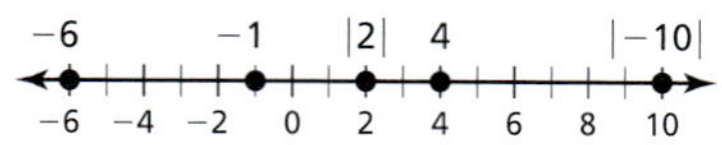

43. 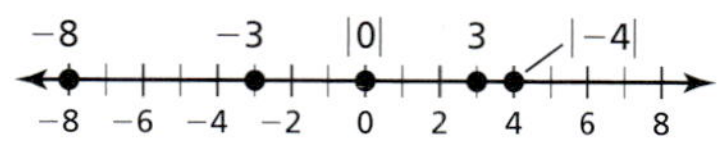

44. 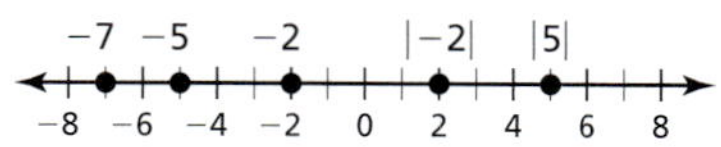

Chapter 12

Record and Practice Journal Fair Game Review

11. $\frac{47}{30}$ **12.** $\frac{1}{3}$

13. $\frac{2}{35}$ **14.** $\frac{5}{27}$

15. $\frac{2}{5}$ **16.** $\frac{14}{11}$

17. $\frac{3}{4}$ **18.** $7\frac{1}{12}$ cups

Section 12.1

Practice and Problem Solving

40. $-5\frac{3}{11} < -5.\overline{2}$ **41.** $-2\frac{13}{16} < -2\frac{11}{14}$

42. *Sample answer:* -0.4, $-0.\overline{45}$

43. Michelle **44.** math quiz

45. No; The base of the skating pool is at -10 feet, which is deeper than $-9\frac{5}{6}$ feet.

47. **a.** when a is negative

b. when a and b have the same sign, $a \neq 0 \neq b$

Section 12.2

Practice and Problem Solving

24. The sum will be positive when the addend with the greater absolute value is positive. The sum will be negative when the addend with the greater absolute value is negative. The sum will be zero when the numbers are opposites.

Section 12.4

Practice and Problem Solving

46. **b.** 0.03 in.;

$$\frac{-0.05 + 0.09 + (-0.04) + (-0.08) + 0.03}{5} = -0.01$$

Chapter 13

Section 13.1

Practice and Problem Solving

24. Solution B is correct.

Solution A: The expression inside the parentheses was simplified first, but 2 and $-5x$ are not like terms.

Solution C: The solution evaluated the expression from left to right by subtracting 4 from 6, instead of using order of operations and distributing the -4 first.

Solution D: The solution did not distribute the negative sign to both of the terms inside the parentheses.

25. $(9 + 3x)\ \text{ft}^2$

26. $(20x + 25y)$ dollars

27. *Sample answer:*

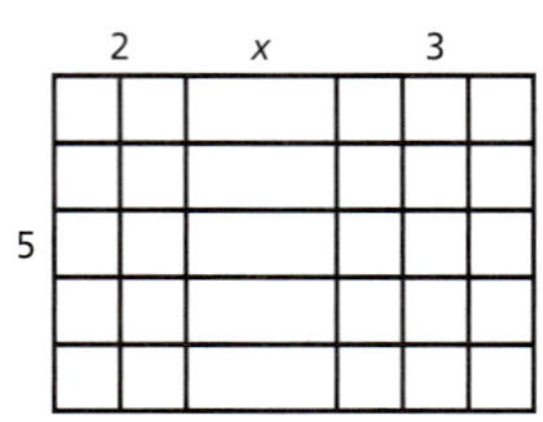

$5x + 25$

28. *Sample answer:*

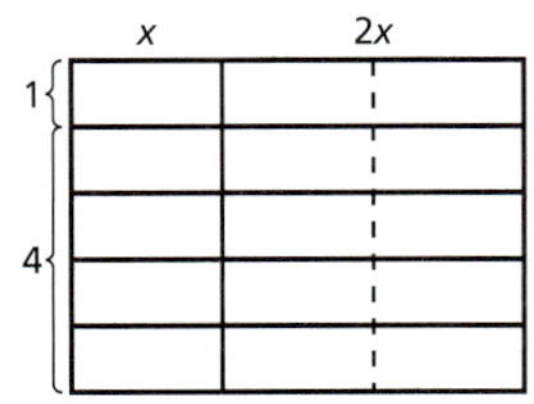

$15x$

Section 13.2

Record and Practice Journal

5. Add—combine like tiles and remove zero pairs

Subtract—remove like tiles, add zero pairs to the first expression as necessary, then remove like tiles

Practice and Problem Solving

25. The -3 was not distributed to both terms inside the parentheses.

$$\begin{aligned}(4m + 9) - 3(2m - 5) &= 4m + 9 - 6m + 15 \\ &= 4m - 6m + 9 + 15 \\ &= -2m + 24\end{aligned}$$

Extension 13.2

Practice

16. **a.** *Sample answer:*

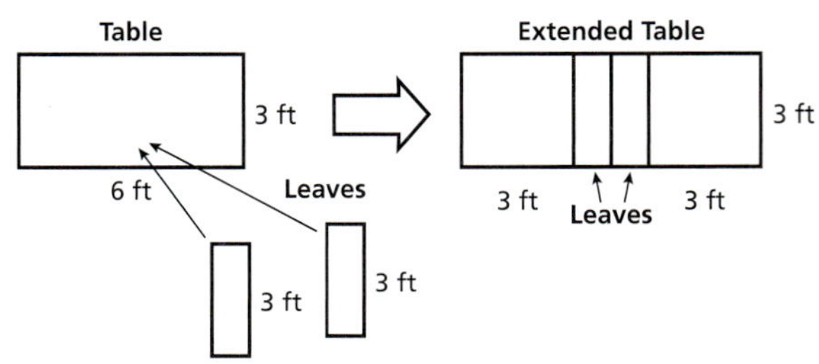

b. $6 + 2x$; shorter dimension of each leaf

Section 13.5

Record and Practice Journal

5. Add enough 1 or -1 tiles to each side to remove the zero pairs and leave only variable tiles on one side. Form as many equal groups of 1 or -1 tiles as there are variable tiles. One of these groups shows the value of the variable.

Chapter 14

Record and Practice Journal
Fair Game Review

7. no **8.** yes

9. yes **10.** no

11. $\frac{6}{29}$ **12.** $d = -48$

13. $x = 21$ **14.** $n = 40$

15. $a = -9$ **16.** $k = -5$

17. $y = -3$ **18.** $w = 18$

19. $z = 90$ **20.** $4p = 35$; $p = \$8.75$

Section 14.1

Record and Practice Journal

1.

Description	Verbal Rate	Numerical
Your running rate in a 100-meter dash	meters per second	$\frac{8 \text{ m}}{\text{sec}}$; $\frac{80 \text{ m}}{\text{sec}}$
The fertilization rate for an apple orchard	pounds per acre	$\frac{150 \text{ lb}}{\text{acre}}$; $\frac{1 \text{ lb}}{\text{acre}}$
The average pay rate for a professional athlete	dollars per year	$\frac{\$3{,}000{,}000}{\text{yr}}$; $\frac{\$3000}{\text{yr}}$
The average rainfall rate in a rain forest	inches per year	$\frac{100 \text{ in.}}{\text{yr}}$; $\frac{5 \text{ in.}}{\text{yr}}$

Section 14.2

Practice and Problem Solving

32. yes; Because Ratio A is equivalent to Ratio B, Ratios A and B simplify to the same ratio. Because Ratio B is equivalent to Ratio C, Ratios B and C simplify to the same ratio. Ratios A and C simplify to the same ratio, so they are equivalent.

Extension 14.2

Practice

3. (0, 0): You earn \$0 for working 0 hours.

(1, 15): You earn \$15 for working 1 hour;

unit rate: $\frac{\$15}{1 \text{ h}}$

(4, 60): You earn \$60 for working 4 hours;

unit rate: $\frac{\$60}{4 \text{ h}} = \frac{\$15}{1 \text{ h}}$

4. (0, 0): The balloon rises 0 feet in 0 seconds.

(1, 5): The balloon rises 5 feet in 1 second;

unit rate: $\frac{5 \text{ ft}}{1 \text{ sec}}$

(6, 30): The balloon rises 30 feet in 6 seconds;

unit rate: $\frac{30 \text{ ft}}{6 \text{ sec}} = \frac{5 \text{ ft}}{1 \text{ sec}}$

5. yes; 5 ft/h **6.** no

7. $y = \frac{4}{3}$

8. a. You

Days	1	2	3	4	5
Cost (dollars)	1.50	2	2.50	3	3.50

Friend

Days	1	2	3	4	5
Cost (dollars)	1.25	2.50	3.75	5	6.25

b. your friend

Record and Practice Journal Practice

1.

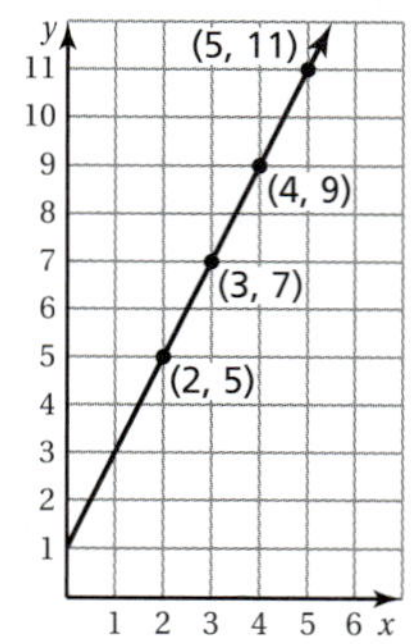

no

2.

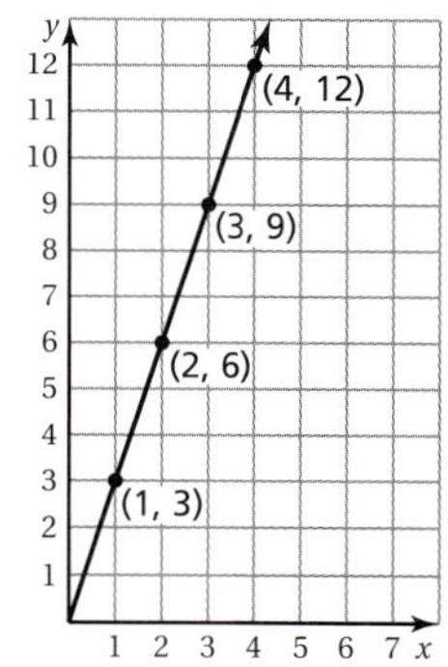

yes

3.

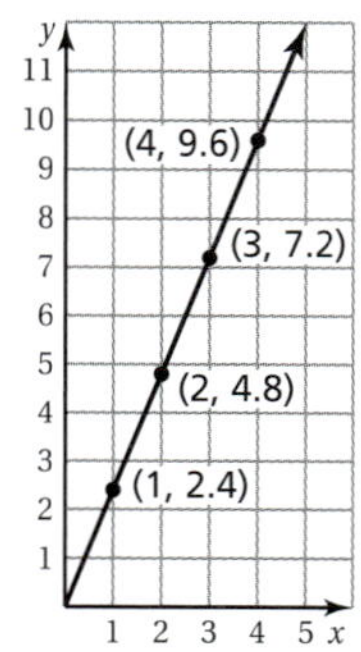

yes

4.

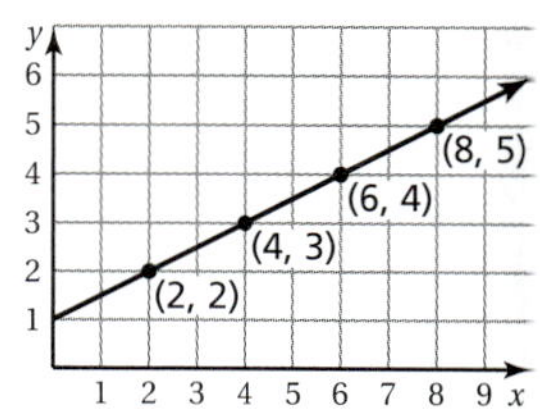

no

5. (0, 0): The car travels 0 miles in 0 hours.
(1, 60): The car travels 60 miles in 1 hour.
(2, 120): The car travels 120 miles in 2 hours.

6. (0, 0): 0 pounds of shrimp costs \$0.
(4, 40): 4 pounds of shrimp costs \$40.
(7, 70): 7 pounds of shrimp costs \$70.

7. (0, 0): You receive 0 emails in 0 days.
(3, 45): You receive 45 emails in 3 days.
(4, 60): You receive 60 emails in 4 days.

8. (0, 0): There are 0 cups of blueberries in 0 pies.
(2, 12): There are 12 cups of blueberries in 2 pies.
(4, 24): There are 24 cups of blueberries in 4 pies.

Section 14.5

Record and Practice Journal

1. **c.** mi/h to ft/sec: Multiply by 5280 to convert miles to feet and divide by 3600 to convert hours to seconds.

   ft/sec to mi/h: Divide by 5280 to convert feet to miles and multiply by 3600 to convert seconds to hours.

Practice and Problem Solving

12. The change in y should be in the numerator. The change in x should be in the denominator.
Slope $= \frac{5}{4}$

13.
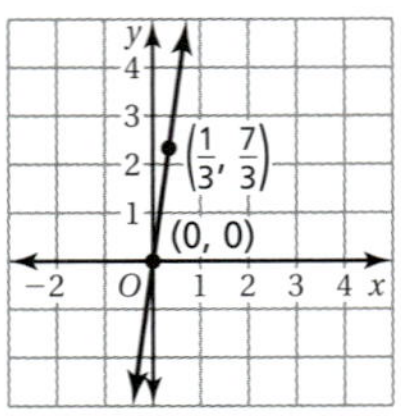

slope = 7

14.
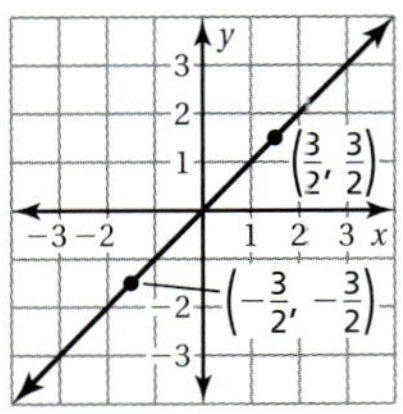

slope = 1

15.
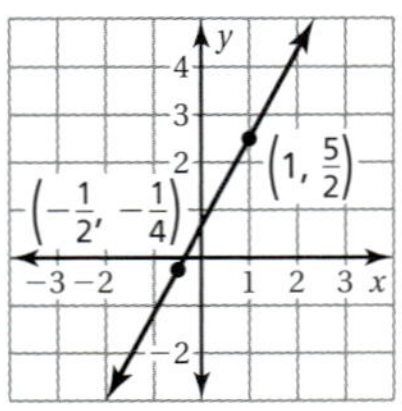

slope $= \frac{11}{6}$

17. **a.**
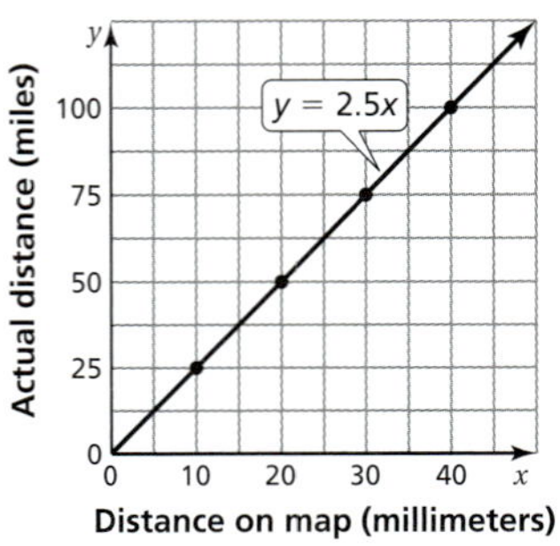

b. 2.5; Every millimeter represents 2.5 miles.

c. 120 mi

d. 90 mm

Section 14.6

On Your Own

4. no; The equation cannot be written as $y = kx$.

5. yes; The equation can be written as $y = kx$.

6. no; The equation cannot be written as $y = kx$.

Practice and Problem Solving

14. yes; The equation can be written as $y = kx$; $k = 1$

15. yes; The equation can be written as $y = kx$; $k = \frac{1}{2}$

16. no; The equation cannot be written as $y = kx$.

17. no; The equation cannot be written as $y = kx$.

18. The line does not pass through the origin, so x and y do not show direct variation.

19.
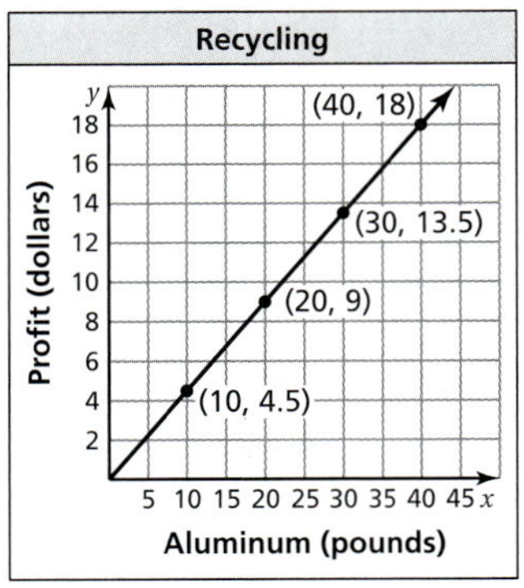

yes; $y = 0.45x$

Chapter 14 Test

12.
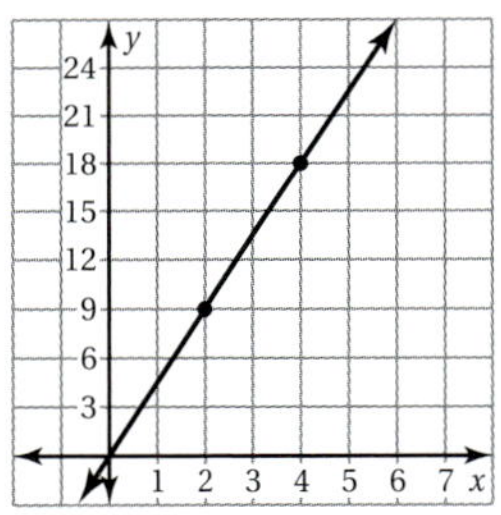

slope $= \frac{9}{2}$

13. no; The equation cannot be written as $y = kx$.

14. no; The equation cannot be written as $y = kx$.

15. yes; The equation can be written as $y = kx$.

16. $58

17. **a.** The number of cycles during the day is greater than the number of cycles during the night.

    b. Day: 40; Night: 30
    The crosswalk cycles 40 times per hour during the day and 30 times per hour during the night.

Chapter 15

Section 15.2

Practice and Problem Solving

20.

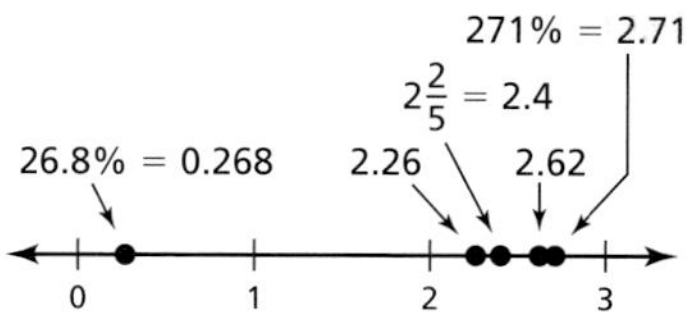

Section 15.3

Practice and Problem Solving

19. 34 represents the part, not the whole.

$$\frac{a}{w} = \frac{p}{100}$$

$$\frac{34}{w} = \frac{40}{100}$$

$$w = 85$$

29. **a.** a scale along the vertical axis

b. 6.25%; *Sample answer:* Although you do not know the actual number of votes, you can visualize each bar as a model with the horizontal lines breaking the data into equal parts. The sum of all the parts is 16. Greg has the least parts with 1, which is 100% ÷ 16 = 6.25%.

c. 31 votes

Section 15.5

Practice and Problem Solving

24. Increasing 20 to 40 is the same as increasing 20 by 20. So, it is a 100% increase. Decreasing 40 to 20 is the same as decreasing 40 by one-half of 40. So, it is a 50% decrease.

25. **a.** about 16.95% increase

b. 161,391 people

29. less than; *Sample answer:* Let x represent the number. A 10% increase is equal to $x + 0.1x$, or $1.1x$. A 10% decrease of this new number is equal to $1.1x - 0.1(1.1x)$, or $0.99x$. Because $0.99x < x$, the result is less than the original number.

Additional Answers

Photo Credits

Front matter

iv Big Ideas Learning, LLC; **viii** *top* ©iStockphoto.com/ALEAIMAGE, ©iStockphoto.com/Ann Marie Kurtz; *bottom* Evok20/Shutterstock.com; **ix** *top* Chiyacat/Shutterstock.com, Zoom Team/Shutterstock.com; *bottom* Sinisa Bobic/Shutterstock.com; **x** *top* ©iStockphoto.com/ALEAIMAGE, ©iStockphoto.com/Ann Marie Kurtz; *bottom* Artpose Adam Borkowski/Shutterstock.com; **xi** *top* Varina and Jay Patel/Shutterstock.com, ©iStockphoto.com/Ann Marie Kurtz; *bottom* ©iStockphoto.com/Aldo Murillo; **xii** *top* Kasiap/Shutterstock.com, ©iStockphoto.com/Ann Marie Kurtz; *bottom* ©iStockphoto.com/kate_sept2004

Chapter 11

475 ©iStockphoto.com/ALEAIMAGE, ©iStockphoto.com/Ann Marie Kurtz; **495** ©iStockphoto.com/RonTech2000; **501** Dmitry Melnikov/Shutterstock.com; **508** *center right* ©iStockphoto.com/Rich Legg; *bottom left* CLFProductions/Shutterstock.com; **511** Liem Bahneman/Shutterstock.com; **512** *center right* ©iStockphoto.com/susaro; *bottom left* © Leonard J. DeFrancisci / Wikimedia Commons / CC-BY-SA-3.0 / GFDL

Chapter 12

516 Chiyacat/Shutterstock.com, Zoom Team/Shutterstock.com; **518** ©iStockphoto.com/Shantell; **527** ultrapro/Shutterstock.com; **528** margouillat photo/Shutterstock.com; **529** Heide Hellebrand/Shutterstock.com; **531** ©iStockphoto.com/Jason Lugo; **547** Pinosub/Shutterstock.com; **548** EdBockStock/Shutterstock.com; **550** Laborant/Shutterstock.com

Chapter 13

552 ©iStockphoto.com/ALEAIMAGE, ©iStockphoto.com/Ann Marie Kurtz; **558** *bottom left* photo25th/Shutterstock.com; *bottom center* ©iStockphoto.com/Don Nichols; **563** Andrew Burgess/Shutterstock.com; **564** Suzanne Tucker/Shutterstock.com; **565** ©iStockphoto.com/Vadim Ponomarenko; **571** MarcelClemens/Shutterstock.com; **575** ©iStockphoto.com/fotoVoyager; **577** *top right* John Kropewnicki/Shutterstock.com; *center left* Yuri Bathan (yuri10b)/Shutterstock.com; *bottom right* Dim Dimich/Shutterstock.com; **595** wacpan/Shutterstock.com

Chapter 14

596 Varina and Jay Patel/Shutterstock.com, ©iStockphoto.com/Ann Marie Kurtz; **599** qingqing/Shutterstock.com; **600** the808/Shutterstock.com; **604** Sergey Peterman/Shutterstock.com; **605** VikaSuh/Shutterstock.com; **610** ©iStockphoto.com/Kemter; **611** ©iStockphoto.com/VikaValter; **619** NASA/Carla Thomas; **622** Jean Tompson; **634** Baldwin Online: Children's Literature Project at www.mainlesson.com; **635** ©iStockphoto.com/Brian Pamphilon; **637** John Kasawa/Shutterstock.com; **640** ©iStockphoto.com/Uyen Le; **646** Peter zijlstra/Shutterstock.com

Chapter 15

648 Kasiap/Shutterstock.com, ©iStockphoto.com/Ann Marie Kurtz; **661** ©iStockphoto.com/Eric Isselée; **666** Image courtesy the President's Challenge, a program of the President's Council on Fitness, Sports and Nutrition; **677** Rob Byron/Shutterstock.com; **678** ©iStockphoto.com/NuStock; **680** ©iStockphoto.com/ARENA Creative; **685** ©iStockphoto.com/amriphoto; **686** ©iStockphoto.com/Albert Smirnov; **687** ©iStockphoto.com/Lori Sparkia; **691** ©iStockphoto.com/anne de Haas; **693** *top right* Big Ideas Learning, LLC; *center left* ©iStockphoto.com/Rui Matos; **694** ©iStockphoto.com/Michael Fernahl; **697** AISPIX by Image Source/Shutterstock.com; **698** ©iStockphoto.com/ted johns

Cartoon illustrations Tyler Stout

Learning Progression

Kindergarten

Counting and Cardinality	– Count to 100 by Ones and Tens; Compare Numbers
Operations and Algebraic Thinking	– Understand and Model Addition and Subtraction
Number and Operations in Base Ten	– Work with Numbers 11–19 to Gain Foundations for Place Value
Measurement and Data	– Describe and Compare Measurable Attributes; Classify Objects into Categories
Geometry	– Identify and Describe Shapes

Grade 1

Operations and Algebraic Thinking	– Represent and Solve Addition and Subtraction Problems
Number and Operations in Base Ten	– Understand Place Value for Two-Digit Numbers; Use Place Value and Properties to Add and Subtract
Measurement and Data	– Measure Lengths Indirectly; Write and Tell Time; Represent and Interpret Data
Geometry	– Draw Shapes; Partition Circles and Rectangles into Two and Four Equal Shares

Grade 2

Operations and Algebraic Thinking	– Solve One- and Two-Step Problems Involving Addition and Subtraction; Build a Foundation for Multiplication
Number and Operations in Base Ten	– Understand Place Value for Three-Digit Numbers; Use Place Value and Properties to Add and Subtract
Measurement and Data	– Measure and Estimate Lengths in Standard Units; Work with Time and Money
Geometry	– Draw and Identify Shapes; Partition Circles and Rectangles into Two, Three, and Four Equal Shares

Grade 3

Operations and Algebraic Thinking – Represent and Solve Problems Involving Multiplication and Division; Solve Two-Step Problems Involving Four Operations

Number and Operations in Base Ten – Round Whole Numbers; Add, Subtract, and Multiply Multi-Digit Whole Numbers

Number and Operations—Fractions – Understand Fractions as Numbers

Measurement and Data – Solve Time, Liquid Volume, and Mass Problems; Understand Perimeter and Area

Geometry – Reason with Shapes and Their Attributes

Grade 4

Operations and Algebraic Thinking – Use the Four Operations with Whole Numbers to Solve Problems; Understand Factors and Multiples

Number and Operations in Base Ten – Generalize Place Value Understanding; Perform Multi-Digit Arithmetic

Number and Operations—Fractions – Build Fractions from Unit Fractions; Understand Decimal Notation for Fractions

Measurement and Data – Convert Measurements; Understand and Measure Angles

Geometry – Draw and Identify Lines and Angles; Classify Shapes

Grade 5

Operations and Algebraic Thinking – Write and Interpret Numerical Expressions

Number and Operations in Base Ten – Perform Operations with Multi-Digit Numbers and Decimals to Hundredths

Number and Operations—Fractions – Add, Subtract, Multiply, and Divide Fractions

Measurement and Data – Convert Measurements within a Measurement System; Understand Volume

Geometry – Graph Points in the First Quadrant of the Coordinate Plane; Classify Two-Dimensional Figures

Mathematics Reference Sheet

Conversions

U.S. Customary
1 foot = 12 inches
1 yard = 3 feet
1 mile = 5280 feet
1 acre = 43,560 square feet
1 cup = 8 fluid ounces
1 pint = 2 cups
1 quart = 2 pints
1 gallon = 4 quarts
1 gallon = 231 cubic inches
1 pound = 16 ounces
1 ton = 2000 pounds
1 cubic foot ≈ 7.5 gallons

U.S. Customary to Metric
1 inch = 2.54 centimeters
1 foot ≈ 0.3 meter
1 mile ≈ 1.61 kilometers
1 quart ≈ 0.95 liter
1 gallon ≈ 3.79 liters
1 cup ≈ 237 milliliters
1 pound ≈ 0.45 kilogram
1 ounce ≈ 28.3 grams
1 gallon ≈ 3785 cubic centimeters

Time
1 minute = 60 seconds
1 hour = 60 minutes
1 hour = 3600 seconds
1 year = 52 weeks

Temperature

$$C = \frac{5}{9}(F - 32)$$

$$F = \frac{9}{5}C + 32$$

Metric
1 centimeter = 10 millimeters
1 meter = 100 centimeters
1 kilometer = 1000 meters
1 liter = 1000 milliliters
1 kiloliter = 1000 liters
1 milliliter = 1 cubic centimeter
1 liter = 1000 cubic centimeters
1 cubic millimeter = 0.001 milliliter
1 gram = 1000 milligrams
1 kilogram = 1000 grams

Metric to U.S. Customary
1 centimeter ≈ 0.39 inch
1 meter ≈ 3.28 feet
1 kilometer ≈ 0.62 mile
1 liter ≈ 1.06 quarts
1 liter ≈ 0.26 gallon
1 kilogram ≈ 2.2 pounds
1 gram ≈ 0.035 ounce
1 cubic meter ≈ 264 gallons

Number Properties

Commutative Properties of Addition and Multiplication

$a + b = b + a$

$a \cdot b = b \cdot a$

Associative Properties of Addition and Multiplication

$(a + b) + c = a + (b + c)$

$(a \cdot b) \cdot c = a \cdot (b \cdot c)$

Addition Property of Zero

$a + 0 = a$

Multiplication Properties of Zero and One

$a \cdot 0 = 0$

$a \cdot 1 = a$

Distributive Property:

$a(b + c) = ab + ac$

$a(b - c) = ab - ac$

Properties of Equality

Addition Property of Equality

If $a = b$, then $a + c = b + c$.

Subtraction Property of Equality

If $a = b$, then $a - c = b - c$.

Multiplication Property of Equality

If $a = b$, then $a \cdot c = b \cdot c$.

Multiplicative Inverse Property

$$n \cdot \frac{1}{n} = \frac{1}{n} \cdot n = 1, n \neq 0$$

Division Property of Equality

If $a = b$, then $a \div c = b \div c$, $c \neq 0$.

Properties of Inequality

Addition Property of Inequality

If $a > b$, then $a + c > b + c$.

Subtraction Property of Inequality

If $a > b$, then $a - c > b - c$.

Multiplication Property of Inequality

If $a > b$ and c is positive, then $a \cdot c > b \cdot c$.

Division Property of Inequality

If $a > b$ and c is positive, then $a \div c > b \div c$.

Perimeter and Area

Square	Rectangle	Parallelogram	Triangle	Trapezoid
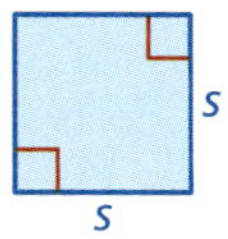	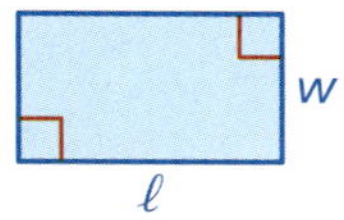	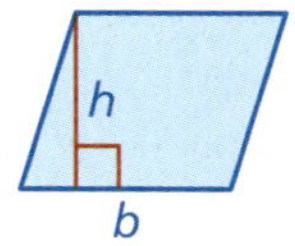	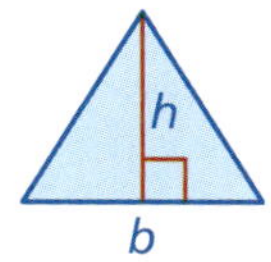	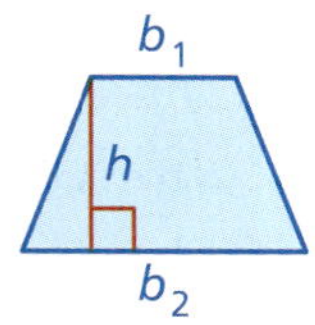
$P = 4s$ $A = s^2$	$P = 2\ell + 2w$ $A = \ell w$	$A = bh$	$A = \frac{1}{2}bh$	$A = \frac{1}{2}h(b_1 + b_2)$

Surface Area

Prism

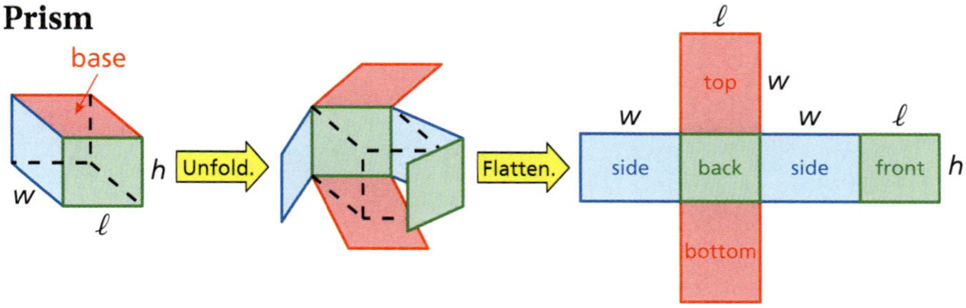

S = areas of bases + areas of lateral faces

Pyramid

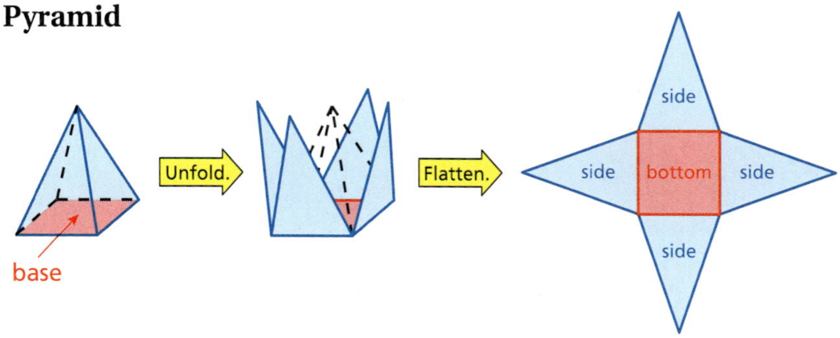

S = area of base + areas of lateral faces

Volume of a Rectangular Prism

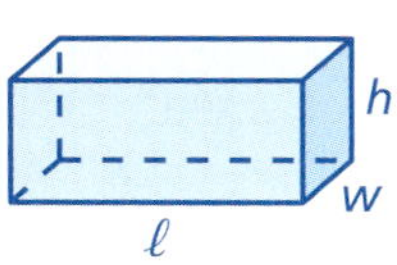

$V = Bh = \ell wh$

Simple Interest

Simple interest formula

$I = Prt$

The Coordinate Plane

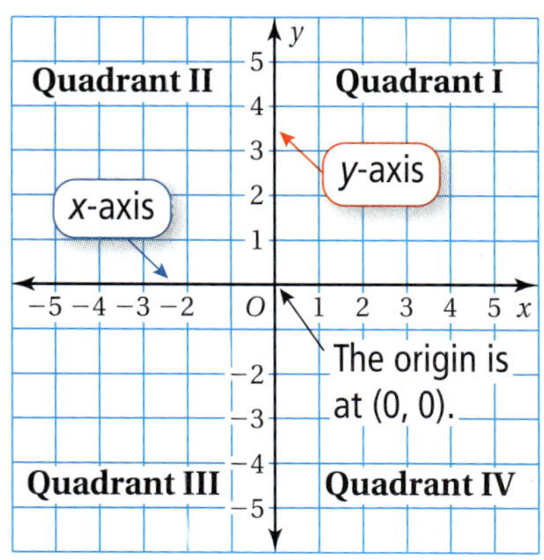